Essential Concepts of Ecological Biodiversity

Essential Concepts of Ecological Biodiversity

Editor: Isaac Hughes

R CALLISTO
REFERENCE

www.callistoreference.com

Callisto Reference,
118-35 Queens Blvd., Suite 400,
Forest Hills, NY 11375, USA

Visit us on the World Wide Web at:
www.callistoreference.com

ISBN: 978-1-63239-937-3 (Hardback)

Cataloging-in-Publication Data

Essential concepts of ecological biodiversity / edited by Isaac Hughes.
    p. cm.
Includes bibliographical references and index.
ISBN 978-1-63239-937-3
1. Ecology. 2. Biodiversity. I. Hughes, Isaac.
QH541 .E87 2018
577--dc23

Table of Contents

Permissions

List of Contributors

Index

Preface

Over the recent decade, advancements and applications have progressed exponentially. This has led to the increased interest in this field and projects are being conducted to enhance knowledge. The main objective of this book is to present some of the critical challenges and provide insights into possible solutions. This book will answer the varied questions that arise in the field and also provide an increased scope for furthering studies.

Biodiversity is defined as the variety of flora and fauna that is present on the Earth. The distribution of distinct species is more in the tropical regions as compared to the polar regions. Conservation of species is primarily carried out through national parks, wildlife sanctuaries, zoological parks, forest reserves, etc. Biodiversity hotspots are areas that contain a large variety of flora and fauna and have been earmarked for conservation. Such selected concepts that redefine biodiversity have been presented in this book. This book, with its detailed analyses and data, will prove immensely beneficial to professionals and students involved in this area at various levels.

I hope that this book, with its visionary approach, will be a valuable addition and will promote interest among readers. Each of the authors has provided their extraordinary competence in their specific fields by providing different perspectives as they come from diverse nations and regions. I thank them for their contributions.

Editor

Biodiversity Effects on Plant Stoichiometry

Maike Abbas[1]*, **Anne Ebeling**[2], **Yvonne Oelmann**[3], **Robert Ptacnik**[1], **Christiane Roscher**[4], **Alexandra Weigelt**[5], **Wolfgang W. Weisser**[2,6], **Wolfgang Wilcke**[7], **Helmut Hillebrand**[1]

1 Institute for Chemistry and Biology of the Marine Environment, Carl von Ossietzky University of Oldenburg, Wilhelmshaven, Germany, 2 Institute of Ecology, Friedrich-Schiller-University Jena, Jena, Germany, 3 Geoecology/Geography, Eberhard Karls-University Tübingen, Tübingen, Germany, 4 UFZ, Helmholtz Centre for Environmental Research, Department of Community Ecology, Halle, Germany, 5 Institute of Biology, University of Leipzig, Leipzig, Germany, 6 Terrestrial Ecology, Department of Ecology and Ecosystem Management, Center of Life and Food Sciences Weihenstephan; Technische Universität München, Freising, Germany, 7 Geographic Institute, University of Berne, Berne, Switzerland

Abstract

In the course of the biodiversity-ecosystem functioning debate, the issue of multifunctionality of species communities has recently become a major focus. Elemental stoichiometry is related to a variety of processes reflecting multiple plant responses to the biotic and abiotic environment. It can thus be expected that the diversity of a plant assemblage alters community level plant tissue chemistry. We explored elemental stoichiometry in aboveground plant tissue (ratios of carbon, nitrogen, phosphorus, and potassium) and its relationship to plant diversity in a 5-year study in a large grassland biodiversity experiment (Jena Experiment). Species richness and functional group richness affected community stoichiometry, especially by increasing C:P and N:P ratios. The primacy of either species or functional group richness effects depended on the sequence of testing these terms, indicating that both aspects of richness were congruent and complementary to expected strong effects of legume presence and grass presence on plant chemical composition. Legumes and grasses had antagonistic effects on C:N (−27.7% in the presence of legumes, +32.7% in the presence of grasses). In addition to diversity effects on mean ratios, higher species richness consistently decreased the variance of chemical composition for all elemental ratios. The diversity effects on plant stoichiometry has several non-exclusive explanations: The reduction in variance can reflect a statistical averaging effect of species with different chemical composition or a optimization of nutrient uptake at high diversity, leading to converging ratios at high diversity. The shifts in mean ratios potentially reflect higher allocation to stem tissue as plants grew taller at higher richness. By showing a first link between plant diversity and stoichiometry in a multiyear experiment, our results indicate that losing plant species from grassland ecosystems will lead to less reliable chemical composition of forage for herbivorous consumers and belowground litter input.

Editor: Andrew Hector, University of Zurich, Switzerland

Funding: The Jena Experiment is funded by the German Science Foundation, FOR 456,1451 (www.dfg.de). The funders had no role in study design, data collection and analysis, decision to publish, or preparation of the manuscript.

Competing Interests: The authors have declared that no competing interests exist.

* E-mail: maike.abbas1@uni-oldenburg.de

Introduction

Recent years have seen the rise of a strong body of literature examining the effects of biodiversity on ecosystem functioning (BDEF), which has been triggered by increasing concerns about potential consequences of loosing species in ecosystems worldwide. Recent syntheses of BDEF research in experimental ecosystems concluded that the loss of biodiversity reduces ecosystem process rates and stability [1–3]. However, there is substantial concern that many previous studies underestimate the strength of BDEF relationships because biodiversity effects become stronger in more complex settings and many of the previous experiments were restricted to a low maximum diversity [4]. Most BDEF studies addressed single ecosystem processes (e.g. primary production), but the loss of biodiversity might impact single processes less than "ecosystem multifunctionality", which is defined as the composite of multiple ecosystem processes [5]. Different species might contribute to different processes in ecosystems, and recent work has shown that the loss of species is more likely to influence multiple processes rather than single processes (e.g., [6–8]).

Ecological stoichiometry (ES) [9] ties multiple processes in ecosystems together as the relation between organisms demand for multiple elements and the availability of these elements in their resources has profound impact on process rates and the relative importance of different processes. The nutrient stoichiometry of plant tissue can be decisive for species interactions with other trophic levels (herbivory, pathogen infestation) and nutrient recycling [10,11]. Consequently, the ratios of carbon:nutrients can be used as a main predictor of the relative role of herbivory and detritus pathways in ecosystems [12,13]. Different plant species may significantly affect the usage of different elements, and processes connected to these elements such as carbon (C) - based total primary productivity or cycling of nitrogen (N) or phosphorus (P), because plants separately consume anorganic elemental resources. Therefore, plants can show trade-offs in resource uptake and storage efficiency for different elements, which leads to higher plasticity in their elemental composition compared to animals, which consume resource packages [9,14].

Unifying ES and BDEF research potentially creates new insights in how communities process available nutrients depending on the number of species involved. However, BDEF research has largely ignored stoichiometric considerations of ecosystem processes, whereas the analysis of ecological stoichiometry has rarely involved biodiversity because ES research often focused upon single species per trophic group or large-scale analyses in certain vegetation types or biomes (e.g., [15,16]). In a pioneering study using algal microcosms, phytoplankton diversity was shown not only to alter primary productivity and P use, but also C:P ratios [17]. Ptacnik et al. [18] suggested on a more general level that plant elemental composition should vary with plant diversity (hypothesis H1), with different outcome depending on how much plant diversity affects resource use efficiency or storage for different elements. If species are highly complementary in their C-acquisition (e.g. different strategies in light acquisition through morphological or physiological traits), but not for mineral nutrients, then C should increase more rapidly with richness than N in community-wide chemical composition and C:nutrient ratios should increase (H1a). In contrast, if plant species show complementarity mainly for the uptake of organic or mineral nutrients but not for C-fixation, then increasing plant diversity should decrease C:nutrient ratios (H1b). The null hypothesis to both is that stoichiometry of community-wide chemical composition is independent of diversity because acquisition of different elements is so strongly coupled that no stoichiometric change is observed or because complementarity in resource acquisition traits is lacking.

It is important to test these potential diversity effects over time as nutrient availability and light limitation might change during community development, leading to time-dependent effects of diversity on stoichiometry (hypothesis H2). In addition to shifting mean ratios, plant diversity will also reduce the variance in chemical composition (hypothesis H3). This can be due to multiple mechanisms underlying diversity effects on resource uptake: More species increase the chance for selection or complementarity effects maximizing nutrient incorporation for each element across the assemblage and thereby lowering the variability of elemental concentrations. Alternatively, the variance in chemical composition can decrease with richness by a statistical averaging effect [19], where more species mask the signature of the stoichiometry of single species.

Here, we tested these three hypotheses on community-wide chemical composition using a grassland biodiversity experiment comprising communities of different species richness (1, 2, 4, 8, 16, and 60) and functional group richness and composition (1 to 4; legumes, grasses, small herb, tall herbs), the Jena Experiment [20]. We analyzed the relationship between plant diversity (species richness, functional group richness and functional composition) and the stoichiometry of plant chemical composition (C, N, P and K) over the first five years of the experiment.

Materials and Methods

Study Site and Experimental Design

The study was conducted in the Jena Experiment, a large biodiversity experiment established in 2002. The experimental site is located on the floodplain of the river Saale in Jena (Thuringia, Germany, $50°55'$ N, $11°35'$ E, 130 m a.s.l.) [20]. The area around Jena has a mean annual temperature of $9.3°C$ and an average annual precipitation of 587 mm [21].

The experimental design is described in detail in Roscher et al. [20]. Briefly, the main experiment comprises 82 plots of 20 m $\times 20$ m size. The soil of the experimental site is an Eutric Fluvisol. Due to flooding dynamics, the soil texture ranges from sandy loam close to the river Saale to silty clay with increasing distance from the river. Species were randomly drawn from a pool of 60 perennial species characteristic for Central European semi-natural, species-rich mesophilic grassland communities (Molinio-Arrhenatheretea [22]). According to the results of a cluster analysis of a literature-based matrix of functional traits, plant species were divided into 4 functional groups (16 grasses, 12 small herbs, 20 tall herbs, and 12 legumes). The experimental design ensures that the presence/absence of each functional group is minimally confounded with species number. Plant communities were established with different levels of species richness increasing on a logarithmic scale (1, 2, 4, 8, 16 and 60). Each species-richness level was replicated with 16 plots with different species composition, only species mixtures with 16 and 60 species were replicated on 14 and 4 plots, respectively. In addition to the main experiment, each experimental species was sown in replicated monocultures resulting in 120 plots of 3.5 m $\times3.5$ m size.

To account for the gradient in soil characteristics, a block design was used with blocks arranged parallel to the river Saale. The plots were mown twice a year, in June and September, and mown material was removed. Additionally, plots were weeded at the beginning of the growing season and after first mowing to maintain the sown species combinations. Weeding was done mostly by hand [23].

Sampling

Aboveground biomass was harvested from 2003 to 2007 at estimated peak standing biomass in late May prior to mowing. Plants were clipped at 3 cm above ground level in four rectangles of 20×50 cm size. In May 2005 only three samples were harvested. Biomass on small monoculture plots was sampled with two replicates. Sample location was selected randomly for each harvest leaving out the outer 70 cm of the plot. After harvest, plant material was sorted into sown species, species which were not sown at a particular plot and detached dead material. Biomass was dried at $70°C$ for at least 48 hours [23].

Biomass samples of the entire plant community per plot were shredded and milled for chemical analyses. Thus, all stoichiometric analyses were done on the mixture of pooled plot biomass. N and C concentrations were determined by an Elemental Analyzer (EA, Vario EL III, Elementar, Germany). Plant material were digested with HNO_3 at $200°C$ using a microwave system (MARS5Xpress, CEM, Germany) to analyze P photometrically after irradiation with UV and oxidation with $K_2S_2O_8$ with a Continuous Flow Analyzer (AutoAnalyzer, Bran&Luebbe, Germany) and K using atomic absorption spectrometry (AAS 240 FS, Varian, Germany) [24].

Statistical Analyses

Bivariate molar ratios of C, N, P and K in the plant community were analyzed across the gradient of plant diversity with different levels of species and functional group richness. We present molar ratios as these are standard in ecological stoichiometry [9]. Since the six different molar ratios (C:N, C:P, C:K, N:P, N:K, P:K) were not independent of each other, a multivariate analysis of variance (MANOVA) was performed with the following factors: block, sown species richness, functional group richness, legume presence, grass presence. Legumes and grasses were explicitly tested because of potential strong impact on N (N_2-fixing) and C (C-storage) concentration. We used the Pillai's trace statistic, which is recommended to test for significant effects on interdependent response variables [25]. In cases of significant effects we used additional univariate tests for each ratio to analyze which ratios responded significantly to the factor. We opted for testing years

Table 1. MANOVA results on bivariate elemental ratios for the years 2003 to 2007.

	May 2003	May 2004	May 2005	May 2006	May 2007
Block	0.543*	0.638***	0.356.	0.415*	0.636***
	(CN,NP,CP,CK,NK)	(CP,CK)	(CP,CK,NK)	(CP)	(CN,CP,CK,NK,PK)
sown diversity	0.079	0.095	0.152.	0.226*	0.301***
			(PK)	(NP,CP,PK)	(NP,CP,PK)
functional group richness	0.147	0.111	0.197*	0.167.	0.296***
			(CP,NK,PK)	(NP,CP)	(CN,NP)
Legume	0.525***	0.287***	0.578***	0.696***	0.706***
	(CN,NP,CK,NK,PK)	(CN,NP,CK,NK,PK)	(all)	(all)	(CN,NP,CK,NK,PK)
Grass	0.200.	0.320***	0.223*	0.385***	0.366***
	(CN)	(CN,CP,PK)	(CN,CP,CK)	(CN,CP,CK)	(CN,CP,CK)

For each factor, the Pillai Trace value and its significance level are given as well as all ratios for which the factor effect was significant at $p<0.05$. Significance levels: $p<0.001 = ***$, $p<0.01 = **$, $p<0.05 = *$, $p<0.1 = .$

separately in order to avoid a repeated measurement MANOVA which would be difficult to interpret.

We made two sensitivity analyses to test the robustness of our results. First, the MANOVA was repeated without the 60 species mixture to test for effects of this less replicated treatment. The results were comparable to those including the 60 species mixtures (Table S1), and changes were restricted to a reduced significance level reflecting the smaller statistical power (6 out of 25 results) and changes in the bivariate ratios becoming significant in the univariate tests of factors significant in the MANOVA (8 out of 25 results). Second, we changed the order of effects and tested functional group richness before testing species richness (see Table S2). Here we found that functional group richness often replaced species richness in importance, reflecting that changes in stoichiometry depended on functional and species richness in a comparable way. We present this alternative model in the supporting online material for comparison.

To analyze how increasing diversity levels effect the bivariate ratios and their variance, we used the software package generalized additive model for location, scale and shape (GAMLSS [26]). We preferred GAMLSS models over ordinary least square regression since they allow for fitting trends in mean and variance simultaneously. Models were fitted assuming normal distribution in the dependent variable. For both, mean and variance, two possible responses (none or linear) result in four possible combinations. The null model assuming no effect on either mean or variance was rejected in every case, such that we present the following models for comparison: Trend in mean, but no trend in variance (model m1), trend in mean and variance (m2) and no trend in mean but trend in variance (m4). The best fitting model per year and ratio was selected by AIC (Akaike's Information Criterion) [27].

The results from MANOVA and GAMLSS were not always congruent for the different bivariate ratios. Generally, the MANOVA detects more significant diversity effects than the GAMLSS, which reflects the fact that the latter tested the direct association between richness and stoichiometry only, whereas the MANOVA extracts additional variation based on block and functional group presence and richness. Thus, in interpreting these outcomes, the MANOVA is the more powerful test for diversity effects on mean ratios, whereas the GAMLSS is the more powerful test for simultaneous changes in the variance.

Additionally, we tested how species diversity affects multi-element stoichiometry and the predictability of chemical composition. For each year, we used a Principal Components Analysis (PCA) on the concentrations of C, N, P, and K, producing two orthogonal axes explaining between 73% and 80% of the total variance in the different years. The first principal component was loaded by C concentration and opposed by P and K concentrations, whereas the second principal component was loaded by N concentrations alone. Factor loadings for PCA analysis (Table S3) were calculated with Statistica 8.0 (Statsoft, Tulsa Oklahoma). We used the origin of the PCA (0;0) as average stoichiometric composition and calculated the stoichiometric distance (SDist) for each of the 82 plots in each year. SDist is a multivariate expression on how deviant the chemical composition of the plant community was from the average across plots and years. An analysis of variance (ANOVA) was performed on SDist as described above.

In order to test whether plant diversity effects on stoichiometry could be explained by mixing plant species only, we compared expected to observed ratios. For the expected ratios, the biomass contribution of each species on the large plots from 2003 to 2006 [23] was multiplied by their element concentrations in monoculture, which were obtained by averaging species replicates from small monoculture plots. Because biomass of the small plots was only analyzed for C and N concentrations from 2003–2006, we lack monoculture information on P-content and thus, the calculation was only done for the C:N ratio in those years. The predicted ratios were plotted against the observed ratios of the years 2003–2006 for plant monocultures and diversity levels of 2, 4, 8 and 16 species mixtures. In addition to perform an ordinary least square regression (OLS), we are assessing the performance of the predicted nutrient ratios by an orthogonal regression (also called 'total least squares'). In OLS, the resulting coefficient is not independent from the choice of predictor and regressor. Conversely, orthogonal regression makes no assumptions regarding the source of the error. The orthogonal regression was estimated using function 'princomp' in R [28]. Confidence intervals for the parameters were estimated from 1000 bootstrap iterations. Additionally, we compared the coefficient of variation (CV) observed in mixtures with a predicted CV derived from species relative cover in mixtures and their monoculture CN ratios.

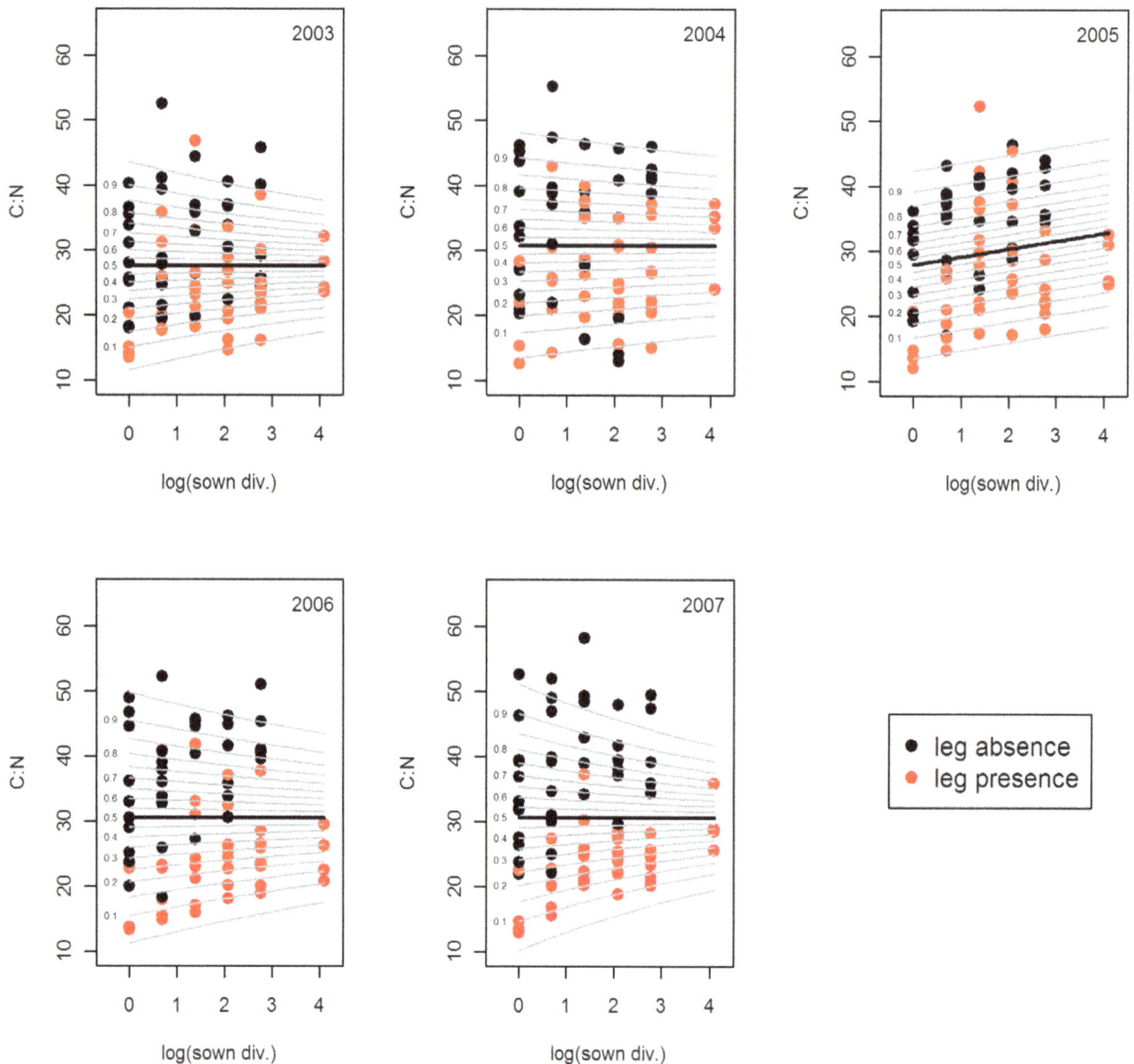

Figure 1. C:N ratio versus plant species richness. GAMLSS (generalized additive model for location scale and shape) model of the molar C:N ratio versus species richness (natural logarithm) of the years 2003–2007. Black line stands for the mean. For better illustration of the variance, percentiles of the standard deviation are given as grey lines. Sown div. = sown diversity, leg = legume.

Results

Effects of Plant Species Richness and Functional Richness on Bivariate Nutrient Ratios

Plant diversity effects on plant stoichiometry started to become significant after three years (from 2005 onwards) for both species and functional group richness (MANOVA, Table 1). The significance of species richness and functional richness in the model partly depended on the sequence of terms (Table 1, Table S1 and S2), but we still found significant influence of functional diversity after accounting for species richness and marginally significant richness effects after accounting for functional group richness.

Plant species richness and functional group richness had similar effects mainly on P-related nutrient ratios, i.e. C:P, N:P and P:K.

In 2007, however, the average C:N ratio increased with functional group richness, but not with species richness (Table 1). GAMLSS detected an increasing average C:N with increasing species richness in 2005 only (Fig. 1). In all other years, the variance in C:N ratios declined with increasing plant species richness, whereas the average ratio remained unchanged (Fig. 1).

The community wide average C:P ratio increased with species richness in 2006 and 2007 (Table 1) and with functional richness in 2005 and 2006. Reversing the order of terms in the model lead to significant increases of C:P with functional group richness from 2005 onwards (see Table S2). GAMLSS detected the same trend only for 2006, when C:P increased by 25% across the richness gradient. Additionally, the C:P ratio became less variable with increasing plant species richness in 2003 (Fig. 2).

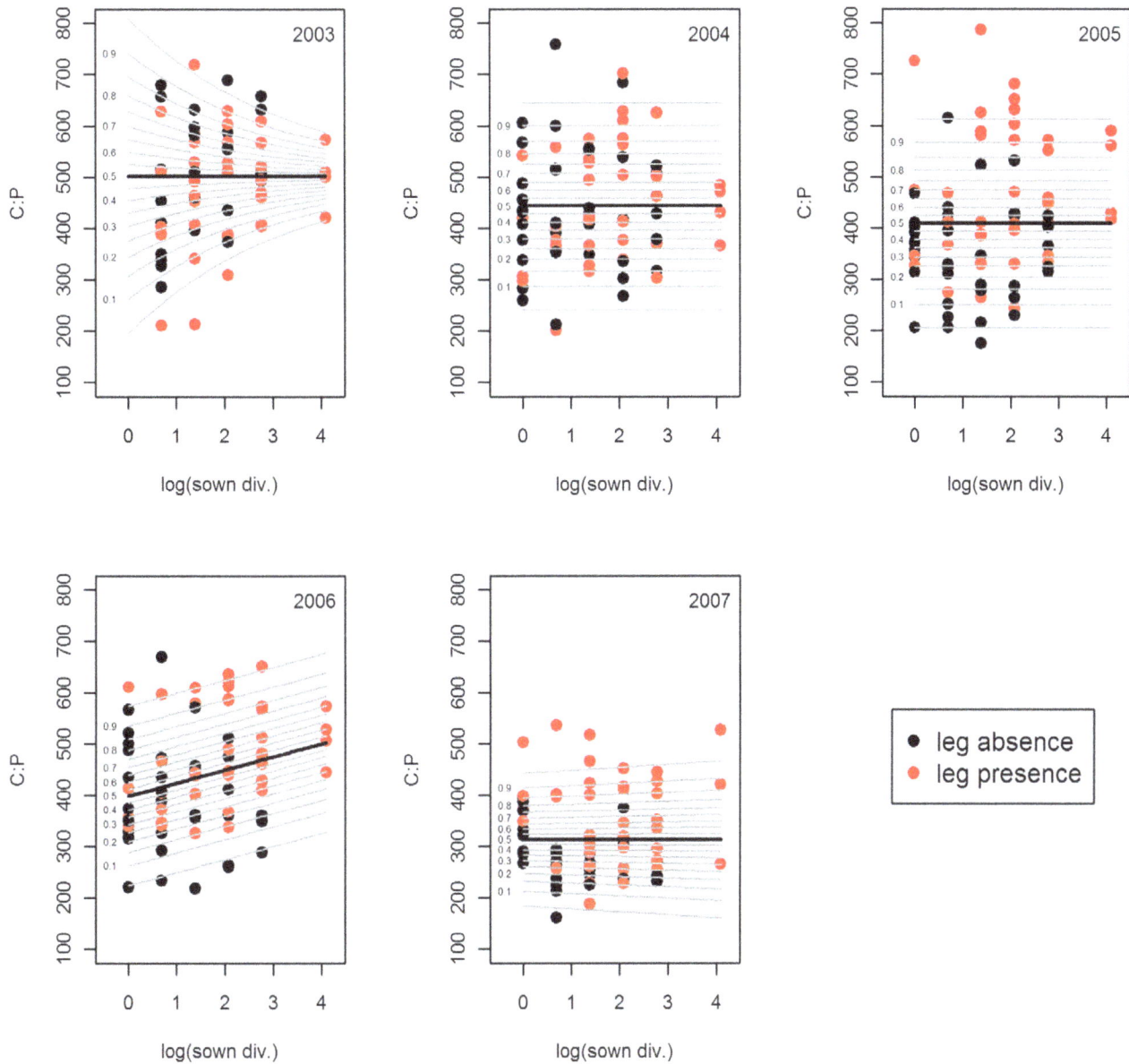

Figure 2. C:P ratio versus plant species richness. GAMLSS (generalized additive model for location scale and shape) model of the molar C:P ratio versus species richness of the years 2003–2007. Black line stands for the mean. For better illustration of the variance, percentiles of the standard deviation are given as grey lines. Sown div. = sown diversity, leg = legume.

Average N:P ratios increased with increasing plant diversity in 2006 and 2007 (both species and functional group richness, Table 1). On average for both years, N:P increased by 26.43% from monocultures to the 60 species-mixture. The same effects were detected in the GAMLSS, which additionally revealed that the variation in N:P ratios decreased with increasing species richness in 2003 and 2005 (see Fig. S1).

The average P:K ratio decreased with increasing plant diversity from 2005 onwards, which was either significant for species richness (Table 1) or functional group richness (see Table S2) depending on term order in the model. The results were consistent in the GAMLSS, which detected a similar negative trend with richness already in 2003. Across the richness gradient, P:K declined by 37.4%. Additionally, the P:K ratios became less variable with increasing richness across all years (Fig. 3).

This decreasing variance with increasing plant diversity was also consistent for the other nutrient ratios containing K (GAMLSS on C:K and N:K, see Figs. S2 and S3), however, these ratios were only marginally shifted by diversity (GAMLSS, C:K, 2006).

Only for C:N were we able to compare the observed community wide ratios to predicted ratios derived from monocultures (see Fig. 4 and 5). The correlation between observed and predicted ratios was significant (linear regression: intercept 8.1448 (SD 1.434), slope 0.756 (SD 0.046)). The slope of the orthogonal regression is not significantly different from 1. As C:N ratios did not show strong richness effects, an analysis of P-related ratios (C:P, N:P, P:K) would have been more informative, but monoculture P-concentrations were not available.

However, we were able to show that the observed coefficient of variation of C:N was not simply reflecting the relative abundances of species, which was used to estimate predicted CV (Fig. 5).

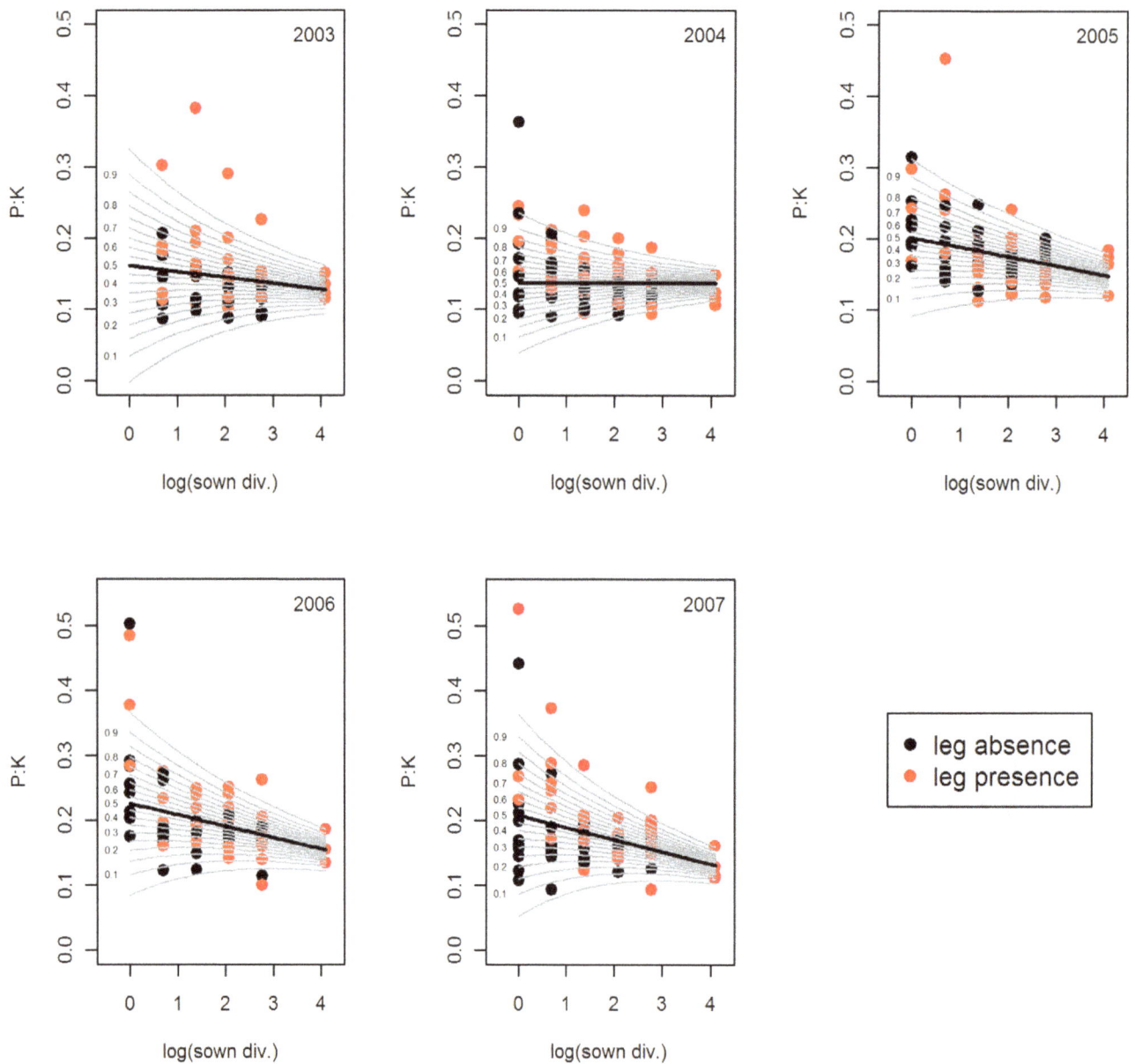

Figure 3. P:K ratio versus plant species richness. GAMLSS (generalized additive model for location scale and shape) model of the molar P:K ratio versus species richness of the years 2003–2007. Black line stands for the mean. For better illustration of the variance, percentiles of the standard deviation are given as grey lines. Sown div. = sown diversity, leg = legume.

Instead, observed CV tended to be higher than predicted CV in the first years and lower in later years. The observed pattern of CV to species richness also differed from the predicted pattern in the different years.

Effect of Legumes and Grasses on Bivariate Nutrient Ratios

Legume presence showed highly significant effects on plant chemical composition of almost all elemental ratios in all years. Grass presence had somewhat weaker effects mainly affecting C-related ratios (Table 1). Legume presence significantly reduced C:N by 27.7% across years (Fig. 1), whereas grass presence increased C:N ratios on average by 32.7%. C:P ratios increased by 26.7% in the presence of legumes in 2005 and 2006 (Fig. 2), and by 21.3% in the presence of grasses from 2004 onwards.

Community-wide N:P ratios increased with legume presence in all years (Fig. S1), showing increasing effect sizes over time (Table 1), with 28.1% increase in 2003 and 90.1% increase in 2007. Grass presence had no effect on N:P ratios. Legume presence increased P:K (Fig. 3) and N:K ratios (Fig. S3) across all years by 7.4% and 80.4% respectively, whereas grass presence decreased C:K on average by 8.1% from 2005–2007.

Shifts in Multivariate Nutrient Ratios

More species-rich assemblages cluster around the origin of the PCA representing all four elements and the species-poor assemblages are more distant from the origin (see Fig. S4). The stoichiometric distance from the origin (*SDist*), which is a measure of stoichiometric imbalance, strongly decreased with increasing plant species richness in all years (Fig. 6). In contrast to the

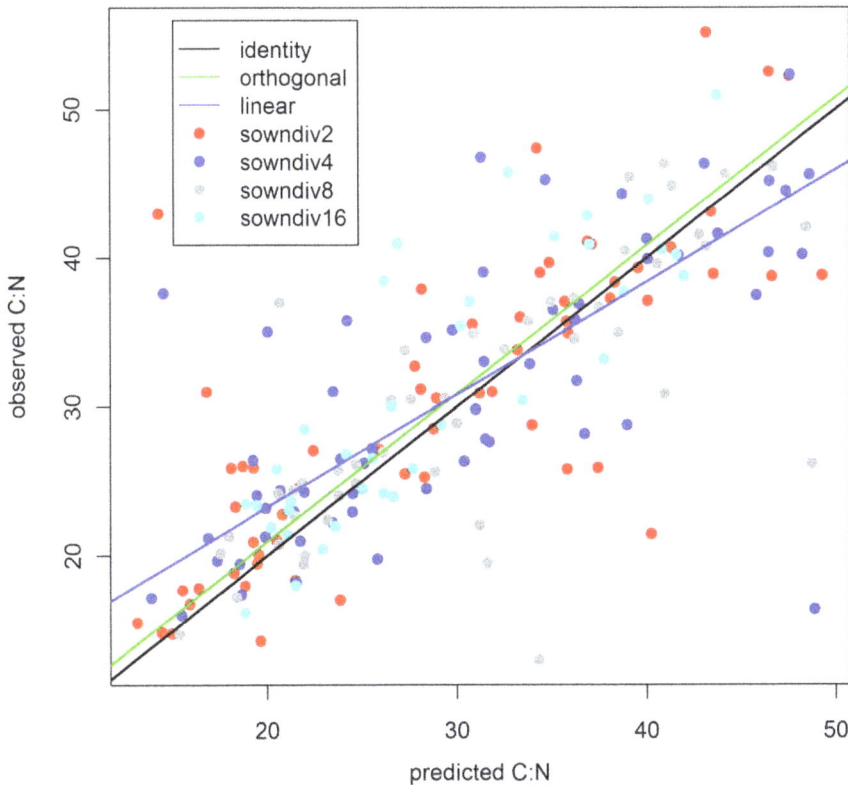

Figure 4. Observed C:N versus predicted C:N ratios. For every measured C:N ratio (2, 4, 8 and 16 species mixtures) the corresponding calculated C:N ratio is shown across years (2003–2006). The green line gives the fit of an orthogonal regression (intercept 1.03 (SD 0.423); estimated C:N 0.999 (SD 0.043). The blue line gives a linear regression (Intercept 8.1448 (SD 1.434); estimated C:N 0.756 (SD 0.046)). For comparison, identity is given by a black line.

analyses on bivariate ratios, the analysis of *SDist* revealed a significant primary effect of plant diversity throughout time, whereas neither block nor grass presence exerted significant influence (see Table S4). On average, *SDist* decreased by 50% from monocultures to the 60 species-mixture. Legume presence increased *SDist* in the first three years of the experiment, but the effect disappeared thereafter. Functional group richness had the same negative effect on *SDist* as species richness, but only in 2004 and 2006. Thus, increasing species richness alone led to less variable multivariate elemental composition of plants.

Discussion

Community-wide plant stoichiometry varied with plant diversity (supporting hypothesis H1). We found significant increases in C:P and N:P with either functional group or species richness from 2005 onwards (supporting H1a). When diversity effects on stoichiometry were present, they increased in strength over time (supporting hypothesis H2, Table 1). In addition to shifting mean ratios, plant diversity also reduced the variance in chemical composition (supporting hypothesis H3), which was seen in both multivariate and bivariate analyses of elemental composition. Plant stoichiometry also strongly responded to the identity of functional groups present, confirming that functional composition of plant communities links to chemical composition. Plant diversity effects on average ratios and stoichiometric variance remained when controlling for the presence of certain functional groups. Whereas previous analyses revealed significant diversity effects on elemental content in plants [29,30] and light use efficiency [31], our study

here is substantially novel as it takes an explicit stoichiometric approach to multiple elements.

Effects of Species Richness on Average Ratios

We found significant effects of species richness or functional group richness on average elemental ratios. The ordering of terms in the model constrained whether species or functional richness effects were dominant, indicating that both aspects were exchangeable. Diversity affected mainly ratios involving P such as C:P and P:K but also C:N. Diversity effects on average stoichiometry showed a lag phase of 3 years, which has also been shown for other responses within the Jena Experiment, e.g. for plant diversity effects on soil microorganisms [32] or soil NO_3-N [33].

Ptacnik et al. [18] suggested that terrestrial plants increase nutrient uptake more efficiently with increasing diversity than their C fixation, because terrestrial plants show little pigment diversity, but high trait variation in nutrient uptake. Therefore, we expected that an increase in species richness reduces C:nutrient ratios in terrestrial plants, consistent with the results of Novotny et al. [34] who observed that plants respond to changes in diversity by decreasing C:P and C:N ratios from monocultures to mixtures. These results rely on data from only one harvest in 2002, the first year of the experiment. In our study, however, we found the opposite trend after a lag phase of three years. Contrary to our expectation, the diversity effects on stoichiometry may rather reflect allocation patterns in the plant than altered resource efficiency. Within the Jena Experiment, increasing diversity increased community biomass, an effect that increased over time

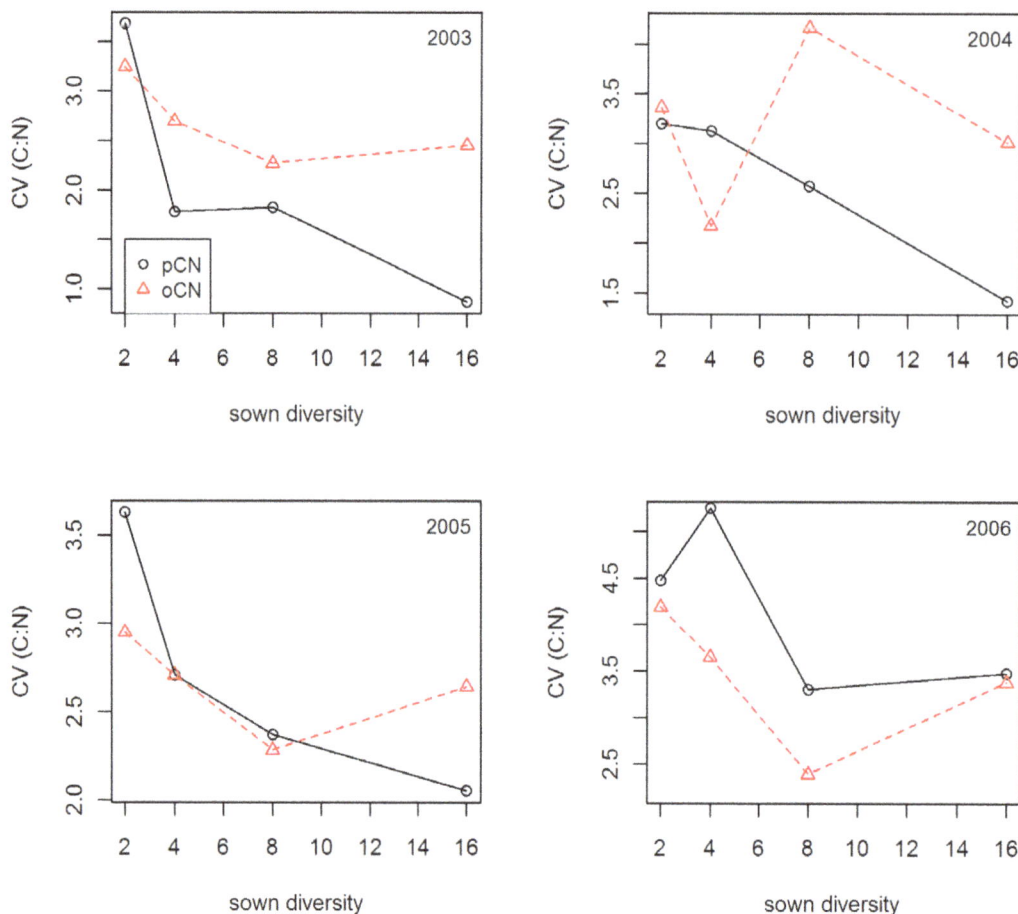

Figure 5. Comparison of the coefficient of variance of observed C:N and predicted C:N ratios. The Coefficient of Variation (CV) of measured (oCN) and calculated C:N (pCN) ratios is shown for different diversity levels (2, 4, 8, and 16 species mixtures) separated by years (2003–2006).

[35]. Plant individuals have to grow taller to successfully compete for light, and in fact this has been shown in the Jena Experiment, where individuals of the same plant species increase in height with increasing diversity [36]. Therefore, plants in more diverse communities have to invest more in rigid stem structure (higher stem:leaf ratios [37,38]), which increases the C-, over proportional to N- and P-concentrations, resulting in the observed increases in C:N and C:P ratios.

The significant correlation between observed C:N ratios and C:N ratios predicted from relative abundances and monoculture C:N indicates that species composition can explain part of community-wide stoichiometric composition. However, this analysis allows little inference with respect to diversity effects, as C:N was not responding to the richness gradient. Thus, we cannot disentangle whether shifts in average ratios or reduced variance were related to mixture effects from species composition or to a more complete resource use as higher richness increases the probability for complementarity effects for multiple elements. Complementarity in elemental specialization would be a special case of the diversity – multifunctionality relationship [5,8], as the number of species affecting elemental concentrations in plants will increase with increasing numbers of elements considered. As in our analysis of stoichiometry, complementarity effects on biomass often show a considerable time lag, which has been shown in the Jena Experiment (see below) and in Cedar Creek [39]. For Jena,

Marquard et al. [35] showed that complementarity effects increased over time and became stronger than selection from 2005 onwards.

Disentangling between statistical averaging and biological complementarity requires the analysis of species-specific elemental concentrations in all mixtures, which have not been analyzed in this experiment or – to the best of our knowledge – any other BDEF experiment. In a new subproject of the Jena Experiment, the trait-based experiment, these analyses will be available from 2011 onwards.

Effects of Functional Groups on Average Ratios

Our analysis corroborated previous findings on the effects of certain functional groups on average nutrient concentrations. In line with the ability of legumes to fix atmospheric N_2 and the associated elevated demand of P, Novotny et al. [34] detected – as we did - that legumes are characterized by lower C:N but higher C:P and N:P ratios than non-legumes (see also [40]). Furthermore, P:K ratios increased significantly across years when legumes were present in the communities, which reflects that communities with legumes had lower concentrations in both P and K, with the reduction being more pronounced with K resulting in higher P per unit K. Within the Jena-Experiment Oelmann et al. [41] found that the absolute aboveground P storage and P exploitation increased if legumes were present in a mixture due to an increased

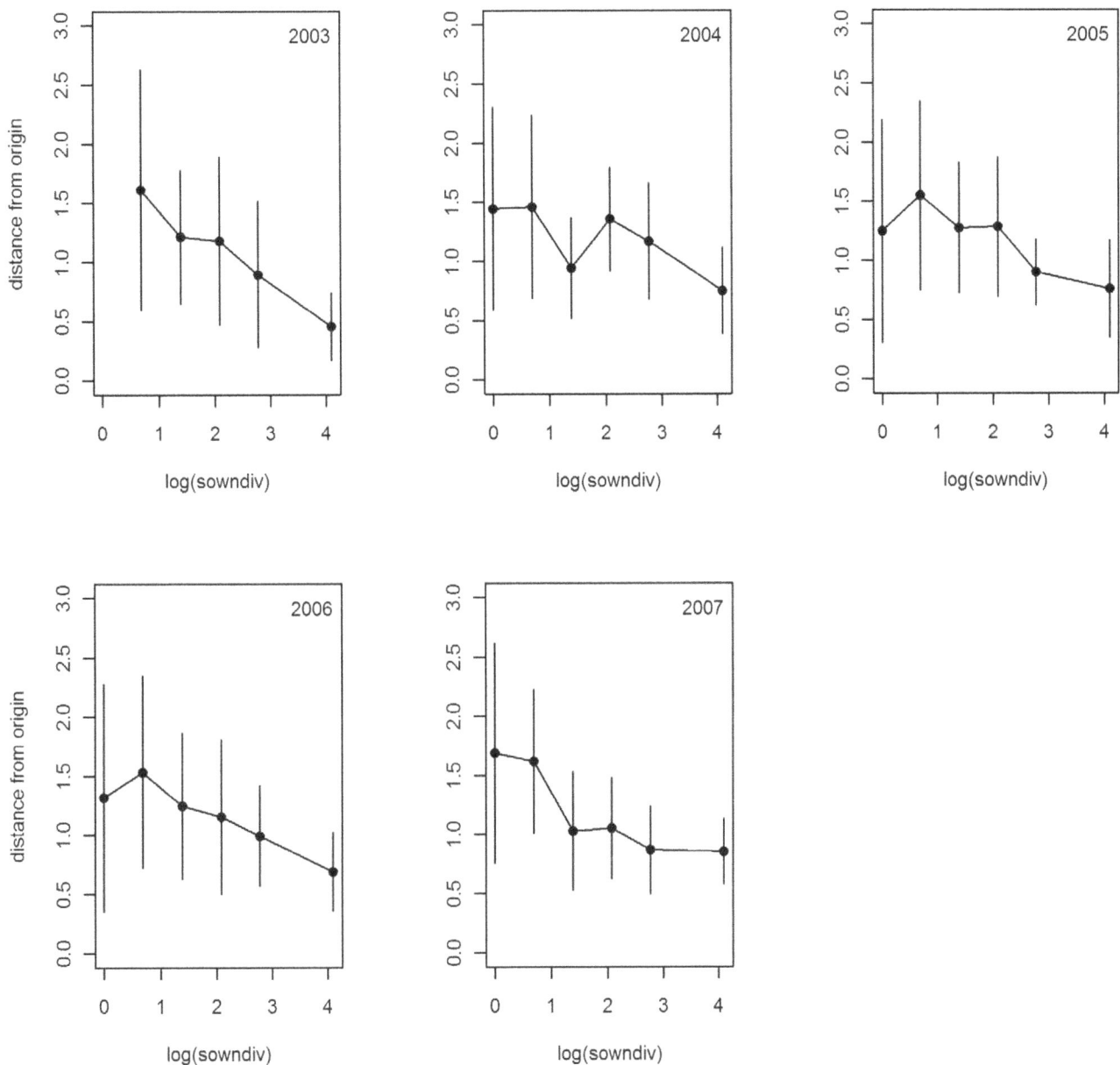

Figure 6. Multielemental stoichiometric distance (SD*ist*) from the origin of the PCA versus sown diversity (log transformed). Error bars are ±1 SD. Axis 1 reflects plant C concentration and opposingly P and K concentration, whereas Axis 2 reflects N alone. The distance from the origin of the PCA (0,0) was used as a measure of stoichiometric imbalance (see methods).

biomass production or due to an increased P demand of legumes for N_2 fixation [42]. Supporting the latter argument, Roscher et al. [43] described a larger requirement of legumes for P, which might reflect higher P requirement for plants fixing N_2 [44]. However, our data rather suggest that P and K are diluted in the average community tissue by legume presence, which indicates that higher N_2 fixation leads to increased biomass production at the expense of reduced tissue concentrations in other elements. The presence of grasses increased the C:nutrient ratios (except for C:K which decreased) across years, corroborating previous findings that the presence of grasses decreases community N and P concentrations because of the lower nutrient demand of monocots compared to dicots [42,45]. Comparing monocultures of grasses and non grasses within the Jena Experiment, we

observed the same pattern as described above but with lower N concentrations.

Effects of Species Diversity on Stoichiometric Variation

Plant diversity was the most important factor constraining stoichiometric variance. There was no effect of grass presence on stoichiometric variance, whereas legume presence had either no or positive effects on variability, i.e. nutrient ratios tended to be less confined with legumes.

We cannot rule out that the diversity effect on variance is based on an averaging effect of species-specific stoichiometry, which would be a statistical effect without biological mechanism. Comparing observed CV of community C:N and predicted CV based on relative abundance and monoculture C:N, we find little evidence that the diversity effect on C:N variance is a statistical averaging effect alone. Especially after a time-lag, observed CV tended to be consistently smaller than predicted CV, indicating that the common resource use of mixtures is less variable than predicted by species composition alone. This might indicate complementarity effects of resource use as discussed above.

Consequences of Diversity-related Changes in Stoichiometry

Irrespective whether plant diversity effects on average stoichiometry and its variance depend on statistical averaging or complementarity, our results indicate that a reduction in plant diversity leads to shifting nutrient ratios and increasing variance in elemental composition of plants. Therefore, we can expect subsequent changes on consumer levels, especially for generalist consumers, as herbivores respond strongly to the elemental concentrations of their food by altering individual consumption and area-specific grazing [10]. Extrapolating from ecosystem comparisons [13], higher C:P ratios in the biomass will result in decreased importance of herbivory but increased importance of decomposition and detritivory.

The enhanced productivity paralleled by elevated C:P ratios allows making different predictions how the loss of plant species might affect secondary production in grasslands, which might increase with increasing food quantity but decreases with decreasing food quality. The net effects of these contrasting changes have to be analyzed. Furthermore, we predict that herbivore body size will mediate the effects of increased variance in plant elemental composition in species-poor grasslands. Large, long-lived herbivores will potentially integrate across the increased stoichiometric variability in low diversity plant assemblages, but for small, short-lived herbivores these assemblages could represent a less reliable food environment with patches differing in food quality.

Supporting Information

Figure S1 N:P ratio versus plant species richness. GAMLSS (generalized additive model for location scale and shape) model of the molar N:P ratio versus species richness of the years 2003–2007. Black line stands for the mean. For better illustration of the variance, percentiles of the standard deviation are given as grey lines. Sown div. = sown diversity, leg = legume.

Figure S2 C:K ratio versus plant species richness. GAMLSS (generalized additive model for location scale and shape) model of the molar C:K ratio versus species richness of the years 2003–2007. Black line stands for the mean. For better

illustration of the variance, percentiles of the standard deviation are given as grey lines. Sown div. = sown diversity, leg = legume.

Figure S3 N:K ratio versus plant species richness. GAMLSS (generalized additive model for location scale and shape) model of the molar N:K ratio versus species richness of the years 2003–2007. Black line stands for the mean. For better illustration of the variance, percentiles of the standard deviation are given as grey lines. Sown div. = sown diversity, leg = legume.

Figure S4 Factor analysis (PCA) of multiple elements across years. For factor loadings see Table S3.

Table S1 MANOVA results on bivariate elemental ratios excluding 60 species mixtures. For each factor, the Pillai Trace value and its significance level are given as well as all ratios for which the factor effect was significant at $p < 0.05$. Significance levels: $p < 0.001 = ***$, $p < 0.01 = **$, $p < 0.05 = *$, $p < 0.1 = .$

Table S2 MANOVA results on bivariate elemental ratios with changed order of effects. For each factor, the Pillai Trace value and its significance level are given as well as all ratios for which the factor effect was significant at $p < 0.05$. Significance levels: $p < 0.001 = ***$, $p < 0.01 = **$, $p < 0.05 = *$, $p < 0.1 = .$

Table S3 Factor loadings for PCA analysis.

Table S4 ANOVA F-values calculating stoichiometric deviance as distance from CA origin.

Acknowledgments

We thank Vicky Temperton (FZ Jülich) for providing CN data of the years 2003/2004, Stephan Rosenkranz for chemical analyses and extractions, Tom Andersen for support with statistical analysis, and the gardening team as well as many student helpers for maintaining the experiment. The manuscript profited from comments by Maren Striebel.

Author Contributions

Conceived and designed the experiments: AE W. Weisser. Performed the experiments: AW CR. Analyzed the data: MA RP HH. Contributed reagents/materials/analysis tools: YO W. Wilcke AW CR. Wrote the paper: MA HH.

References

1. Balvanera P, Pfisterer AB, Buchmann N, He J-S, Nakashizuka T, et al. (2006) Quantifying the evidence for biodiversity effects on ecosystem functioning and services. Ecology letters 9: 1146–1156. Available: http://www.ncbi.nlm.nih.gov/pubmed/16972878. Accessed 5 July 2011.
2. Cardinale BJ, Srivastava DS, Duffy JE, Wright JP, Downing AL, et al. (2006) Effects of biodiversity on the functioning of trophic groups and ecosystems. Nature 443: 989–992. Available: http://www.ncbi.nlm.nih.gov/pubmed/17066035. Accessed 6 July 2011.
3. Hooper DU, Chapin FS, Ewel JJ, Hector A, Inchausti P, et al. (2005) Effects of Biodiversity on Ecosystem Functioning: A consensus of current knowledge. Ecological Monographs 75: 3–35.
4. Hillebrand H, Matthiessen B (2009) Biodiversity in a complex world: consolidation and progress in functional biodiversity research. Ecology letters 12: 1405–1419. Available: http://www.ncbi.nlm.nih.gov/pubmed/19849711. Accessed 5 July 2011.
5. Gamfeldt L, Hillebrand H, Jonsson PR (2008) Multiple functions increase the importance of biodiversity for overall ecosystem functioning. Ecology 89: 1223–1231.
6. Hector A, Bagchi R (2007) Biodiversity and ecosystem multifunctionality. Nature 448: 188–190. Available: http://www.ncbi.nlm.nih.gov/pubmed/17625564. Accessed 7 July 2011.
7. Mouillot D, Villéger S, Scherer-Lorenzen M, Mason NWH (2011) Functional structure of biological communities predicts ecosystem multifunctionality. PloS one 6: e17476. Available: http://www.pubmedcentral.nih.gov/articlerender.fcgi?artid=3053366&tool=pmcentrez&rendertype=abstract. Accessed 21 August 2011.
8. Isbell F, Calcagno V, Hector A, Connolly J, Harpole WS, et al. (2011) High plant diversity is needed to maintain ecosystem services. Nature 477: 199–202. Available: http://www.nature.com/doifinder/10.1038/nature10282. Accessed 11 August 2011.
9. Sterner RW, Elser JJ (2002) Ecological stoichiometry: the biology of elements from molecules to the biosphere. USA: Princeton University Press, USA.
10. Hillebrand H, Borer ET, Bracken MES, Cardinale BJ, Cebrian J, et al. (2009) Herbivore metabolism and stoichiometry each constrain herbivory at different organizational scales across ecosystems. Ecology Letters 12: 516–527. Available: http://www.ncbi.nlm.nih.gov/pubmed/19392711. Accessed 10 June 2011.

11. Elser JJ, Fagan WF, Denno RF, Dobberfuhl DR, Folarin A, et al. (2000) Nutritional constraints in terrestrial and freshwater food webs. Nature 408: 578–580. Available: http://www.ncbi.nlm.nih.gov/pubmed/11117743.

12. Cebrian J, Shurin JB, Borer ET, Cardinale BJ, Ngai JT, et al. (2009) Producer nutritional quality controls ecosystem trophic structure. PloS one 4: e4929. Available: http://www.pubmedcentral.nih.gov/articlerender. fcgi?artid = 2654170&tool = pmcentrez&rendertype = abstract. Accessed 20 September 2011.

13. Cebrian J, Lartigue J (2004) Patterns of herbivory and decomposition in aquatic and terrestrial ecosystems. Ecological Monographs 74: 237–259.

14. Persson J, Fink P, Goto A, Hood JM, Jonas J, et al. (2010) To be or not to be what you eat: regulation of stoichiometric homeostasis among autotrophs and heterotrophs. Oikos 119: 741–751. Available: http://doi.wiley.com/10.1111/j. 1600-0706.2009.18545.x. Accessed 25 July 2011.

15. Ågren GI (2008) Stoichiometry and Nutrition of Plant Growth in Natural Communities. Annual Review of Ecology, Evolution, and Systematics 39: 153–170. Available: http://www.annualreviews.org/doi/abs/10.1146/annurev. ecolsys.39.110707.173515. Accessed 20 July 2011.

16. Reich PB, Oleksyn J, Wright IJ, Niklas KJ, Hedin L, et al. (2010) Evidence of a general 2/3-power law of scaling leaf nitrogen to phosphorus among major plant groups and biomes. Proceedings Biological sciences/The Royal Society 277: 877–883. Available: http://www.pubmedcentral.nih.gov/articlerender. fcgi?artid = 2842731&tool = pmcentrez&rendertype = abstract. Accessed 26 August 2011.

17. Striebel M, Behl S, Stibor H (2009) The coupling of biodiversity and productivity in phytoplankton communities: consequences for biomass stoichiometry. Ecology 90: 2025–2031.

18. Ptacnik R, Moorthi SD, Hillebrand H (2010) Hutchinson Reversed, or Why There Need to Be So Many Species. Advances In Ecological Research. Vol. 43. 1–43. doi:10.1016/S0065-2504(10)43001-9.

19. Doak DF, Bigger D, Harding EK, Marvier MA, O'Malley RE, et al. (1998) The statistical inevitability of stability-diversity relationships in community ecology. American Naturalist 151: 264–276. doi:10.1086/286117.

20. Roscher C, Schumacher J, Baade J, Wilcke W, Gleixner G, et al. (2004) The role of biodiversity for element cycling and trophic interactions: an experimental approach in a grassland community. Basic and Applied Ecology 121: 107–121.

21. Kluge G, Müller-Westermeier G (2000) Das Klima ausgewählter Orte der Bundesrepublik Deutschland: Jena. Berichte des Deutschen Wetterdienstes 213.

22. Ellenberg H, Leuschner C (2010) Vegetation Mitteleuropas mit den Alpen. Ulmer Verlag, Stuttgart.

23. Weigelt A, Marquard E, Temperton VM, Roscher C, Scherber C, et al. (2010) The Jena Experiment: six years of data from a grassland biodiversity experiment. Ecology 91: 929.

24. Oelmann Y, Potvin C, Mark T, Werther L, Tapernon S, et al. (2010) Tree mixture effects on aboveground nutrient pools of trees in an experimental plantation in Panama. Plant Soil 326: 199–212. Available: http://www. springerlink.com/index/10.1007/s11104-009-9997-x. Accessed 16 November 2011.

25. Scheiner SM (1993) MANOVA: multiple response variables and multispecies interactions. Design and analysis of ecological experiments Chapman and Hall, New York, New York, USA: 94–112.

26. Stasinopoulos DM, Rigby RA (2007) Generalized additive models for location scale and shape (GAMLSS) in R. Journal Of Statistical Software 23: 1–64.

27. Johnson JB, Omland KS (2004) Model selection in ecology and evolution. Trends in Ecology & Evolution 19: 101–108. doi:10.1016/j.tree.2003.10.013.

28. R Development Core Team (2010) R: A language and environment for statistical computing. The R Development Core Team R Foundati.

29. Oelmann Y, Wilcke W, Temperton VM, Buchmann N, Roscher C, et al. (2007) Soil and Plant Nitrogen Pools as Related to Plant Diversity in an Experimental Grassland. Soil Science Society of America Journal 71: 720. Available: https:// www.soils.org/publications/sssaj/abstracts/71/3/720. Accessed 16 November 2011.

30. van Ruijven JV, Berendse F (2005) Diversity – productivity relationships: Initial effects, long-term patterns, and underlying mechanisms. Proceedings of the National Academy of Sciences 102: 695–700.

31. Roscher C, Kutsch WL, Schulze E-D (2011) Light and nitrogen competition limit Lolium perenne in experimental grasslands of increasing plant diversity. Plant biology (Stuttgart, Germany) 13: 134–144. Available: http://www.ncbi. nlm.nih.gov/pubmed/21143734. Accessed 12 June 2011.

32. Eisenhauer N, Bessler H, Engels C, Gleixner G, Habekost M, et al. (2010) Plant diversity effects on soil microorganisms support the singular hypothesis. Ecology 91: 485–496. Available: http://www.ncbi.nlm.nih.gov/pubmed/20392013.

33. Oelmann Y, Buchmann N, Gleixner G, Habekost M, Roscher C, et al. (2011) Plant diversity effects on aboveground and belowground N pools in temperate grassland ecosystems: Development in the first 5 years after establishment. Global Biogeochemical Cycles 25: 1–11. Available: http://www.agu.org/pubs/ crossref/2011/2010GB003869.shtml. Accessed 21 December 2011.

34. Novotny AM, Schade JD, Hobbie SE, Kay AD, Kyle M, et al. (2007) Stoichiometric response of nitrogen-fixing and non-fixing dicots to manipulations of CO2, nitrogen, and diversity. Oecologia 151: 687–696. Available: http://www.ncbi.nlm.nih.gov/pubmed/17106721. Accessed 17 July 2011.

35. Marquard E, Weigelt A, Temperton VM, Roscher C, Schumacher J, et al. (2009) Plant species richness and functional composition drive overyielding in a six-year grassland experiment. Ecology 90: 3290–3302. Available: http://www. ncbi.nlm.nih.gov/pubmed/20120799.

36. Schmidtke A, Rottstock T, Gaedke U, Fischer M (2010) Plant community diversity and composition affect individual plant performance. Oecologia 164: 665–677. Available: http://www.ncbi.nlm.nih.gov/pubmed/20617445. Accessed 23 August 2011.

37. Gubsch M, Buchmann N, Schmid B, Schulze E-D, Lipowsky A, et al. (2011) Differential effects of plant diversity on functional trait variation of grass species. Annals of botany 107: 157–169. Available: http://www.pubmedcentral.nih. gov/articlerender. fcgi?artid = 3002477&tool = pmcentrez&rendertype = abstract. Accessed 16 November 2011.

38. Roscher C, Schmid B, Buchmann N, Weigelt A, Schulze E-D (2011) Legume species differ in the responses of their functional traits to plant diversity. Oecologia 165: 437–452. Available: http://www.ncbi.nlm.nih.gov/pubmed/ 20680645. Accessed 3 October 2011.

39. Fargione J, Tilman D, Dybzinski R, Lambers JHR, Clark C, et al. (2007) From selection to complementarity: shifts in the causes of biodiversity-productivity relationships in a long-term biodiversity experiment. Proceedings Biological sciences/The Royal Society 274: 871–876. Available: http://www. pubmedcentral.nih.gov/articlerender. fcgi?artid = 2093979&tool = pmcentrez&rendertype = abstract. Accessed 17 July 2012.

40. Güsewell S (2004) N: P ratios in terrestrial plants: variation and functional significance. New Phytologist 164: 243–266. Available: http://doi.wiley.com/ 10.1111/j.1469-8137.2004.01192.x. Accessed 4 August 2011.

41. Oelmann Y, Richter AK, Roscher C, Rosenkranz S, Temperton VM, et al. (2011) Does plant diversity influence phosphorus cycling in experimental grasslands? Geoderma 167–168: 178–187. Available: http://linkinghub.elsevier. com/retrieve/pii/S0016706111002795. Accessed 17 November 2011.

42. Oelmann Y, Kreutziger Y, Temperton VM, Buchmann N, Roscher C, et al. (2007) Nitrogen and phosphorus budgets in experimental grasslands of variable diversity. Journal of environmental quality 36: 396–407. Available: http://www. ncbi.nlm.nih.gov/pubmed/17255627. Accessed 25 July 2011.

43. Roscher C, Thein S, Weigelt A, Temperton VM, Buchmann N, et al. (2011) N2 fixation and performance of 12 legume species in a 6-year grassland biodiversity experiment. Plant and Soil 341: 333–348. Available: http://www.springerlink. com/index/10.1007/s11104-010-0647-0. Accessed 16 November 2011.

44. Vitousek PM, Cassman K, Cleveland C, Field CB, Grimm NB, et al. (2002) Towards an ecological understanding of biological nitrogen fixation. Biogeochemistry 57/58: 1–45.

45. Broadley MR, Bowen HC, Cotterill HL, Hammond JP, Meacham MC, et al. (2004) Phylogenetic variation in the shoot mineral concentration of angiosperms. Journal of experimental botany 55: 321–336. Available: http://www.ncbi.nlm. nih.gov/pubmed/14739259. Accessed 16 November 2011.

2

Taxonomic and Numerical Resolutions of Nepomorpha (Insecta: Heteroptera) in Cerrado Streams

Nubia França da Silva Giehl[1,4]*, Karina Dias-Silva[2], Leandro Juen[3], Joana Darc Batista[4], Helena Soares Ramos Cabette[4,5]

1 Programa de Pós-graduação em Ecologia e Conservação, Universidade do Estado de Mato Grosso (UNEMAT), Nova Xavantina, Mato Grosso, Brasil, 2 Programa de Pós-graduação em Ciências Ambientais, ICB1, Universidade Federal de Goiás, Goiânia, Goiás, Brasil, 3 Laboratório de Ecologia e Conservação, Instituto de Ciências Biológicas, Universidade Federal do Pará (UFPA), Belém, Pará, Brasil, 4 Laboratório de Entomologia, Universidade do Estado de Mato Grosso (UNEMAT), Nova Xavantina, Mato Grosso, Brasil, 5 Departamento de Biologia, Universidade do Estado de Mato Grosso (UNEMAT), Nova Xavantina, Mato Grosso, Brasil

Abstract

Transformations of natural landscapes and their biodiversity have become increasingly dramatic and intense, creating a demand for rapid and inexpensive methods to assess and monitor ecosystems, especially the most vulnerable ones, such as aquatic systems. The speed with which surveys can collect, identify, and describe ecological patterns is much slower than that of the loss of biodiversity. Thus, there is a tendency for higher-level taxonomic identification to be used, a practice that is justified by factors such as the cost-benefit ratio, and the lack of taxonomists and reliable information on species distributions and diversity. However, most of these studies do not evaluate the degree of representativeness obtained by different taxonomic resolutions. Given this demand, the present study aims to investigate the congruence between species-level and genus-level data for the infraorder Nepomorpha, based on taxonomic and numerical resolutions. We collected specimens of aquatic Nepomorpha from five streams of first to fourth order of magnitude in the Pindaíba River Basin in the Cerrado of the state of Mato Grosso, Brazil, totaling 20 sites. A principal coordinates analysis (PCoA) applied to the data indicated that species-level and genus-level abundances were relatively similar (>80% similarity), although this similarity was reduced when compared with the presence/absence of genera (R = 0.77). The presence/absence ordinations of species and genera were similar to those recorded for their abundances (R = 0.95 and R = 0.74, respectively). The results indicate that analyses at the genus level may be used instead of species, given a loss of information of 11 to 19%, although congruence is higher when using abundance data instead of presence/absence. This analysis confirms that the use of the genus level data is a safe shortcut for environmental monitoring studies, although this approach must be treated with caution when the objectives include conservation actions, and faunal complementarity and/or inventories.

Editor: Fabio S. Nascimento, Universidade de São Paulo, Faculdade de Filosofia Ciências e Letras de Ribeirão Preto, Brazil

Funding: The authors thank the Coordenação de Aperfeiçoamento de Pessoal de Nível Superior (CAPES) for a masters scholarship and PELD/CNPq for a DTI-C Scholarship; KDS thanks the Conselho Nacional de Desenvolvimento Científico e Tecnológico (CNPq) for a doctoral scholarship; JDB thanks PELD/CNPq for PDJ scholarship; to the Fundação de Amparo a Pesquisa do Estado de Mato Grosso (FAPEMAT) (Proc. 98/04), Programa de Apoio à Pos-Graduação (PROAP) and Programa de Cooperação Técnica (PROCAD #109/2007) for financial support for data collection. The funders had no role in study design, data collection and analysis, decision to publish, or preparation of the manuscript.

Competing Interests: The authors have declared that no competing interests exist.

* Email: nubiagiehl@gmail.com

Introduction

Human activities generate serious impacts on natural aquatic environments, especially by replacing the vegetation with plantations and pastures [1], constructing dams and reservoirs, and diverting the natural course of waterways [2,3]. These processes, either alone or in synergy, result in decreased habitat heterogeneity, increased input of sediments into the channel, and the loss of aquatic biodiversity [4,5].

Due to major environmental problems and limited resources for evaluating diversity, researchers have resorted to relatively fast and inexpensive methods to evaluate and monitor aquatic ecosystems [6–9]. In particular, while invertebrates are widely used for the monitoring of freshwater systems, they are prone to a "Linnean deficit" [10], in which closely-related species have been diagnosed based on subtle or even subjective morphological characters, and many have still not been described [11]. The lack of appropriate

identification keys may also contribute to the lack of an accurate taxonomic resolution in many studies, few of which are capable of working at the species level. More often than not, then, higher taxonomic levels have been analyzed, and this has been justified based on arguments such as an effective cost-benefit ratio, the time needed for processing samples [8], limited resources [12], a lack of taxonomists and/or data on the ranges and ecological requirements of the species [11,13].

Given the difficulties and demands for rapid methods for bioevaluation, a priority issue is understanding the level of taxonomic identification required for studies of anthropogenic impacts and the monitoring of aquatic systems [14]. The question of numerical resolution–whether abundance or presence/absence data are more effective for the analysis of patterns of biodiversity and environmental quality–has also been evaluated [9,15]. The use of numerical resolution may be an alternative that reduces sample processing time and allows a comparison of results using different methods of inventory and monitoring, and across different regions [16].

Insects of the order Heteroptera (Nepomorpha and Gerromorpha) have aquatic and semi-aquatic habits, are widely distributed, and vary considerably in form and niche, allowing them to occupy a diversity of habitats, including both lentic and lotic bodies of water [17]. Most heteropterans live exclusively in water from the nymphal stage until becoming adults, as in the case of the Nepomorpha. These organisms are ecologically-important predators, and provide a variety of ecological services, including: (i) the biological control of disease vectors, such as the larvae of *Anopheles*, *Culex*, and *Aedes* mosquitoes [18,19] and *Biomphalaria* snails [20]; (ii) the predation of aquatic and terrestrial organisms at the water/air interface; (iii) serving as prey for fish, amphibians, and birds [21]; and (iv) being able to respond to environmental disturbances [17,22,23].

Nepomorpha have only recently become the subjects of environmental evaluation or conservation studies [e.g. 22–26]. In brasilian studies, difficulties have been encountered with regard to the identification of species, which are exacerbated by the lack of reference material, which is often dispersed and inaccessible. On the other hand, identification at the generic level is relatively simple, being based on well-defined characters. The time necessary for the identification of genera tends to be much shorter, the results more reliable, and there is no need to submit the material to experts.

Based on these premises, this study aims to investigate the level of congruence in the data matrices of Nepomorpha genera versus species, by applying taxonomic and numerical resolutions and analyzing whether congruence remains similar when analyzing altered and preserved sites separately, in order to answer to the following questions: a) how much information is lost when the analyses are conducted with a genus-level data matrix in comparison with a species-level one; and b) whether the ordinating patterns obtained from the incidence (presence/absence) of genera and/or species are similar to those obtained from abundance data. A marked congruence of genus-level data with those for species is assumed to support the use of the former for environmental-monitoring studies, enabling quicker decision-making by managers of affected populations, and scholars, when dealing with areas with high habitat-conversion rates.

Materials and Methods

The last author, HSRC, has a permanent license to collect scientific specimens of aquatic insects (14457-1) granted by IBAMA/SISBIO, an organ of the Brazilian Environment Ministry, according to federal legislation. The areas where the specimens were collected are privately owned (Appendix S1), but prior permission was obtained from the owners or managers. None of the specimens collected in work represented IACUC-registered or endangered species.

Study area

The Pindaíba River Basin is a right-bank tributary of the middle Mortes River, located in the southwestern state of Mato Grosso, Brazil, and its basin includes the municipalities of Barra do Garças, Araguaiana, Cocalinho, and Nova Xavantina. The region's climate is Aw in the Köppen classification, with two clearly distinct seasons: dry and rainy [27]. Mean annual rainfall ranges from 1500 mm to 1800 mm, and temperatures from 18.9°C to 33.7°C [28].

The study was carried out in stream sections ranging from the first to the fourth order (classification proposed by [29]). In this classification, first-order watercourses have no tributaries, and when two first order tributaries are connected, a second-order stream is formed. Thus two combined *n* order streams form a river of (n+1) order. A total of 20 sampling sites (Figure 1), with different conservation status were sampled in the Cachoeirinha, Caveira, Da Mata, Papagaio and Taquaral streams. The sites were visually evaluated using the Habitat Integrity Index (HII) [30]. This index, adapted from Petersen's protocol [31] for small brazilian streams, is composed of twelve questions that describe the pattern of land use outside the riparian forest, riparian forest conditions, and stream channel characteristics. The HII provides values from 0 to 1, and the closer the value is to 1, the more pristine is the stream.

Environments with a HII of less than 0.70 presented considerable deforestation on one bank and the riparian vegetation, when it existed, was no more than a narrow strip. Land use outside the riparian vegetation consisted of pastures, with periodic access of cattle to the streams, resulting in banks with excavations or deep rutting. The most disturbed stream had a weir just upstream from the section sampled. Environments with HII≥0.70 varied from areas with clearly defined riparian vegetation (from 5 to 30 m wide) to environments with intact forest without gaps in their canopy. Sites with indices closer to 0.70 had extensive pasture or soybean plantations outside the riparian vegetation.

Collection and identification

The specimens of Heteroptera (Nepomorpha) were collected in the rainy (January), dry (June/July), and early rainy (October/November) seasons in 2005, except in the Caveira Stream, which was sampled in 2008 during the same periods of the year. For sampling, we demarcated a 100 m transect in each of the streams, subdivided into twenty 5-meter segments. In each segment, a 18 cm sieve with a 250 μm mesh were applied three times, constituting a sample with 20 sites per stream/order in each season [24].

Specimens were preserved in 85% ethyl alcohol and identified by means of dichotomous keys [32–37]. Whenever necessary, specimens were analyzed by experts from the Federal University of the Pampa (UNIPAMPA) and the Federal University of Minas Gerais (UFMG). Samples were deposited in the "James Alexander Ratter" Zoo-Botanical Collection at the Nova Xavantina *campus* of Mato Grosso State University (UNEMAT), Mato Grosso, Brazil.

Data analysis

In order to minimize data-normality problems, we used the logarithmic transformation log (x+1) in the abundance matrices. To test whether the data matrices were congruent, the analyses were initially conducted including all sites (r_{all}, 20 sites with HII between 0.51 and 0.96) and, then, in order to determine whether congruence was sustained, we repeated the analyses separating the environments classified as "altered" (r_{alter}; 10 sites, HII<0.70) from those identified as "preserved" ($r_{preserv}$; 10 sites, HII≥0.70).

A Principal Coordinates Analysis (PCoA) was used to order sampling sites. The Bray-Curtis distance matrix was used for the abundance data and the Jaccard similarity matrix for presence/absences data (incidence), generating eigenvectors which were subsequently used in the Procrustes analysis [38]. This method was used to evaluate the degree of congruence of the Nepomorpha community, comparing the abundance of species and genera (taxonomic resolution), species abundance *versus* the presence/absence of species and genera, and finally, the abundance of genera *versus* their incidence (numerical resolution).

This is an overlapping method that compares each ordination pair using a rotational algorithm (rotational-fit), which finds the

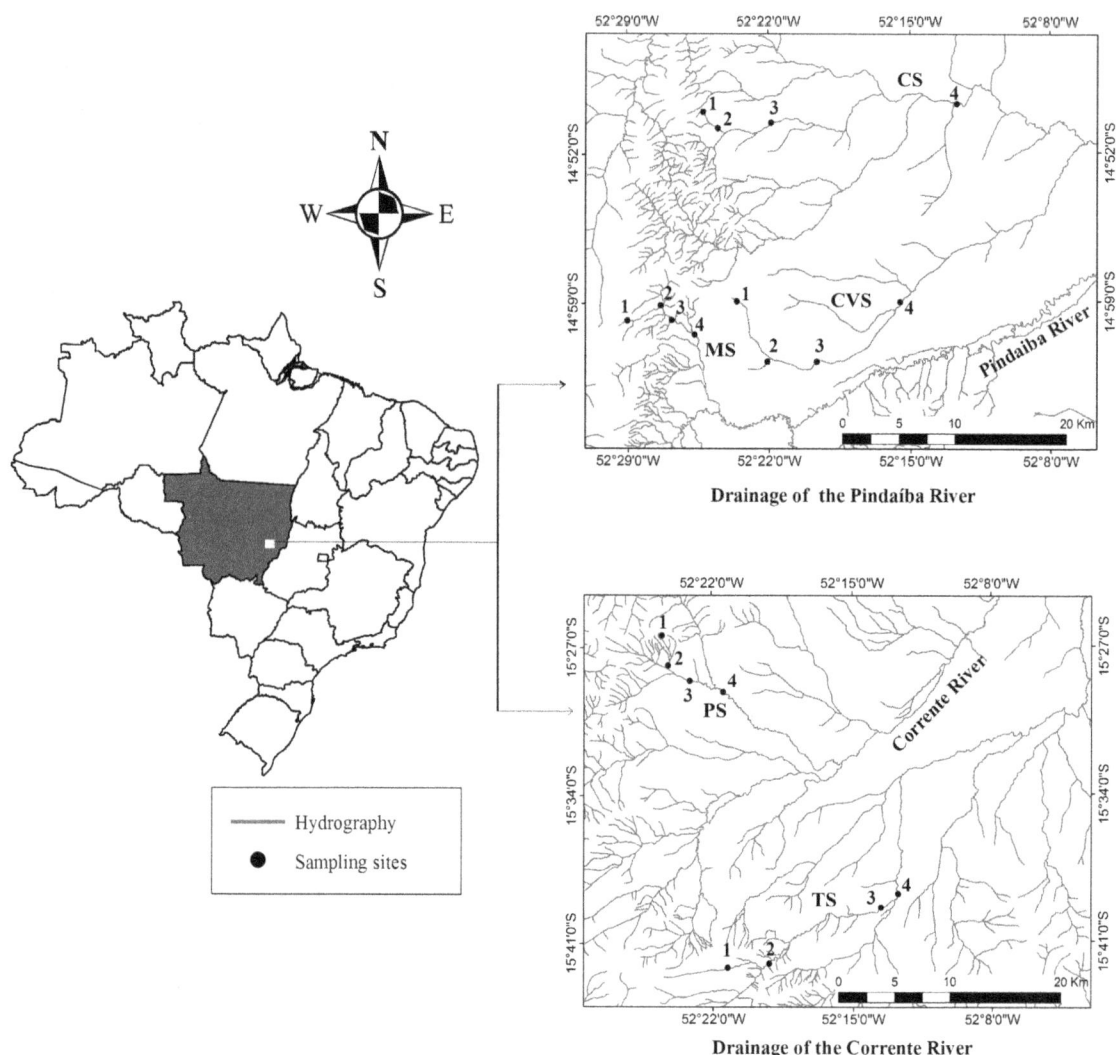

Figure 1. Sampling sites in the drainage basins of the Pindaíba and Corrente rivers, Mato Grosso, Brazil. (CS–Cachoeirinha Stream; CVS–Caveira Stream; MS–Mata Stream; PS–Papagaio Stream; TS–Taquaral Stream; 1st to 4th–stream orders).

best fit between the corresponding geometric ordinations and, provides a correlation value, r (the square root of $1-m^2$). Values closer to 0 indicate a greater difference between ordination patterns and a value of 1 indicates complete overlap between the matrices [38,39]. The fits were quantified and tested for statistical significance (p<0.05) using the Monte Carlo test (with 10,000 permutations). By using all the PCoA axes, we sought to represent the total variance of the system and enable the visualization of all the dimensions of the composition of the community. This permitted the direct comparison of pairs of matrices. In this study, the complementary part to a value of r = 1.0 (100%) was used as an indication of a loss of information. All analyzes were performed through the R Software, using the protest function in the vegan package [40].

Results

We collected 465 Nepomorpha specimens representing six families, 13 genera, and 43 species. When the incidence of Nepomorphan at the different sites was ordered by the HII, 10 species occurred only at sites with an integrity of >0.7 and 12 only where HII was <0.7 (Figure 2). Most of these species were

relatively rare (n≤4 specimens), except for *Belostoma estevezae* Ribeiro & Alecrim, 2008 (n = 17) and *B. ribeiroi* De Carlo, 1993 (n = 9).

Taxonomic resolution – genera *versus* species

The ordination of the abundance data for genera was similar to that for species, with >80% similarity for the whole data set ($r_{all} = 0.84$, p<0.05) and similar values for the two types of habitat ($r_{alter} = 0.92$ and $r_{preserv} = 0.81$, p<0.05). There was thus a 16% loss of information when analyzing all sites together, 8% for altered sites only and 19% for preserved sites (Table 1).

Numerical resolution – abundance *versus* presence/absence

The ordination patterns of the species presence/absence data were highly congruent (>94% similarity) when compared to the data on species abundance. As a consequence, the loss of information was reduced: 6, 5 and 3% ($r_{all} = 0.94$, $r_{alter} = 0.95$, and $r_{preserv} = 0.97$, p<0.05), respectively. On the other hand, when considering the incidence matrix for genera, despite a degree of congruence, there was an increased loss of information in

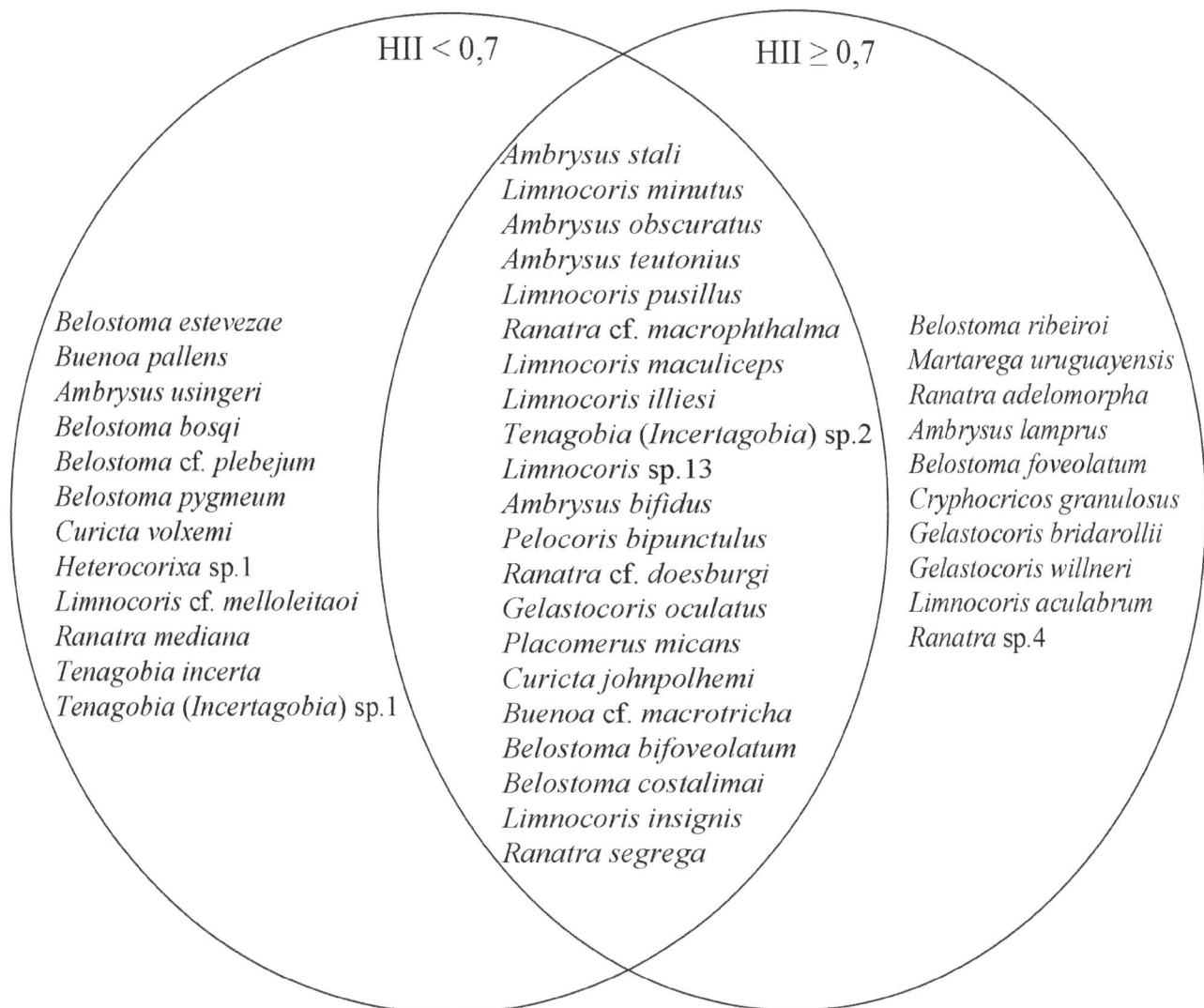

HII < 0,7   **HII ≥ 0,7**

Belostoma estevezae
Buenoa pallens
Ambrysus usingeri
Belostoma bosqi
Belostoma cf. *plebejum*
Belostoma pygmeum
Curicta volxemi
Heterocorixa sp.1
Limnocoris cf. *melloleitaoi*
Ranatra mediana
Tenagobia incerta
Tenagobia (Incertagobia) sp.1

Ambrysus stali
Limnocoris minutus
Ambrysus obscuratus
Ambrysus teutonius
Limnocoris pusillus
Ranatra cf. *macrophthalma*
Limnocoris maculiceps
Limnocoris illiesi
Tenagobia (Incertagobia) sp.2
Limnocoris sp.13
Ambrysus bifidus
Pelocoris bipunctulus
Ranatra cf. *doesburgi*
Gelastocoris oculatus
Placomerus micans
Curicta johnpolhemi
Buenoa cf. *macrotricha*
Belostoma bifoveolatum
Belostoma costalimai
Limnocoris insignis
Ranatra segrega

Belostoma ribeiroi
Martarega uruguayensis
Ranatra adelomorpha
Ambrysus lamprus
Belostoma foveolatum
Cryphocricos granulosus
Gelastocoris bridarollii
Gelastocoris willneri
Limnocoris aculabrum
Ranatra sp.4

Figure 2. Species and morphospecies occurring in the streams surveyed in the Pindaíba River basin, Mato Grosso, Brazil.

comparison with the species data, i.e. 26, 11 and 20% ($r_{all} = 0.74$, $r_{alter} = 0.89$, and $r_{preserv} = 0.80$, p<0.05), results similar to those obtained for the comparison of the species abundance matrix with the genus presence/absence data ($r_{all} = 0.77$, $r_{alter} = 0.91$, and $r_{preserv} = 0.84$, p<0.05).

Discussion

Effects of taxonomic resolution

The results of the taxonomic resolution indicated a high degree of congruence in the ordination of matrices generated for genera and species, showing that the distribution of the Nepomorpha community was represented adequately when using the genus level, with >80% similarity. This congruence may be explained by

Table 1. Taxonomic and numerical resolutions between data sets (abundance and incidence) for nepomorphan genera and species in Cerrado streams in the Pindaíba River basin, Mato Grosso, Brazil.

Data sets	r_{all}	p	r_{alter}	p	$r_{preserv}$	p
Species abundance *vs.* Genus abundance	0.84	<0.001	0.92	<0.001	0.81	=0.011
Species abundance *vs.* Species incidence	0.94	<0.001	0.95	<0.001	0.97	<0.001
Species abundance *vs.* Genus incidence	0.77	<0.001	0.91	<0.001	0.84	=0.031
Genus abundance *vs.* genus incidence	0.74	<0.001	0.89	<0.001	0.80	=0.012

the small number of species found in each genus, given that more than half of the genera were represented by only one or two species, and that the genera with the highest species richness were also those that were most abundant and most amply distributed. This reduced diversity is characteristic of predator taxa, such as the Plecoptera, for example. A previous study of macroinvertebrate communities [41] regarded congruence values of over 0.75 to be indicators of good levels of congruence, and in our study, which analyzed only the Nepomorpha, all the values were always similar to or above this threshold. Thus, our results indicate that it is safe to use genus-level, rather than species data, in monitoring studies, although caution is a recommended when congruence values are lower than 0.75.

This taxonomic resolution approach has been tested and applied in a number of different types of study, including the monitoring of communities, the evaluation of pollution in freshwater and seawater habitats, as well as the analysis of environmental gradients, demonstrating congruence between different taxonomic levels, even when focusing on different taxonomic groups (e.g. macrobenthos [14]; invertebrates and diatoms [42]; macroinvertebrates [43–45]; and phytoplankton [9,46].

A number of recent studies analyzing different levels of taxonomic resolution in macroinvertebrates under bio-evaluation have found that higher taxonomic levels provide relevant information on environmental status [47,26], and when species are aggregated on a higher taxonomic level, some environmental or biological information is lost [45]. By contrast, taxonomic difficulties can lead to errors of identification, which may result in major problems of interpretation [11,48]. The balance between these opposing tendencies will depend on the purpose of the study.

In the present study, we observed that, in the Nepomorpha, there was reduced taxonomic congruence when sites were more preserved. This may be due to the fact that these sites have suffered little environmental change, and that any change would be reflected more effectively at the species level. However, the speed at which environmental changes are occurring in tropical aquatic ecosystems, and the delay in reaching the species level for data analysis, justifies the use of a genus-level taxonomic resolution. The analysis of genera rather than species should nevertheless be considered with caution, given that, while this level of analysis may be suitable for one ecosystem or region, it may not be equally effective elsewhere [41,49]. It may also vary according to the index used, the type of analysis adopted, and the group of organisms [50,51].

In any case, identification to the species level is not always possible, given the paucity of taxonomic information (Linnean deficit) and the lack of geographic distribution data, that is, a Wallacean deficit [10,52,53], as well as the inadequate number of taxonomists, identification keys, and reference collections. Moreover, the reliable morphological identification of species is often only possible for adult male and larvae in the last aquatic instar stage [48]. As a result, many of the specimens remain unidentified and are excluded from the study. According to [54], there is a payoff between the taxonomic resolution adopted and the clarity of the pattern or effect, which must be taken into account before determining the level of data resolution most appropriate for the study. Given the congruence found between the data matrices and the difficulty of identifying species reliably, our results indicate that a genus-level approach should be used in monitoring studies, considering costs and benefits (8), lower demands for time and resources, and reduced taxonomic errors compared to a species–level approach. The resources saved by choosing an optimal taxonomic (and numerical) resolution may be allocated to other purposes, such as ensuring temporal continuity and increasing spatial coverage in biomonitoring programs [9,39], something which may increase the reliability of the results, given that communities may change depending on a wide range of environmental factors [55].

Effects of numerical resolution

In the present study, we found a lower degree of similarity between the abundance matrices (for species or genera) and the presence/absence of genera when compared with that of species, indicating that monitoring studies based on a genus-level approach must use data on the abundance of taxa to guarantee reliability. Incidence data would nevertheless be sufficient for a species-level study. This is probably due to the fact that the community was made up of many rare species and a few dominant ones, a typical pattern in many ecosystems [56].

However, studies of aquatic macroinvertebrates, [43] have shown that data on the abundance of taxa provide the basis for more reliable interpretations than simply recording the presence/absence of taxa, although in monitoring studies, it is often difficult to measure abundance accurately. In general, these studies are based on rapid surveys, which means that reporting occurrence may be safer and more reliable. Presence/absence data also permit comparisons between different taxonomic [16] or inventories with different sampling methods. According to [57], many monitoring studies are based on presence/absence data, rather than abundance estimates, due to their vulnerability to sampling errors.

Conclusion

The data matrix of Nepomorpha species was similar to that of the genera, and the incidence matrices of genera and species were similar to those of abundance. There was a loss of information of the order of 11–19% when the abundance matrix of genera was used instead of that of species. The congruence between the genus and species matrices indicates that, a genus-level approach is adequate for monitoring studies of nepomorphans that aim to identify environmental impacts quickly and at low cost. The potential loss of information incurred by considering only the presence/absence of taxa, supports the use of abundance data to guarantee the reliability of analyses.

Taxonomic congruence decreased at better preserved sites. This may be due to the fact that these sites presented little environmental modification and that the detection of changes would require a more refined level of taxonomic identification, that is, species rather than genus.

Supporting Information

Appendix S1 List of the sites surveyed in Mato Grosso state, Brazil (2005–2007/08), their respective acronyms, and geographic coordinates. The numbers 1–4 after an acronym refer to the classification of the river, following [29]. Privately-owned areas [O = Owner of the farm; M = farm manager; Faz = Farm].

Acknowledgments

We thank the Graduate Program in Ecology and Conservation and Entomology Laboratory of the State University of Mato Grosso (UNEMAT), and Biologist L.A. Castro provided logistic support, Drs. A.L. Melo (UFMG) and. J.R.I. Ribeiro (UNIPAMPA) provided confirmation of biological material, Drs. R.K. Umetsu (UNEMAT) and F.M. Carneiro (UEG) reviewed the manuscript, and Dr. S.F. Ferrari (UFS) reviewed the English.

Author Contributions

Conceived and designed the experiments: NFSG LJ HSRC. Performed the experiments: NFSG KDS JDB HSRC. Analyzed the data: NFSG KDS LJ.

Contributed reagents/materials/analysis tools: HSRC. Wrote the paper: NFSG JDB KDS LJ HSRC.

References

1. Schiesari L, Waichman A, Brock T, Adams C, Grillitsch B (2013) Pesticide use and biodiversity conservation in the Amazonian agricultural frontier. Phil Trans R Soc B 368: 1–9.
2. Goulart MDC, Callisto M (2003) Bioindicadores de qualidade de água como ferramenta em estudos de impacto ambiental. Revista FAPAM 1: 1–9.
3. Couceiro SRM, Hamada N, Forsberg BR, Padovesi-Fonseca C (2011) Trophic structure of macroinvertebrates in Amazonian streams impacted by anthropogenic siltation. Austral Ecol 36: 628–637.
4. Vörösmarty CJ, Mcintyre PB, Gessner MO, Dudgeon D, Prusevich A, et al. (2010) Global threats to human water security and river biodiversity. Nature 467: 555–561.
5. Couceiro SR, Hamada N, Forsberg BR, Pimentel TP, Luz SL (2012) A macroinvertebrate multimetric index to evaluate the biological condition of streams in the Central Amazon region of Brazil. Ecol Indic 18: 118–125.
6. Rodrigues ASL, Brooks TM (2007) Shortcuts for biodiversity conservation planning: the effectiveness of surrogates. Annu Rev Ecol Evol Syst 38: 713–37.
7. Heino J (2010) Are indicator groups and cross-taxon congruence useful for predicting biodiversity in aquatic ecosystems? Ecol. Indic 14: 112–117.
8. Kallimanis AS, Mazaris AD, Tsakanikas D, Dimopoulos P, Pantis JD, et al. (2012) Efficient biodiversity monitoring: which taxonomic level to study? Ecol Indic 15: 100–104.
9. Carneiro FM, Bini LM, Rodrigues LC (2010) Influence of taxonomic and numerical resolution on the analysis of temporal changes in phytoplankton communities. Ecol Indic 10: 249–255.
10. Whittaker RJ, Araujo MB, Jepson P, Ladle RJ, Watson JEM, et al. (2005) Conservation biogeography: assessment and project. Diversity Distrib 11: 3–23.
11. Jones FC (2008) Taxonomic sufficiency: the influence of taxonomic resolution on freshwater bioassessments using benthic macroinvertebrates. Environ Rev 16: 45–69.
12. Curry CJ, Zhou X, Baird DJ (2012) Congruence of biodiversity measures among larval dragonflies and caddisflies from three Canadian rivers. Freshwater Biol 57: 628–639.
13. Schmidt-Kloiber A, Nijboer RC (2004) The effect of taxonomic resolution on the assessment of ecological water quality classes. Hydrobiologia 516: 269–283.
14. Sánchez-Moyano JE, Fa DA, Estacio FJ, Garcia-Gómez JC (2006) Monitoring of marine benthic comunities and taxonomic resolution: an approach through diverse habitats and substrates along the Southern Iberian coastline. Helgol Mar Res 60: 243–255.
15. Melo AS (2005) Effects of taxonomic and numeric resolution on the ability to detect ecological patterns at a local scale using stream macroinvertebrates. Arch Hydrobiol 164: 309–323.
16. Hortal J, Borges PAV, Gaspar C (2006) Evaluating the performance of species richness estimators: sensitivity to sample grain size. J Anim Ecol 75: 274–287.
17. Souza MAA, Melo AL, Vianna GJC (2006) Heterópteros aquáticos oriundos do município de Mariana, MG. Neotrop Entomol 35: 803–810.
18. Quiroz-Martínez H, Rodríguez-Castro A (2007) Aquatic insects as predators of mosquito larvae. J Am Mosq Control Assoc 23: 110–117.
19. Kweka EJ, Zhou G, Munga S, Lee M-C, Atieli HE, et al. (2012) Anopheline Larval Habitats Seasonality and Species Distribution: A Prerequisite for Effective Targeted Larval Habitats Control Programmes. PLoS ONE 7(12): e52084. doi:10.1371/journal.pone.0052084.
20. Armúa De Reyes CA, Estévez AL (2006) Predation on Biomphalaria sp. (Mollusca: Planorbidae) by three species of the genus Belostoma (Heteroptera: Belostomatidae). Braz J Biol 66: 1033–1035.
21. Papacek M (2001) Small aquatic and ripicolous bugs (Heteroptera: Nepomorpha) as predators and prey: the question of economic importance. Eur J Entomol 98: 1–12.
22. Karaouzas I, Gritzalis KC (2006) Local and regional factors determining aquatic and semi-aquatic bug (Heteroptera) assemblages in rivers and streams of Greece. Hydrobiologia 573: 199–212.
23. Lock K, Adriaens T, Van De Meutter F, Goethals P (2013) Effect of water quality on waterbugs (Hemiptera: Gerromorpha & Nepomorpha) in Flanders (Belgium): results from a large-scale field survey. Ann Limnol-Int J Lim 49: 121–128.
24. Cabette HSR, Giehl NF, Dias-Silva K, Luen L, Batista JD (2010) Distribuição de Nepomorpha e Gerromorpha (Insecta: Heteroptera) da Bacia Hidrográfica do Rio Suiá-Miçu, MT: Riqueza relacionada à qualidade do hábitat. In: Santos JE, Galbiati C, Moschini LE, editores. Gestão e educação ambiental, água, biodiversidade e cultura. São Carlos: Rima, v.2. 113–137.
25. Dias-Silva K, Cabette HSR, Juen L, De Marco P Jr (2010) The influence of habitat integrity and physical-chemical water variables on the structure of aquatic and semi-aquatic Heteroptera. Zoologia 27: 918–930.
26. Dudgeon D (2012) Responses of benthic macroinvertebrate communities to altitude and geology in tributaries of the Sepik River (Papua New Guinea): the influence of taxonomic resolution on the detection of environmental gradients. Freshwater Biol 57: 1794–1812.
27. Peel MC, Finlayson BL, Mcmahon TA (2007) Updated world map of the Köppen-Geiger climate classification. Hydrol Earth Syst Sci 11: 1633–1644.
28. INMET (2009) Instituto Nacional de Metereologia. Ministério de Agricultura, Pecuária e Abastecimento – MAPA, Brasil. http://www.inmet.gov.br/html/clima.php. Accessed 6 August 2009.
29. Strahler HN (1957) Quantitative analysis of watershed geomorphology. Trans Am Geophys Union 38: 913–920.
30. Nessimian JL, Venticinque EM, Zuanon J, De Marco P Jr., Gordo M, et al. (2008) Land use, habitat integrity, and aquatic insect assemblages in Central Amazonian streams. Hydrobiologia 614: 117–131.
31. Petersen RC Jr. (1992) The RCE: A riparian, channel, and environmental inventory for small streams in agricultural landscape. Freshwater Biology 27: 295–306.
32. Nieser N, Melo AL (1997) Os heterópteros aquáticos de Minas Gerais: guia introdutório com chave de identificação para as espécies de Nepomorpha e - Gerromorpha. Belo Horizonte: UFMG. 177 p.
33. Estévez AL, Polhemus JT (2001) The small species of Belostoma (Heteroptera, Belostomatidae). I. Key to species groups and a revision of the Denticolle group. Iheringia Sér Zool 91: 151–158.
34. Nieser N, Lopez-Ruf M (2001) A review of Limnocoris Stål (Heteroptera: Naucoridae) in southern South America? East of the Andes. Entomol 144: 261–328.
35. Nieser N (1975) The water bugs (Heteroptera: Nepomorpha) of the Guyana Region. Stud. Fauna Suriname & other Guyanas, 16: 1–308.
36. Ribeiro JRI (2007) A review of the species of Belostoma Latreille, 1807 (Hemiptera: Heteroptera: Belostomatidae) from the four Southeastern Brazilian states. Zootaxa 1477: 1–70.
37. Heckman CW (2011) Encyclopedia of South American aquatic insects: Hemiptera Heteroptera. Illustrated keys to known families, genera, and species in South America. Olympia: Springer. 680p.
38. Jackson DA (1995) PROTEST: a PROcrustean Randomization TEST of community environment concordance. Ecoscience 2: 297–303.
39. Carneiro FM, Nabout JC, Vieira LCG, Lodi S, Bini LM (2013) Higher Taxa Predict Plankton Beta-diversity Patterns Across an Eutrophication Gradient. Natureza & Conservação 11: 43–47.
40. R Development Core Team (2012) R: A language and environment for statistical computing. Vienna: R Foundation for Statistical Computing. Available: http://www.R-project.org.
41. Lovell S, Hamer M, Slotow R, Herbert D (2007) Assessment of congruency across invertebrate taxa and taxonomic levels to identify potential surrogates. Biological Conservation 139: 113–125.
42. Heino J, Soininen J (2007) Are higher taxa adequate surrogates for species-level assemblage patterns and species richness in stream organisms? Biological Conservation 137: 78–89.
43. Marshall JC, Steward AL, Harch BD (2006) Taxonomic resolution and quantification of freshwater macroinvertebrate samples from an Australian dryland river: the benefits and costs of using species abundance data. Hydrobiologia 572: 171–194.
44. Buss DF, Vitorino AS (2010) Rapid Bioassessment Protocols using benthic macroinvertebrates in Brazil: evaluation of taxonomic sufficiency. J N Am Benthol Soc 29: 562–571.
45. Jiang X, Xiong J, Song Z, Morse JC, Jones FC, et al. (2013) Is coarse taxonomy sufficient for detecting macroinvertebrate patterns in floodplain lakes? Ecol Indic 27: 48–55.
46. Gallego I, Davidson TA, Jeppesenb E, Pérez-Martínez C, Sánchez-Castillo P, et al. (2012) Taxonomic or ecological approaches? Searching for phytoplankton surrogates in the determination of richness and assemblage composition in ponds. Ecol Indic 18: 575–585.
47. Cortelezzi A, Armendáriz LC, Oosterom MVLV, Cepeda R, Capítulo AR (2011) Different levels of taxonomic resolution in bioassessment: a case study of Oligochaeta in lowland streams. Acta Limnol Bras 23: 412–425.
48. Yoshimura M (2012) Effects of forest disturbances on aquatic insect assemblages. Entomol. Science 15: 145–154.
49. Waite IR, Herlihy AT, Larsen DP, Urquhart NS, Klemm DJ (2004) The effects of macroinvertebrate taxonomic resolution in large landscape bioassessments: an example from the Mid-Atlantic Highlands, U.S.A. Freshwater Biol 49: 474–489.
50. Resh VC, McElravy EP (1993) Contemporary quantitative approaches to biomonitoring using benthic macroinvertebrates. In: Rosenberg DM, Resh VH, editors. Freshwater biomonitoring and benthic macroinvertebrates. New York: Chapman and Hall. 159–194.
51. Heino J (2008) Influence of taxonomic resolution and data transformation on biotic matrix concordance and assemblage-environment relationship in stream macroinvertebrates. Boreal Environ Res 13: 359–369.
52. Brown JH, Lomolino MV (1998) Biogeography, 2nd edn. Massachusetts: Sinauer Press. 691 p.

53. Bini LM, Diniz-Filho JAF, Rangel TFLVB, Bastos RP, Pinto MP (2006) Challenging Wallacean and Linnean shortfalls: knowledge gradients and conservation planning in a biodiversity hotspot. Diversity Distrib 12: 475–482.

54. Landeiro VL, Bini LM, Costa FRC, Franklin E, Nogueira A, et al. (2012) How far can we go in simplifying biomonitoring assessments? An integrated analysis of taxonomic surrogacy, taxonomic sufficiency and numerical resolution in a megadiverse region. Ecol Indic 23: 366–373.

55. Wagenhoff A, Townsend CR, Matthaei CD (2012) Macroinvertebrate responses along broad stressor gradients of deposited fine sediment and dissolved nutrients: a stream mesocosm experiment. J Appl Ecol 49: 892–902.

56. Siqueira T, Bini LM, Roque FO, Couceiro SRM, Trivinho-Strixino S, et al. (2012) Common and rare species respond to similar niche processes in macroinvertebrate metacommunities. Ecography 35: 183–192.

57. Metzeling L, Chessman B, Hardwick R, Wong V (2003) Rapid assessment of rivers using macroinvertebrates: the role of experience, and comparisons with quantitative methods. Hydrobiologia 510: 39–52.

3

Plant Diversity Impacts Decomposition and Herbivory via Changes in Aboveground Arthropods

Anne Ebeling[1]*, **Sebastian T. Meyer**[2], **Maike Abbas**[3], **Nico Eisenhauer**[1], **Helmut Hillebrand**[3], **Markus Lange**[4], **Christoph Scherber**[5], **Anja Vogel**[1], **Alexandra Weigelt**[6], **Wolfgang W. Weisser**[1,2]

1 Institute of Ecology, University of Jena, Jena, Germany, 2 Department of Ecology and Ecosystem Management, Center for Food and Life Sciences Weihenstephan, Technische Universität München, Freising, Germany, 3 Institute for Chemistry and Biology of the Marine Environment, Carl-von-Ossietzky-University Oldenburg, Wilhelmshaven, Germany, 4 Max Planck Institute for Biogeochemistry, Jena, Germany, 5 DNPW, Agroecology, Georg-August University Göttingen, Göttingen, Germany, 6 Department for Systematic Botany and Functional Biodiversity, University of Leipzig, Leipzig, Germany

Abstract

Loss of plant diversity influences essential ecosystem processes as aboveground productivity, and can have cascading effects on the arthropod communities in adjacent trophic levels. However, few studies have examined how those changes in arthropod communities can have additional impacts on ecosystem processes caused by them (e.g. pollination, bioturbation, predation, decomposition, herbivory). Therefore, including arthropod effects in predictions of the impact of plant diversity loss on such ecosystem processes is an important but little studied piece of information. In a grassland biodiversity experiment, we addressed this gap by assessing aboveground decomposer and herbivore communities and linking their abundance and diversity to rates of decomposition and herbivory. Path analyses showed that increasing plant diversity led to higher abundance and diversity of decomposing arthropods through higher plant biomass. Higher species richness of decomposers, in turn, enhanced decomposition. Similarly, species-rich plant communities hosted a higher abundance and diversity of herbivores through elevated plant biomass and C:N ratio, leading to higher herbivory rates. Integrating trophic interactions into the study of biodiversity effects is required to understand the multiple pathways by which biodiversity affects ecosystem functioning.

Editor: Mari Moora, University of Tartu, Estonia

Funding: The study was funded by the Deutsche Forschungsgemeinschaft (FOR 1451). The funders had no role in study design, data collection and analysis, decision to publish, or preparation of the manuscript.

Competing Interests: The authors have declared that no competing interests exist.

* Email: anne.ebeling@uni-jena.de

Introduction

Extensive studies have identified biodiversity loss as one of the main drivers of ecosystem change [1], showing that the relation between biodiversity and ecosystem processes is mainly positive [2,3]. For a long time biodiversity research has focused mainly on processes directly relating to the trophic level manipulated (e.g. aboveground productivity in plant diversity experiments [4,5]; but see [6]). More recent studies have shown that plant diversity and associated changes in plant species composition also affect ecosystem processes governed by higher trophic levels, such as decomposition and herbivory [7–10].

Decomposition of organic matter is a key process in ecosystems, replenishing the pool of plant available soil nutrients and releasing photosynthetically fixed carbon back to the atmosphere by the activity of several groups of soil organisms. Invertebrate macro-fauna, such as isopods, diplopods and earthworms, fragment dead plant material and thereby pave the way for further microbial decay and mineralization. Consequently, changes in their density and diversity should lead to altered decomposition rates [11,12]. They may also affect nutrient cycling and plant productivity of grassland ecosystems. Herbivores can decrease plant productivity by decreasing plant performance, but contrary to that, they could also increase plant productivity by recycling nutrients and triggering compensatory growth [13,14]. Arthropod density and diversity were discussed as drivers of plant diversity effects on decomposition and herbivory, but empirical evidence is scarce [8,15]. Studies in experimental grasslands have generally reported higher diversity and abundances of herbivores, detritivores, pollinators, carnivores and parasitoid arthropods at higher plant diversity [16–23], although the strength of the effect differed between the groups.

Plant-animal, plant-ecosystem process, and animal-ecosystem process relationships have often been studied separately, but have rarely been linked. To understand how plants, arthropods, and arthropod-related processes such as decomposition and herbivory are linked, there is a strong need to conduct studies considering multiple trophic levels and ecosystem processes simultaneously. There is a large number of competing hypotheses linking consumer-resource dynamics in diverse communities (summarized in Fig. 1), applicable to the relationship between arthropod and plant communities. Hypotheses emphasizing effects of resources on consumers have focused on biomass (More Individuals Hypothesis- MIH, Productivity Hypothesis- PH), and stoichiometry (C:N- Mechanism 3 in Fig. 1), both variables being strongly affected by changing plant species richness [24,25] but also on

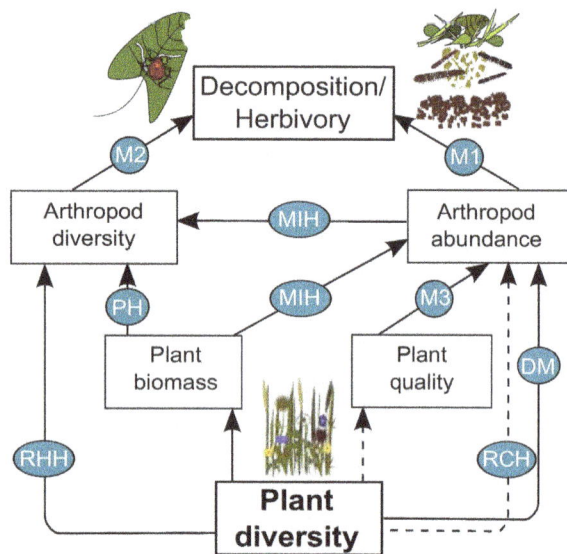

Figure 1. Possible mechanisms driving a positive plant diversity-ecosystem process relationship. The consumer-related ecosystem processes studied here are herbivory and decomposition. Based on existing literature we hypothesize different mechanisms driving the positive relationship between plant diversity and herbivory/ decomposition, represented here by the different arrows. For details on the hypotheses see Introduction. Solid lines indicate expected positive correlations, dotted lines expected negative correlations. M1-3 = Mechanisms 1-3, DM = Dietary Mixing, MIH = More Individuals Hypothesis, PH = Productivity Hypothesis, RCH = Resource Concentration Hypothesis, RHH = Resource Heterogeneity Hypothesis. M1-3 are "unnamed" mechanisms, which are reported in the literature.

species richness per se (Dietary Mixing- DM, Resource Concentration Hypothesis- RCH, Resource Heterogeneity Hypothesis- RHH).

MIH and PH hypothesize a positive relationship between plant biomass and consumer diversity, but they differ in the underlying mechanisms. The MIH [26] applies to communities, which are limited in productivity and predicts higher consumer abundances in more productive sites as well as consumer diversity as an increasing function of total abundance. Following the PH, a higher overall resource level in diverse plant communities directly attracts more generalist consumer species [27]. The stoichiometry of resources is also reported to be a strong predictor of consumer abundances as higher food quality (e.g. higher N content of plants) increases arthropod fitness (e.g. fecundity) and hence their abundance [28]. Following the literature, the way and direction how plant species richness affects consumer abundance strongly depends on their food specialization. According to the RCH [29] we would expect lower abundances of specialized herbivorous pests in more diverse habitats due to lower densities of the respective host plant. For generalist herbivores the opposite pattern may be true (DM): the ability of some generalist herbivores (e.g., grasshoppers) to mix their diet to an optimal combination of nutrients or to dilute toxins could increase their fitness and hence their abundance, indicated by a positive relation between plant diversity and arthropod abundance [30]. For consumer species richness the RHH [31] predicts that due to a higher number of different resources with increasing plant diversity, the number of specialist consumer species increases.

While the theoretical framework of how plant diversity affects consumer (arthropod) abundance and diversity is quite elaborated,

the effects of arthropod communities on herbivory and decomposition are less clear. Consumer effects on related processes could be driven by arthropod abundance (Mechanism 1 in Fig. 1) or species richness (Mechanism 2 in Fig. 1). If specialist herbivores dominate, then increasing herbivore diversity is likely to increase community-wide herbivory rates, while a higher number of herbivores will only increase herbivory on those plants that are attacked. By contrast, if herbivory is mainly due to generalists, then increasing herbivore abundance rather than diversity will increase herbivory. For decomposition, functional complementarity among decomposers has been shown in a number of studies [11], yet decomposition may well be limited by decomposer abundance. Thus, various chains of effects are possible emphasizing the need for empirical approaches. All the hypotheses and mechanisms (Fig. 1) are not mutually exclusive, however, and many of the hypothesized pathways can have interactive and direct or indirect effects. The relative importance of these pathways may change when considering herbivores as opposed to decomposers because some mechanisms like compensatory plant growth would only result from herbivory, but not from decomposition.

Here, we integrate consumers into our models of plant diversity effects on ecosystem processes by focusing on three components of plant communities (food quantity: aboveground biomass; food quality: C:N ratio of plants, and species richness- hereafter: plant diversity), two components of consumer communities (arthropod abundance, arthropod species richness), and two processes (decomposition and herbivory). We test (1) if plant diversity effects on arthropods and decomposition/herbivory found in former studies can be confirmed, (2) if effects of plant diversity on decomposition and herbivory are mediated by altered arthropod community composition in terms of overall abundance, and (3) which mechanisms link plants to arthropods (plant diversity, quantity or quality), and arthropods to decomposition and herbivory (abundance or species richness).

Material and Methods

Ethic statement

Arthropod sampling was conducted with the permission of the city council of Jena, Germany.

Experimental field site

The study was conducted in the framework of The Jena-Experiment (Thuringia, Germany, 50°55′ N, 11°35′ E; 130 m above sea level). This grassland biodiversity experiment uses sixty species native to regional floodplains. Plant communities were sown in 82 plots of 20×20 m with a gradient of species richness (1, 2, 4, 8, 16 and 60) and plant functional group richness (1–4 functional groups: grasses, small herbs, tall herbs, legumes) [32]. In 2009, the plot size was reduced to 100 m², and two of the monocultures were abandoned due to poor cover of the target species. The particular species mixtures were randomly selected with some constraints [32]. A gradient in abiotic soil properties was taken into account by arranging plots in four blocks perpendicular to the river Saale [32]. All plots are mown twice a year, a management regime common for such meadows. Plots are weeded twice a year to maintain the experimental diversity gradient. As sown and realized number of target species are highly correlated ($R^2 = 0.97$), we used the number of sown plant species for our analyses. The study site was established in May 2002 allowing for plenty of time for colonization by arthropods and making the established plant diversity gradient particularly well suited for examining consumer effects on ecosystem functioning [20].

Plant community analyses

In our path analyses we used aboveground plant biomass as measure of resource availability for arthropods, because herbivores directly feed on plants and decomposers on plant litter material, which is reported to be highly correlated with plant biomass [19]. In 2010, aboveground plant community biomass was harvested during peak standing biomass in late May and August on all plots, in the frame of the long-term measurement of aboveground productivity. This was done by clipping the vegetation three cm above the soil surface in two rectangles of 0.2×0.5 m. The harvested biomass was separated into sown species, cleaned from weeds, dried at 70°C for 72 h and weighed. As a general measure of plant quality we used plant molar C:N ratio, because it is the most practicable measure of plant quality across 60 different plant species, applicable to many different herbivore and decomposer species. Community carbon (C) and nitrogen (N) concentrations of the pooled, dried and milled plant material were measured separately for May and August 2010 with an elemental analyzer (Flash EA 112 Thermo).

Arthropod sampling

Arthropods were sampled in 2010 according to methods used in previous years with the permission of the city council of Jena. We sampled decomposers from May until September 2010, using two pitfall traps of 4.5 cm diameter per plot. The use of pit fall traps is quite common to sample litter- and surface-dwelling macro-arthropods like isopods and diplopods [33]. Traps were emptied every two weeks, but closed for two weeks during the mowing campaigns in June and September. In total, this resulted in 125 sampling days over the whole vegetation period. During the sampling periods, the field traps were filled with 3% formalin, and after emptying the traps, animals were stored in 70% ethanol. The resulting activity abundance data will be termed 'abundance' hereafter. Our data on decomposers comprise surface active decomposer macrofauna only. Less mobile decomposers, such as earthworms, may be important agents in surface litter decomposition [34], but are underrepresented in these samples.

Herbivorous arthropods were collected by suction sampling in June and July using a modified commercial vacuum cleaner (Kärcher A2500, Kärcher GmbH, Winnenden, Germany). Three subplots of 0.75 m×0.75 m were randomly chosen within each plot, covered quickly by a cage with gauze coat of the same size, and sampled until no arthropods were spotted anymore. The sampling was carried out between 9 a.m. and 4 p.m. within two 4-day sampling periods. Pit fall samples and suction samples were sorted to order level and most orders containing decomposers and herbivores were further identified to species level. For herbivores only Auchenorrhyncha, Coleoptera, and Heteroptera were identified to species level, and these orders covered 55% of the potential herbivores we sampled. Aphids and Diptera, which made together 30.2% of the sampled potential herbivores were not identified, because of their patchy distribution within a single plot (Aphids) and due to identification difficulties (Diptera).

Data of the different sampling campaigns were pooled.

Litter decomposition

We measured litter decomposition using the standard litterbag technique [11]. We placed one litterbag in each of the 80 plots of The Jena-Experiment between 17th June and 24th August 2009, as successfully done in previous studies [9,10,35]. The fact that we have a very high number of replicates within each plant diversity level (across plots) compensates for the low replication within plots (one litterbag each) since we were more interested in the effects of plant diversity than of those of specific plant mixtures. To allow access for decomposer macrofauna (e.g. earthworms, isopods, diplopods) we constructed litterbags with 4 mm mesh and anchored them to the soil surface. We filled each litterbag with ~3 g of wheat straw (N = 0.4%, C = 45.2%, C:N = 111.5) of ~3 cm length. At the end of the experiment, the remaining litter was cleaned, dried (70°C, 48 h) and weighed to determine mass loss. The use of wheat straw acts as a standard method to measure potential decomposition rates caused by decomposers and affected by microclimatic conditions, without including effects of litter quality [9,10,36]. This method focusses on microenvironmental effects on decomposition caused by changes in plant communities instead of combined effects of microenvironment and litter quality, composition or structure.

Herbivory

Arthropod herbivory was measured in the plant communities during peak biomass in August 2010. A maximum of 30 fully developed leaves were sampled randomly for each species from each plot where it occurred in the plant biomass samples. The minimum number of leaves for a species in a plot was one, and the average number was 21 per species and plot. On average 65% (SE±2.1%) of the sown species per plot occurred in the biomass sampled, therefore the measure can be regarded representative of the sown plant communities. The damaged surface area caused by herbivores of each leaf (in mm^2) was estimated visually by comparing the damaged leaf area to circular and square templates ranging in size from 1 mm^2 to 500 mm^2. The leaf area was measured with a leaf-area meter (LI-3000C Area Meter, LI-COR Biosciences, Lincoln, USA), and the proportional herbivory damage (herbivory rates) was calculated as the damaged area divided by measured leaf area. The community herbivory rates for each plot were calculated by summing the species-specific herbivory rates weighted by species-specific leaf biomass in the plot. Mammals were excluded from the experimental field site by a fence, thus only invertebrates were responsible for the observed herbivore damage. Slugs contributed to the measured leaf damage, but were not included in herbivore analyses. For details in the methods see [8].

Statistical analysis

We analyzed our data using (i) linear models and (ii) path analyses. Linear models served as affirmation of the results arising from the path analyses, as our sample size (N = 80) was somewhat lower than usually recommended for path analyses [37,38].

Linear models were used to test for effects of plant community properties on arthropods and ecosystem processes. These models were fitted using the statistical software package R (R development Core Team, http://www.R-project.org. version 2.13.1). Arthropod abundance, arthropod species richness (decomposer and herbivores), decomposition and herbivory were used as response variables. For each of the response variables, we fitted a model including block (factor), plant diversity (numeric), and functional group richness (numeric; number of plant functional groups occurring in the respective plot- see experimental design section) as explanatory variables. We omitted plant functional group richness from our further models, as it had no consistent effect on any of the response variables (see Table S1).

Path analyses were calculated in Amos v.20.0 (IBM SPSS Amos, http://www.spss.com/amos/) and were used to infer the hypothesized causal links between plant diversity and ecosystem processes, as ecosystem process responses are likely indirectly mediated by physical, structural or chemical attributes of plants and/or consumer groups studied. We constructed two models (one for decomposition and one for herbivory), based on our hypotheses extracted from the published literature (see Introduction, and Fig. 1), and did not perform any model simplification. Models included arthropod abundance and species richness, aboveground plant biomass as a measure of food quantity, and plant community C:N ratio as measure of food quality as endogenous variables. Because of our low overall sample size (N = 80), we did not construct latent variables but assumed direct paths among each pair of endogenous variables. As herbivores were sampled after the first biomass harvest we used herbivory, biomass and C:N data of the data collection in August 2010. For decomposition we used mean values of the first and second data collection for aboveground biomass and C:N ratio of plants.

For all analyses (linear models and path analyses) measures of plant diversity, decomposer abundance, herbivore abundance and decomposer species richness were log transformed, aboveground plant biomass was square-root transformed, and herbivory was logit transformed [39] to account for non-normality of errors and heteroscedasticity, and to linearize the relationships among variables. All data used in this publication are deposited in the project database (www.the-jena-experiment.de).

Results

Effects of plant diversity on the decomposer community and decomposition

Overall, the final dataset contained 1,316 macrofauna decomposer individuals belonging to 12 species (four Isopoda and eight Diplopoda species). Isopoda dominated the macrofauna decomposer community, representing on average 92.8% of the total community.

Decomposer abundance and species richness increased significantly with increasing plant diversity, but remained unaffected by the number of plant functional groups (Table S1, Fig. S1). The number of individuals increased from 78.6 per plot (SE±19.2) in monocultures to 244 per plot (SE±30.5) in 60-plant-species-mixtures. Mean decomposer species richness increased from 3.6 (SE±0.5) per plot in plant monocultures to 6.3 (SE±0.4) in 60-plant-species-mixtures. Decomposition rate increased with increasing plant species richness from 3 mg mg^{-1} d^{-1} (SE±0.0002) in monocultures up to 5 mg mg^{-1} d^{-1} (SE±0.0004) in 60-plant-species-mixtures. The number of plant functional groups, available in the respective plant communities (Table S1), did not significantly affect decomposition.

Mechanisms driving altered decomposition rates

A chi-squared test indicated that our path analysis model cannot be rejected as a potential explanation of the observed covariance matrix ($\chi^2_4 = 4.49$, $P = 0.344$). The model explained 27% of the variance in decomposition. The positive effect of plant diversity on decomposition (Table S2 and S3, Fig. 2a) was mediated by aboveground plant biomass, which increased with increasing plant diversity and which positively affected decomposer abundance (Fig. 3a) and species richness (marginally). Although decomposer abundances and species richness were strongly correlated (Fig. 3b; Table S1), only species richness increased litter decomposition (Fig. 3c; Table S1). While increasing plant diversity increased plant C:N ratio, there was no significant effect of plant C:N ratio on decomposers or decomposition. A chi-squared test for the same path analysis excluding species richness and abundance of decomposers indicated that it need to be rejected as a potential explanation of the observed covariance matrix ($\chi^2_6 = 15.60$, $P = 0.016$).

Effects of plant diversity on the herbivore community and herbivory rate

Overall, the final dataset contained 12,829 herbivores, belonging to 127 species (38 Auchenorrhyncha, 62 Coleoptera, 27 Heteroptera). On average the herbivore community was dominated by Coleoptera (48.6%), followed by Auchenorrhyncha (43.6%), and Heteroptera (7.8%). Plant diversity, but not plant functional group richness, strongly affected the abundance and species richness of herbivorous insects and herbivory rate (Table S1, Fig. S2). Herbivore abundance increased from 65.2 (SE±13.7) individuals per plot in monocultures to 251 (SE±36.7) individuals per plot in 60-plant-species-mixtures. Herbivore communities in monocultures contained on average 15.6 (SE±1.0) species per plot, whereas in 60-plant-species-mixtures 23.5 (SE±0.8) species per plot occurred. Plant diversity also led to a higher herbivory rate, nearly doubling from 2% (SE±0.4%) in monocultures to 3.5% (SE±0.9%) in 60-plant-species-mixtures.

Mechanisms driving altered herbivory

A chi-squared test indicated that the path analysis explained 22% of the variance in herbivory, and cannot be rejected as a potential explanation of the observed covariance matrix ($\chi^2_4 = 7.50$, $P = 0.112$). The positive relationship between plant diversity and herbivory rate (Table S4 and S5, Fig. 2b), was not driven by aboveground plant biomass, as there was no significant correlation between aboveground biomass and herbivore abundance and species richness. Herbivore abundance and species richness increased significantly with increasing plant diversity (Fig. S2). Increasing plant diversity increased plant C:N ratio, which, in turn, was positively correlated with herbivore abundance (Fig. 4a). Herbivore abundances increased herbivore species richness (Fig. 4b), but only higher abundances of herbivorous arthropods increased herbivory rate (Fig. 4c). A chi-squared test for the same path analysis excluding species richness and abundance of herbivores indicated that it need to be rejected as a potential explanation of the observed covariance matrix ($\chi^2_6 = 24.70$, $P = 0.001$).

Discussion

Plant diversity influenced the abundance and diversity of decomposers and herbivores, and in turn, these effects propagated to decomposition and herbivory rates. Effects of plant diversity on process rates could be better explained when including information on the consumer community into the explanatory model for both, decomposition and herbivory. Abundance of decomposers was positively associated with increased resource (plant) biomass, whereas herbivore abundance increased with higher C:N ratio of the plant material and plant species richness. However, herbivore richness increased directly with higher plant diversity, whereas decomposer richness increased indirectly (via plant biomass and decomposer abundance). Accordingly, we provide support for the More Individuals Hypothesis (MIH) to drive altered decomposer communities, whereas we explain changes in herbivore communities by Dietary Mixing (DM) and the Resource Heterogeneity Hypothesis (RHH). Decomposer species richness enhanced decomposition of litter material (Mechanism 2 in Fig. 1), whereas higher herbivory was mainly driven by the higher abundance of herbivores (Mechanism 1 in Fig. 1).

Decomposers and decomposition

Notably, the discussion of plant community effects (biomass and C:N ratio) on the decomposer community is based on to the use of standard litter material in our study instead of plot-specific litter

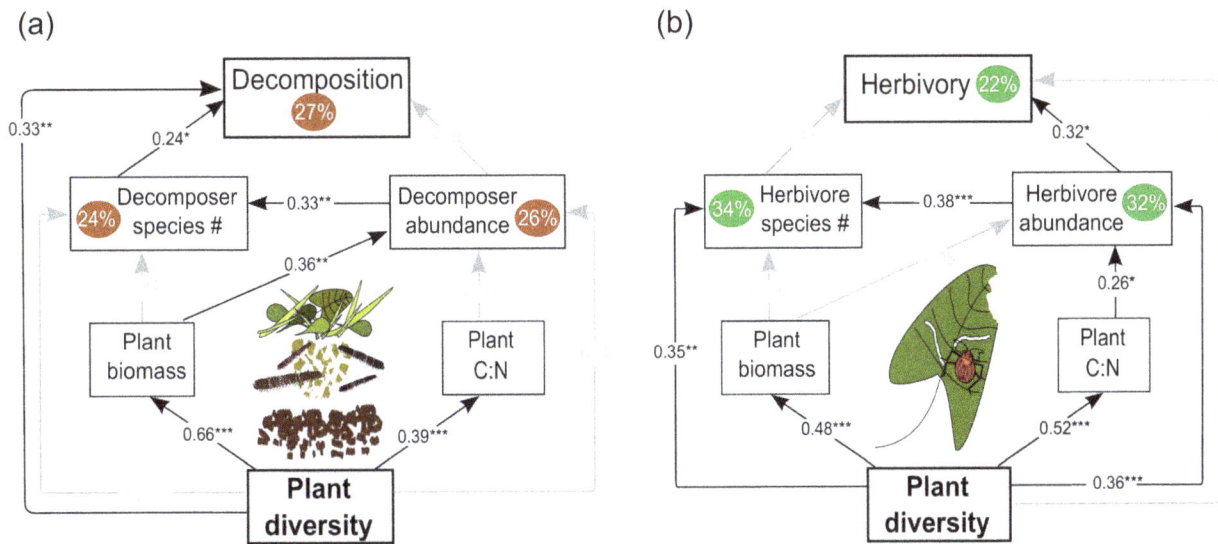

Figure 2. Path diagram explaining plant community effects on decomposition and herbivory. Models relate plant community variables (diversity, quantity and quality), species richness and abundance of (a) decomposer arthropods to decomposition, and (b) herbivorous insects to herbivory. Standardised path coefficients are given on top of the path arrows with significances indicated by *, $P<0.05$; **, $P<0.01$; ***, $P<0.001$. Non-significant paths are given in grey.

material. Consequently, we here discuss indirect effects between plant variables and decomposer communities (plants affect the environment for decomposers, which in turn leads to changes in the decomposer community) [9,10].

Decomposition was positively correlated with plant diversity as has been shown in some studies using standard litter [9,40–42] but not all [10,34]. Reich and colleagues [3] recently showed that plant diversity effects on plant biomass production increased over time, and it was suggested that this increase in plant diversity effects may be linked to delayed soil feedback effects of slowly assembling decomposer communities [43]. Indeed, the density and diversity of macro-decomposers and the density of meso-decomposers increased significantly with increasing plant diversity only four (aboveground) and six years (belowground) after establishment of The Jena-Experiment [20,23]. Consequently, the lack of plant diversity effects in other studies may be due to the short-term character of most previous experiments. Positive effects of plant diversity on decomposition processes can possibly be attributed to higher productivity of more diverse plant communities and thus elevated food availability for decomposers (see MIH in Fig. 1), as the amount of available litter material is positively correlated to aboveground biomass in grasslands [44].

Our path analysis identified aboveground biomass as an important driver of changes in the decomposer abundance, which supports the MIH. Plant biomass production increased with increasing plant diversity and this, in turn, increased decomposer species richness via abundance. Resulting increases in decomposition rates were mediated by increased decomposer species richness but not abundance, indicating a more diverse decomposer community to be functionally complementary in fragmenting litter material (according to M2 in Fig. 1).

Our results on decomposition are in accordance with a study of Heemsbergen et al. [36], who found functional dissimilarity among decomposer species being the best predictor for changes in leaf litter mass loss and soil respiration. Similar results are reported for benthic detritivores in streams, where decomposition rates also increase with detritivore richness [45]. In these systems, potential mechanisms involve increasing faciliation of resource use among

species and reduction of inter-specific competiton [46]. Such facilitation can arise if the processing of dead organic matter by one species enhances the suitability of this resource for another species. The positive correlation between decomposer species richness and decomposition could also occur due to a sampling effect (higher chance of having a functionally dominant decomposer species in a species-rich community than in a species-poor one); however, we are not able to disentangle both possible effects here.

Even if we used wheat straw as litter material in our study we expected a negative effect of plant C:N ratio on the environment for decomposers (lower nutrient availability in litter and soil), thereby indirectly affecting their fitness (in our case: abundance). In contrast to our hypothesis we detected no significant effect of aboveground plant quality (C:N ratio) on decomposer abundance and species richness, maybe indicating that the C:N ratio of the fresh plant material is a poor predictor of the C:N ratio of the litter material.

In our path analysis a significant direct link between plant diversity and decomposition remained, indicating that part of the plant diversity effect on decomposition was independent of consumer-mediated effects. Potential environmental variables that could contribute to this link might include soil moisture of the top soil layer, vegetation density, and shade which are important predictors of soil dwelling decomposer community composition. All three variables are documented to increase with increasing plant diversity [10,47]. Further, our analysis was restricted to aboveground living decomposers, but we know that belowground decomposer groups are strongly affected by plant diversity [19,48]. Therefore, we suggest that also belowground decomposition might show an imprint of plant species richness via decomposer abundance or diversity. As data on soil dwelling decomposer species were not available from the the study year and could therefore not be included in the analyses this is a natural extension for future studies.

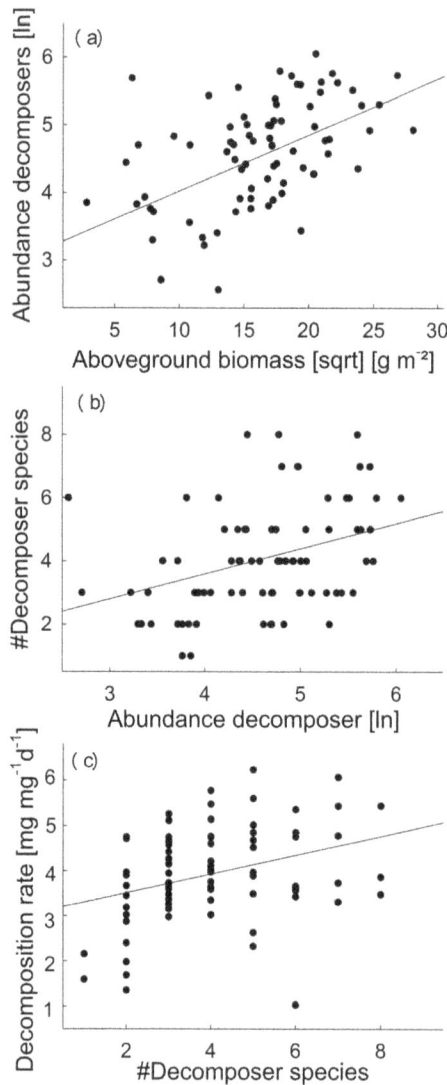

Figure 3. Pairwise correlations visualizing the significant links detected in the path analysis relating plants, decomposers and decomposition. We show the relationships between aboveground plant biomass and decomposer abundances (a), between decomposer abundances and their species richness (b) and decomposer species richness and decomposition (c). For statistics, see Table S2.

Figure 4. Pairwise correlations visualizing the significant links detected in the path analysis relating plants, herbivores and herbivory. We show the relationships between plant C:N ratio and herbivore abundance (a), herbivore species richness and their abundance (b), and between herbivore abundance and herbivory rate (c). For statistics, see Table S4.

Herbivores and herbivory

Plant diversity increased both herbivore abundance and richness. Plant diversity effects on herbivore richness could not be explained by variations in plant biomass or plant C:N, which indicates that it may be a direct link between resource diversity and consumer diversity. Such relationships are expressly predicted for specialist consumers, as more diverse resource assemblages enable the presence of a higher number of specialist consumer species (e.g. [16,17,21,31]; see RHH in Fig. 1). However, generalist consumers can also benefit from more diverse food plants, if these provide a nutritionally more balanced or temporally less variable food [17,30]. In contrast to findings of McNaughton et al. [49] and Borer et al. [22], who found higher aboveground plant biomass at higher plant species richness providing the food for a larger abundance of herbivores (see MIH in Fig. 1), increases in herbivore abundance in this study were not mediated by plant biomass.

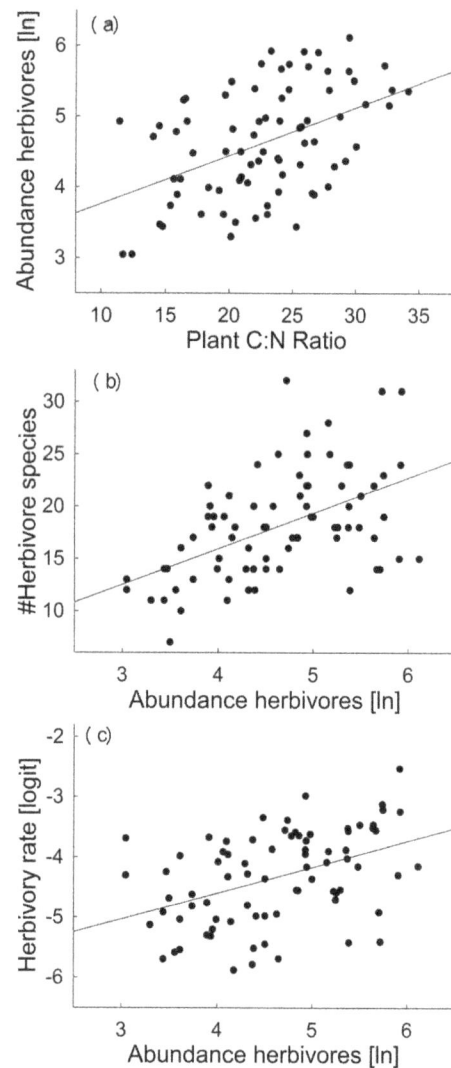

Food plant quality (here C:N ratio) affected the herbivore abundance and thereby indirectly herbivory. As previously shown [25,50], plant C:N ratios increased with plant species richness. A higher food C:N ratio should lead to lower fitness and abundance of herbivores [28,51,52] if the plant C:N is above the consumers' threshold elemental ratios [53]. Consequently, community-wide herbivory was shown to decrease with poorer autotroph quality although individuals compensated for poor food quality by ingesting more [54]. If compensatory feeding in a low nutrition environment is effective, a neutral relationship between food quality and community herbivory might occur [51,55]. Yet, it is hard to imagine how compensatory feeding could lead to increased herbivore abundance at higher plant C:N as observed in our analysis, unless C:N ratio is correlated with other non-nutritional aspects of these trophic interactions. Increased investment in vertical growth and thus C-rich stem-material in the more species-rich plots is a promising candidate for such a relationship, which would increase the habitable volume for arthropods and thereby increase abundance. This interpretation is supported by the fact that the average height in summer of all plants occuring in a communitiy was a significant predictor of herbivory levels when modelling community herbivory based on plant functional traits [56].

Overall, plant diversity increased both herbivore abundance and richness, but only herbivore abundance then increased herbivory rates (confirming M1 in Fig. 1). The obvious interpretation of this relationship is that at higher plant diversity a larger abundance of herbivores consumes more plant biomass. Consequently, an increase of herbivory at higher plant diversity [8,57], can potentially be explained by an increase in herbivore abundance with increasing plant diversity, as documented in this study. Integrating data on feeding type or food specialization could help for a detailed understanding of the mechanisms responsible for the observed positive relation between herbivore abundance and herbivory rate. Herbivore species richness was indirectly linked to herbivory rates via its positive correlation to herbivore abundance.

Conclusion

Our results demonstrate the importance of interactions across trophic levels for ecosystem process rates of green and brown food webs. Decomposition was driven by arthropod species richness via arthropod abundance, and herbivory depended on the number of herbivores. Excluding data on consumer levels from our analysis (thus measuring only the directly plant related links from plant diversity and composition to rates of decomposition or herbivory) resulted in no explained variance (model rejection, see result section) for decomposition and herbivory. In particular, our results strongly support the conclusion that integrating trophic interactions into the study of biodiversity effects is required to understand the multiple pathways by which biodiversity affects ecosystem functioning [58]. In addition, we could show that the functional role of an important part of overall biodiversity – in our case the arthropod community – cannot always be deduced *a priori*, but must be unraveled using a mechanistic approach. While functional biodiversity research has shown the general importance of biodiversity for many ecosystem processes, understanding the functional role of biodiversity requires analyzing the mechanisms underlying BEF relationships. Given that decomposers and herbivores interact with communities of natural enemies exerting a top-down control in addition to the bottom-up control by the plant community [59] a consequent step forward would be to include higher trophic levels (predators, parasites, parasitoids) in the analysis of plant diversity effects on arthropod-mediated ecosystem processes.

Supporting Information

Figure S1 Effects of plant diversity on the abundance and species richness of decomposer arthropods, and decomposition.

Figure S2 Effects of plant diversity on the abundance and species richness of herbivorous arthropods, and herbivory rate.

Table S1 Results of linear models testing plant diversity effects on decomposing and herbivorous arthropods, and decomposition (mg mg^{-1} d^{-1}) and herbivory rate (%, logit transformed).

Table S2 Results of the path analysis linking plants, decomposers and decomposition.

Table S3 Standardized total effects of the structural equation model analysing plant diversity effects on decomposition.

Table S4 Results of the path analysis linking plants, herbivores and herbivory rate.

Table S5 Standardized total effects of the structural equation model analysing plant diversity effects on herbivory rate.

Acknowledgments

The study was funded by the Deutsche Forschungsgemeinschaft (FOR 1451). We thank the technical staff for their work in maintaining the field site and also many student helpers for weeding of the experimental plots and support during measurements. Hannah Loranger and Enrica De Luca are greatly acknowledged for providing data on herbivory rates and plant biomass respectively. We thank Jes Hines for helpful comments on this manuscript. Additional financial support was received by the AquaDiva @Jena project financed by the state of Thuringia, Germany.

Author Contributions

Conceived and designed the experiments: AE STM NE HH AV AW WWW. Performed the experiments: STM NE AV AE ML MA HH AW. Analyzed the data: AE STM CS. Contributed reagents/materials/analysis tools: STM CS NE AE HH AW WWW. Wrote the paper: AE STM MA NE HH ML AV AW WWW.

References

1. Hooper DU, Adair EC, Cardinale BJ, Byrnes JEK, Hungate BA, et al. (2012) A global synthesis reveals biodiversity loss as a major driver of ecosystem change. Nature 486: 105–108.
2. Cardinale BJ, Duffy JE, Gonzalez A, Hooper DU, Perrings C, et al. (2012) Biodiversity loss and its impact on humanity. Nature 486: 59–67.
3. Reich PB, Tilman D, Isbell F, Mueller K, Hobbie SE, et al. (2012) Impacts of biodiversity loss escalate through time as redundancy fades. Science (80-) 336: 589–592.
4. Cardinale BJ, Srivastava DS, Duffy JE, Wright JP, Downing AL, et al. (2006) Effects of biodiversity on the functioning of trophic groups and ecosystems. Nature 443: 989–992.
5. Marquard E, Weigelt A, Temperton VM, Roscher C, Schumacher J, et al. (2009) Plant species richness and functional composition drive overyielding in a six-year grassland experiment. Ecology 90: 3290–3302.
6. Naeem S, Thompson LJ, Lawler SP, Lawton JH, Woodfin RM (1994) Declining biodiversity can alter the performance of ecosystems. Nature 368: 734–737.

7. Scherber C, Milcu A, Partsch S, Scheu S, Weisser WW (2006) The effects of plant diversity and insect herbivory on performance of individual plant species in experimental grassland. J Ecol 94: 922–931.
8. Loranger H, Weisser WW, Ebeling A, Eggers T, Luca E De, et al. (2013) Invertebrate herbivory increases along an experimental gradient of grassland plant diversity. Oecologia:
9. Hector A, Beale AJ, Minns A, Otway SJ, Lawton JH (2000) Consequences of the reduction of plant diversity for litter decomposition: effects through litter quality and microenvironment. Oikos 90: 357–371.
10. Scherer-Lorenzen M (2008) Functional diversity affects decomposition processes in experimental grasslands. Funct Ecol 22: 547–555.
11. Hattenschwiler S, Gasser P (2005) Soil animals alter plant litter diversity effects on decomposition. Proc Natl Acad Sci U S A 102: 1519–1524.
12. Eisenhauer N (2012) Aboveground-belowground interactions as a source of complementarity effects in biodiversity experiments. Plant Soil 351: 1–22.

13. Gruner DS, Smith JE, Seabloom EW, Sandin SA, Ngai JT, et al. (2008) A cross-system synthesis of consumer and nutrient resource control on producer biomass. Ecol Lett 11: 740–755.

14. Allan E, Crawley MJ (2011) Contrasting effects of insect and molluscan herbivores on plant diversity in a long-term field experiment. Ecol Lett 14: 1246–1253.

15. Mulder CPH, Koricheva J, Huss-Danell K, Hogberg P, Joshi J (1999) Insects affect relationships between plant species richness and ecosystem processes. Ecol Lett 2: 237–246.

16. Koricheva J, Mulder CPH, Schmid B, Joshi J, Huss-Danell K (2000) Numerical responses of different trophic groups of invertebrates to manipulations of plant diversity in grasslands. Oecologia 125: 271–282.

17. Ebeling A, Klein AM, Schumacher J, Weisser WW, Tscharntke T (2008) How does plant richness affect pollinator richness and temporal stability of flower visits? Oikos 117: 1808–1815.

18. Haddad NM, Crutsinger GM, Gross K, Haarstad J, Knops JMH, et al. (2009) Plant species loss decreases arthropod diversity and shifts trophic structure. Ecol Lett 12: 1029–1039.

19. Scherber C, Eisenhauer N, Weisser WW, Schmid B, Voigt W, et al. (2010) Bottom-up effects of plant diversity on multitrophic interactions in a biodiversity experiment. Nature 468: 553–556.

20. Eisenhauer N, Milcu A, Sabais A, Bessler H, Brenner J, et al. (2011) Plant diversity surpasses plant functional groups and plant productivity as driver of soil biota in the long term. PLoS One 6: e16055.

21. Sabais ACW, Scheu S, Eisenhauer N (2011) Plant species richness drives the density and diversity of Collembola in temperate grassland. Acta Oecologica-International J Ecol 37: 195–202.

22. Borer ET, Seabloom EW, Tilman D, Novotny V (2012) Plant diversity controls arthropod biomass and temporal stability. Ecol Lett 15: 1457–1464.

23. Rzanny M, Voigt W (2012) Complexity of multitrophic interactions in a grassland ecosystem depends on plant species diversity. J Anim Ecol 81: 614–627.

24. Cardinale BJ, Wright JP, Cadotte MW, Carroll IT, Hector A, et al. (2007) Impacts of plant diversity on biomass production increase through time because of species complementarity. Proc Natl Acad Sci U S A 104: 18123–18128.

25. Abbas M, Ebeling A, Oelmann Y, Ptacnik R, Roscher C, et al. (2013) Biodiversity effects on plant stoichiometry. PLoS One 8: e58179.

26. Srivastava DS, Lawton JH (1998) Why more productive sites have more species: An experimental test of theory using tree-hole communities. Am Nat 152: 510–529.

27. Abrams PA (1995) Monotonic or unimodal diversity productivity gradients - what does competition theory predict. Ecology 76: 2019–2027.

28. Awmack CS, Leather SR (2002) Host plant quality and fecundity in herbivorous insects. Annu Rev Entomol 47: 817–844.

29. Root RB (1973) Organization of a plant-arthropod association in simple and diverse habitats - fauna of collards (Brassica-Oleracea). Ecol Monogr 43: 95–120.

30. Bernays EA, Bright KL (1993) Mechanisms of dietary mixing in grasshoppers - a review. Comp Biochem Physiol a-Physiology 104: 125–131.

31. Hutchinson GE (1959) Homage to Santa-Rosalia or why are there so many kinds of animals. Am Nat 93: 145–159.

32. Roscher C, Schumacher J, Baade J, Wilcke W, Gleixner G, et al. (2004) The role of biodiversity for element cycling and trophic interactions: an experimental approach in a grassland community. Basic Appl Ecol 5: 107–121.

33. Coleman DC, Crossley DA, Hendrix PF (2004) Fundamentals in soil ecology. Elsevier Academic Press.

34. Milcu A, Partsch S, Scherber C, Weisser WW, Scheu S (2008) Earthworms and legumes control litter decomposition in a plant diversity gradient. Ecology 89: 1872–1882.

35. Vogel A, Eisenhauer N, Weigelt A, Scherer-Lorenzen M (2013) Plant diversity does not buffer drought effects on early-stage litter mass loss rates and microbial properties. Glob Chang Biol 19: 2795–2803.

36. Heemsbergen DA, Berg MP, Loreau M, van Haj JR, Faber JH, et al. (2004) Biodiversity effects on soil processes explained by interspecific functional dissimilarity. Science (80-) 306: 1019–1020.

37. Arbuckle JL, Wothke W (1995) Amos 4.0 User's guide. Chicago: Small waters.

38. Grace J (2006) Structural equation modeling and natural systems. Cambridge: Cambridge University Press.

39. Warton DI, Hui FKC (2011) The arcsine is asinine: the analysis of proportions in ecology. Ecology 92: 3–10.

40. Bardgett RD, Shine A (1999) Linkages between plant litter diversity, soil microbial biomass and ecosystem function in emperate grasslands. Soil Biol Biochem 31: 317–321.

41. Spehn EM, Hector A, Joshi J, Scherer-Lorenzen M, Schmid B, et al. (2005) Ecosystem effects of biodiversity manipulations in european grasslands. Ecol Monogr 75: 37–63.

42. Cardinale BJ, Matulich KL, Hooper DU, Byrnes JE, Duffy E, et al. (2011) The functional role of producer diversity in ecosystems. Am J Bot 98: 572–592.

43. Eisenhauer N, Reich PB, Scheu S (2012) Increasing plant diversity effects on productivity with time due to delayed soil biota effects on plants. Basic Appl Ecol 13: 571–578.

44. Long ZT, Mohler CL, Carson WP (2003) Extending the resource concentration hypothesis to plant communities: Effects of litter and herbivores. Ecology 84: 652–665.

45. Jonsson M, Malmqvist B (2000) Ecosystem process rate increases with animal species richness: evidence from leaf-eating, aquatic insects. Oikos 89: 519–523.

46. Jonsson M, Malmqvist B (2003) Mechanisms behind positive diversity effects on ecosystem functioning: testing the facilitation and interference hypotheses. Oecologia 134: 554–559.

47. Rosenkranz S, Wilcke W, Eisenhauer N, Oelmann Y (2012) Net ammonification as influenced by plant diversity in experimental grasslands. Soil Biol Biochem 48: 78–87.

48. Hooper DU, Bignell DE, Brown VK, Brussaard L, Dangerfield JM, et al. (2000) Interactions between aboveground and belowground biodiversity in terrestrial ecosystems: Patterns, mechanisms, and feedbacks. Bioscience 50: 1049–1061.

49. McNaughton SJ, Oesterheld M, Frank DA, Williams KJ (1989) Ecosystem-level patterns of primary productivity and herbivory in terrestrial habitats. Nature 341: 142–144.

50. Van Ruijven J, Berendse F (2005) Diversity-productivity relationships: initial effects, long-term patterns, and underlying mechanisms. Proc Natl Acad Sci U S A 102: 695–700.

51. Berner D, Blanckenhorn WU, Korner C (2005) Grasshoppers cope with low host plant quality by compensatory feeding and food selection: N limitation challenged. Oikos 111: 525–533.

52. Ebeling A, Allan E, Heimann J, Köhler G, Scherer-Lorenzen M, et al. (2013) The impact of plant diversity and fertilization on fitness of a generalist. Basic Appl Ecol 14: 246–254.

53. Frost PC, Benstead JP, Cross WF, Hillebrand H, Larson JH, et al. (2006) Threshold elemental ratios of carbon and phosphorus in aquatic consumers. Ecol Lett 9: 774–779.

54. Hillebrand H, Borer ET, Bracken MES, Cardinale BJ, Cebrian J, et al. (2009) Herbivore metabolism and stoichiometry each constrain herbivory at different organizational scales across ecosystems. Ecol Lett 12: 516–527.

55. Schuldt A, Bruelheide H, Durka W, Eichenberg D, Fischer M, et al. (2012) Plant traits affecting herbivory on tree recruits in highly diverse subtropical forests. Ecol Lett 15: 732–739.

56. Loranger J, Meyer ST, Shipley B, Kattge J, Loranger H, et al. (2013) Predicting invertebrate herbivory from plant traits: polycultures show strong nonadditive effects. Ecology 94: 1499–1509.

57. Schuldt A, Baruffol M, Böhnke M, Bruelheide H, Härdtle W, et al. (2010) Tree diversity promotes insect herbivory in subtropical forests of south-east China. J Ecol 98: 917–926.

58. Duffy JE, Cardinale BJ, France KE, McIntyre PB, Thebault E, et al. (2007) The functional role of biodiversity in ecosystems: incorporating trophic complexity. Ecol Lett 10: 522–538.

59. Rzanny M, Kuu A, Voigt W (2013) Bottom-up and top-down forces structuring consumer communities in an experimental grassland. Oikos 122: 967–976.

4

Documenting Biogeographical Patterns of African Timber Species Using Herbarium Records: A Conservation Perspective Based on Native Trees from Angola

Maria M. Romeiras[1,2]*, **Rui Figueira**[1,3], **Maria Cristina Duarte**[1,3], **Pedro Beja**[3], **Iain Darbyshire**[4]

1 Tropical Botanical Garden, Tropical Research Institute (IICT), Lisbon, Portugal, **2** Centre for Biodiversity, Functional and Integrative Genomics (BIOFIG), Faculty of Sciences, University of Lisbon, Lisbon, Portugal, **3** CIBIO - Research Center in Biodiversity and Genetic Resources/InBIO, University of Porto, Vairão, Portugal, **4** Royal Botanic Gardens, Kew. Richmond, United Kingdom

Abstract

In many tropical regions the development of informed conservation strategies is hindered by a dearth of biodiversity information. Biological collections can help to overcome this problem, by providing baseline information to guide research and conservation efforts. This study focuses on the timber trees of Angola, combining herbarium (2670 records) and bibliographic data to identify the main timber species, document biogeographic patterns and identify conservation priorities. The study recognized 18 key species, most of which are threatened or near-threatened globally, or lack formal conservation assessments. Biogeographical analysis reveals three groups of species associated with the enclave of Cabinda and northwest Angola, which occur primarily in Guineo-Congolian rainforests, and evergreen forests and woodlands. The fourth group is widespread across the country, and is mostly associated with dry forests. There is little correspondence between the spatial pattern of species groups and the ecoregions adopted by WWF, suggesting that these may not provide an adequate basis for conservation planning for Angolan timber trees. Eight of the species evaluated should be given high conservation priority since they are of global conservation concern, they have very restricted distributions in Angola, their historical collection localities are largely outside protected areas and they may be under increasing logging pressure. High conservation priority was also attributed to another three species that have a large proportion of their global range concentrated in Angola and that occur in dry forests where deforestation rates are high. Our results suggest that timber tree species in Angola may be under increasing risk, thus calling for efforts to promote their conservation and sustainable exploitation. The study also highlights the importance of studying historic herbarium collections in poorly explored regions of the tropics, though new field surveys remain a priority to update historical information.

Editor: Giovanni G. Vendramin, CNR, Italy

Funding: This work was supported by the Portuguese Foundation for Science and Technology with the FCT/Ciência 2008 to MMR and FCT/Ciência 2007 to RF, EDP Biodiversity Chair (CIBIO) to PB and BioFIG PEst-OE/BIA/UI4046/2011. The funders had no role in study design, data collection and analysis, decision to publish, or preparation of the manuscript.

Competing Interests: The authors have declared that no competing interests exist.

* Email: mromeiras@yahoo.co.uk

Introduction

Legacy data from natural history collections contain invaluable information about biodiversity in the recent past, providing a baseline for detecting change and forecasting future trends [1]. In the case of plants, specimens have accumulated for hundreds of years in herbaria, and these may be used as the basis for identifying threatened or declining species, guiding future research and monitoring programs, and establishing conservation priorities [2]. For instance, the IUCN Sampled Red List Index for plants was driven in its first iteration almost solely by herbarium specimen data [3]. Data from herbaria are particularly important in poorly explored regions of the tropics, where the lack of continuous field-based botanical research has emphasized the pivotal role of herbaria in documenting plant diversity and species distributions [4–6]. The interest in herbaria for undertaking conservation biology research has thus grown in recent years,

though less than about 2% of the herbarium specimens have been used to answer biogeographical or environmental questions [6].

Establishing baselines is particularly important for those tropical tree species that are exploited commercially and have come under increasing pressure from the global timber trade [7–8]. Over-exploitation has resulted in declining populations of the most valuable timber species and it is one of the foremost causes for the loss and degradation of tropical forests [9], with utmost negative consequences for the conservation of biodiversity and ecosystem services [10–11]. In recent decades, efforts have been made to increase the sustainability of tropical timber exploitation, through for instance the outright ban on or severe restrictions to the trade of endangered species, or the implementation of certification schemes for timber harvested sustainably [8–12]. These approaches face several problems, however, including uncertainties related to the conservation status of many exploited species due to insufficient knowledge of their distribution, abundance and

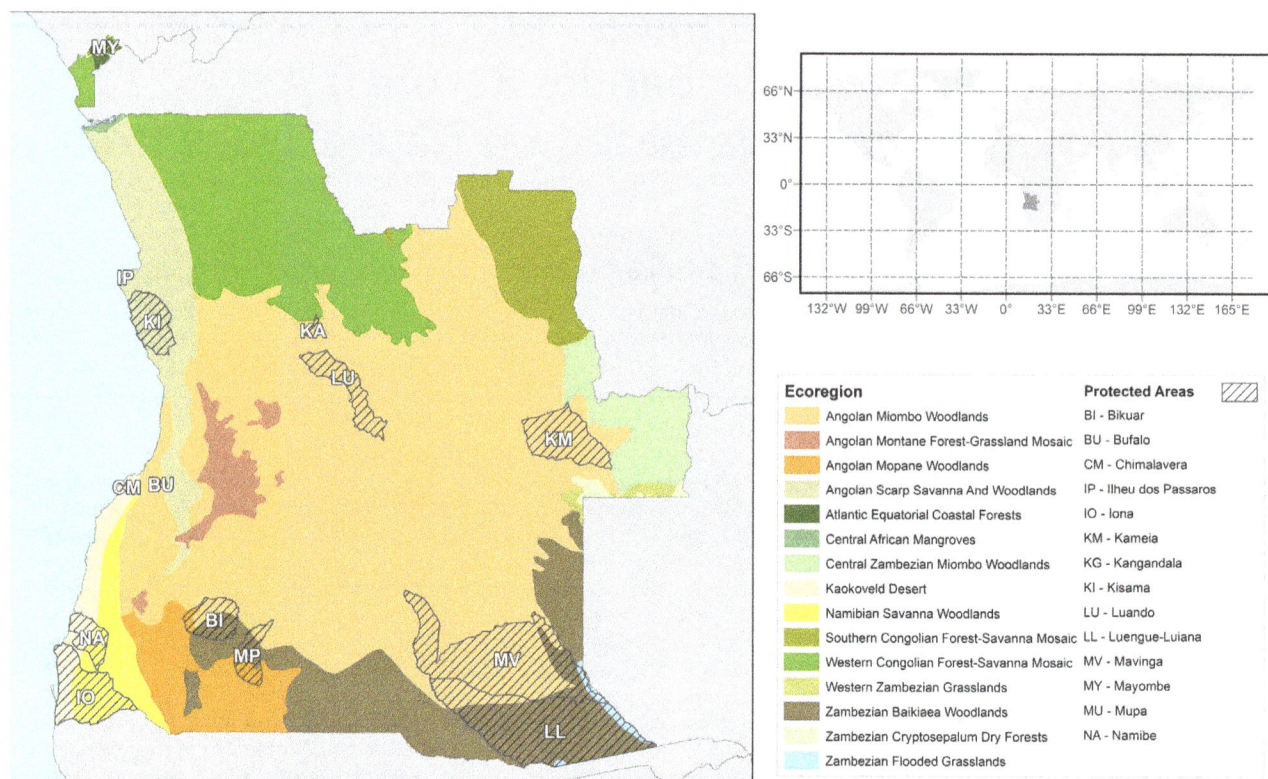

Figure 1. Map of Angola. The 15 WWF ecoregions represented in Angola are displayed together with the network of protected areas (see text for details).

population trends [13–15]. Although this type of information has become increasingly available for tropical forests of Central and South America [16–17] and Asia [18–19], data are still very limited for most African forests [20]. Considering that Africa still holds some of the most important tropical forests in the world [21–22] and that these have been increasingly exploited [23], information on the conservation status of its timber species is urgently required [24].

Angola is one of the African countries for which basic data on timber tree species are most severely lacking, though the country has a forested area of about 40–60 million hectares largely administered by the government [25–26]. Deforestation rates in Angola are among the highest in Sub-Saharan Africa [27], which is likely a consequence of wood extraction for firewood and charcoal, slash-and-burn cultivation, urban expansion, and logging [25]. Illegal logging of valuable timber is considered one of the potential causes of forest degradation, but there is no information on the extent of this problem [25]. Despite some early studies [28–32], botanical data on the forests of Angola are scarce because most of the country was inaccessible to researchers during the war of independence (1961–1974), and the subsequent civil war (1975–2002). Despite increases in safety during the first decade of the 21st century, field biodiversity research has remained very limited, thereby making historic herbarium specimens the main source of data for studying the distribution patterns of tree species exploited commercially in Angola. This information is urgently required because Angola is currently experiencing rapid economic and human population growth, which is likely to place further pressure on its forest resources, with negative consequences for biodiversity, ecosystem services, and ultimately for human well-being [25]. Data on timber trees is also

required to inform ongoing initiatives to improve the protected area network of Angola [33].

The present study focuses on the timber trees of Angola, combining herbarium and bibliographic data to assess biogeographical patterns and conservation priorities, thereby providing baseline information required for their conservation management and sustainable exploitation. Specifically, the study aims (i) to inventory the timber tree species of Angola based on a thorough review of literature and data held in herbaria, (ii) to document biogeographical patterns of the timber species in relation to WWF ecoregions [34]; and (iii) to estimate species conservation priorities based on distribution patterns, representation in protected areas and deforestation rates.

Materials and Methods

Study area

The Republic of Angola (Fig. 1) is the largest country in southern Africa (1.24 million km^2), encompassing a variety of climatic characteristics, which correspond to five climate types by the Köppen–Geiger system [35]. The phytogeographic study of Grandvaux-Barbosa [36] identified 32 vegetation units in the country, ranging from rainforests in the northwest to the desert in the southwest. The global ecoregions map of World Wildlife Fund (WWF) [34] recognises the presence of 15 biogeographic units in Angola (Fig. 1), of which the most widespread are the miombo woodlands of the central plateau, and the western Congolian forest-savanna mosaics in the north. Other important but less widespread forest types include the Atlantic Equatorial coastal forests in Cabinda, the mopane (*Colophospermum mopane*) woodlands, and the Namibian savanna woodlands in the southwest (see Fig. 1). The network of protected areas was mainly established in colonial times to protect large ungulates, and it has been considered too limited to adequately protect most biodiversity components, notably vascular plants [25,33].

Species data

Data on the timber species of Angola were obtained through a combination of bibliographic sources and the study of 2670 herbarium records: 417 of Angolan specimens (see Table S1) and 2253 records from 62 providers available through the GBIF data portal (Tables S2 and S3). First, a thorough literature review was undertaken, focusing on studies of the flora of Angola [36–44], and on studies documenting the use of afro-tropical timber trees [28–32]. Based on this information, we selected for further analysis the subset of timber trees that are: (i) known to be native in Angola; (ii) documented in the Angolan literature to be exploited for timber in colonial times or at present; and (iii) traded in international timber markets. Many of these timber species are important components of the upper forest layer, above 25 m, and all are known for their economic value, thus making them interesting from both ecological and conservation standpoints. For each species selected, we compiled information on their distribution in Angola and across Africa, their habitat, ecology, timber value and characteristics, and their global conservation status based on the IUCN Red List of Threatened Species [45].

Second, a thorough study of herbarium specimen data was undertaken for all timber species selected. The research was concentrated on herbaria holding the largest collections of Angola vascular plants, including LISC (Tropical Research Institute), LISU (University of Lisbon), COI (University of Coimbra), BM (Natural History Museum, London), and K (Royal Botanic Gardens Kew). The collecting locality for each specimen was georeferenced wherever possible, using 1:100,000 cartographic maps and geographic gazetteers [46], and data was compiled in a geographic database prepared in ArcGIS Arcinfo ver. 10.0 [47]. Further information about the global native distribution of each selected timber species in Africa was gathered from the GBIF data portal. Although it is recognised that GBIF does not contain all known records of the species studies, it is deemed adequate to provide a first approximation of their geographic range.

Biogeographic patterns

Patterns of timber tree species distribution in Angola were analysed in relation to the 15 WWF ecoregions identified in the country [34]. We focused on WWF ecoregions because they have been produced mainly as a utility tool for conservation planning [48], and so it was considered important to examine whether they could be used as meaningful spatial units for conservation prioritization and management of Angolan timber tree species.

Analyses were based on a presence/absence matrix, which indicated whether or not each timber tree species had been recorded within each WWF ecoregion. Presence/absence was used instead of the number of records, to reduce the bias associated with geographic variation in sampling effort. Although this approach does not avoid the problem of false absences (i.e. absence due to lack of sampling rather than a true record of absence), we believe that this problem has been minimised by using a small number of spatial units, each covering a large geographic area and encompassing many species records. Hierarchical clustering was then carried out, using the Jaccard index as a measure of similarity between species distribution, and the Ward agglomerative procedure [49]. The Jaccard index was used because it does not consider double absences [49]. Several agglomerative methods were tested (e.g., UPGMA, WPGMA),

but they produced largely similar results. Clusters identified at different levels of the dendrogram were mapped and checked for spatial consistency, i.e., whether each group was associated with a well-defined spatial region, and we selected the number of clusters that maximised spatial interpretability [49]. Quantitative approaches such as the L-Method [50] were also tested but the number of clusters produced was excessively large and with no spatial consistency. Analyses were carried out using 'dist' and 'hclust' functions implemented in R version 3.02 [51].

The spatial distribution of the species groups emerging from the cluster analysis was overlapped with the WWF ecoregions map, and spatial consistency between species groups and ecoregions was visually inspected. A similar investigation was carried out by overlapping the spatial distribution of species groups and the climate classification map of Köppen–Geiger [35].

Species conservation priorities

Estimating conservation priorities from herbaria data is difficult, because a species may no longer exist in localities where it was historically recorded, and because collectors may be biased towards or against certain species or regions [2,5]. To overcome these problems, we used a combination of three relatively coarse criteria, which were judged useful in helping to guide future conservation efforts, despite some potential shortcomings and limitations.

A first approximation for conservation prioritization was obtained by computing the extent of occurrence (EOO) of each species, assuming that the highest priority should be given to species with a small EOO in Angola, and to species with a large proportion of its global EOO concentrated in the country. EOO was computed from the georeferenced locality data for each species, using the minimum convex hull polygon method [52], implemented in GEOCAT [53]. Computations were carried out at the scale of the African Continent and that of Angola, and we calculated Angola's contribution to the overall EOO for each species. Areas offshore from the African continent were calculated using ArcGIS Arcinfo ver. 10.0 [47] and were excluded from the EOO polygon. Although the area of occupancy (AOO) is an important parameter to assess species conservation status [52], it was not estimated because large gaps in species distribution are likely to be due primarily to the lack of comprehensive field surveys or lack of data reporting by herbaria to GBIF, rather than resulting from true species absences.

A second indicator of conservation priority was based on the occurrence of herbarium specimens' locations in national parks and reserves, assuming that a higher conservation risk should be attributed to the species poorly represented within protected areas. We considered both the number of locations recorded within protected areas, and the percentage of the EOO that is included in protected areas. Although we recognise that it is uncertain whether a given species occurs at any particular location within its EOO, we assumed that the overlap between EOO and protected areas could be taken as a coarse approximation of the relative representation of a species within the protected area network. The geographical limits of protected areas were obtained in GIS shape file format from WDPA [54]. New protected areas unavailable in WDPA were digitised in ArcGIS ArcInfo ver. 10.0 [37] from maps published in the official journal of The Republic of Angola (law n° 38/11 of December, 29 2011, p. 6340).

Finally, conservation priority was also evaluated by estimating rates of forest loss between 2000 and 2012 around the georeferenced localities for each species. We assumed that higher conservation priority should be given to species occurring in areas with low forest cover, and where the recent deforestation rate is highest. Forest cover was estimated for each georeferenced specimens location using raster maps provided by Hansen et al. [27], by multiplying the percent tree (crown) cover per pixel and the pixel area (30-m resolution), and then summing across all pixels extracted in a 5-km buffer of the location. Deforestation rate was calculated by estimating the area of pixels showing forest loss, and then expressing it as a percentage of total tree cover in 2000. Similar analyses were carried out using 1, 2.5 and 10-km buffers, but the results were much the same, and so they were not considered further.

Results

From the literature review and the study of herbarium specimens, we identified eighteen native timber species occurring in Angola (Table 1), which have a high commercial value due to the quality of their timber (Table S4). Available herbarium data are rather old, corresponding primarily to specimens collected in 1850–1860, 1910–1920, and 1950–1975 (Fig. S1). Most species (> 80%) belong to the Fabaceae and Meliaceae families, and they are associated with tropical rainforests (11 species), evergreen forests and woodlands (2 species), and mainly with dry forests and savannas (5 species) (Table 1, Table S4). Half the species are either classified as threatened (7) or near-threatened (2) by IUCN at the global scale, whereas the conservation status of eight species has not yet been evaluated.

The cluster analysis of timber trees identified four groups of species (Fig. 2a). The first and second groups have similar spatial patterns, with most occurrences concentrated in the small enclave of Cabinda. The second group, however, is also represented in north-western regions of Angola. For eight of the eleven species in these two groups, Cabinda is the southern limit of wider distributions concentrated in the Guineo-Congolian rainforests (Fig. 3a). The third group includes only two species and it has a distribution concentrated in north-western regions of Angola, though it is absent from Cabinda. The fourth group includes five species, and it occupies most of the Angolan territory, with the exception of Cabinda.

There is a poor match between the spatial distribution of the four species groups and the WWF ecoregions, as each group occurs in several ecoregions (Fig. 2b). Overlay with the climate classification map of Köppen–Geiger suggests a rough association between groups 1–3 and a single climate type (Aw - Equatorial savanna with dry winter), while group 4 occurs in a wide range of climate types (Fig. 2c).

Most of the timber species have a large global extent of occurrence - EOO (Table 2). Smaller EOO values are found for *Entandrophragma spicatum* and *Guibourtia arnoldiana* ($\approx 1 \times 10^5$ km^2), but they are still one order of magnitude above the threshold for species qualifying as threatened under IUCN criterion B (i.e., $> 2 \times 10^4$ km^2) (Table 2). Within Angola, however, there are nine species with a restricted EOO ($< 2 \times 10^4$ km^2) and thus potentially qualifying as threatened at the national level. Three species have more than 15% of their global EOO concentrated in Angola (Fig. 3b–d), reaching >50% in the case of *E. spicatum* and *G. coleosperma* (Table 2). Few of the historical herbarium specimens were collected from within current protected areas, with ≤5 localities for all species evaluated (Table 2). More than 10% of the EOO of eight species overlaps with protected areas, whereas there was no overlap for another six species (Table 2).

Forest cover, in 2000, around the location of collection localities ranged from <20% in the case of *Diospyros mespiliformis*, *E.*

spicatum and *G. coleosperma*, to >70% in the case of *Entandrophragma candollei*, *Milicia excelsa*, *Oxystigma oxyphyllum* and *Terminalia superba* (Table 3). Variation among species in deforestation rate (2000–2012) was less marked, but particularly high values (>10%) were recorded in the occurrence areas of *G. coleosperma* and *Pterocarpus angolensis*.

Discussion

This study recognized the presence of 18 key timber tree species in Angola, most of which are widely used as timber trees elsewhere in Africa [26]. These species are important components of woody vegetation communities and are known for their economic value, thus making them important from both ecological and economic perspectives. Several of these species are highly valued in international timber markets and they have been historically exploited in Angola, including the African mahoganies (*Entandrophragma* spp.), the agba (*G. balsamiferum*), and the tchitola (*O. oxyphyllum*) [43], hence they are under increasing pressure in the country.

Biogeographical patterns

This study revealed striking differences in biogeographic patterns of timber species in Angola, recognizing different groups associated with regions with relatively homogeneous climatic conditions (i.e. tropical rainforests; evergreen forests and woodlands; and dry forests, woodlands and savannas). Three of the four clusters identified were associated with Cabinda and the north-western regions of Angola, showing a close matching with the Aw climate category of Köppen–Geiger and with the Congolian region identified in the recent bioregionalization study of Linder et al. [56]. The timber species included in the first two clusters (ca. 60%) corresponding largely to Guineo-Congolian rainforest species [57], where the rainy season lasts for six months or more, and the relative air humidity is above 80%, with some areas having persistent and dense fogs (locally known as *cacimbo*). In Angola they occur in Cabinda's Maiombe forest (extending through Congo, Democratic Republic of the Congo and Angola), which are dense moist forest formations, with high ecological and floristic diversity [58,59]. Most of the species found in Guineo-Congolian rainforests have their south-western range limit in Cabinda (see Fig. 3a), where they may face drier climatic conditions than within their core range. These peripheral populations may have unique adaptations to specific environmental conditions that are absent from other populations [60], and so may be particularly valuable in a warming scenario due to climate change [61].

The third cluster included just two species (*K. anthotheca* and *P. tinctorius*), and it was restricted to woodlands and evergreen forests of northwest Angola. This region is characterized by a rainy season lasting about six months and relative air humidity of about 75% [36]. Finally, the fourth cluster comprised five species and is widespread throughout the country, albeit little represented in the north-western regions of Angola. Species included in this group occur mainly in dry forests and savannas, and sometimes their distribution reach the semi-arid and arid regions of southern Angola, characterized by xerophytic vegetation. Among the studied timber species, the legumes *G. coleosperma* (see Fig. 3b) and *P. angolensis* (see Fig. 3c) are the most widespread in Angola, and are mainly found in miombo woodlands, which is the dominant forest component of Angola and one of the major dry forest-savanna biomes of the world [34].

The spatial distribution of the four clusters retrieved from the biogeographical analysis showed little concordance with the WWF

Table 1. Geographical distribution in Angola and in the African continent, main types of ecosystems and global conservation status [45] for each species considered in the present study. Nomenclature according to The Plant List [55].

| Main types of ecosystems | Geographical distribution | | Conservation status and criteria |
Species	Angola (district)	Africa	
Tropical Rainforests			
Bobgunnia fistuloides (Harms) J.H. Kirkbr. & Wiersema	Cabinda; Malanje; Zaire	Nigeria, Cameroon, Gabon, Angola, D.R. Congo, Mozambique.	Least Concern
Entandrophragma angolense (Welw.) C. DC.	Cabinda; Cuanza Norte; Cuanza Sul; Malanje	From Guinea to Uganda, Kenya and Angola	Vulnerable A1cd
Entandrophragma candollei Harms	Cabinda	From the Ivory Coast to Angola and D.R. Congo	Vulnerable A1cd
Entandrophragma cylindricum (Sprague) Sprague	Cabinda	From Sierra Leone to Cabinda and Uganda	Vulnerable A1cd
Entandrophragma utile (Dawe & Sprague) Sprague	Cabinda	From the Ivory Coast to Angola, D.R. Congo and Uganda	Vulnerable A1cd
Gossweilerodendron balsamiferum (Vermoesen) Harms	Cabinda	From the south of Nigeria and Cameroon to D.R. Congo and Angola	Endangered A1cd
Guibourtia arnoldiana (De Wild. & T. Durand) J. Léonard	Cabinda; Zaire	Gabon, Congo, Angola (Cabinda), D. R. Congo (Maiombe)	Not Evaluated
Khaya ivorensis A. Chev.	Cabinda	From the Ivory Coast to Angola (Cabinda)	Vulnerable A1cd
Milicia excelsa (Welw.) C.C. Berg	Cabinda; Cuanza Norte	Widely distributed in Africa, from Senegal to Angola, D.R. Congo, East Africa and Mozambique	Near Threatened
Oxystigma oxyphyllum (Harms) J. Léonard	Cabinda	From Nigeria to Angola (Cabinda)	Not Evaluated
Terminalia superba Engl. & Diels	Cabinda	From Guinea to D.R. Congo (Maiombe)	Not Evaluated
Evergreen Forests and Woodlands			
Khaya anthotheca (Welw.) C. DC.	Bengo; Cuanza Norte; Malanje	From Sierra Leone to Uganda and Tanzania, and central Angola, Zambia, Malawi, Mozambique and Zimbabwe; also in the Ivory Coast, the Gold Coast, Nigeria and Cameroon.	Vulnerable A1cd
Pterocarpus tinctorius Welw.	Bengo; Cuanza Norte; Cuanza Sul; Luanda; Malanje; Zaire	Congo, Angola, Zambia, Zimbabwe, Tanzania and Mozambique	Not Evaluated
Dry Forests, Woodlands and Savannas			
Afzelia quanzensis Welw.	Benguela; Bie; Cuando Cubango; Cuanza Norte; Cuanza Sul; Cunene; Huila; Malanje; Namibe	In Angola, Namibia, D.R. Congo, Zambia, Zimbabwe and Botswana, and from Somalia to South Africa	Not Evaluated
Diospyros mespiliformis Hochst. ex A. DC.	Bengo; Cuando Cubango; Cuanza Norte; Cuanza Sul; Cunene; Huila; Luanda; Namibe	From Senegal to Sudan, and southwards to Namibia. It can be found from the mouth of the Zaire river to the Transvaal and South Mozambique, but not in the Guineo-Congolian rainforests	Not Evaluated
Entandrophragma spicatum (C. DC.) Sprague	Benguela; Cunene; Huila; Namibe	Southern Angola and Namibia	Not Evaluated
Guibourtia coleosperma (Benth.) J. Léonard	Bengo; Bie; Cuando Cubango; Cunene; Huambo; Huila; Lunda Norte; Lunda Sul; Moxico	D.R. Congo, Angola, Namibia, Botswana, Zambia, Zimbabwe	Not Evaluated
Pterocarpus angolensis DC.	Benguela; Bie; Cuando Cubango; Cuanza Norte; Cuanza Sul; Cunene; Huambo; Huila; Lunda Norte; Lunda Sul; Malanje; Moxico; Namibe; Uige	From Congo to Namibia and from Tanzania to Swaziland	Near Threatened

ecoregions. The mismatch was particularly notable in the case of the Cabinda forests, which have a very unique set of timber tree species that are not adequately captured by WWF ecoregions. In fact, although one of the two regions dominating the enclave of Cabinda also occurs in a larger area in the northwest of Angola (western Congolian forest-savanna mosaic), most timber species characteristic of the former region were not found elsewhere. Reasons for the mismatches are uncertain, but they are probably related to the operation of climatic and historical factors that are not adequately captured by the WWF ecoregion definitions. Irrespective of the reason, however, these results suggest that

WWF ecoregions may not provide an adequate operational basis for conservation planning exercises targeting timber tree species in Angola.

It is suggested that the biogeographic patterns observed in this study, might be better explained by the climatic classification of Köppen–Geiger [35], which suggests that the native range of these timber species is conditioned by large scale patterns. These regions are broadly similar to those recently proposed by Linder et al. [56], that demonstrate the existence of only seven well-defined and consistent biogeographical regions in sub-Saharan Africa, proposing that the best approach might be to recognize, as White [57]

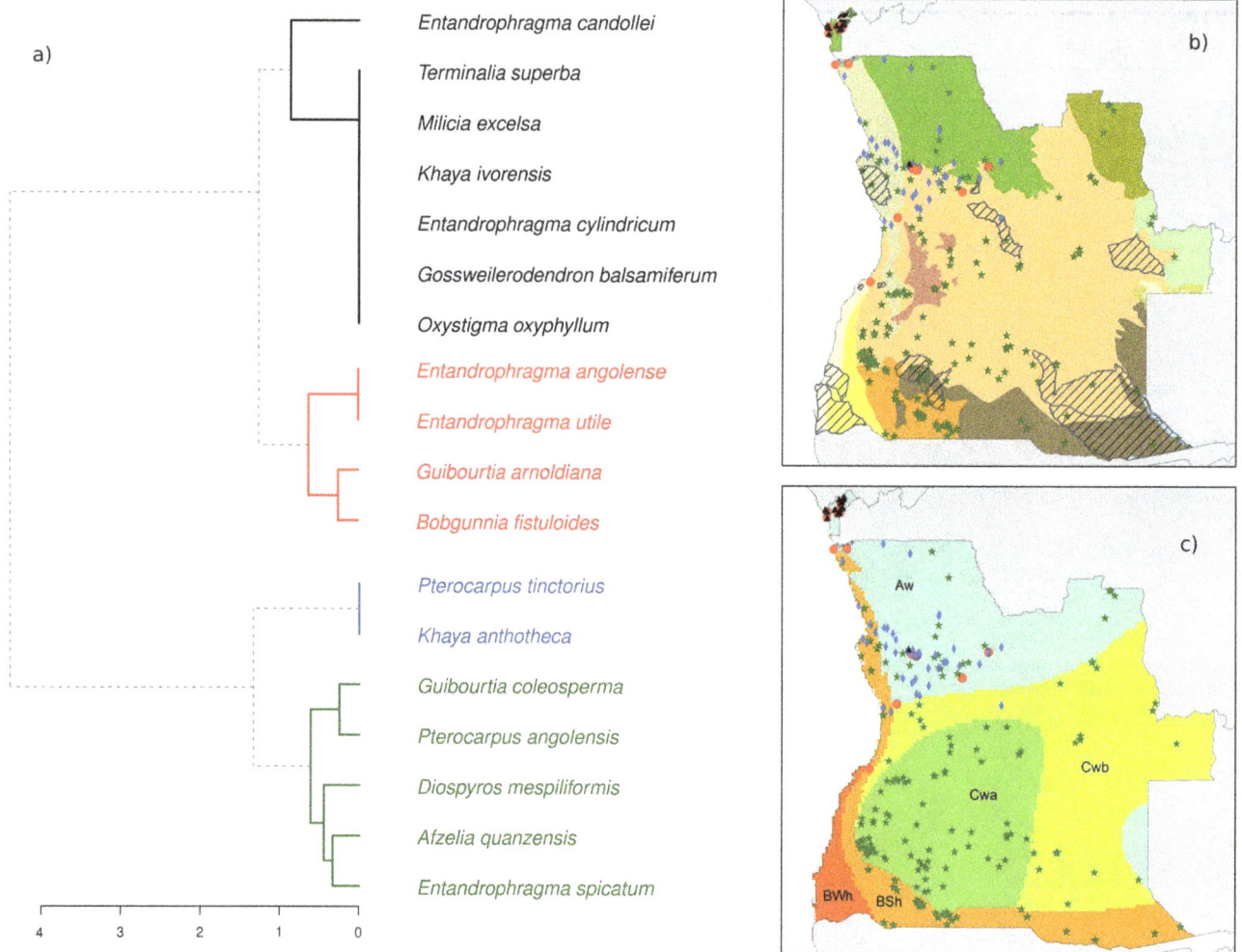

Figure 2. Biogeographical patterns of Angolan forest species selected for this study. a) Dendrogram of a cluster analysis based on the distribution (presence/absence) of timber tree species in each of the ecoregions. Distribution of collection localities of the four species groups identified in this study, in relation to: **b)** protected areas and the 15 WWF ecoregions represented in Angola; and **c)** Köppen–Geiger climate classification. Clustering was based on the Jaccard index of similarity and on the Ward agglomeration algorithm.

did, a small number of very broad biogeographical regions in Africa that can reflect the patterns found in both vertebrates and plants [56].

Species conservation priorities

Our results suggest that at least 11 of the species evaluated should be given high conservation priority in Angola. These include: (1) globally threatened or near-threatened species with small ranges in Angola and largely restricted to Cabinda or, to a much lesser extent, the north-western regions (*E. angolense, E. candollei, E. cylindricum, E. utile, G. balsamiferum, K. anthotheca, K. ivorensis* and *M. excelsa*); and (2) species from dry forests with a large proportion of their global range concentrated in Angola, and occurring in areas that are affected at present by high deforestation rates (*E. spicatum, G. coleosperma* and *P. angolensis*).

From the first group, all but two species (*E. angolense* and *K. anthotheca*) are concentrated in Cabinda's Maiombe forest, with a very small EOO in Angola, though they are widely distributed elsewhere in Guineo-Congolian rainforests. The herbarium collections studied of all these species were made in locations still retaining a relatively extensive forest cover (46.7 to 84.4%), and

where deforestation rates between 2000 and 2012 were lower than in other areas of Angola. The overlap between the EOO and protected areas for these species is negligible (<0.5%), except in the case of *G. balsamiferum* (10.5%). These species may thus remain largely unprotected in Angola despite the recent creation of the Maiombe National Park, which was specifically designed to protect the Cabinda's forest. This is worrying in view of ongoing logging activities in Cabinda, where these species may be under increasing pressure [25,58].

The second group of conservation priority species includes timber trees that are mainly found in dry forests, and that are particularly important from a conservation perspective because their Angolan range represents a large proportion (17.7–55.4%) of their global range. Although these species have a relatively large overlap between their EOO and protected areas in Angola (12.9–16.4%), they occur in areas where tree cover is among the lowest for timber trees in Angola (9.3–20.4%). Further, the collection localities of these species have suffered high recent deforestation rates, amounting to >10% in 12 years for *G. coleosperma* and *P. angolensis*. These species should therefore merit national conservation attention, though only *P. angolensis* has been considered

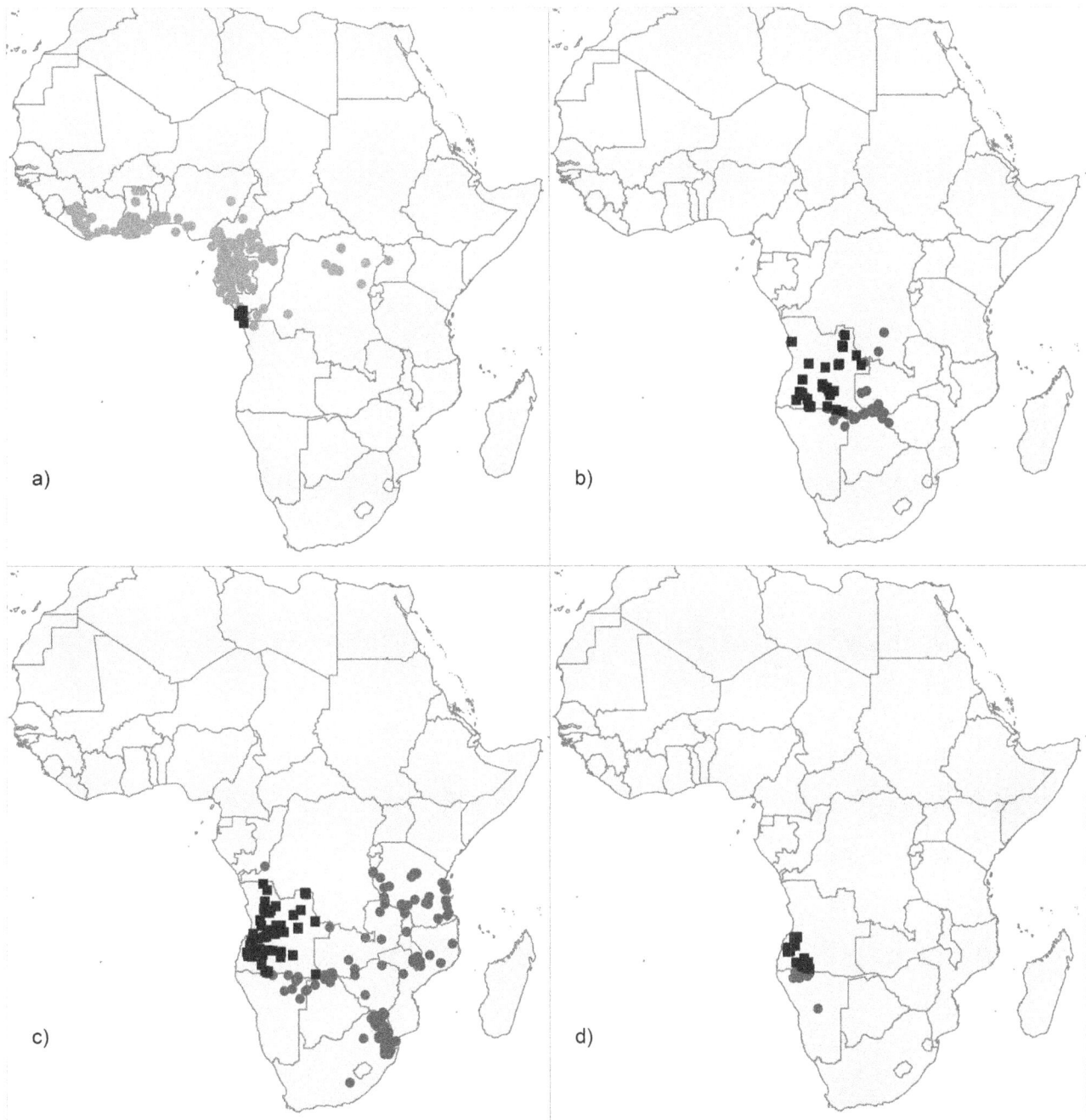

Figure 3. Geographic distribution in Africa of representative timber trees from Angola. (a) Species typical of Guineo-Congolian rainforests that reach their southern limit in Cabinda (*E. cylindricum; E. utile; G. arnoldiana; G. balsamiferum; K. ivorensis; O. oxyphyllum; T. superba*); and species with >15% of their global range concentrated in Angola, including (**b**) *G. coleosperma*, (**c**) *P. angolensis*, (**d**) *E. spicatum*. Black squares: studied specimens from Angola, housed in LISC, LISU, COI, BM and K; grey circles: data obtained via the GBIF portal.

near-threatened in IUCN global assessment [45], while the evaluation of *E. spicatum* and *G. coleosperma* is lacking.

The conservation priorities identified in this study are limited because herbarium data may not reflect current distribution, the ecological information for most species is scarce and there are virtually no data on present logging pressure. Notwithstanding, we believe that our approach provides a first approximation for timber tree species prioritization in Angola, which may be a useful guide to conservation decisions until more detailed information becomes available.

Conservation implications

Our study clearly underlines the need to take urgent action to protect the Cabinda's Maiombe forest, where there is a significant concentration of threatened timber species of high conservation priority. At present, these forests may be better conserved than similar Guineo-Congolian forests in neighbouring countries (see Fig. S2 in Supporting Information) where deforestation rates are high and concessions for industrial logging are expanding [27,62]. However, the Cabinda's forest may be under increasing legal and illegal logging pressure, though data to quantify this problem are

Table 2. Number of specimen collection localities and estimates of the global and national (Angola) extent of occurrence (EOO) for the selected timber tree species.

Species	Number of localities			Extent of Occurrence			%EOO	%EOO protected
	Global	Angola	Protected areas	Global (×10⁶ km²)	Angola (×10³ km²)	Protected areas (×10³ km²)		
Afzelia quanzensis	107	34	3	5.2	429.6	56.7	8.3	13.2
Bobgunnia fistuloides	47	5	1	0.8	0.2	0.0	0.0	0.0
Diospyros mespiliformis	272	37	3	12.4	464.5	96.5	3.8	20.8
Entandrophragma angolense	67	8	1	2.7	169.6	0.5	6.2	0.3
Entandrophragma candollei	42	1	1	1.6	a	a	a	a
Entandrophragma cylindricum	50	3	1	1.9	0.01	0.0	0.0	0.0
Entandrophragma spicatum	13	21	1	0.1	75.1	12.3	52.5	16.4
Entandrophragma utile	46	2	1	1.6	a	a	a	a
Gossweilerodendron balsamiferum	30	9	3	1.0	1.4	0.1	0.1	10.5
Guibourtia arnoldiana	10	5	0	0.1	10.1	0.0	11.2	0.0
Guibourtia coleosperma	36	32	5	1.3	697.0	89.6	55.4	12.9
Khaya anthotheca	76	8	0	6.0	58.4	0.0	1.0	0.0
Khaya ivorensis	49	5	0	1.0	0.6	0.0	0.1	0.0
Milicia excelsa	156	3	1	7.1	0.1	0.0	0.0	0.0
Oxystigma oxyphyllum	34	6	2	1.3	1.1	0.3	0.1	28.1
Pterocarpus angolensis	135	61	2	4.8	856.2	113.3	17.7	13.2
Pterocarpus tinctorius	45	38	1	2.1	158.6	10.3	7.5	6.5
Terminalia superba	98	6	2	2.0	0.7	0.1	0.0	14.3

% EOO is the percentage contribution of Angola to the global extent of occurrence; % EOO protected is the percentage of the EOO within Angola which is included in protected areas network.
[a]EOO not estimated due to insufficient data (≤2 locations).

Table 3. Forest cover (in 2000) and forest cover changes (2000–2012) estimated in 5-km buffers around the herbarium collection localities for each Angolan timber tree species.

Species	N	Tree Cover (%)	Forest Gain (%)	Forest Loss (%)	Deforested (%)
Afzelia quanzensis	6	23.8	0.1	2.3	9.6
Bobgunnia fistuloides	8	60.6	0.8	4.5	7.4
Diospyros mespiliformis	18	13.6	0.1	0.6	4.6
Entandrophragma angolense	9	62.1	0.9	5.5	8.9
Entandrophragma candollei	13	77.2	1.6	6.6	8.6
Entandrophragma cylindricum	10	68.6	1.4	3.8	5.6
Entandrophragma spicatum	11	9.3	0.0	0.8	8.2
Entandrophragma utile	12	46.7	1.1	3.5	7.6
Gossweilerodendron balsamiferum	1	69.0	0.6	2.5	3.6
Guibourtia arnoldiana	7	55.9	0.8	1.5	2.8
Guibourtia coleosperma	3	18.9	0.0	5.1	27.1
Khaya anthoteca	14	52.8	0.4	4.9	9.2
Khaya ivorensis	15	48.8	0.5	1.7	3.6
Milicia excelsa	16	84.4	0.9	4.2	5.0
Oxystigma oxyphyllum	2	76.8	0.7	2.8	3.6
Pterocarpus angolensis	4	20.4	0.0	3.0	14.7
Pterocarpus tinctorius	5	42.1	0.2	3.3	7.8
Terminalia superba	17	72.8	0.3	1.6	2.2

Tree cover, forest gain and forest loss are percentages expressed in relation to total buffer area. Deforestation rate is computed as the percentage of forest loss in 2012, in relation to total tree cover in 2000. Estimates were based on data extracted from Hansen et al. [27].

scarce [25]. According to Buza et al. [58], Cabinda is the largest producer of timber from Angola, being responsible for 33.9% of total timber exportations between 1990 and 1995; from 1996 to 2000 the external market consumed 85% of the logged timber. An important step towards the conservation of these forests has been the recent creation of the Maiombe National Park in a transfrontier conservation area, as a result of an international cooperation between Angola, Congo and the Democratic Republic of the Congo [63]. Despite its value, however, this new protected area may provide an incomplete representation of priority timber species, as most of the historically known populations are located outside the Park boundaries. Eventual refinements to the limits of the Park may thus be desirable, which, together with prevention of illegal logging, could greatly assist in the protection of threatened timber tree species in Cabinda's Maiombe forest.

Urgent consideration should also be given to dry forests of Angola, where there are at least three timber species of high conservation priority, and where deforestation rates are increasing rapidly [27]. This is in line with the growing perception that tropical dry forests should be given high conservation priority, as they have a high biodiversity value in Sub-Saharan Africa [64,65]. The current exploitation of dry forest trees of high conservation priority is unknown in Angola, but in the past they were all valued timber species [43], and they are exploited elsewhere in Africa [24]. Although some of these species may be represented in protected areas in the south of Angola, the degree of on-the-ground protection that these areas presently afford is very uncertain. It is thus recommended that particular conservation attention should be given to timber trees from dry forests, in the context of ongoing efforts to strengthen the network of protected areas in Angola [25,33].

Lack of recent information is one of the key problems affecting biodiversity conservation in Angola. Shortage of data is probably more serious in Angola than in most places elsewhere in Africa, because of the prolonged war of independence (1961–1974) and post-independence civil war (1975–2002) which left the country largely inaccessible to most researchers. In these circumstances, studies such as the present one, based on accumulated historical information, may provide initial guidance on the identification of conservation priorities and problems, providing useful insights that would otherwise be very difficult to obtain [1,2,5,6,66]. However, it is now more than one decade after the end of the conflicts in Angola and, given the growing prosperity and development in the country, it is essential that new field surveys are undertaken to document contemporary species distributions and conservation challenges. In the particular case of tree species, it is also essential to collect quantitative information on species identity and places of origin of timber exports, which can then be used to guide new surveys and conservation assessments. Collecting this information would be essential to provide a solid basis for the conservation and sustainable use of forest resources in Angola.

Supporting Information

Figure S1 Temporal profile of herbarium records of the selected timber species of Angola. For each species, the temporal range of herbarium specimens housed in the selected herbaria is indicated (grey horizontal line).

Figure S2 Raster data of Maiombe forest cover in the lower Congo basin, showing Cabinda (Angola) and adjacent areas of Congo (upper) and Democratic Republic of the Congo (lower). The pin bullet indicates a transition where a change in the forest cover density across the border is identified, with the higher density being on the Cabinda side. Maps were produced using data available on-line from: http://earthenginepartners.appspot.com/science-2013-global-forest ([27] Hansen et al. 2013. High-Resolution Global Maps of 21st-Century Forest Cover Change. Science 342. 850–853).

Table S1 Georeferenced vouchers or bibliographic records for selected timber species in Angola. Data from the LISC Herbarium is available through GBIF at http://www.gbif.org/dataset/231c5bcf-1b56-4905-a398-6d0e18f6de1a.

Table S2 Sixty-two datasets from GBIF providers queried for the 18 timber species, producing 2253 records. Accessed 29 April 2013.

Table S3 Data provider per species accessed through GBIF data portal for each of the 18 timber species considered in this study. Accessed 29 April 2013.

Table S4 Characteristics of the 18 timber species studied, including information about their family, synonymy, common names, habit and ecology, and timber characteristics and uses.

Acknowledgments

We thank to the Academic Editor, Prof. Giovanni Vendramin, and to the reviewers (Prof. Berthold Heinze and one anonymous reviewer) for the valuable comments and suggestions, that improved the manuscript. We are also grateful to Patrícia Rodrigues for helping with GIS editing and Filipa Monteiro for technical assistance. We dedicate this paper to Prof. Maria Salomé Pais (Full Professor - BioFIG/FCUL) who made significant contributions to our understanding on Plant Biology and always encouraged the first author to work on the fascinating plant groups of tropical regions.

Author Contributions

Conceived and designed the experiments: MR. Analyzed the data: RF. Wrote the paper: MR RF MCD PB ID.

References

1. Krishtalka L, Humphrey PS (2000) Can natural history museums capture the future? BioScience 50: 611–617.
2. Rivers MC, Taylor L, Brummitt NA, Meagher TR, Roberts DL, et al. (2011) How many herbarium specimens are needed to detect threatened species? Biol Conserv 144: 2541–2547.
3. Brummitt N, Bachman S (2010) Plants under pressure a global assessment. The first report of the IUCN sampled red list index for plants. London: Natural History Museum. Available: http://www.kew.org/ucm/groups/public/documents/document/kppcont_027709.pdf. Accessed 2013 Dec 10.
4. Bebber DP, Carine MA, Wood JRI, Wortley AH, Harris DJ, et al. (2010) Herbaria are a major frontier for species discovery. Proc Natl Acad Sci USA 107: 22169–22171.
5. Costion CM, Liston J, Kitalong AH, Iida A, Lowe AJ (2012) Using the ancient past for establishing current threat in poorly inventoried regions. Biol Conserv 147: 153–162.
6. Lavoie C (2013) Biological collections in an ever changing world: herbaria as tools for biogeographical and environmental studies. Perspect Plant Ecol Evol Syst 15: 68–76.
7. Rands MR, Adams WM, Bennun L, Butchart SH, Clements A, et al. (2010) Biodiversity conservation: challenges beyond 2010. Science 329: 1298–1303.
8. Putz FE, Zuidema PA, Synnott T, Peña-Claros M, Pinard MA, et al. (2012) Sustaining conservation values in selectively logged tropical forests: the attained and the attainable. Conserv Lett 5: 296–303. http://dx.doi.org/10.1111/j.1755-263X.2012.00242.x.
9. FAO (2005) State of the World's forests 2005. Food and Agriculture Organization of the United Nations, Rome.
10. Balmford A, Bond W (2005) Trends in the state of nature and their implications for human well-being. Ecol Lett 8: 1218–1234.
11. Wallace KJ (2007) Classification of ecosystem services: problems and solutions. Biol Conserv 139: 235–246.
12. Giurca A, Jonsson R, Rinaldi F, Priyadi H (2013) Ambiguity in timber trade regarding efforts to combat illegal logging: potential impacts on trade between South-East Asia and Europe. Forests 4: 730–750.
13. Barrett MA, Brown JL, Morikawa MK, Labat JN, Yoder AD (2010) CITES designation for endangered rosewood in Madagascar. Science 328: 1109–1110.
14. Grogan J, Blundell AG, Landis RM, Youatt A, Gullison RE, et al. (2010) Over-harvesting driven by consumer demand leads to population decline: big-leaf mahogany in South America. Conserv Lett 3: 12–20.
15. Cerrillo RMN, Agote N, Pizarro F, Ceacero CJ, Palacios G (2013) Elements for a non-detriment finding of *Cedrela* spp. in Bolivia - A CITES implementation case study. J Nat Conserv 21: 241–252.
16. Pitman NC, Terborgh JW, Silman MR, Nunez PV, Neill DA, et al. (2001) Dominance and distribution of tree species in upper Amazonian terra firme forests. Ecology 82: 2101–2117.
17. Toledo M, Peña-Claros M, Bongers F, Alarcón A, Balcázar J, et al. (2012) Distribution patterns of tropical woody species in response to climatic and edaphic gradients. J Ecol 100: 253–263.
18. Lai J, Mi X, Ren H, Ma K (2009) Species-habitat associations change in a subtropical forest of China. J Veg Sci 20: 415–423.
19. Li L, Huang Z, Ye W, Cao H, Wei S, et al. (2009) Spatial distributions of tree species in a subtropical forest of China. Oikos 118: 495–50.
20. Schmitt CB, Denich M, Demissew S, Friis I, Boehmer HJ (2010) Floristic diversity in fragmented Afromontane rainforests: altitudinal variation and conservation importance. Appl Veg Sci 13: 291–304. doi: 10.1111/j.1654-109X.2009.01067.x.
21. Blom B, Cummins I, Ashton MS (2012) Large and intact forests: drivers and inhibitors of deforestation and forest degradation. In: Ashton MS, Tyrrell ML, Spalding D, Gentry B, editors. Managing forest carbon in a changing climate. Springer Press. pp. 285–304.
22. Dauby G, Hard OJ, Leal M, Breteler F, Stévart T (2014) Drivers of tree diversity in tropical rain forests: new insights from a comparison between littoral and hilly landscapes of Central Africa. J Biogeogr 41: 574–586. doi: 10.1111/jbi.12233.
23. Bodart C, Brink AB, Donnay F, Lupi A, Mayaux P, et al. (2013) Continental estimates of forest cover and forest cover changes in the dry ecosystems of Africa between 1990 and 2000. J Biogeogr 40: 1036–1047.
24. Chidumayo EN, Gumbo DJ (2010) The dry forests and woodlands of Africa: managing for products and services. London: Earthscan.
25. USAID (2008) 118/119 Biodiversity and tropical forest assessment for Angola. Biodiversity Analysis and Technical Support (BATS) Program. Washington, DC.
26. Blaser J, Sarre A, Poore D, Johnson S (2011) Status of tropical forest management 2011, ITTO Technical Series 38. International Tropical Timber Organization, Yokohama, Japan.
27. Hansen MC, Potapov PV, Moore R, Hancher M, Turubanova SA, et al. (2013) "High-Resolution Global Maps of 21st-Century Forest Cover Change." Science 342: 850–853.
28. Ferreirinha MP (1954) Notas sobre as madeiras do Ultramar - 1ª Série. Estudos e Informação 21. Lisboa: Direcção Geral dos Serviços Florestais e Aquícolas. 12 p.
29. Ferreirinha MP (1959) Madeiras do Ultramar português. Garcia de Orta 7: 363–365.
30. Ferreirinha MP (1962) Madeiras de Angola - 2ª Série. Garcia de Orta 10: 111–123.
31. Ferreirinha MP, Reis JEB (1969) Madeiras de Angola - 3ª Série. Garcia de Orta 17: 289–298.
32. Freitas MC (1961) Madeiras de Angola - 1ª Série. Garcia de Orta 9: 699–712.
33. Huntley BJ, Matos EM (1994) Botanical biodiversity and its conservation in Angola. Strelitzia 1: 53–74.
34. Olson DM, Dinerstein E, Wikramanayake ED, Burgess ND, Powell GV, et al. (2001) Terrestrial ecoregions of the world: a new map of life on earth. BioScience 51: 933–938.
35. Peel MC, Finlayson BL, McMahon TA (2007) Updated world map of the Köppen-Geiger climate classification, Hydrol. Earth Syst Sci 11: 1633–1644, doi:10.5194/hess-11-1633-2007.
36. Grandvaux-Barbosa LA (1970) Carta fitogeográfica de Angola. Luanda: Instituto de Investigação Científica de Angola. 323 p.
37. Exell AW, Mendonça FA (1951) Meliaceae. In: Exell AW, Mendonça FA, editors. Conspectus Florae Angolensis 1. Lisboa: Junta de Investigações do Ultramar. pp. 305–320.
38. Gossweiler J (1953) Nomes indígenas de plantas de Angola. Agronomia Angolana 7: 1–587.
39. Torre AR, Hillcoat D (1956) Caesalpinioideae. In: Exell AW, Mendonça FA, editors. Conspectus Florae Angolensis 2. Lisboa: Junta de Investigações do Ultramar. pp. 162–253.

40. Sousa EP (1966) Papilionoideae Tribo VIII-Dalbergieae. In: Exell AW, Fernandes A, editors. Conspectus Florae Angolensis 3. Lisboa: Junta de Investigações do Ultramar. pp. 344–372.

41. Exell AW, Garcia JG (1970) Combretaceae. In: Exell AW, Fernandes A, Mendes EJ, editors. Conspectus Florae Angolensis 4. Lisboa: Junta de Investigações do Ultramar. pp. 44–93.

42. Mambo A (1990) Taxonomia florestal de Cabinda. Técnicas de herbariologia. Lisboa: Instituto de Investigação Científica Tropical. 98 p.

43. Diniz AC (1991) Angola. O meio físico e potencialidades agrárias. Lisboa: Instituto para a Cooperação Económica. 189 p.

44. Barreto LS (1963) Madeiras ultramarinas. Lourenço Marques: Instituto de Investigação Científica de Moçambique. 52 p.

45. IUCN (2012) IUCN Red List of threatened species. Version 2012.2. Available: www.iucnredlist.org. Accessed 2013 Oct 20.

46. Straw HT (1956) Gazetteer n° 20 Angola. United States Board on Geographic Names. Washington. 234 p.

47. Environmental Systems Research Incorporated (2011) ArcGIS 10.0. Environmental Systems Research Incorporated, Redlands, CA. USA.

48. Ladle RJ, Whittaker RJ (editors) (2011) Conservation biogeography. Oxford: Wiley-Blackwell. 320 p.

49. Legendre P, Legendre L (1998) Numerical ecology, 2nd English edition. Amsterdam: Elsevier.

50. Salvador S, Chan P (2004) Determining the number of clusters/segments in hierarchical clustering/segmentation algorithms. Proceedings of the 16th IEEE – International Conference on Tools with Artificial Inteligence, pp. 576–584. Institute of Electrical and Electronics Engineers, Piscataway, NJ.

51. R Development Core Team (2013) R: A language and environment for statistical computing. R Foundation for Statistical Computing. Vienna, Austria. Available: http://www.R-project.org/

52. IUCN (2001) IUCN Red List Categories and Criteria: Version 3.1. IUCN Species Survival Commission. IUCN, Gland, Switzerland and Cambridge, UK. 30 p.

53. Bachman S, Moat J, Hill AW, de la Torre J, Scott B (2011) Supporting Red List threat assessments with GeoCAT: geospatial conservation assessment tool. In: Smith V, Penev L, editors. e-Infrastructures for data publishing in biodiversity science. ZooKeys 150: 117–126.

54. IUCN and UNEP-WCMC (2013) The world database on protected areas (WDPA) [On-line]. Cambridge, UK: UNEP- WCMC. Available: www.protectedplanet.net. Accessed 2013 Dec 1.

55. The Plant List (2010) Version 1. Published on the Internet; http://www.theplantlist.org/. Accessed 10 Dec 2013.

56. Linder HP, de Klerk HM, Born J, Burgess ND, Fjeldså J, et al. (2012) The partitioning of Africa: statistically defined biogeographical regions in sub-Saharan Africa. J Biogeogr 39: 1189–1205. doi: 10.1111/j.1365-2699.2012.02728.x.

57. White F (1983) The vegetation of Africa. Paris: UNESCO. 356 p.

58. Buza AG, Tourinho MM, Silva JN (2006) Caracterização da colheita florestal em Cabinda, Angola. Rev Ciênc Agrár Belém 45: 59–78.

59. Ijang TP, Cleto N, Ewane NW, Chicaia A, Tamar R (2012) Transboundary dialogue and cooperation: first lessons from igniting negotiations on joint management of the Maiombe forest in the Congo Basin. Int J Agric For 2: 121–131.

60. Sexton JP, McIntyre PJ, Angert AL, Rice KJ (2009) Evolution and ecology of species range limits. Annu. Rev Ecol Evol Syst 40:415–36.

61. Hulme M, Doherty R, Ngara T, New M, Lister D (2001) African climate change: 1900–2100. Climate research 17: 145–168.

62. Laporte NT, Stabach JA, Grosch R, Lin TS, Goetz SJ (2007) Expansion of industrial logging in Central Africa. Science 316: 1451–1451.

63. Kuedikuenda S, Xavier MNG (2009) Framework report on Angola's biodiversity. Luanda: Republic of Angola, Ministry of Environment. 60 p.

64. Miles L, Newton AC, DeFries RS, Ravilious C, May I, et al. (2006) A global overview of the conservation status of tropical dry forests. J Biogeogr 33: 491–505.

65. Rudel TK (2013) The national determinants of deforestation in sub-Saharan Africa. Phil Trans R Soc B: 368(1625), 20120405.

66. Pyke GH, Ehrlich PR (2010) Biological collections and ecological/environmental research: a review, some observations and a look to the future. Biol Rev Camb Philos Soc 85: 247–266. doi: 10.1111/j.1469-185X.2009.00098.x.

5

Invasiveness Does Not Predict Impact: Response of Native Land Snail Communities to Plant Invasions in Riparian Habitats

Jitka Horáčková[1,2]*, Lucie Juřičková[2], Arnošt L. Šizling[3], Vojtěch Jarošík[1,4]†, Petr Pyšek[4,1]

1 Department of Ecology, Charles University in Prague, Faculty of Science, Prague 2, Czech Republic, 2 Department of Zoology, Charles University in Prague, Faculty of Science, Prague 2, Czech Republic, 3 Center for Theoretical Study, Charles University and the Academy of Sciences of the Czech Republic, Prague 1, Czech Republic, 4 Department of Invasion Ecology, Institute of Botany, Academy of Sciences of the Czech Republic, Průhonice, Czech Republic

Abstract

Studies of plant invasions rarely address impacts on molluscs. By comparing pairs of invaded and corresponding uninvaded plots in 96 sites in floodplain forests, we examined effects of four invasive alien plants (*Impatiens glandulifera*, *Fallopia japonica*, *F. sachalinensis*, and *F.×bohemica*) in the Czech Republic on communities of land snails. The richness and abundance of living land snail species were recorded separately for all species, rare species listed on the national Red List, and small species with shell size below 5 mm. The significant impacts ranged from 16–48% reduction in snail species numbers, and 29–90% reduction in abundance. Small species were especially prone to reduction in species richness by all four invasive plant taxa. Rare snails were also negatively impacted by all plant invaders, both in terms of species richness or abundance. Overall, the impacts on snails were invader-specific, differing among plant taxa. The strong effect of *I. glandulifera* could be related to the post-invasion decrease in abundance of tall nitrophilous native plant species that are a nutrient-rich food source for snails in riparian habitats. *Fallopia sachalinensis* had the strongest negative impact of the three knotweeds, which reflects differences in their canopy structure, microhabitat humidity and litter decomposition. The ranking of *Fallopia* taxa according to the strength of impacts on snail communities differs from ranking by their invasiveness, known from previous studies. This indicates that invasiveness does not simply translate to impacts of invasion and needs to be borne in mind by conservation and management authorities.

Editor: Brock Fenton, University of Western Ontario, Canada

Funding: This work was supported by grant of the Grant Agency of Charles University in Prague no. 40007 and institutional resources of Ministry of Education, Youth and Sports of the Czech Republic (http://www.msmt.cz/). PP was supported by long-term research development project no. RVO 67985939 (Academy of Sciences of the Czech Republic) and acknowledges the support by Praemium Academiae award from the Academy of Sciences of the Czech Republic (http://www.cas.cz/). The funders had no role in study design, data collection and analysis, decision to publish, or preparation of the manuscript.

Competing Interests: The authors have declared that no competing interests exist.

* Email: jitka.horackova@gmail.com

† Deceased.

Introduction

Invasive species are one of the major biotic stressors in native ecosystems all over the world [1–3], affecting the diversity of resident biota at various scales [4–10]. Plants are the most frequently studied group of invaders [11,12] and in the last decades, extensive literature has accumulated on how they impact ecosystem structure, functioning and services [13–20].

The majority of studies on impacts of plant invasions focus on the same trophic level, i.e., what effects invasive species have on the performance of populations, species and communities of resident plants. In their global review of available data on impact, Pyšek et al. ([21]; their Table 1) found that effects of plant invasions on plant diversity are addressed about twice as frequently as those on animal diversity e.g., [22–25]; see [9,26] for meta-analyses. However, invasive plants may alter interactions between trophic groups via the co-introduction of alien pollinators, seed dispersers, herbivores and predators, that cause profound disruptions to plant reproductive mutualisms [27], and by changing the biotic environment they may also impact reproduc-

tive output and population status of animal species [28,29]. Invasions not only have major implications for biodiversity, but by forging novel functions in resident ecosystems, they also limit the effectiveness of restoration efforts that can be followed by unpredictable responses [18,30].

Studies addressing the impacts of plant invasions on macroinvertebrates mostly reported significant reductions in species abundance, richness and diversity of arthropod communities [31–38] although in some studies this effect was restricted only to some groups [39]. Rarely, the studies reported shifting in food guilds [37]. However, studies exploring the impact of invasive plants on the abundance, species richness and diversity of molluscs, one of the model groups of herbivore generalists, are rather rare [40–46]. It has been shown that, for example, mollusc abundance decreased in areas invaded by *Tamarix ramosissima* in the southwestern United States [42] and in the riverine *Fallopia* stands in western Germany [43] and Switzerland [45]. Additionally, the litter of alien grasses from the genera *Avena* and *Bromus* reduced the number of snails in the Mediterranean biome of Australia [41]. However, the abundance of molluscs was not

Table 1. Quantitative summary of the effects of the four invasive plants studied on species numbers and abundances of land snail communities separated into groups (see text for criteria).

Invading plant	Snail category					
	Total		Small		Rare	
	Species number	Abundance	Species number	Abundance	Species number	Abundance
Fallopia sachalinensis	↓ 41.7	↓ 69.6	↓ 48.0			↓ 89.6
F. japonica			↓ 48.0			↓ 65.3
F. ×bohemica			↓ 48.0			↑ 19.5
Impatiens glandulifera	↓ 16.3		↓ 48.0			↓ 28.8

Species numbers indicate percentage reduction in invaded compared to control plots, the arrow a decrease or increase in invaded plots. Empty cells refer to non-significant effects. Abundance is expressed as the number of living snail individuals.

significantly affected in vegetation invaded by *Spartina anglica* in Australia [40] and both gastropod species richness and abundance even increased following invasion by *I. glandulifera* in northern Switzerland [46]. The results are thus rather scarce and contradictory and none of the studies compared the impact of several invasive plants on multiple criteria of mollusc performance. Such impacts are, however, likely to differ; mollusc assemblages were shown to respond strongly to the change in vegetation, with associated changes in calcium content and humidity being the most important factors determining their occurrence [46–48]. Therefore, the close dependence of land-snail assemblages on soil and vegetation, resulting from their food preferences, makes this group of invertebrates a promising model for studying the impact of plant invasion on higher trophic levels. It can be assumed that invasive plants differing in stature, canopy structure, and chemical composition of tissues would exert different impacts on the structure and composition of land-snail communities.

Here we examine the effects of four invasive alien plants on communities of land snails inhabiting invaded stands. The plants studied are all highly invasive in the Czech Republic [49] and include representatives of contrasting life forms: clonal perennials (three taxa of the genus *Fallopia*) versus an annual species (*I. glandulifera*). The impact of these invaders on plant diversity has been thoroughly documented (see below), but there is a lack of information on changes they induce in the species richness of terrestrial snail communities. To get insight into this issue we address the following questions: (1) Do invasive alien plants exert impacts on species richness and abundance of land snail communities? (2) If so, do the impacts differ with respect to particular invasive plant taxa? Finally, using the three *Fallopia* congeners for which there is a thorough knowledge of mechanisms of invasion in central Europe that makes it possible to rank them according to their invasiveness e.g., [50,51], we ask (3) whether their ranking according to invasiveness corresponds to that based on the strength of impact on land snail communities?

Materials and Methods

Ethics statement

No permits and approvals were required for the field work, as sampling sites were under neither nature nor law protection.

Invasive plants studied

Fallopia japonica (Houtt.) Ronse Decr. var. *japonica* and *F. sachalinensis* (F. Schmidt) Ronse Decr. (Polygonaceae) are stout rhizomatous perennials native to East Asia, introduced to Europe (the former as a single female clone that spread across the continent) as garden ornamentals and fodder plants in the 19th century [52,53]. In the Czech Republic, both species are classified as invasive [10] and the genus *Fallopia* is represented also by the invasive hybrid *F. ×bohemica* (Chrtek and Chrtková) J.P. Bailey, that is likely to have arisen on this continent several times independently and is also known from the native range of the parental species [53]. The first record of *F. japonica* var. *japonica* in the wild is from 1902, that of *F. sachalinensis* from 1921, and the earliest record of the hybrid *F. ×bohemica* is from 1950. The invasion occurred in the second half of the 20th century, the hybrid lagged behind the two parental species but proceeded faster [54] due to its competitive superiority over the parents [50,51]. In the early 2000s, *F. japonica* var. *japonica* was recorded from 1335 localities, *F. sachalinensis* from 261 and the hybrid from 382 [54]. Their dispersal is mainly vegetative through regeneration from rhizome and stem segments transported with contaminated soil and water [51,55]. All three taxa became invasive (sensu [56,57]) in a number of habitats including riparian, where they reach high covers and reduce species richness and diversity of invaded vegetation [49]. The invasion by *Fallopia* taxa exhibits the most severe impact on species richness and diversity among central-European alien plants, reducing the number of species present prior to invasion by 66–86%, depending on the taxon [23]. *Fallopia* taxa affect infrastructure by damaging roads and flood-prevention structures, and increasing the erosion potential of rivers [58,59].

Impatiens glandulifera Royle (Balsaminaceae) is an annual species, up to 2.5 m tall, native to the Himalayas, introduced as a garden ornamental to Europe in 1839 and first recorded as escaped in 1855 [60]. In the Czech Republic, it was first recorded outside cultivation in 1896 [10], but rapid invasion only started in the mid-20th century [61]. *Impatiens glandulifera* is a dominant species of nitrophilous herbaceous fringes of rivers, willow galleries of loamy and sandy riverbanks and of riverine reed vegetation [49]. The species produces higher biomass than its congeners and is plastic in terms of response to nutrient availability and shading, but it also exhibits some genetically based population differentiation [62,63]. Due to its massive spread and extensive populations in riparian habitats, it is considered a conservation problem [64]. However, despite forming populations with a high cover of up to 90%, it does not markedly reduce the numbers of species co-occurring in invaded stands, although invasion does alter species composition in favour of ruderal species [23,65], but see [66]. *Impatiens glandulifera* was also shown to reduce the availability of pollinators for co-occurring native species [67].

The number of plant taxa included in the study was constrained by the fact that they represent a complete set of widespread invasive aliens in riparian habitats in the Czech Republic, with the only additional species being *Helianthus tuberosus*, that, however, invades different vegetation types than natural floodplain forests addressed in our study [49]. The massive invasion of all taxa under study on the rivers in the Czech Republic started at comparable times, around the mid-20th century [61,68], with some local differences [69]; in study sites the invasive species were permanently present for at least 10–15 years therefore it is unlikely that possible differences in residence times among localities affected the results.

Study area and field sampling

Field work was conducted from 2006 to 2011. In total, 96 sites with a maximum distance of 279.5 km were located in floodplain forests, in alluvia of six rivers of the lower Elbe catchment area in the western part of the Czech Republic, Central Europe (see Supporting Information, Table S1). For each of the four invasive plant taxa, one pair of 10×10 m plots was established in each of the sampling sites (*I. glandulifera*, n = 16 paired plots at 32 sites; *F. japonica*, n = 10 paired plots at 20 sites; *F. sachalinensis*, n = 10 paired plots at 20 sites; *F.×bohemica*, n = 12 paired plots at 24 sites). One plot of the pair was located in invaded vegetation where the cover of the invader was 70–100%, the second non-invaded (control) plot was placed in a close vicinity to ensure that the habitat conditions matched as closely as possible to the invaded part [23]. Both plots in the studied pair were sampled once only on the same day. Most of the plots within pairs were placed within the distance of 200 m (median value 159 m) and with four exceptions, all plots were paired within one kilometre. In a few cases, the invader occurred in the non-invaded plot, but its cover range of 1–2% could not have any effect on vegetation or land snail species. That the invasion was the main factor in which plots within the pair differed was confirmed by direct measurements in both plots of the pair of environmental characteristics that might be important predictors of land snail species richness and composition [70,71]. Of these we controlled for elevation (as a surrogate for climate), soil pH, and soil Ca content.

Land snail communities were sampled in the same plot as vegetation using a standard sampling procedure [72]. To document the presence of large and especially dendrophilous species (that rarely occur in litter samples), one person searched by eye for half an hour in all appropriate microhabitats within the whole plot, from which the litter sample could not be taken (e.g., dead wood, stones, tree trunks). Slugs were not included in data analysis because their activity depends mostly on weather conditions [73], and our sampling method was not suitable to record slugs quantitatively. The leaf litter samples with topsoil, twigs and vegetation were taken from four randomly selected quadrats (each measured 25×25 cm^2) at each plot. These subsamples were amalgamated, air-dried and all shells were sorted out using sieves of different mesh size. All empty shells, including their fragments, were excluded from analyses in order to reduce potential bias caused by (1) including species that were not living in the locality but that had their shells redeposited by floods, (2) not including species living in the locality whose accumulated empty shells were removed by accidental flooding [74], and (3) a different length of shell degradation time in various floodplain forest types, which depends mainly on humidity [70,75] and topsoil calcium content [76]. For these reasons in our analyses we only used the total numbers of living land snail species (further referred to as "total species") and the total number of individuals per species. Species included in any of the four threat categories used in the

Red List of molluscs of the Czech Republic [77] were labelled as "rare species" and considered as indicators of the state of molluscan assemblages. Species with shells smaller than 5 mm [78] were classified as "small species". We distinguish this category because of biologically important variation of snail size in relationship to ecological [79] and geographical [80] scales. Specifically, large snails are often associated with moist conditions and low latitudes and thus their representation in communities is not uniform. Total number of species, the total number of individuals and their categorization into rare and small species, are shown in Table S2.

Statistical analysis

Because the data were hierarchically structured in the sense that the invaded and non-invaded plots were nested within locations, to account for the spatial dependencies within locations statistical models were constructed by introducing a random effect for the locations. Invasion status of the plots (invaded/non-invaded) and plant taxon (*F. sachalinensis*, *F. japonica*, *F.×bohemica* and *I. glandulifera*; further termed 'plant species') were fixed factors and location a random intercept, implicitly introducing the compound symmetrical correlation structure [81]. To check whether the models adequately accounted for the spatial dependences in the data, meaning that the models did not violate the basic assumption of the independence of errors of the observations due to spatial autocorrelation [82,83], we used a spline correlogram with 1000 resamples for bootstrap [84–86] based on Moran's I [87,88], to investigate residuals of the models [89].

Numbers of total and small snail species were square-root or square-root+1 transformed, and numbers of total and small individuals log or log+1 transformed to normalize the data e.g., [90], and analyzed by linear mixed models (LMMs) using the function *lme* [91]. Numbers of rare species and individuals could not be transformed to normal distribution due to a large number of zero counts, and these data were therefore analyzed by generalized linear mixed models (GLMMs) with Poisson errors, using the functions *glmmPQL* [92] and *lmer* [93]. LMMs and the function *glmmPQL* also made it possible to calculate intra-class correlation, i.e., the associations between non-invaded and invaded plots within locations, and distinguish, after explaining the part of variance due to differences between the paired plots, the part of residual variance within paired plots at a particular location from the part of residual variance among the locations. Fitted models were checked by plotting appropriate residuals against fitted values and predictors, and by Q-Q plots e.g., [81]. Calculations were done in R 2.12.1 [94].

Finally, a binomial test across all snail species for all four invasive plant species was performed in order to assess the effect of the invaders on each snail species separately.

Results

The effect of spatial autocorrelations was eliminated. This was so for all linear mixed models, and generalized linear mixed models obtained by application of *glmmPQL* function for rare species and *lmer* function for rare individuals (see Figure S1). This means that the explanatory variables were properly included in the models and their effects on the recorded mollusc species adequately measured, successfully accommodating for the spatial autocorrelation within the invaded and non-invaded plots.

High values of associations between non-invaded and invaded plots within sites, as well as relatively low residual variance within paired plots at each site and, at the same time, relatively high residual variance among sites (Table S3), indicate that the invaded and non-invaded plots within each site were appropriately selected. This is so because these results suggest that there was a relatively high similarity of the environmental factors listed above within the pairs of invaded and non-invaded plots within each site.

A simple pairwise test indicated significant reduction of plant and snail species richness at the invaded sites (p<0.01, df = 57, for details see legend to the Fig. 1A), but none or insignificant difference in elevation, soil pH, and soil Ca content between the invaded and non-invaded sites (Fig. S2). Moreover the snail species richness was independent from plant species richness in both the invaded and non-invaded plots as well as when using pooled data, where all the plots were analyzed together (Fig. 1B). Hence we conclude that neither the reduction of plant species richness nor difference in environmental factors between the invaded and non-invaded sites can be a direct driver of snail species richness. We therefore interpret the reduction of snail species richness as a consequence of the focal plant invaders presence/absence at the sites.

Except for small snails, the effect of individual plant taxa on the numbers of all snail species (i.e., including small and rare) and individuals was statistically different, as indicated by the significant plant species×invasion status interactions (Tables 2 and 3). *Fallopia sachalinensis* had the greatest negative effect on snail communities, significantly decreasing the total number of species and individuals, and the number of rare individuals. *Impatiens glandulifera* had a significant negative effect on the total number of species and on rare individuals. *Fallopia japonica* significantly decreased the number of rare individuals. Surprisingly, *F.×bohemica* significantly increased the number of rare individuals. All invasive species had the same, significant negative effect on the number of small snail species (Fig. 2, Tables 1 and 4).

Detailed binomial analyses showed, after the Bonfferoni correction applied at the significance level of 0.01, that *I. glandulifera* decreased abundances of 11 species and increased those of eight species of the 51 in total; *F.×bohemica* decreased and increased abundances of 11 and six species, respectively, of 54

in total; *F. japonica* decreased abundances of seven species and increased abundances of two species of 50 in total; and *F. sachalinensis* decreased abundances of 19 of 43 species. The snails whose abundances were significantly higher at sites with the invasive plant present compared to non-invaded ones belonged mostly to small, leaf litter-dwelling species (see Table S4).

Discussion

Impacts on snail communities are invader-specific

Our study shows that invasive plants in temperate riparian habitats significantly affect species composition and structure of land-snail communities, and that these impacts vary with respect to the ecological groups of snails (i.e., they depend on woodland, open-country, mesic and/or aquatic character of particular snail species). Overall, the significant impacts range from 16 to 48% reduction in terms of species numbers, and 29–90% reduction in abundance. However, unlike in previous studies that mostly addressed the impacts of a single invasive plant species on mollusc communities e.g., [40,42,43,45,46], but see [41], our results provide insights into how impacts differ with respect to the identity of the invader.

That *I. glandulifera* was the plant with the second strongest impact on snail communities in the study, the only one besides *F. sachalinensis* that decreased total snail numbers, is rather surprising. This plant forms less homogenous and less dense cover than *Fallopia* taxa, and was reported to exert relatively minor impact on species richness of native plants following invasion in the Czech Republic [23,65]. This plant species has been documented to cause an increase in the richness and abundance of gastropods in deciduous forests in Switzerland, attributed to higher humidity in invaded sites [46]. It needs to be noted, however, that a greater impact on native plant species richness than that recorded in the Czech Republic was reported from the UK [22]. The strong effect of this invader could be related to the decreased abundance, following invasion, of tall nitrophilous

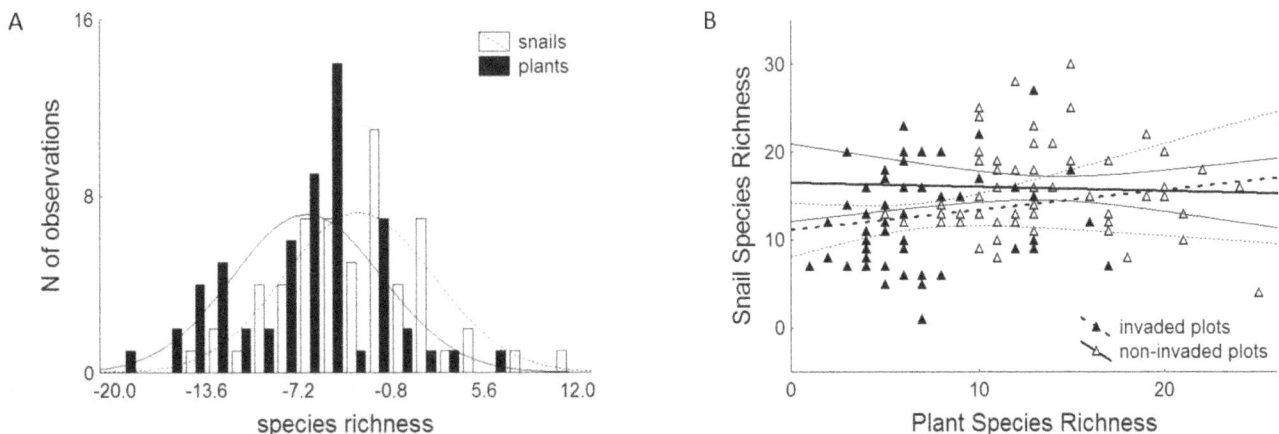

Figure 1. Frequency distribution of the pairwise residuals between species richness of invaded and non-invaded plots across the whole dataset. A) Frequency distribution of the pairwise residuals between species richness of invaded and non-invaded plots (richness of an invaded plot minus richness of the non-invaded plot) across the whole dataset. As the mean values of residuals in both the taxa (−6.7, and −3.1 for plants and snails, respectively) lies below zero value, and their two-sided 99% confidence intervals ([−8.5; −4.8] and [−4.9; −1.3], N = 58, df = 57) does not overlap zero, we conclude that the presence of the focal plant invaders reduce simultaneously plant and snail species richness. **B)** The lack of significant relationship (straight lines – mean trends, curved lines 95% confidence intervals) between the plant and snail species richness (in both the invaded-full lines, open symbols- and non-invaded plots-dashed lines, full symbols- as well as when using pooled data, where all the plots were analyzed together-is not shown) suggests that the reduction of plant species richness at the invaded plots is not a direct driver of the observed snail species richness reduction.

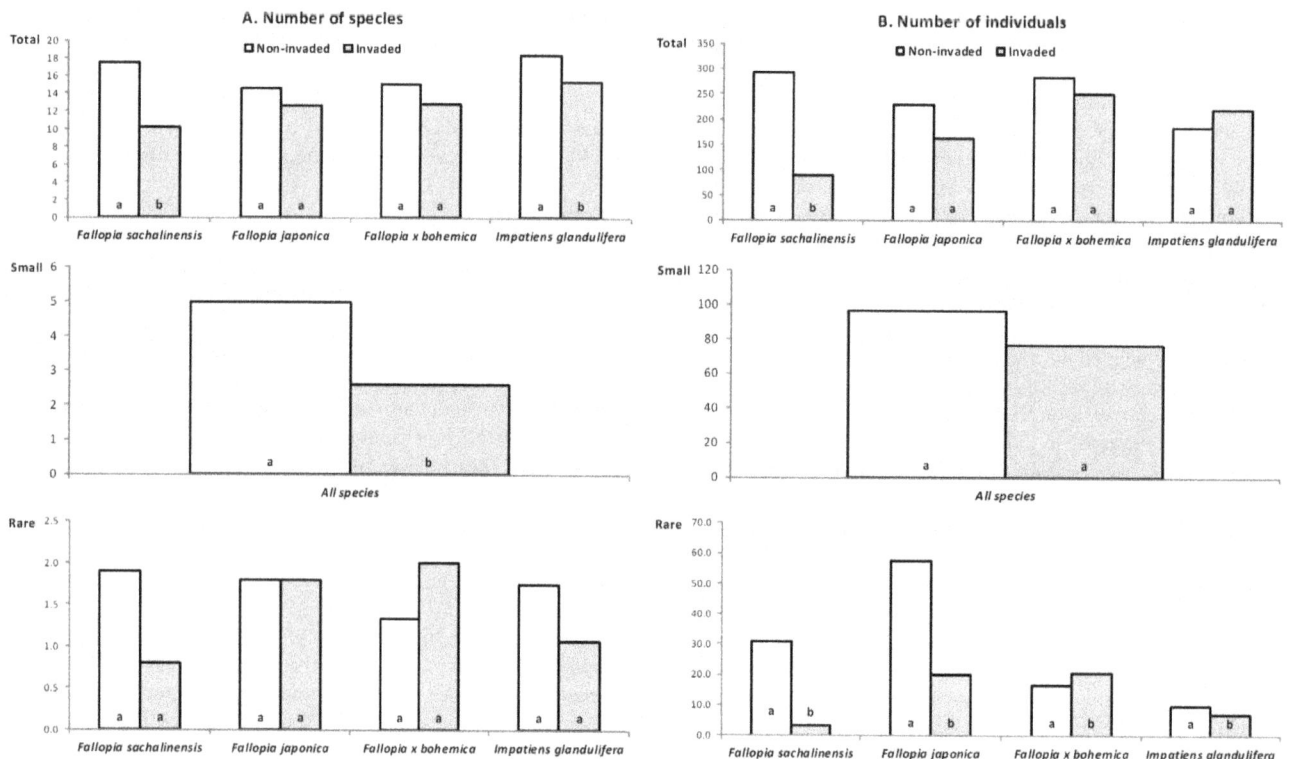

Figure 2. Average numbers of snail species and individuals from 48 paired invaded and non-invaded plots. Average numbers of total, small and rare snail species (A) and individuals (B) from 48 paired plots at individual sites for the species of invasive plants studied (*Fallopia sachalinensis*, *F. japonica*, *F.×bohemica* and *Impatiens glandulifera*). Counts for small snail species and individuals are shown together for all invasive plants as these numbers changed consistently for all plant species (non-significant invasion status×plant species interaction in Table 2). Paired columns followed by different letters differ significantly (*P*<0.05). Full statistics are given in Table 4.

native species (e.g., *Urtica dioica* and *Aegopodium podagraria*) that are a nutrient-rich food source for snails in riparian habitats. *Urtica*-dominated stands are characteristic of the understory of native floodplain forests in the study area, and harbour typical woodland snail fauna that includes a number of rare species.

Fallopia sachalinensis had the strongest negative impact on land-snail communities. Invasion by this species markedly decreased total species number and abundance of snails, as well as the number of small species and abundance of rare species. This is in contrast with the recorded impacts of the other two *Fallopia* taxa, which were much less profound and did not affect the snail community as a whole; their effects were only evident with regard to small and/or rare species. Interestingly, Stoll et al. [45] who examined the impact of a single *Fallopia* species, *F. japonica*, on snail communities in northern Switzerland arrived to opposite conclusions. This invasion reduced average snail richness but the impact differed with respect to shell size; the decreases in species richness were even more pronounced in large, long-lived species as compared to slugs and small, short-lived snails. Moreover, in our study there was a positive effect of *F.×bohemica*, *F. japonica*, and *I. glandulifera* on numbers of individuals of rare snail species. Nevertheless, the overall pattern of impacts markedly differing among the three closely related taxa is surprising if compared to how they affect the plant species richness of invaded communities. The degree to which plant species richness is reduced following invasion is rather high and consistent for all three *Fallopia* taxa. They exhibit one of the most severe impacts on species richness and diversity among central-European alien plants, reducing the

number of species present prior to invasion by 66–86%, depending on the taxon [23]; see also [45] for *F. japonica*.

Laboratory experiments may shed a light on the differences among the three *Fallopia* taxa in respect with the impact on snail communities. Laboratory experiments have shown that there exists a pronounced phytotoxic effect of *Fallopia* leaf extracts on seed germination. *Fallopia sachalinensis* exerts the largest negative effect on germination of *Urtica dioica*, the most abundant native species commonly growing in floodplain habitats invaded by *Fallopia* taxa in the studied area, while *F.×bohemica* consistently has the lowest inhibitory effect [97]. Although these results do not provide direct evidence for differential effect of the individual *Fallopia* taxa on snail communities, they clearly show that litter quality differs among the *Fallopia* species and their hybrid. Importantly, this difference in phytotoxicity of leaf litter for seed germination is consistent with the different impact of the individual *Fallopia* taxa on snail community; *F. sachalinensis* had consistently the strongest negative impact on land snail communities, while for *F.×bohemica* there was a positive effect. In addition, the high negative effect of *F. sachalinensis* leaf litter on germination of native plant species can further exacerbate the negative effect of this species on snail communities by an indirect way, via the suppression of the important native food plant *U. dioica*.

Body size affects the response of snails to invasion

Only snail community characteristics for which the impact was not invader-specific is the proportion of small species relative to

Table 2. ANOVA tables for the numbers of total and small snail species and individuals.

| Source of variation | Species | | | | | | Individuals | | | | | |
| | Total | | | Small | | | Total | | | Small | | |
	Df	F	P	Df	F	P	Df	F	P	Df	F	P
Site status	1, 44	25.707	<0.0001	1, 44	6.589	<0.05	1, 44	8.181	<0.01	1, 44	3.159	<0.1
Plant species	3, 44	1.499	NS	3, 44	0.935	NS	3, 44	1.129	NS	3, 44	0.207	NS
Status×Plant	**3, 44**	**3.115**	**<0.05**	3, 44	1.640	NS	**3, 44**	**6.800**	**<0.001**	3, 44	1.844	NS

ANOVA tables for the numbers of total and small snail species and individuals, analyzed by linear mixed models with plot invasion status (invaded/non-invaded) and invading plant species (*Fallopia sachalinensis*, *F. japonica*, *F.×bohemica* and *Impatiens glandulifera*) as fixed effects and sites with paired invaded/non-invaded plots as random intercepts. Significant invasion status×plant species interactions are in bold. Rare species and individuals were analyzed by generalized linear models (GLMMs) for which ANOVAs are not available. Results of t-tests for the fixed effects of these GLMMs are in Table 3.

Table 3. Results of t-tests for the numbers of rare snail species and individuals.

| Source of variation | Rare species | | | | | Rare individuals | | | |
	Value	Std. Error	Df	t-value	P	Estimate	Std. Error	z-value	P
Intercept	0.144	0.294067	44	0.489	NS	**2.0096**	**0.484**	**4.154**	**<0.001**
Status invasive	0.405	0.242285	44	1.674	NS	**0.2171**	**0.096**	**2.273**	**<0.05**
Fallopia japonica	0.257	0.424453	44	0.607	NS	1.1191	0.713	1.569	NS
Fallopia sachalinensis	0.270	0.425662	44	0.634	NS	-0.1536	0.733	-0.209	NS
Impatiens glandulifera	0.204	0.382983	44	0.534	NS	-0.4884	0.646	-0.757	NS
Invasive×*Fallopia japonica*	-0.405	0.348307	44	-1.164	NS	**-1.2764**	**0.126**	**-10.126**	**<0.001**
Invasive×*Fallopia sachalinensis*	**-1.270**	**0.398501**	**44**	**-3.188**	**<0.01**	**-2.4814**	**0.209**	**-11.875**	**<0.001**
Invasive×*Impatiens glandulifera*	**-0.90446**	**0.33462**	**44**	**-2.70294**	**<0.01**	**-0.5574**	**0.157**	**-3.556**	**<0.001**

Results of t-tests for the numbers of rare snail species and individuals, analyzed by generalized linear mixed models with plot invasion status (invaded/non-invaded) and invading plant species (*Fallopia sachalinensis*, *F. japonica*, *F.×bohemica* and *Impatiens glandulifera*) as fixed effects and sites with paired invaded/non-invaded plots as random intercepts. Rare species were analyzed using the function *glmmPQL* and rare individuals using the function *glmer* in R [94]. Results were obtained by the function *summary* and show fixed effects based on treatment contrasts where the intercept is for the plot invasion status 'non-invaded' and invading plant species *F.×bohemica*. Significant invasion status×plant species interactions are in bold.

Table 4. Results of full statistical analyses describing numbers of total, small and rare snail species and individuals between non-invaded and invaded plots.

Numbers of species	Total				Small				Rare			
	Difference	F	Df	P	Difference	F	Df	P	Difference	t-value	Df	P
Fallopia sachalinensis	**-1.070 ±0.189** (-7.3)	32.519	1, 9	**<0.001**	**-0.229 ±0.091** (-2.4)	6.332	1, 47	**<0.05**	-0.865 ±0.438 (-1.1)	1.473	9	NS
Fallopia japonica	-0.298 ±0.256 (-2.0)	1.350	1, 9	NS	**-0.229 ±0.091** (-2.4)	6.332	1, 47	**<0.05**	0.000 ±0.150 (0.0)	0.000	9	NS
Fallopia×bohemica	-0.315 ±0.177 (-2.3)	3.176	1, 11	NS	**-0.229 ±0.091** (-2.4)	6.332	1, 47	**<0.05**	0.405 ±0.206 (0.7)	1.964	11	NS
Impatiens glandulifera	**-0.390 ±0.169** (-3.0)	5.359	1, 15	**<0.05**	**-0.229 ±0.091** (-2.4)	6.332	1, 47	**<0.05**	-0.499 ±0.236 (-0.7)	2.117	15	NS

Numbers of individuals	Total				Small				Rare			
	Difference	F	Df	P	Difference	F	Df	P	Difference	z-value		P
Fallopia sachalinensis	**-1.408 ±0.309** (-203.9)	20.821	1, 9	**<0.001**	-0.437 ±0.252 (-17.3)	2.998	1, 47	<0.1	**-2.264 ±0.186** (-27.6)	**12.185**		**<0.001**
Fallopia japonica	-0.556 ±0.367 (-66.3)	2.291	1, 9	NS	-0.437 ±0.252 (-17.3)	2.998	1, 47	<0.1	**-1.059 ±0.082** (-37.5)	**12.876**		**<0.001**
Fallopia×bohemica	-0.188 ±0.250 (-31.5)	0.568	1, 11	NS	-0.437 ±0.252 (-17.3)	2.998	1, 47	<0.1	**0.217 ±0.095** (4)	**2.273**		**<0.05**
Impatiens glandulifera	0.229 ±0.186 (36.2)	1.526	1, 15	NS	-0.437 ±0.252 (-17.3)	2.998	1, 47	<0.1	**-0.340 ±0.124** (-2.8)	**2.74**		**<0.01**

Results of full statistical analyses describing numbers of total, small and rare snail species and individuals between non-invaded plots and plots invaded by plant species *Fallopia sachalinensis*, *F. japonica*, *F.×bohemica* and *Impatiens glandulifera*. Difference is a change in counts between non-invaded and invaded plots, with negative values indicating a decrease in snail numbers on invaded sites. Values ± standard errors are on transformed scales, numbers in parentheses are original counts. Significant differences between invaded and non-invaded plots for the individual plant species are in bold.

the total snail community. Our results indicate that small snail species are a group especially prone to reduction in species richness resulting from plant invasions; their numbers in invaded plots consistently decreased regardless of the identity of the invader (but see [45]). This holds also for plots invaded by *F. japonica* and *F.×bohemica*, where total snail numbers were not affected, but the proportion of small snails decreased by 48% (Table 1). Stoll et al. [45] argued that small snail species in *F. japonica* invaded plots feed on algae, fungi and leaf litter, hence are less impacted by invasion than herbivorous large snails suffering from low palatability of knotweed tissues caused by high concentrations of phenolic compounds and lignin [96,97]. On the other hand, slow decomposition of knotweed litter [97,98] most likely results in a limited availability of food for small snail species, causing their reduction in invaded plots.

The post-invasion shift in snail species size hierarchies can be also linked to large snail species controlling a greater proportion of available resources than the smaller ones [80]. It can be hypothesized that under deteriorated conditions and namely reduced diversity of available food after the invasions, small snails are more affected than large ones that are superior in utilization of the limited resources [99]. An additional explanation could be that large snails inhabiting riparian vegetation are capable of profiting from the presence of tall invasive plants due to their climbing behaviour which is not the case of epigeic small snails.

High invasiveness does not automatically translate into strong impact

The three *Fallopia* taxa addressed in our study represent a thoroughly investigated study system for which there is detailed information on the history of invasion, ecology and traits conferring invasiveness in the invaded range in Europe. Previous research consistently points to an increased invasiveness of the hybrid compared to both parental species. The hybrid was reported to spread faster [54], and its abundance in the landscape can be related to better regeneration capacity from rhizome fragments [51]; it is also more difficult to control [50,55] and was a superior competitor to both parents when grown together in an experimental garden (P. Pyšek et al. unpublished data). In other studies, one of the parents performed poorly, such as with *F. sachalinensis* in a field study addressing the establishment of the three taxa [100] or *F. japonica* in a laboratory study investigating phytotoxic effects on germination of native species [95]; however, in both studies the hybrid was, together with the other parent, superior to the poorly performing one.

That the ranking of *Fallopia* taxa according to the strength of impacts on snail communities markedly differs from that according to their invasiveness as measured in the above studies, points to the fact that invasiveness does not simply translate to impacts. This is in accordance with conclusions of Ricciardi and Cohen [7] who found no correlations between invasiveness of alien plants, mammals, fishes, invertebrates, amphibians and reptiles, and their impact on biodiversity on a broad scale. Although the issue requires further study the possibility that the mechanisms of invasion and impact may not be strongly linked needs to be taken into account by managers. For our study it needs to be borne in mind that the impact on snail communities is only one particular type of impacts of plant invasions. Therefore, our results also emphasize the necessity of employing a variety of response measures when studying impacts of invasive species, as what we measure to a large extent determines whether or not the impact of a particular invasion appears serious [21,20].

The results of our study convey an important message for conservation authorities in the Czech Republic. Riparian habitats serve as refugia for many snail species that lost the majority of their natural habitats in the fragmented, intensively used landscape. Invasions of riparian zones by alien plants are an important factor further contributing to deterioration of snail habitats, and knotweeds are among the major invaders of these habitats. Focusing management effort on the hybrid, as the taxon with the greatest potential to spread [54], and paying the least widespread parent, *F. sachalinensis*, less attention, would be justified if one was primarily concerned with plant diversity. Without knowledge of impacts on snails, as documented in our study, this might seem the best strategy in general. However, based on a more comprehensive picture of taxon-specific impacts that vary with respect to the affected group of biota, and with specific conservation goals in mind, our results may help to inform conservation policy in a given area. For example, in regions with high land snail diversity and conservation value, allocation of resources to *Fallopia* control should reflect the ranking of taxa according to impact on snails.

Supporting Information

Figure S1 Spline autocorrelation statistics for residuals of models describing the numbers of total, small and rare snail species and individuals.

Figure S2 Frequency distribution of residuals between environmental parameters of the invaded and non-invaded plots.

Table S1 Overview of non-invaded and invaded sites used in this study.

Table S2 Overview of all recorded land snail species.

Table S3 Results of full statistical analyses describing intraclass correlations between plots and partition of residual variance for numbers of total, small and rare snail species and individuals between invaded and non-invaded plots.

Table S4 Results of binomial analyses describing differences in total abundances of each snail species in the invaded and non-invaded plots.

Acknowledgments

We thank Christina Alba for improving our English, and commenting on the manuscript.

Author Contributions

Conceived and designed the experiments: LJ JH. Performed the experiments: JH. Analyzed the data: VJ ALŠ. Contributed reagents/materials/analysis tools: JH. Wrote the paper: PP VJ ALŠ JH LJ.

References

1. Sala OE, Chapin FS, Armesto JJ, Berlow E, Bloomfield J, et al. (2000) Global biodiversity scenarios for the year 2100. Sci 287: 1770–1774.
2. Millenium Ecosystem Assessment (2005) Ecosystems and Human Well-being: Synthesis. Washington, DC.: Island Press. 160p.
3. Pimentel D, Zuniga R, Morrison D (2005) Update on the environmental and economic costs associated with alien-invasive species in the United States. Ecol Econ 52: 273–288.
4. Wilcove DS, Rothstein D, Dubow J, Phillips A, Losos E (1998) Quantifying threats to imperiled species in the United States. BioSci 48: 607–615.
5. Mack RN, Simberloff D, Lonsdale WM, Evans H, Clout M, et al. (2000) Biotic invasions: causes, epidemiology, global consequences, and control. Ecol Appl 10: 689–710.
6. Gurevitch J, Padilla D (2004) Are invasive species a major cause of extinctions? Trends Ecol Evol 19: 470–474.
7. Ricciardi A, Cohen J (2007) The invasiveness of an introduced species does not predict its impact. Biol Inv 9: 309–315.
8. Powell KI, Chase JM, Knight TM (2011) A synthesis of plant invasion effects on biodiversity across spatial scales. Am J Bot 98: 539–548.
9. Vilà M, Espinar JL, Hejda M, Hulme PE, Jarošík V, et al. (2011) Ecological impacts of invasive alien plants: a meta-analysis of their effects on species, communities and ecosystems. Ecol Lett 14: 702–708.
10. Pyšek P, Danihelka J, Sádlo J, Chrtek Jr J, Chytrý M, et al. (2012) Catalogue of alien plants of the Czech Republic (2nd edition): checklist update, taxonomic diversity and invasion patterns. Preslia 84: 155–255.
11. Pyšek P, Richardson DM, Pergl J, Jarošík V, Sixtová Z, et al. (2008) Geographical and taxonomic biases in invasion ecology. Trends Ecol Evol 23: 237–244.
12. Vilà M, Basnou C, Pyšek P, Josefsson M, Genovesi P, et al. (2010) How well do we understand the impacts of alien species on ecosystem services? A pan-European, cross-taxa assessment. Front Ecol Environ 8: 135–144.
13. Gordon DR (1998) Effects of invasive, non-indigenous plant species on ecosystem processes: lessons from Florida. Ecol Appl 8: 975–989.
14. Parker IM, Simberloff D, Lonsdale WM, Goodell K, Wonham M, et al. (1999) Impact: toward a framework for understanding the ecological effects on invaders. Biol Invasions 1: 3–19.
15. Tilman D (1999) The ecological consequences of changes in biodiversity: a search for general principles. Ecol 80: 1455–1474.
16. Dangles O, Jonsson M, Malmqvist B (2002) The importance of detritivore species diversity for maintaining stream ecosystem functioning following the invasion of a riparian plant. Biol Invasions 4: 441–446.
17. Levine JM, Vilà M, D'Antonio CM, Dukes JS, Grigulis K, et al. (2003) Mechanisms underlying the impacts of exotic plant invasions. Proc R Soc Lond B 270: 775–781.
18. Pyšek P, Richardson DM (2010) Invasive species, environmental change and management, and health. Annu Rev Environ Res 35: 25–55.
19. Gaertner M, Richardson DM, Privett SDJ (2011) Effects of alien plants on ecosystem structure and functioning and implications for restoration: insights from three degraded sites in South African Fynbos. Environ Manage 48: 57–69.
20. Hulme PE, Pyšek P, Jarošík V, Pergl J, Schaffner U, et al. (2013) Bias and error in current knowledge of plant invasions impacts. Trends Ecol Evol 28: 212–218.
21. Pyšek P, Jarošík V, Hulme PE, Pergl J, Hejda M, et al. (2012) A global assessment of invasive plant impacts on resident species, communities and ecosystems: the interaction of impact measures, invading species' traits and environment. Glob Change Biol 18: 1725–1737.
22. Hulme PE, Bremner ET (2006) Assessing the impact of *Impatiens glandulifera* on riparian habitats: partitioning diversity components following species removal. J Appl Ecol 43: 43–50.
23. Hejda M, Pyšek P, Jarošík V (2009) Impact of invasive plants on the species richness, diversity and composition of invaded communities. J Ecol 97: 393–403.
24. Cushman JH, Gaffney KA (2010) Community-level consequences of invasion: impacts of exotic clonal plants on riparian vegetation. Biol Invasions 12: 2765–2776.
25. Hladyz S, Åbjörnsson K, Giller PS, Woodward G (2011) Impacts of an aggressive riparian invader on community structure and ecosystem functioning in stream food webs. J Appl Ecol 48: 443–452.
26. Gaertner M, Breeyen AD, Hui C, Richardson DM (2009) Impacts of alien plant invasions on species richness in Mediterranean-type ecosystems: a meta-analysis. Progr Phys Geogr 33: 319–338.
27. Traveset A, Richardson DM (2006) Biological invasions as disruptors of plant–animal reproductive mutualisms. Trends Ecol Evol 21: 208–216.
28. Leslie AJ, Spotila JR (2001) Alien plants threaten Nile crocodile (*Crocodylus niloticus*) breeding in Lake St. Lucia, South Africa. Biol Conserv 98: 347–355.
29. Dibble KL, Meyerson LA (2012) Tidal flushing restores the physiological condition of fish residing in degraded salt marshes. PLoS ONE 7: e46161.
30. Milton SJ, Wilson JRU, Richardson DM, Seymour CL, Dean WRJ, et al. (2007) Invasive alien plants infiltrate bird-mediated shrub nucleation processes in arid savanna. J Ecol 95: 648–661.
31. Slobodchikoff CN, Doyen JT (1977) Effects of *Ammophila arenaria* on sand dune arthropod communities. Ecol 58: 1171–1175.

32. Lambrinos JG (2000) The impact of the invasive alien grass *Cortaderia jubata* (Lemoine): stapf on an endangered mediterranean-type shrubland in California. Divers Distrib 6: 217–231.

33. Herrera AM, Dudley TL (2003) Reduction of riparian arthropod abundance and diversity as a consequence of giant reed (*Arundo donax*) invasion. Biol Invasions 5: 167–177.

34. Greenwood H, O'Dowd DJ, Lake PS (2004) Willow (*Salix×rubens*) invasion of the riparian zone in south-eastern Australia: reduced abundance and altered composition of terrestrial arthropods. Divers Distrib 10: 485–492.

35. Ernst CM, Cappuccino N (2005) The effect of an invasive alien vine, *Vincetoxicum rossicum* (Asclepiadaceae), on arthropod populations in Ontario old fields. Biol Invasions 7: 417–425.

36. Gratton C, Denno RF (2006) Arthropod food web restoration following removal of an invasive wetland plant. Ecol Appl 16: 622–631.

37. Topp W, Kappes H, Rogers E (2008) Response of ground-dwelling beetle (Coleoptera) assemblages to giant knotweed (*Reynoutria* spp.) invasion. Biol Invasions 10: 381–390.

38. Wilkie L, Cassis G, Gray M (2007) The effects on terrestrial arthropod communities of invasion of a coastal heath ecosystem by the exotic weed bitou bush (*Chrysanthemoides monilifera* ssp. *rotundata* L.). Biol Invasions 9: 477–498.

39. Durst SL, Theimer TC, Paxton EH, Sogge MK (2008) Temporal variation in the arthropod community of desert riparian habitats with varying amounts of saltcedar (*Tamarix ramosissima*). J Arid Environ 72: 1644–1653.

40. Hedge P, Kriwoken LK (2000) Evidence for effects of *Spartina anglica* invasion on benthic macrofauna in Little Swanport estuary, Tasmania. Austral Ecol 25: 150–159.

41. Lenz TI, Moyle-Croft JL, Facelli JM (2003) Direct and indirect effects of exotic annual grasses on species composition of a South Australian grassland. Austral Ecol 28: 23–32.

42. Kennedy TA, Finlay JC, Hobbie SE (2005) Eradication of invasive *Tamarix ramosissima* along a desert stream increases native fish density. Ecol Appl 15: 2072–2083.

43. Kappes H, Lay R, Topp W (2007) Changes in different trophic levels of litter-dwelling macrofauna associated with giant knotweed invasion. Ecosyst 10: 734–744.

44. Gerber E, Krebs C, Murrell C, Moretti M, Rocklin R, et al. (2008) Exotic invasive knotweeds (*Fallopia* spp.) negatively affect native plant and invertebrate assemblages in European riparian habitats. Biol Conserv 141: 646–654.

45. Stoll P, Gatzsch K, Rusterholz R, Baur B (2012) Response of plant and gastropod species to knotweed invasion. Basic Appl Ecol 13: 232–240.

46. Ruckli R, Rusterholz H, Baur B (2013) Invasion of *Impatiens glandulifera* affects terrestrial gastropods by altering microclimate. Acta Oecol 47: 16–24.

47. Martin K, Sommer M (2004) Relationships between land snail assemblage patterns and soil properties in temperate-humid forest ecosystems. J Biogeogr 31: 531–545.

48. Juřičková L, Horsák M, Cameron RDA, Hylander K, Míkovcová A, et al. (2008) Land snail distribution patterns within a site: the role of different calcium sources. Eur J Soil Biol 44: 172–179.

49. Pyšek P, Chytrý M, Pergl J, Sádlo J, Wild J (2012) Plant invasions in the Czech Republic: current state, introduction dynamics, invasive species and invaded habitats. Preslia 84: 576–630.

50. Bímová K, Mandák B, Pyšek P (2001) Experimental control of *Reynoutria* congeners: a comparative study of a hybrid and its parents. In: Brundu G, Brock J, Camarda I, Child l, Wade M, editors. Plant Invasions: Species Ecology and Ecosystem Management. Leiden: Backhuys Publishers. pp. 283–290.

51. Pyšek P, Brock JH, Bímová K, Mandák B, Jarošík V, et al. (2003) Vegetative regeneration in invasive *Reynoutria* (Polygonaceae) taxa: the determinant of invasibility at the genotype level. Am J Bot 90: 1487–1495.

52. Beerling DJ, Bailey JP, Conolly AP (1994) *Fallopia japonica* (Hout.) Ronse Decraene (*Reynoutria japonica* Houtt., *Polygonum cuspidatum* Sieb. & Zucc.). J Ecol 82: 959–979.

53. Bailey JP, Conolly AP (2000) Prize-winners to pariahs: a history of Japanese knotweed s.l. (Polygonaceae) in the British Isles. Watsonia 23: 93–110.

54. Mandák B, Pyšek P, Bímová K (2004) History of the invasion and distribution of *Reynoutria* taxa in the Czech Republic: a hybrid spreading faster than its parents. Preslia 76: 15–64.

55. Bímová K, Mandák B, Pyšek P (2003) Experimental study of vegetative regeneration in four invasive *Reynoutria* taxa. Plant Ecol 166: 1–11.

56. Richardson DM, Pyšek P, Rejmánek M, Barbour MG, Panetta FD, et al. (2000) Naturalization and invasion of alien plants: concepts and definitions. Divers Distrib 6: 93–107.

57. Blackburn TM, Pyšek P, Bacher S, Carlton JT, Duncan RP, et al. (2011) A proposed unified framework for biological invasions. Trends Ecol Evolut 26: 333–339.

58. Beerling DJ (1991) The effect of riparian land use on the occurrence and abundance of Japanese knotweed (*Reynoutria japonica*) on selected rivers in South Wales. Biol Cons 55: 329–337.

59. Reinhardt F, Herle M, Bastiansen F, Streit B (2003) Economic Impact of the Spread of Alien Species in Germany. Frankfurt am Main: Biological and Computer Sciences Division, Dept. of Ecology and Evolution. 229p.

60. Beerling DJ, Perrins DM (1993) Biological flora of the British Isles: *Impatiens glandulifera* Royle (*Impatiens Roylei* Walp.). J Ecol 81: 367–381.

61. Pyšek P, Prach K (1993) Plant invasions and the role of riparian habitats: a comparison of four species alien to central Europe. J Biogeogr 20: 413–420.

62. Skálová H, Moravcová L, Pyšek P (2011) Germination dynamics and seedling frost resistance of invasive and native *Impatiens* species reflect local climatic conditions. Persp Pl Ecol Evol Syst 13: 173–180.

63. Skálová H, Havlíčková V, Pyšek P (2012) Seedling traits, plasticity and local differentiation as strategies of invasive species of *Impatiens* in central Europe. Ann Bot 110: 1429–1438.

64. DAISIE (eds.) (2009) Handbook of Alien Species in Europe. Berlin: Springer. 381 p.

65. Hejda M, Pyšek P (2006) What is the impact of *Impatiens glandulifera* on species diversity of invaded riparian vegetation? Biol Conserv 132: 143–152.

66. Bremner ET, Hulme PE (2006) Assessing the impact of *Impatiens glandulifera* on riparian habitats: partitioning diversity components following species removal. J Appl Ecol 43: 43–50.

67. Chittka L, Schürkens S (2001) Successful invasion of a floral market. Nature 411: 653–654.

68. Pyšek P, Prach K (1995) Invasion dynamics of *Impatiens glandulifera*: a century of spreading reconstructed. Biol Conserv 74: 41–48.

69. Malíková L, Prach K (2010) Spread of alien *Impatiens glandulifera* along rivers invaded at different times. Ecohydrol Hydrobiol 10: 81–85.

70. Čejka T, Horsák M, Némethová D (2008) The composition and richness of danubian floodplain forest land snail faunas in relation to forest type and flood frequency. J Mollusc Stud 74: 37–45.

71. Dvořáková J, Horsák M (2012) Variation of snail assemblages in hay meadows: disentangling the predictive power of abiotic environment and vegetation. Malacologia 55: 151–162.

72. Cameron RAD, Pokryszko BM (2005) Estimating the species richness and composition of land mollusc communities: problems, consequences and practical advice. J Conchol 38: 529–548.

73. Rollo CD (1991) Endogenous and exogenous regulation of activity in *Deroceras reticulatum*, a weather-sensitive terrestrial slug. Malacol 33: 199–220.

74. Ilg Ch, Foeckler F, Deichner O, Henle K (2009) Extreme flood events favour floodplain mollusc diversity. Hydrobiol 621: 63–73.

75. Horáčková J, Horsák M, Juřičková L (2014) Land snail diversity and composition in relation to ecological variations in Central European floodplain forests and their history. Community Ecol 15: 44–53.

76. Cernohorsky N, Horsák M, Cameron RAD (2010) Land snail species richness and abundance at small scales: the effects of distinguishing between live individuals and empty shells. J Conchol 40: 233–241.

77. Beran L, Juřičková L, Horsák M (2005) Mollusca. In: Farkač J, Král D, Škorpík M, editors. Red List of Threatened Species in the Czech Republic - Invertebrates. Prague: AOPK ČR. pp. 67–69.

78. Kerney MP, Cameron RAD, Jungbluth JH (1983) Die Landschnecken Nord-und Mitteleuropas. Hamburg and Berlin: Verlag Paul Parey. 384 p.

79. Goodfriend GA (1986) Variation in land-snail shell form and size and its causes: a review. Syst Biol 35:204–223.

80. Hausdorf B (2007) The interspecific relationship between abundance and body size in central European land snail assemblages. Basic Appl Ecol 8: 125–134.

81. Zuur AF, Ieno EN, Waker NJ, Saveliev AA, Smith GM (2009) Mixed Effects Models and Extensions in Ecology. New York: Springer. 574p.

82. Legendre P (1993) Spatial autocorrelation: Trouble or new paradigm? Ecol 74: 1659–1673.

83. Lichstein JW, Simons TR, Shriner SA, Franzreb KE (2002) Spatial autocorrelation and autoregressive models in ecology. Ecol Monogr 72: 445–463.

84. Cliff AA, Ord JK (1981) Spatial processes: models and applications. London: Pion.

85. Bjørnstadt ON, Falck W (2001) Nonparametric spatial covariance functions: estimation and testing. Environ Ecol Stat 8: 53–70.

86. Bjørnstadt ON (2008) Ncf: Spatial Nonparametric Covariance Functions. R package version 1.1-1. Available: http://onb.ent.psu.edu/onb1/R. Accessed 29 May 2012.

87. Sokal RR, Oden NL (1978) Spatial autocorrelation in biology: 1. Methodology. Biol J Linn Soc 10: 199–228.

88. Legendre P, Legendre L (1998) Numerical Ecology. Amsterdam: Elsevier. 853 p.

89. Rhodes JR, McAlpine CA, Zuur AF, Smith GM, Ieno EN (2009) Chapter 21. GLMM applied on the spatial distribution of koalas in a fragmented landscape. In: Zuur AF, Ieno EN, Walker NJ, Saveliev AA, Smith GM, editors. Mixed Effects Models and Extensions in Ecology. New York: Springer. pp. 469–492.

90. Sokal RR, Rohlf FJ (1995) Biometry: the principles and practise of statistics in biological research, third ed. New York: Freeman. 887p.

91. Pinheiro J, Bates D, DebRoy S, Sarkar D, the R Core team (2009) Nlme: Linear and Nonlinear Mixed Effects Models. R package version 3: 1–93.

92. Venables WN, Ripley BD (2002) Modern Applied Statistics with S, fourth ed. New York: Springer. 497p.

93. Bates D, Maechler M, Bolker B (2011) lme4: Linear Mixed-Effects Models Using S4 Classes. R package version 0.999375-39. Available: http://CRAN.R-project.org/package=lme4. Accessed 15 April 2013.

94. R Development Core Team (2009) R: A Language and Environment for Statistical Computing. Vienna: R Foundation for Statistical Computing. Available: http://www.R-project.org. Accessed January 2012.

95. Moravcová L, Pyšek P, Jarošík V, Zákravský P (2011) Potential phytotoxic and shading effects of invasive *Fallopia* (Polygonaceae) taxa on the germination of dominant native species. NeoBiota 9: 31–47.

96. Vrchotová N, Šerá B, Tříska J (2007) The stilbene and catechin content of the spring sprouts of *Reynoutria* species. Acta Chromatogr 19: 21–28.

97. Pálková K (2007) Palatabilita druhů, jejich bionomické vlastnosti a rychlost rozkladu detritu. Master's Thesis, České Budějovice: University of South Bohemia in České Budějovice. 68 p.

98. Mincheva T, Barni E, Varese GC, Brusa G, Cerabolini B et al. (2014) Litter quality, decomposition rates and saprotrophic mycoflora in *Fallopia japonica* (Houtt.) Ronse Decraene and in adjacent native grassland vegetation. Acta Oecol 54: 2–35.

99. Baur B, Baur A (1990) Experimental evidence for intra- and interspecific competition in two species of rock-dweling land snails. J Anim Ecol 59: 301–315.

100. Brabec J, Pyšek P (2000) Establishment and survival of three invasive taxa of the genus *Reynoutria* (Polygonaceae) in mesic mown meadows: a field experimental study. Folia Geobot 35: 27–42.

6

Impacts of *Carpobrotus edulis* (L.) N.E.Br. on the Germination, Establishment and Survival of Native Plants: A Clue for Assessing Its Competitive Strength

Ana Novoa*¤, Luís González

Departamento de Bioloxía Vexetal e Ciencias do Solo, Facultade de Bioloxía, Universidade de Vigo, Vigo, Spain

Abstract

Does *Carpobrotus edulis* have an impact on native plants? How do *C. edulis'* soil residual effects affect the maintenance of native populations? What is the extent of interspecific competition in its invasion process? In order to answer those questions, we established pure and mixed cultures of native species and *C. edulis* on soil collected from invaded and native areas of Mediterranean coastal dunes in the Iberian Peninsula. We examined the impact of the invader on the germination, growth and survival of seeds and adult plants of two native plant species (*Malcolmia littorea* (L.) R.Br, and *Scabiosa atropurpurea* L.) growing with ramets or seeds of *C. edulis*. Residual effects of *C. edulis* on soils affected the germination process and early growth of native plants in different ways, depending on plant species and density. Interspecific competition significantly reduced the germination and early growth of native plants but this result was soil, density, timing and plant species dependent. Also, at any density of adult individuals of *C. edulis*, established native adult plants were not competitive. Moreover, ramets of *C. edulis* had a lethal effect on native plants, which died in a short period of time. Even the presence of *C. edulis* seedlings prevents the recruitment of native species. In conclusion, *C. edulis* have strong negative impacts on the germination, growth and survival of the native species *M. littorea* and *S. atropurpurea*. These impacts were highly depended on the development stages of native and invasive plants. Our findings are crucial for new strategies of biodiversity conservation in coastal habitats.

Editor: Meng-xiang Sun, Wuhan University, China

Funding: The authors received no specific funding for this work.

Competing Interests: The authors have declared that no competing interests exist.

* Email: ANAN@sun.ac.za

¤ Current address: Centre for Invasion Biology, Department of Botany and Zoology, Stellenbosch University, Stellenbosch, South Africa

Introduction

Invasive alien plants are considered as one of the greatest threats to the diversity, structure and functioning of natural ecosystems around the world [1,2]. They can exert significant impacts on many ecological variables [3]. In particular, Mediterranean coastal dune ecosystems are highly sensitive to invasion by exotic plants since the disruption caused by the movement of sand constantly produces open spaces that are susceptible to colonization by alien species [4]. In those open spaces, competitive interactions between invasive and native species are extremely important [5]. As Mediterranean coastal dune ecosystems present a high cultural and ecological value, and support many threatened and endemic species [6], efforts on their alien species management strategies are critically needed.

One of the major invaders of Mediterranean coastal dune ecosystems is *Carpobrotus edulis* (L.) N.E.Br., a perennial clone plant native to South Africa [7–9]. Since its introduction, this invasive succulent now dominates millions of kilometres of Mediterranean dune ecosystems, leading to loss of species and irreversible changes on the substrate [7–10]. However, there is little information about the competitive interactions between this invasive species with native plants, even though it is a crucial

aspect for prioritizing Mediterranean coastal dune ecosystems management.

Impacts of invasive species through competition with native plants is a primary ecological process limiting restoration success [11]. The phenologic stage of each species is decisive in these competitive relationships [12]: plants pass through different physiological stages and their development processes and competition occurs not only within species, but also within and between stages of different species [13]. However, most competition studies are focused on plants at the same development stage.

This study deals with the relative competitive ability of native seeds, seedlings and adult plants. It was designed to address the following questions: (i) the presence of *C. edulis* causes changes to soil characteristics (see Novoa et al. (2013a) for further information and detailed data). How do these changes affect the maintenance of native populations during each development stage? And (ii) To what extent are the ecological impacts caused by *C. edulis* based on different development stages of native plants? Despite the fact that the plant invasion process is a result of multiple interacting factors [14,15], to the best of our knowledge this study is the first reporting simultaneous examination of multiple mechanisms: competition, density, timing of sowing, plant developmental stage

Figure 1. Collection of adult plants of *C. edulis.*

and residual soil effects on the limitation of native flora by an invasive plant.

Moreover, since *C. edulis* can reproduce both vegetatively and sexually [16], understanding the competitive relationships established between clones or seeds of *C. edulis* and native species during the germination, establishment and growth processes are crucial for the conservation of the high biodiversity of Mediterranean coastal habitats [17].

Materials and Methods

1. Plant material

The two main dominant native species in the study area were selected: a typical semi-fixed dunes species (*Malcolmia littorea* (L.) R.Br.) and a common fixed-dune and rocky species (*Scabiosa atropurpurea* L.) both of them Chamaephytes. Seeds of the native species *M. littorea* and *S. atropurpurea* and the invasive *C. edulis* were collected between 10 Sep and 10 Oct 2011 from at least 15 plants from 20 different populations of each species located along 20 km of the coast of Pontevedra, Spain (between 42°29′56.17″N 8°52′16.22″O and 42°20′16.22″N 8°49′41.17″O). The land

accessed is not privately owned or protected. The seeds were separated from the rest of the fruit and its accessory dispersion parts and stored in the dark at 4°C until assay. Seeds were surface-sterilized for 5 min in 0.1% sodium hypochlorite, rinsed 3 times in distilled water and dried at room temperature.

On the 19 Nov 2010, adult plants of native species (*M. littorea* and *S. atropurpurea*) and apical ramets of the invasive *C. edulis* -as well as a volume of 1 L of sand around each plant- were collected in the same area and immediately transplanted to 1 L sand pots for greenhouse acclimatization, integrating the plant stock. As *C. edulis* presents clonal growth, we could obtain individuals with approximately the same developmental state from the ends of the branches (Figure 1).

2. Soil collection

On the 19 Nov 2010, soil was collected up to 10 cm depth in those dunes from where the seeds, adult plants and ramets had previously been collected. The top soil layer from 20 randomly selected points (1×1 m) was collected in an area invaded by *C. edulis* and in an adjacent native area. The adjacent native area

Table 1. Methodological scheme for greenhouse experiment.

		Ramets of *C. edulis*				
		0	1	2	3	4
N° of native species seeds or seedlings	0	•				X
	10/1	•			X	
	15/2	•		X		
	20/3	•	X			
	25/4	X				

(X) Represent classic replacement series design. (•) Represent modified replacement series. N = 5.

Table 2. Methodological scheme for growth chamber experiment.

	Day 0	Day 5		Day 10	
	5 days before sowing native species	*Sowing date of native seeds*		*5 days after sowing native species*	
	Seeds number				
	C. edulis	native	C. edulis	native	C. edulis
Assay 1	10	+	10		
	10	+	30		
	30	+	10		
Assay 2		10	+	10	
		30	+	10	
		10	+	30	
Assay 3		10		+	10
		30		+	10
		10		+	30
Monocultures		10			10
		30			30

Assay 1: Interspecific competition, *C. edulis* seeds sowed before native seeds. Assay 2: Interspecific competition, *C. edulis* seeds sowed at the same time that native seeds. Assay 3: Interspecific competition, *C. edulis* seeds sowed after native seeds. Monocultures: intraspecific competition for native and *C. edulis* seeds. N = 5.

Table 3. Intraspecific competition effect on the germination indices (Gt: Total germination and AS: Cumulative rate of germination), seedling shoot and root length and survival of the native species *Malcolmia littorea* in both invaded and native soil.

Seed number	Invaded soil				Native soil			
	10	15	20	25	10	15	20	25
Gt	82.5^{a}*	81.7^{a}*	57.0^{b}	36.0^{c}*	51.2	62.3	53.0	65.6
	(9.8)	(9.6)	(6.8)	(5.7)	(4.0)	(3.4)	(6.4)	(3.4)
AS	105.8^{ab}*	112.9^{a}	84.3^{b}*	47.8^{c}*	67.2	82.7	54.8	88.5
	(14.1)	(10.5)	(6.3)	(3.8)	(8.6)	(10.7)	(12.6)	(43)
Survival (%)	71.0	94.1	82	77.9	72^{b}	81^{b}	79^{b}	92^{a}
	(12.7)	(11.9)	(14.3)	(17.6)	(2.4)	(9.6)	(6.3)	(2.8)
Shoot length (cm)	1.2*	1.5*	1.4*	1.3*	0.9	1.1	0.9	1.1
	(0.1)	(0.2)	(0.1)	(0.1)	(0.1)	(0.1)	(0.1)	(0.1)
Root length (cm)	2.3*	2.3*	2.3*	2.3*	1.4	1.6	1.4	1.5
	(0.2)	(0.3)	(0.2)	(0.2)	(0.1)	(0.1)	(0.1)	(0.1)

Different letters mean significant difference of 5% between treatments (10, 15, 20 or 25 seeds).*: indicates significant difference of 5% between invaded and native soils on each treatment. Numbers in parentheses indicate the standard error. N=5.

Table 4. Intraspecific competition effect on the germination indices (Gt: Total germination and AS: Cumulative rate of germination), seedling shoot and root length and survival of the native species *Scabiosa atropurpurea* in both invaded and native soil.

Seed number	Invaded soil				Native soil			
	10	15	20	25	10	15	20	25
Gt	64.0	53.3	43.0	45.6	44.0	49.3	40.0	48.0
	(11.7)	(9.2)	(8.6)	(2.7)	(10.3)	(6.5)	(8.2)	(7.2)
AS	51.0	41.7	33.6	40.1	42.7	39.9	35.8	46.1
	(5.7)	(4.0)	(5.0)	(2.9)	(5.1)	(3.4)	(6.5)	(1.9)
Survival (%)	80.2	87.7	90.2	93.5	68.8	84.7	89.0	94.5
	(11.1)	(4.8)	(4.9)	(1.7)	(12.8)	(4.6)	(3.5)	(2.5)
Shoot length (cm)	0.9^{b}	1.2^{ab}	1.1^{ab}	1.4^{a}	0.79	1.1	1.0	1.2
	(0.1)	(0.2)	(0.2)	(0.2)	(0.1)	(0.2)	(0.1)	(0.2)
Root length (cm)	5.3^{b}	5.6^{b}	5.3^{b}	6.7^{a}	4.5^{b}	5.2^{b}	5.4^{b}	7.1^{a}
	(0.7)	(0.4)	(0.2)	(0.6)	(0.9)	(0.4)	(0.3)	(0.7)

Different letters mean significant difference of 1% between treatments (10, 15, 20 or 25 seeds). Numbers in parentheses indicate the standard error. N=5.

was sufficiently separated from the invaded area, to affirm that there was no effect of *C. edulis* on the soil. The soil taken from each area (invaded or native) was homogenized (approx. 100 Kg) and reserved for the establishment of the crops, as explained later. See [8] for physicochemical results.

3. Competition between *C. edulis* ramets and native species

In order to look for competitive relationships established between *C. edulis* and native plants in different stages of native plant's development (germination, seedling and adult plant), pot culture experiments were established in both invaded and native soil following the principle of replacement or substitution [18], modified for our purposes (Table 1). Competition experiments were carried out on two soils with different origins (native and invaded soil). To avoid interference in the replacement series due to physico-chemical and biological characteristics of the soil, a concomitant treatment checking intra-specific competition density was established. Four experimental trials were then established to study the competitive interaction between the invasive *C. edulis* and native species: (a) intra-specific competition between native seeds/seedlings (b) ramets of *C. edulis* vs native seeds/seedlings, (c) intra-specific competition between native adult plants and (d) ramets of *C. edulis* vs adult native plants. Cultures were established in 1 L pots filled with soil from native and invaded zones and replicated five times. A total of 320 pots were used.

3.1. Intra-specific competition between native seeds and seedlings. 80 pots with cultures containing 10 15, 20 or 25 seeds of the native species (*M. littorea* or *S. atropurpurea*) were established (Table 1), which we refer to in this paper as "pure seed cultures." Total germination rate (Gt), cumulative rate of germination (AS) [19], survival and early growth were determined. The number of germinated seeds and plant survival were recorded daily for ten weeks.

3.2. Ramets of *C. edulis* vs native seeds and seedlings. In order to check inter-specific competition, mixed cultures were established in 80 pots, combining 3, 2, 1 and 0 ramets of the invasive *C. edulis* with 10, 15, 20 and 25 seeds of native species, referred to as "mixed seed/ramet cultures" (10/3, 15/2, 20/1 and 25/0). Total germination rate (Gt), cumulative rate of germination (AS) [19], survival and early growth were determined. The number of germinated seeds and plant survival were recorded daily for ten weeks.

3.3. Intra-specific competition between native adult plants. In order to take into account intra-specific competition, cultures containing 4, 3, 2 or 1 adult native plants (*M. littorea* or *S. atropurpurea*) were established (Table 1) in 80 pots, referred to as "pure adult cultures". Leaf number, height and survival were recorded every three days for two weeks.

3.4. Ramets of *C. edulis* vs adult native plants. In order to check inter-specific competition, mixed cultures with 3, 2, 1 and 0 ramets of the invasive *C. edulis* combined with 1, 2, 3 or 4 adult native plants were cultivated in 80 pots, referred to as "adult/ramets cultures" (1/3, 2/2, 3/1 and 4/0). Leaf number, height and survival were recorded every three days for two weeks.

4. Competition between *C. edulis* seeds and native seeds

Seeds of *C. edulis* and *M. littorea* or *S. atropurpurea* were sowed at different densities and times, following the scheme proposed by Tielbörger and Prasse [5]. Five replicates of the following seed mixture were established: 10 seeds of each native species plus 10 seeds of *C. edulis*, 10 seeds of each native species plus 30 seeds of *C. edulis*, 30 seeds of each native species plus 10 seeds of *C. edulis* and pure crops of 30 or 10 seeds of each species.

Table 5. Inter-specific competition effect on the germination indices (Gt: Total germination and AS: Cumulative rate of germination), seedling shoot and root length and survival of the native species *Malcolmia littorea* on both invaded and native soil.

Seed N°/ramets	Invaded soil				Native soil			
	10/3	15/2	20/1	25/0	10/3	15/2	20/1	25/0
Gt	46.0*	33.3*	37.0	36.0*	15.8[c]	26.7[bc]	45[ab]	65.6[a]
	(13.3)	(9.7)	(12.9)	(5.7)	(5.5)	(3.8)	(13.5)	(3.4)
AS	51.5*	42.1	38.0	47.8*	12.8[c]	23.6[bc]	44.9[b]	88.5[a]
	(10.8)	(8.6)	(10.5)	(3.8)	(7.4)	(5.8)	(15.2)	(43)
Survival (%)	32.5[b]*	18.6[b]	51.8[ab]*	77.9[a]	0[b]	0[b]	92[a]	(2.8)
	(12.8)	(9.9)	(16.0)	(17.6)				
Shoot length (cm)	-	-	-	-	-	-	-	-
Root length (cm)	-	-	-	-	-	-	-	-

Different letters mean significant difference of 5% between treatments (10/3, 15/2, 20/1 or 25/0 seeds/ramets). *: indicates significant difference of 5% between invaded and native soils on each treatment. Numbers in parentheses indicate the standard error. N = 5.

Table 6. Inter-specific competition effect on the germination indices (Gt: Total germination and AS: Cumulative rate of germination), seedling shoot and root length and survival of the native species Scabiosa atropurpurea on both invaded and native soil.

Seed N°/ramets	Invaded soil				Native soil			
	10/3	15/2	20/1	25/0	10/3	15/2	20/1	25/0
Gt	30.0	41.3	37.1	45.6	14.0	24.0	37.0	48.0
	(8.9)	(3.9)	(3.7)	(2.7)	(9.8)	(7.5)	(3.0)	(7.2)
AS	23.2*	28.3	32.8	40.1*	3.8	19.8	30.4	46.1
	(7.0)	(2.7)	(4.7)	(2.9)	(7.9)	(6.1)	(1.7)	(1.9)
Survival (%)	31.1*	68.3	79.8	93.5	70.0	51.4	51.7	94.5
	(18.1)	(18.1)	(7.5)	(1.7)	(20.0)	(16.4)	(21.5)	(2.5)
Shoot length (cm)	1.2	1.5	1.1	1.4	1.7	1.2	1.2	1.3
	(0.2)	(0.5)	(0.2)	(0.2)	(0.3)	(0.2)	(0.2)	(0.2)
Root length (cm)	4.1	5.0	5.8	6.4	5.4	4.4	4.9	6.3
	(1.1)	(1.4)	(0.7)	(1.4)	(0.5)	(0.9)	(0.3)	(0.9)

Different letters mean significant difference of 5% between treatments (10/3, 15/2, 20/1 or 25/0 seeds/ramets). *: indicates significant difference of 5% between invaded and native soils on each treatment. Numbers in parentheses indicate the standard error. N = 5. None of the post-hoc ANOVA results were significant.

With the aim of testing the effect of time, the seeds of C. edulis were sowed at different date ranges (5 days before the native species, at the same time, or 5 days later than the native species). Competition conditions were established in Petri dishes filled with soil from invaded and from native areas. A total of 240 petri dishes were used. The experimental design (Table 2) therefore had 5 independent factors: neighbour density (10 vs 30), target species (two native species), neighbour species (one invasive species), soil type (native or invaded) and timing of sowing (5 days before, at the same time or 5 days latter).

The Petri dishes were incubated in germination chambers with periods of 12 hours of light and 25°C/15°C, temperatures and light regimes similar to those in the field. Substrate moisture in sandy soils is one of the most limiting factors of plant growth [20]. Therefore, all of the seeds were watered every two days, as previous trials have indicated that this procedure permits maximum germination despite the limited amount of substrate. Percolation of the water through holes in the bottom of the dishes was allowed, avoiding the formation of a salt crust. The number of germinated seeds and plant survival were recorded daily for ten weeks. Total germination rate (Gt), and the cumulative rate of germination (As) were determined [19]. After approximately ten weeks of watering, no further germination was observed and the length of leaf, stem and roots of 7 seedlings per plate and species were measured.

5. Statistical analysis

The Kolmogorov–Smirnov test and Levene's test were used to ensure the normality assumption and the homogeneity of variances, respectively. Two-way analysis of variance (ANOVA) was performed to assess the significance of the effects of soil characteristics and density as well as of their interaction on the studied parameters of germination, growth and survival. Tukey's test was applied for all post-hoc comparisons between groups.

A two-way ANOVA involving soil characteristics and density as factors was carried out to detect significant differences between treatments for each native species. All statistical analyses were performed using the IBM SPSS Statistic 19.0 (IBM, Armonk, NY, USA) software package.

Results

1. Competition of C. edulis ramets vs native species

1.1. Intra-specific competition between native seeds and seedlings. In invaded soil, pure seed cultures of M. littorea (10, 15, 20 and 25 seeds) showed a significant decrease in the total germination percentage (up to 56.4%) and cumulative rate of germination (up to 55.1%) with the increase in density (Table 3). However, in native soil there were no significant differences between treatments (seed number). It is quite remarkable that at the highest density (25 seeds), the total germination and cumulative rate of germination were significantly greater on native soil compared with invaded soil ($P \leq 0.05$), while at low seed densities (10, 15 and 20 seeds) we observed the opposite trend (Table 3). The survival percentage of seedlings of M. littorea showed a significant increase (21.74%) at high densities in native soil. The shoot and root length of M. seedlings in both soils showed no significant effect of intra-specific competition between treatments (number of individuals). Despite this, the growth values from invaded soil are significantly higher than the lengths in the native soil ($P \leq 0.05$).

No significant effect of intra-specific competition was found between S. atropurpurea seedlings affecting the growth, survival or germination rates (Table 4).

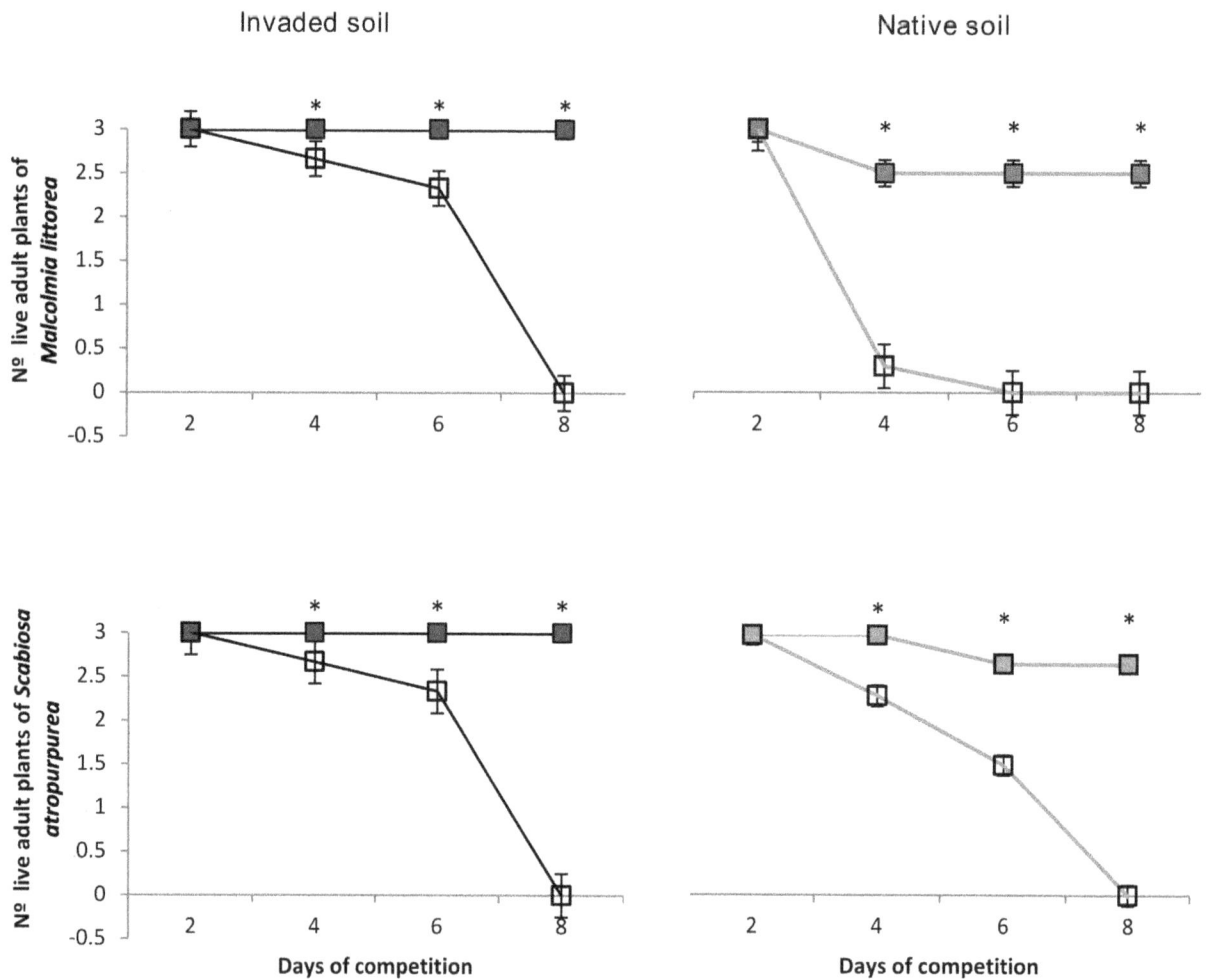

Figure 2. Survival of the native plants *Malcolmia littorea* **and** *Scabiosa atropurpurea* **in pure adult cultures (3 adult native plants) and mixed adult/ramets cultures (3 adult native plants against 1 ramet of** *C. edulis***) on invaded and native soil.** *: indicate differences ($P \leq$ 0.05) between monocultures (filled squares) and mixed (empty squares).

1.2. Ramets of *C. edulis* vs native seeds and seedlings. The increase in density of *C. edulis* per pot did not affect the germination process of *M. littorea* in invaded soil. However, in native soil, the higher density of *C.* caused a decrease in the percentage and speed of germination of *M. littorea* (78% and 87% respectively). The survival of *M. littorea* seedlings was lower as the density of *C. edulis* increased, and was null in native soil (Table 5). Most of the treatments with a high density of *C. edulis* showed higher germination rates and survival percentages in invaded than in native soil. The opposite situation was shown for germination rates when *C. edulis* was not present. It was impossible to test the early seedling response of *M. littorea* experimentally, due to the low or null survival percentage.

The germination process and early growth of *S. atropurpurea* were not affected by the increase in density of *C. edulis* and slightly affected by the type of soil (Table 6).

The interaction between the dependent factors (soil origin and density) was studied by a two-way ANOVA. The interaction was not significant, indicating that the internal relationship between soil origin and density in the germination process and early growth of native plants is very small, or does not exist.

1.3. Intra-specific competition between native adult plants. In native pure adult cultures there were no significant

differences in leaf number, height and survival between plant densities or any soil effect on the plant traits recorded (data not shown).

1.4. Ramets of *C. edulis* vs adult native plants. Mixed adult/ramet cultures revealed that the presence of the invasive plant *C. edulis* had a lethal effect on the adult plants of *M. littorea* and *S. atropurpurea*. All native plants died within 8 days of competition, even when the ratio for the invasive plant *C. edulis* was minimal (3 adult native plants against 1 ramet of *C. edulis*) (Figure 2).

2. *C. edulis* seeds *vs* native seeds

2.1. Intra-specific competition and soil effects. Considering pure seed cultures, we found no density effect ($P \leq 0.05$) and almost not soil effect (Table 7). However, the Gt and AS indices of were in general higher in Petri dishes filled with invaded soil than in those filled with native soil while the shoot growth seemed higher on non-invaded soils (Table 7).

2.2. Inter-specific competition and soil effects. In relation to time of sowing, the germination indices for *M. littorea* seeds were generally significantly lower when the seeds of *C. edulis* were sown 5 days before. Values were generally similar in assay 2 and 3 and in monocultures.

Table 7. Effect of soil on the germination and early growth of the invader *C. edulis* and the native species *Malcolmia littorea* and *Scabiosa atropurpurea* in monocultures.

	N° seeds	Gt		AS		Shoot growth		Radicle growth	
		I	N	I	N	I	N	I	N
C. edulis	10	30.0	23.3	0.19	0.16	0.49*	0.30	1.7*	2.6
		(10.0)	(8.8)	(0.09)	(0.09)	(0.08)	(0.07)	(0.1)	(0.3)
	30	38.9	22.2	0.21	0.12	0.53	0.38	2.1	2.6
		(4.0)	(5.9)	(0.04)	(0.01)	(0.07)	(0.04)	(0.2)	(0.38)
Malcolmia littorea	10	60.0	53.3	0.78	0.53	0.24	0.24	1.79	1.9
		(10.0)	(8.8)	(0.20)	(0.10)	(0.02)	(0.04)	(0.2)	(0.1)
	30	65.6	56.7	1.05*	0.54	0.25	0.22	2.1	2.5
		(1.9)	(10.0)	(0.02)	(0.09)	(0.02)	(0.01)	(0.1)	(0.4)
Scabiosa atropurpurea	10	46.7	73.3	0.73	0.99	0.35	0.33	8.1*	10.2
		(12.0)	(13.3)	(0.24)	(0.19)	(0.02)	(0.02)	(1.2)	(0.8)
	30	65.6	70.0	0.94	0.90	0.44*	0.29	12.0	10.4
		(5.9)	(5.1)	(0.09)	(0.10)	(0.04)	(0.01)	(0.6)	(0.7)

(Gt: Total germination and AS: Cumulative rate of germination). *indicate significant differences at 5% level between the seeds sown in Petri dishes filled with invaded soil and those sowed in native soil. Numbers in parentheses indicate the standard error. I: Invaded soil, N: Native soil. N = 5.

Table 8. Effect of the timing of sowing on the germination of the invader *C. edulis* and the native species *Malcolmia littorea* and *Scabiosa atropurpurea littorea*.

		Gt				AS			
		A1	A2	A3	M	A1	A2	A3	M
CM	I	36.7[a] (3.3)	26.7[b] (3.3)	20.0[b] (10.0)	30.0[a] (10.0)	0.50 (0.04)	0.24 (0.09)	0.14 (0.09)	0.19 (0.09)
	N	30.0[a] (0.0)	3.3[b] (3.3)	3.3[b] (3.3)	23.3[a] (8.8)	0.43[a] (0.06)	0.02[b] (0.02)	0.01[b] (0.01)	0.16[a] (0.09)
CS	I	43.3 (3.3)	23.3 (3.3)	30.0 (5.8)	30.0 (10.0)	0.39 (0.01)	0.14 (0.07)	0.07 (0.01)	0.19 (0.09)
	N	20.0 (5.8)	13.3 (3.3)	10.0 (5.8)	23.3 (8.8)	0.22 (0.11)	0.09 (0.05)	0.05 (0.02)	0.16 (0.09)
ML	I	33.3[b] (8.8)	70.0[a] (17.3)	66.7[a] (12.0)	60.0[a] (10.0)	0.36[b] (0.15)	1.01[a] (0.14)	0.86[a] (0.15)	0.78[a] (0.20)
	N	20.0[b] (5.8)	30.0[a] (2.8)	40.0[a] (7.5)	53.3[a] (8.8)	0.13[b] (0.03)	0.30[a] (0.05)	0.46[a] (0.09)	0.53[a] (0.10)
SA	I	46.7 (6.7)	63.3 (3.3)	66.7 (6.7)	46.7 (12.0)	0.59 (0.07)	0.89 (0.01)	0.98 (0.06)	0.73 (0.24)
	N	70.0 (15.3)	70.0 (15.3)	66.7 (12.0)	73.3 (13.3)	0.77 (0.19)	1.05 (0.20)	0.92 (0.10)	0.99 (0.19)

Different letters means significant differences at 5% level among treatments (A1, A2, A3, M).
A1: Assay 1, *C. edulis* seeds sown before native seeds. A2: Assay 2: *C. edulis* seeds sowed at the same time that native seeds. A3: Assay 3: *C. edulis* seeds sown after native seeds. M: native monocultures. I: Invaded soil. N: Native soil.
CM: *C. edulis* + *M. littorea*. CS: *C. edulis* + *S. atropurpurea*. ML: *M. littorea*. SA: *S. atropurpurea*. Numbers in parentheses indicate the standard error. N = 5.

Table 9. The effect of timing of sowing on the early growth of the native species *Malcolmia littorea* and *Scabiosa atropurpurea*.

		Shoot growth				Radicle growth			
		A1	A2	A3	M	A1	A2	A3	M
ML	I	0.20[b]	0.20[b]	0.17[b]	0.24[a]	2.1	1.79	1.7	1.79
		(0.00)	(0.00)	(0.02)	(0.02)	(0.5)	(0.2)	(0.2)	(0.2)
	N	0.24[b]	0.23[b]	0.17[b]	0.30[a]	1.9	1.9	1.7	1.9
		(0.00)	(0.03)	(0.02)	(0.04)	(0.1)	(0.3)	(0.3)	(0.1)
SA	I	0.34	0.35	0.40	0.35	10.9	10.0	10.5	8.1
		(0.03)	(0.02)	(0.03)	(0.02)	(1.3)	(0.6)	(1.0)	(1.2)
	N	0.35	0.32	0.28	0.33	9.0	11.1	10.7	10.2
		(0.03)	(0.02)	(0.03)	(0.02)	(1.1)	(1.1)	(1.0)	(0.8)

Different letters means significant differences at 5% level among treatments (A1, A2, A3, M). A1: Assay 1, C. *edulis* seeds sown before native seeds. A2: Assay 2: *C. edulis* seeds sowed at the same time that native seeds. A3: Assay 3: *C. edulis* seeds sown after native seeds. M: native monocultures. I: Invaded soil. N: Native soil. ML: *M. littorea*. SA: *S. atropurpurea*. Numbers in parentheses indicate the standard error. N = 5.

C. edulis seeds reacted the same way in the presence of *M. littorea* seeds in native soils. The germination indices were generally significantly lower when the seeds of *M. littorea* were sown 5 days before (80–90%) but not in the presence of *S. atropurpurea* (Table 8). The germination process of *S.* did not appear to be influenced by any of the treatments (Table 8).

The growth results in relation to the time of sowing factor only refer to native species, as *C. edulis* had different growth periods depending on the treatment (assay 1, 2 or 3). *M. littorea* seedlings showed a decrease in shoot growth with the presence of *C. edulis* seeds, especially in native soils, although the radicle growth of *M. littorea* showed no differences (Table 9). No effects were observed in *S. atropurpurea* growth in relation to the timing factor (Table 9).

Despite the effects of the time of sowing, there were no significant differences ($P \leq 0.05$) between pure and mixed cultures in any treatment (results not shown).

Discussion

1. Competition between ramets of *C. edulis* and native species

1.1. Intra-specific competition between native seeds and seedlings. Soil from habitats invaded by *C. edulis* had a markedly species-dependent effect. *M. littorea*, a typical semi-fixed dunes species, is a specialist plant of poor and slightly saline soils [21]. Its germination and growth process is significantly influenced by soil changes induced by *C. edulis* [7,8]. In general, we found a reduction in the germination process as the density increased in invaded soil. The advantage of this behaviour of density-dependent germination in invaded soils is difficult to understand. It is possible that for these plants there is an advantage of not germinating under competitive conditions, so they may exploit more favourable conditions in later years [5]. Shoot and root length was always greater in invaded soils, probably due to a greater availability of nutrients in these soils [9,22] and salt concentration [23] that stimulate growth. In native soil, the survival percentage increased at the highest density, which may be due to the nurse plant phenomenon [20] taking into account not differences between life forms.

Germination was not density dependent for *S. atropurpurea* in either invaded or native soils. As previously mentioned *S. atropurpurea* presented a greater plasticity to adapt to different soil conditions and seems to be less of a specialist than *M. littorea*.

1.2. Ramets of *C. edulis* vs native seeds and native seedlings. Inter-specific competition is species-dependent and determines what species can coexist [24]. *C. edulis* differentially affected germination of the selected native plants. Soil characteristics became as one of the factors that directly affect the competitive potential of the exotic [25].

M. littorea showed a significant decrease in the germination process caused by the presence of *C. edulis* adult plants. However, the most dramatic effect found on *M. littorea* is in the survivorship, which is soil dependent. In invaded soil, the survival percentage decreases with the proportion of *C. edulis* adult plants in the mixed cultures, while in native soils, none of the germinated seedlings survived at the end of the bioassay. Inter-specific competition seems to be responsible for these results [26,27]. Thus, although barriers to native plant germination could be overcome, *M. littorea* seedlings would not establish viable populations in the presence of *C. edulis* adult plants.

Soil characteristics are not determinant in the response of *S. artropurpurea* to the presence of *C. edulis* (except to mixed seed/ramets cultures 10/3), once again probably due to its plasticity.

Although *C. edulis* threatens the establishment of *S. atropurpurea* in both soils in different ways, seedlings are phenotypically plastic in their allocation of biomass into roots and shoots [28].

1.3. Intra-specific competition between natives adult plants. Our results indicate that the residual effect of *C. edulis* on dune soil does not affect the development of native adult plants. Adult native plants seem to be better adapted to soil changing conditions than seedlings, suffering stress from the residual effect in the soil.

1.4. Ramets of *C. edulis* vs adult native plants. The presence of *C. edulis* in the replacement series had a deleterious effect on native plants independently of density and soil type. Different authors have indicated that allelopathy is a fairly common invasion mechanism (29). Ens et al. [30] proposed that the eventual dominance of invasive species could be explained by direct or indirect chemical inhibition of the establishment of indigenous plants, which was confirmed by Novoa et al. [7] for *C. edulis*.

2. Seed competition of *C. edulis* vs seeds of native species

2.1. Soil effects. Dune species germinate in autumn, the rainy season, when in addition to having more water in the soil, the salt content decreases [20]. Water softens the seed coat so that the root can emerge more easily and also solubilizes nutrients [31]. A high salt content can block the germination process by the osmotic effect, drawing water from seeds [32]. During the assay, all of the seeds were watered every two days. However, invaded soils have a higher level of organic matter than native soils [7,8], so they can hold water for a longer period of time. As a result, the Gt and AS indices of *M.* were higher in Petri dishes filled with invaded soil than in those filled with native soil. Also, in the presence of *C. edulis*, *M. littorea* has greater shoot growth in native soils. This could be due to an allelopathic effect of *C. edulis* seeds on *M. littorea* [7].

Once again *S. atropurpurea* did not show any differences in the germination process between soils. However, the radicle growth of *S. atropurpurea* seedlings seemed to be stimulated by native soil, while shoot growth seemed to be stimulated by invaded soil. The nutrient content of sand (especially nitrogen and phosphorus) positively or negatively affects the growth of dune species depending on the species [20]. The radical growth of *S. atropurpurea* decreases and its shoot growth increases with an increase in the nutrient level [8]. This could explain the differences

observed in the growth of *S. atropurpurea* between invaded and native soil.

Finally, when *C. edulis* invades coastal habitats, it modifies the conditions of the substrate and suffers from difficulties as a result of tissue decomposition [33]. This feature could have evolved as a mechanism to facilitate recolonization when the clones die, and it influences the germination process of the invasive species. As a result, the germination process of *C. edulis* depends on the nutrient level of the soil [7]. Therefore, the Gt and AS indices of *C. edulis* were higher in Petri dishes filled with invaded soil than in those filled with native soil. Also, the radicle of *C. edulis* grows more at high pH levels [7], and so, it grew more in native than in invaded soils.

2.2. Timing of sowing. We observed that the timing of sowing affects the establishment of both *C. edulis* and *M. littorea*, although there are no competitive interactions if they are sown at the same time. Therefore, they could have some mechanisms that allow their seeds to evaluate the conditions of their neighbours prior to emergence and to plastically respond to them [5]. *C. edulis* could also have an allelopathic effect on *M. littorea* and vice versa, which should be explored in future assays. These are important results to be taken into account for restoration actions; but also in conservation actions. If *C. edulis* is a new invader arriving from seeds into a native area and *M. littorea* is present, according to our results it appears that M. littorea might have the ability to supress *C. edulis* germination.

Conclusion

Our results show that the impacts of *C. edulis* on native plants are highly dependent on its development stages as well as on the development stages of native plants. These findings are crucial for new strategies of biodiversity conservation in coastal habitats.

Acknowledgments

We thank Paula Lorenzo for valuable comments and Jorge Tabares for technical assistance.

Author Contributions

Conceived and designed the experiments: AN LG. Performed the experiments: AN. Analyzed the data: AN. Contributed reagents/materials/analysis tools: LG. Contributed to the writing of the manuscript: AN LG.

References

1. Kowarik I (2003) Biologische Invasionen-Neophyten und Neozoen in Mitteleuropa. Stuttgart: Ulmer.
2. Simberloff D, Parker IM, Windle PN (2005) Introduced species policy, management, and future research needs. Frontiers in Ecology and the Environment 3: 12–20.
3. Vilà M, Espinar JL, Hejda M, Hulme PE, Jarošík V, et al. (2011) Ecological impacts of invasive alien plants: a meta-analysis of their effects on species, communities and ecosystems. Ecology letters 14: 702–708. Available: http://www.ncbi.nlm.nih.gov/pubmed/21592274. Accessed 25 May 2014.
4. Ley C, Gallego J, Vidal C (2007) Manual de restauración de dunas costeras. Ministerio de medio ambiente.
5. Tielbörger K, Prasse R (2009) Do seeds sense each other? Testing for density-dependent germination in desert perennial plants. Oikos 118: 792–800. Available: http://doi.wiley.com/10.1111/j.1600-0706.2008.17175.x. Accessed 6 November 2012.
6. Council Directive 92/43/EEC (1992): 7.
7. Novoa A, González L, Moravcová L, Pyšek P (2012) Effects of Soil Characteristics, Allelopathy and Frugivory on Establishment of the Invasive Plant Carpobrotus edulis and a Co-Occuring Native, Malcolmia littorea. Plos One 7: 1–11. doi:10.1371/journal.pone.0053166.
8. Novoa A, González L, Moravcová L, Pyšek P (2013) Constraints to native plant species establishment in coastal dune communities invaded by Carpobrotus edulis: Implications for restoration. Biological Conservation 164: 1–9. Available:

http://linkinghub.elsevier.com/retrieve/pii/S0006320713001031. Accessed 20 June 2013.
9. Novoa A, Rodríguez R, Richardson D, González L (2014) Soil quality: a key factor in understanding plant invasion? The case of Carpobrotus edulis (L.) N.E.Br. Biological Invasions 16: 429–443. Available: http://link.springer.com/10.1007/s10530-013-0531-y. Accessed 30 June 2014.
10. Santoro R, Jucker T, Carranza ML, Acosta ATR (2011) Assessing the effects of Carpobrotus invasion on coastal dune soils. Does the nature of the invaded habitat matter? Community ecology 12: 234–240. doi:10.1556/ComEc.12.2011.2.12.
11. Brown CS, Anderson VJ, Claassen VP, Stannard ME, Wilson LM, et al. (2008) Restoration Ecology and Invasive Plants in the Semiarid West. Invasive plant science and management 1: 399–413.
12. Lorenzo P, Palomera-Pérez A, Reigosa MJ, González L (2011) Allelopathic interference of invasive Acacia dealbata Link on the physiological parameters of native understory species. Plant Ecology 212: 403–412. doi:10.1007/s11258-010-9831-9.
13. Mangla S, Sheley RL, James JJ, Radosevich SR (2011) Intra and interspecific competition among invasive and native species during early stages of plant growth. Plant Ecology 212: 531–542. Available: http://link.springer.com/10.1007/s11258-011-9909-z. Accessed 19 June 2014.
14. Lorenzo P, Rodríguez-Echeverría S, González L, Freitas H (2010) Effect of invasive Acacia dealbata Link on soil microorganisms as determined by PCR-

DGGE. Applied Soil Ecology 44: 245–251. Available: http://linkinghub. elsevier.com/retrieve/pii/S0929139310000065. Accessed 7 November 2012.

15. Thuiller W, Richardson DM, Rouget M, Thuiller W, Richardson DM, et al. (2006) Interactions between Environment, Species Traits, and Human Uses Describe Patterns of Plant Invasions. Ecology 87: 1755–1769.

16. D'Antonio C, Mahall BE (1991) Root profiles and competition between the invasive, exotic perennial, Carpobrotus edulis, and two native shrub species in California coastal scrub. American Journal of Botany 78: 885–894.

17. Combs JK, Reichard SH, Groom MJ, Wilderman DL, Camp PA (2011) Invasive competitor and native seed predators contribute to rarity of the narrow endemic Astragalus sinuatus Piper. Ecological Applications 21: 2498–2509.

18. De Wit CT, Van den Bergh JP (1965) Competition between herbage plants. Journal of Agricultural Science 13: 212–221.

19. Hussain MI, González-Rodríguez L, Reigosa MJ (2008) Germination and growth response of four plant species to different allelochemicals and herbicides. Allelopathy Journal 22: 101–110.

20. Maun MA (2009) The biology of coastal sand dunes. Oxford University Press.

21. Del Vecchio S, Giovib E, Izzia CF, Abbateb G, Acosta ATR (2012) Malcolmia littorea: The isolated Italian population in the European context. Journal for Nature Conservation 20: 357–363. doi:10.1016/j.jnc.2012.08.001.

22. Del Vecchio S, Marbà N, Acosta A, Vignolo C, Traveset A (2013) Effects of Posidonia oceanica beach-cast on germination, growth and nutrient uptake of coastal dune plants. PloS one 8: e70607. Available: http://www.pubmedcentral. nih.gov/articlerender.fcgi?artid=3720909&tool=pmcentrez&rendertype=abstract. Accessed 5 August 2014.

23. Lee JA, Caporn SJM (1998) Ecological effects of atmospheric reactive nitrogen deposition on semi-natural terrestrial ecosystems. New Phytologist 139: 127–134.

24. Tschirhart J (2002) Resource competition among plants: from maximizing individuals to community structure. Ecological Modelling 148: 191–212. Available: http://linkinghub.elsevier.com/retrieve/pii/S0304380001004392.

25. Huangfu CH, Zhang TR, Chen DQ, Wang N, Yang DL (2011) Residual effects of invasive weed yellowtop (Flaveria bidentis) on forage plants for ecological restoration. Allelopathy Journal 27: 55–64.

26. Ammondt S, Litton CM (2012) Competition between Native Hawaiian Plants and the Invasive Grass Megathyrsus maximus: Implications of Functional Diversity for Ecological Restoration. Restoration Ecology 20: 638–646. Available: http://doi.wiley.com/10.1111/j.1526-100X.2011.00806.x. Accessed 28 November 2012.

27. Holdredge C, Bertness MD (2010) Litter legacy increases the competitive advantage of invasive Phragmites australis in New England wetlands. Biological Invasions 13: 423–433. Available: http://www.springerlink.com/index/10. 1007/s10530-010-9836-2. Accessed 3 November 2012.

28. Shadel WP, Molofsky J (2003) Habitat and population effects on the germination and early survival of the invasive weed, Lythrum salicaria L. (purple loosestrife). Biological Invasions 4: 413–423.

29. Ren M-X, Zhang Q-G (2009) The relative generality of plant invasion mechanisms and predicting future invasive plants. Weed Research 49: 449–460. Available: http://doi.wiley.com/10.1111/j.1365-3180.2009.00723.x. Accessed 6 November 2012.

30. Ens EJ, French K, Bremner JB (2009) Evidence for allelopathy as a mechanism of community composition change by an invasive exotic shrub, Chrysanthe-moides monilifera spp. rotundata. Plant and Soil 316: 125–137. Available: http://www.springerlink.com/index/10.1007/s11104-008-9765-3. Accessed 7 January 2013.

31. Khurana E, Singh JS (2004) Germination and seedling growth of five tree species from tropical dry forest in relation to water stress: impact of seed size. Journal of Tropical Ecology 20: 385–396. Available: http://www.journals.cambridge.org/abstract_S026646740400135X. Accessed 26 November 2012.

32. Bubel N (1988) The new seed-starters handbook. Rodale Press, Inc.

33. Conser C, Connor EF (2008) Assessing the residual effects of Carpobrotus edulis invasion, implications for restoration. Biological Invasions 11: 349–358. Available: http://www.springerlink.com/index/10.1007/s10530-008-9252-z. Accessed 4 November 2012.

7

Policy Development for Environmental Licensing and Biodiversity Offsets in Latin America

Ana Villarroya[1]*, **Ana Cristina Barros**[2], **Joseph Kiesecker**[3]

1 The Nature Conservancy, Boulder, Colorado, United States of America, **2** The Nature Conservancy, Brasilia, Brazil, **3** The Nature Conservancy, Fort Collins, Colorado, United States of America

Abstract

Attempts to meet biodiversity goals through application of the mitigation hierarchy have gained wide traction globally with increased development of public policy, lending standards, and corporate practices. With interest in biodiversity offsets increasing in Latin America, we seek to strengthen the basis for policy development through a review of major environmental licensing policy frameworks in Argentina, Brazil, Chile, Colombia, Mexico, Peru and Venezuela. Here we focused our review on an examination of national level policies to evaluate to which degree current provisions promote positive environmental outcomes. All the surveyed countries have national-level Environmental Impact Assessment laws or regulations that cover the habitats present in their territories. Although most countries enable the use of offsets only Brazil, Colombia, Mexico and Peru explicitly require their implementation. Our review has shown that while advancing quite detailed offset policies, most countries do not seem to have strong requirements regarding impact avoidance. Despite this deficiency most countries have a strong foundation from which to develop policy for biodiversity offsets, but several issues require further guidance, including how best to: (1) ensure conformance with the mitigation hierarchy; (2) identify the most environmentally preferable offsets within a landscape context; (3) determine appropriate mitigation replacement ratios; and (4) ensure appropriate time and effort is given to monitor offset performance.

Editor: Clinton N. Jenkins, Instituto de Pesquisas Ecológicas, Brazil

Funding: Funding was provided by the Grantham Foundation, the Anne Ray Charitable Trust and The Nature Conservancy. The funders had no role in study design, data collection and analysis, decision to publish, or preparation of the manuscript.

Competing Interests: The authors have declared that no competing interests exist.

* Email: ana.villarroya@tnc.org

Introduction

Over the next two decades, governments and companies will invest unprecedented sums – well over 20 trillion dollars – in development projects around the world, from Argentina to Zambia. Rapidly developing countries are making trillion dollar investments in infrastructure. For example, Latin America is in the midst of unprecedented and sustained growth in development as worldwide demand for the region's mineral, agricultural, and energy wealth grows [1,2]. The region will need to construct more roads, energy facilities, and mines as this economic development continues. To be sustainable it is important to ascertain how this development can be done in a way that minimizes impacts and maximizes the benefits to nature and people. It will require that we find ways to balance the seemingly conflicting goals of improving infrastructure, increasing food production, and expanding access to reliable energy and housing while also preserving and protecting the biodiversity and ecosystem services of the region. To simultaneously achieve these goals will be challenging and require that development is complemented by public and private investments to prevent the loss of biodiversity and ecosystem services.

Environmental licensing processes, such as Environmental Impact Assessment (EIA) play a critical role in controlling the way development projects result in damage to the environment. In most countries developers are required to get an environmental license before development activities can be implemented, and currently EIA has been legally adopted in almost all countries in the world [3]. Obtaining such permit usually depends on the way the predicted negative impacts will be mitigated, or depends on the fulfillment of additional requirements set by the licensing authority. EIA is a systematic, iterative process that examines the environmental consequences of planned developments and emphasizes prediction and prevention of environmental damage [4]. The mitigation of environmental impacts is thus a key stage of the environmental impact assessment process and lies at its core [5]. Practitioners seek to reduce impacts through application of the mitigation hierarchy: avoid, minimize, restore, and offset [6]. To avoid impacts on biodiversity, measures are taken to prevent creating impacts from the outset, such as careful spatial or temporal placement of elements of infrastructure. In minimization, measures are taken to reduce the duration, intensity, and/or extent of impacts that cannot be completely avoided. In restoration, measures are taken to rehabilitate degraded ecosystems or restore cleared ecosystems after impacts that cannot be completely avoided and/or minimized. To offset impacts measures are taken to compensate for any residual adverse impacts that cannot be avoided, minimized, and/or restored. Offsets can take the form of positive management interventions such as restoration of degraded habitat, arrested degradation or averted risk, or protecting areas where there is imminent or projected loss of biodiversity [7,8]. Attempts to meet biodiversity goals through application of the mitigation hierarchy have gained wide traction globally with increased development of public policy, lending standards, and corporate policy. In the public policy sector there are approximately 45 compensatory mitigation programs for biodiversity impacts worldwide, with another 27 programs in development [9]. In the financial sector, major institutions including the International Finance Corporation (IFC) and more than 70 Equator Principles financial institutions that base their requirements on IFC's Performance Standards are requiring projects they finance to adhere to the mitigation hierarchy. This means they should seek to avoid impacts on biodiversity and ecosystem services or - where this is not possible - to minimize or restore them. In critical habitats, this also means achieving net gains of biodiversity values for which these habitats have been designated. European Bank for Reconstruction and Development (EBRD) also has similar requirements [10–12]. As new performance standards and public policies drive mitigation biodiversity goals from a voluntary objective into the sphere of compliance, businesses (especially mining companies) are increasingly adopting it into corporate biodiversity management policies and mitigation practices as a normal way/cost of doing business [13–15].

With interest in biodiversity offsets increasing in Latin America, we seek to strengthen the basis for policy development through a review of major environmental licensing policy frameworks in Argentina, Brazil, Chile, Colombia, Mexico, Peru and Venezuela (Figure 1). We focused on these countries because they represent ~85% of the area of all Central and South America and ~80% of the population of the region. Since we relied mainly on colleagues to identify and interpret policy documents we focused on countries where The Nature Conservancy has country level programs and staff available. We recognize the limitation of this approach but also do not consider this sample of countries to be a random sample intended to capture broader patterns in other countries found in the region. By comparing the goals, approaches, and key issues highlighted in these frameworks, and distilling important commonalities and differences, our aim is to provide guidance to countries that have not yet developed frameworks and to support improvements in existing policies. The frameworks selected for review include both established offset programs and rapidly emerging policies. With this analysis we sought to explore and analyze the role mitigation hierarchy and, more specifically, offsets are given in different Latin-American legal frameworks. First we conducted a broad review of policies related to the environmental licensing process, because the ecological effectiveness of mitigation depends heavily on the existence of strong environmental laws and regulations [16]. We then reviewed current offset frameworks from the selected countries. Finally we highlight negative and positive aspects of each countries mitigation frameworks as a guide to improve existing tools or proposal of new ones.

Methods

We focused our review on national (federal) level policies to assess how current requirements would affect implementation of the mitigation hierarchy and promote positive environmental outcomes. State and provincial policies have not been included in this study. Although they are necessary to respond to local environmental contexts, the paper focuses on national policies because (a) the constitution of Chile, Colombia, Peru only allow for all laws to be established at the national level [17–19]. In addition in Venezuela environmental laws are only made at the national level [20]. Decisions on environmental licensing in Mexico are context dependent with large scale impacts (e.g. oil and gas, large hydropower, forest clearing, roads and railways) regulated at the federal level while localized environment impacts

(e.g. urban expansion, small hydropower) are made at the state level [21,22]. (b) National/federal policies often establish a common base for more specific documents such as state or provincial policies [23,24]. (c) Infrastructure projects are often large and may affect more than one province or state. Thus we decided to focus on national level policies that would govern these types of projects. We sought to include three primary sources to gather and assess existing policies: 1. Official websites of each country's Ministry of the Environment (or equivalent agency), and any official agencies involved with the country's environmental licensing processes. 2. Published articles and reports about EIA and offset procedures in the selected countries. 3. Interviews with persons directly engaged in the mitigation agenda in each country. These interviews also helped ensure we interpreted the legal texts correctly. The interviews also helped confirm that all relevant legal texts had been selected and that we were not missing any information. For a complete list of sources used in our analysis see Table S1 and Appendix S1. We focused our analysis on existing policies and laws but we also included the new offsets law in Peru, that is about to be signed into law.

While a policy analysis may provide interesting and relevant information, it also has limitations that cannot be overlooked [25]. We acknowledge that environmental policies are numerous, varied and constantly changing, and the information they contain can be sometimes misinterpreted. Thus, even though all effort has been made to find and comprehensively review all relevant policies, we acknowledge that there is a chance that some regulations or information were missed. In those cases, we state that "no information has been found" instead of "no information exists on the subject". Also, we must keep in mind that policies are in most aspects qualitative and difficult to compare and/or evaluate in a standard way that leaves little place to subjective interpretation. We have tried to overcome this handicap as much as possible by setting a list of specific and well defined questions to answer when reviewing the selected texts (see below).

Review of Policies Related to Application of the Mitigation Hierarchy

To assess how a countries environmental licensing process would promote positive conservation outcomes we have reviewed national legal texts related to EIA processes and mitigation for infrastructure projects. We focused on general environmental policies, such as environmental acts as these laws often make provisions for EIA, or establish how mitigation activities are carried out. We also paid attention to sector-specific policies, since it is common to find specifications on how EIA shall be carried out for certain types of development (i.e. mining) projects, or require specific mitigation measures for particular types of development. In addition we also examined habitat/area-specific policies (e.g. wetlands), since sometimes they include provisions related to impact mitigation [26].

Although our analysis examined some aspects of the EIA process it was not intended as a detailed review of these procedures. Impact assessments are highly technical processes, whose success depends on the quality of the regulatory requirements, availability of analytical tools and technical capacity. Here we focused on aspects of EIA that we think influence implementation of the mitigation hierarchy. Moreover several publications have conducted broader analysis of EIA process in the region, including operational and implementation issues [27–31]. In addition a recent review by Reid et al. (in prep.) analyzes the SEA procedures in the region. To evaluate mitigation frameworks as the basis for offset practices, our assessment is focused on regulatory features directly related to impact mitigation, mainly:

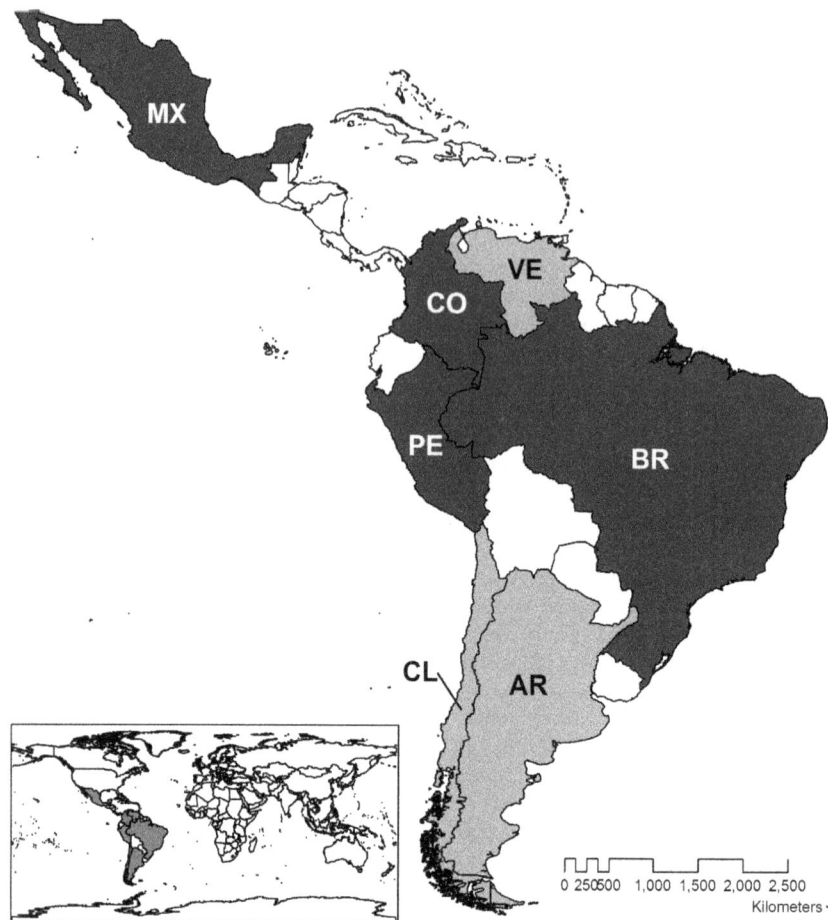

Figure 1. Countries selected for the study (in color). In dark grey, countries for which offset frameworks have been established. Countries' names have been abbreviated to the codes set by ISO 3166.

impact evaluation and the use of the mitigation hierarchy. The aim is to assess to what extent the reviewed policies set requirements that may eventually promote solid mitigation practices. This portion of our analysis serves as the starting point for the more detailed review of offset frameworks, see below.

To standardize the review as much as possible, we have defined a set of questions that have been answered for each country on a yes/no basis, depending on the contents of their laws and regulations (Appendix S2). We grouped policies for the review as follows: General (policies that apply to all projects: environmental acts, general EIA, and habitat-specific laws and regulations), and Sector-specific (mining, hydrocarbons, energy (electricity), transport infrastructure (i.e. roads, railways, airports and ports), and waste management). All the projects covered under these policies also have to follow the requirements set by general laws and regulations, so the sectorial provisions supplement the general ones.

Review of Offset Specific Policies and Laws

Our assessment of the environmental licensing processes of the seven countries identified four (Brazil, Colombia, Mexico and Peru) that have developed specific policies that regulate offset implementation (Table 1). For these four countries we focused on the following laws that dictated offset usage:

In Brazil projects subject to environmental licensing must offset their impacts on environmental assets. Impacts on Protected Areas

(Law 9985 of 2000), caves (Decree 6640 of 2006) and coastal native vegetation (Decree 5300 of 2004) shall always be offset, although the environmental authority (IBAMA) may require the developer to offset any other residual impacts identified. In addition to these laws, Law 12651 of 2012 (on the Protection of Native Vegetation) regulates offsets for impacts to native vegetation, although these are not required for obtaining an environmental license so we will not examine it in this paper. In this case, our review will focus on the framework first set by Law 9985 of 2000 (see Table 1).

In Colombia projects subject to EIA must offset their impacts on terrestrial ecosystems (as regulated by Resolution 1517 of 2012) and freshwater (Law 99 of 1993, Decree 1900 of 2006 and Decree 1933 of 1994). In addition there are some offset requirements for impacts to forests (Decree 1791 of 1996) as well as several specific activities (Resolutions that implement TORs for elaborating EISs, see Table S1). However, these latter policies only address a few aspects related to offset implementation, so they cannot be considered equal to the 2012 law focused on terrestrial ecosystems. For this country we will focus the review on this 2012 framework (see Table 1).

In Mexico the Sustainable Forest Development Act of 2003 requires offsets for impacts that result in land-use change to forested areas. The recently enacted Environmental Liability Act (2013) also requires offsets, but only when impacts are not predicted or approved in the EIA and are deemed an environ-

Table 1. Current policies reviewed for the selected offset frameworks.

Country	Year	Document reference	What it regulates
Brazil	2000	Law 9985	Sets the obligation for projects subject to environmental licensing of offsetting impacts by making payments to support the National System of Protected Areas
	2002	Decree 4340	Regulates calculation of offset payments, sets the need of an Offsets Chamber, and establishes how to use offset funds
	2004	Direct action of unconstitutionality 3378	Partially modifies Art.36 § 1° of Law 9985 (original one declared partially unconstitutional)
	2006	CONAMA Resolution 371/06	Sets guidelines for the environmental authority to calculate, collect, use, approve and manage offset funds related to Law 9985
	2006	Decree 5746	Regulates offsets for impacts to Natural Heritage Reserves
	2009	Decree 6848	Modifies Decree 4340
	2010	Ordinance 416	Creates the Environmental Offsets Federal Chamber (CFCA)
	2010	Ordinance 458	Designates the representatives of each organization that compound the Environmental Offsets Federal Chamber (CFCA)
	2011	Ordinance 10	Regulates the selection of environmental non-governmental organizations that will be part of the Environmental Offsets Federal Chamber (CFCA)
	2011	Ordinance 225	Creates the Environmental Offsets Federal Committee (CCAF)
	2011	Normative Instruction 8	Regulates the Environmental Offsets procedure set in Decree 4340 and modified by Decree 6848
	2011	Normative Instruction 20	Regulates the administrative procedures for setting the terms of commitment regarding offsets
	2011	IBAMA Ordinance 16	Sets the bylaws of the Environmental Offsets Federal Committee (CCAF)
Colombia	2010	Resolution 1503	Sets the obligation to follow the instructions of the "Manual for allocating offsets for loss of biodiversity" for implementing offsets in projects subject to EIA
	2012	Resolution 1517	Approves the Manual for allocating offsets for loss of biodiversity
Mexico	2003	General Law on Sustainable Forestry	Sets the obligation of making offset payments for land-use change of forest areas
	2005	Regulation of the General Law on Sustainable Forestry	Sets the basis for regulating offset payments for land-use change of forest areas
	2005	Agreement on offsets equivalency	Sets the method for calculating the required offsets area
	2011	Agreement on offsets costs	Sets the reference costs for calculating the required offset payments
Peru	2014?	Offsets law [to be passed]	Sets the basis for offsetting impacts to biodiversity in projects subject to EIA (categories II and III)

mental offence. Since this is not a part of the environmental licensing process we have not included it in our assessment. The general law on ecological balance and environmental protection of 1988 (most commonly known as the LGEEPA) also enables the use of offsets as does the Official Mexican Rule NOM-120-SEMARNAT-2011 which makes some provisions for offsets related to mining projects. Neither of these two laws can be considered a specific offset framework since they only enable the use of offsets but do not make any specific requirements or guidance for when or how they should be used.

In Peru a new law about to be passed requires offsets for certain projects subject to EIA, and provides details on how such measures shall be implemented. Although the law is not currently enacted, we have included it in this study since it establishes a new offset framework that is different from the other country level programs.

Offset design is a complex process that entails multiple challenges. Several principles have been outlined to guide this process, the most widespread being the ones set by the Business and Biodiversity Offsets Programme (BBOP) [32]. However, applying theoretical guidance into practice often proves difficult,

as when trying to translate best practice principles into effective policy requirements. Several challenges, which may be especially tricky for policy making, have been identified and discussed in the scientific literature (see [33,34]). We want to contribute to this discussion by evaluating how the selected policies deal with these challenges, and how the theory we know may help improving legal frameworks.

Following the approach outlined in McKenney and Kiesecker 2010 [33] and Bull et al. 2013 [34] we distilled a set of criteria that constitute main current challenges and at the same time are key to policies which seek to ensure that offsets provide the following values: (1) they provide additional replacement for unavoidable negative impacts of human activity on biodiversity, (2) they involve measurable, equivalent biodiversity losses and gains, and (3) they achieve, as a minimum, no net loss of biodiversity. Following these principles we identified twelve criteria (which include most of the ones listed by the above cited references, plus two additional ones) that we used to assess the current state of offset frameworks in our four target countries (Table 2).

Table 2. List of criteria used for the assessment of the reviewed offset frameworks.

Criterion	Description	Discussion and Recommendations
Offset goal	Setting a target outcome (i.e. no-net-loss) and requirements for demonstrating achievement of biodiversity goal	Offset framework should set specific measureable target goals and goals should be measured against dynamic baseline, incorporating trends. Ideally net-gain, but at least no-net-loss, of biodiversity should be required [32]
Thresholds	Requirements to determine threshold for which biodiversity offset are not acceptable	Offset frameworks should acknowledge there are things that cannot be offset and thus define criteria for when the use offsets is not appropriate and avoidance or minimization should be applied [32]. These criteria could include the irreplaceability of biological resources or the irreversibility of the impacts [34]
Offset currency	Metrics for measuring biodiversity	Offset valuation should use multiple or compound metrics and incorporate measure of ecological function as well as biodiversity [34]
Equivalence	Requiring equivalence between biodiversity losses and gains	Offset should not allow 'out of kind' trading unless this involves 'trading up' from losses that have little or no conservation value. Adherence to the "like-for-like or better" principle is recommended [10]
Offset timing	Deciding in which moment offsets should be implemented	Ideally, offsets should be implemented in advance of the project so that their benefits are already in place when impacts occur [51,68]
Time lag	Deciding whether an additional offset for the temporal loss is required in case there is a temporal gap between impact & offset gains	There is no way of completely offsetting the possible negative consequences of time lags. However, where offset benefits cannot be delivered prior to impacts it is often recommended that offset value should be discounted to account for temporal loss [56]
Offset longevity	Deciding how long offset schemes should endure	Offsets should last at least as long as the impacts of development and should be adaptively managed for change. Ideally, they should be permanent [32,33]
Uncertainty	Establishing requirements for managing for uncertainties throughout the offset process	Uncertainty may be avoided by implementing offsets in advance. When this proves not feasible increasing offset ratios may minimize uncertainty over offset gains, although the effectiveness of this approach is still being discussed ([51]
Additionality	Ensuring that offset actions result in additional conservation outcomes that would not have occurred without the use of an offset	Ideally all offset actions should seek to provide additionality [32]. Policies should require project developers to demonstrate the gains achieved through offsets.
Link to landscape-level conservation goals	Ensuring offsets benefit broader landscape level conservation goals	Offsets should seek to complement landscape level conservation goals [42]
Offset monitoring	Requiring post implementation monitoring to track progress of projected offset benefits	Offset frameworks should always seek to monitor projected returns for a period long enough to ensure the offset values have reached maturity

Results

Policies Related to Application of the Mitigation Hierarchy

All the surveyed countries have national-level EIA laws or regulations that cover all the habitats present in their territories. In addition some have also developed specific EIA or environmental management regulations for particular types of development, e.g. energy and mining (see Table S1). As Figure 2 shows, most environmental policies related to licensing processes in the reviewed Latin-American countries have been enacted in the last ten years. None of the countries have explicitly established a general goal of no-net-loss or net-gain for the EIA process. Only the general EIA regulations of Chile, Colombia and Mexico specifically mention the complete mitigation hierarchy (avoid, minimize, restore, offset) although none of them explicitly requires adherence to it.

Cumulative/Indirect Impacts and Impact Significance: Several countries make provisions for evaluating strategic development plans under their general EIA regulations (Chile, Decree 40 of 2013; Peru, Supreme Decree 019-2009-MINAM; Mexico, LGEEPA of 1988; Venezuela, Decree 1257 of 1996). Some sector-specific policies require the assessment of impacts from a landscape perspective (see Table S1), but for most the scale of impact assessment is not clearly stated. Only Brazil, Chile and Peru include provisions for assessing indirect impacts as part of their general EIA policies, although Argentina and Colombia add

that requirement in some of their sectorial policies (roads and hydrocarbons, respectively). Cumulative impacts are required in all EIAs in all countries except for Argentina and Chile. Although Argentina includes assessment of cumulative impacts under Law 26331 of 2007 on native forests and some sector-specific policies. When it comes to how to evaluate impact significance Argentina, Brazil, Colombia and Peru provide some guidance, although only Colombia and Peru include that in their general EIA policies. In most cases, this guidance consists of a list of environmental assets that should be tackled in impact evaluation (e.g. soils, wildlife), or a list of impact characteristics that should be evaluated (e.g. positive/negative, medium/long term). However, more detailed guidance can be found in some sector-specific regulations, especially in the case of Argentina (see Table S2 for details).

Avoidance: Our results indicate that environmental licensing provisions targeted at the hydrocarbon sector have the strongest requirements for avoidance of impacts followed by provisions targeted at all energy-related development. These sectorial policies frequently include guidance and recommend activities to avoid impacts, although these requirements vary greatly among countries. The rest of sectorial policies do not seem strong regarding avoidance (Figure 3), and in some countries specific policies for certain sectors have not been found (Table S1). Only in two cases (Resolution 1604/2007 on environmental assessment and management for road projects in Argentina, and Resolution 1288 of 2006 on the TOR for EIS of electric lines in Colombia) do laws clearly state that avoidance shall be prioritized over all other forms

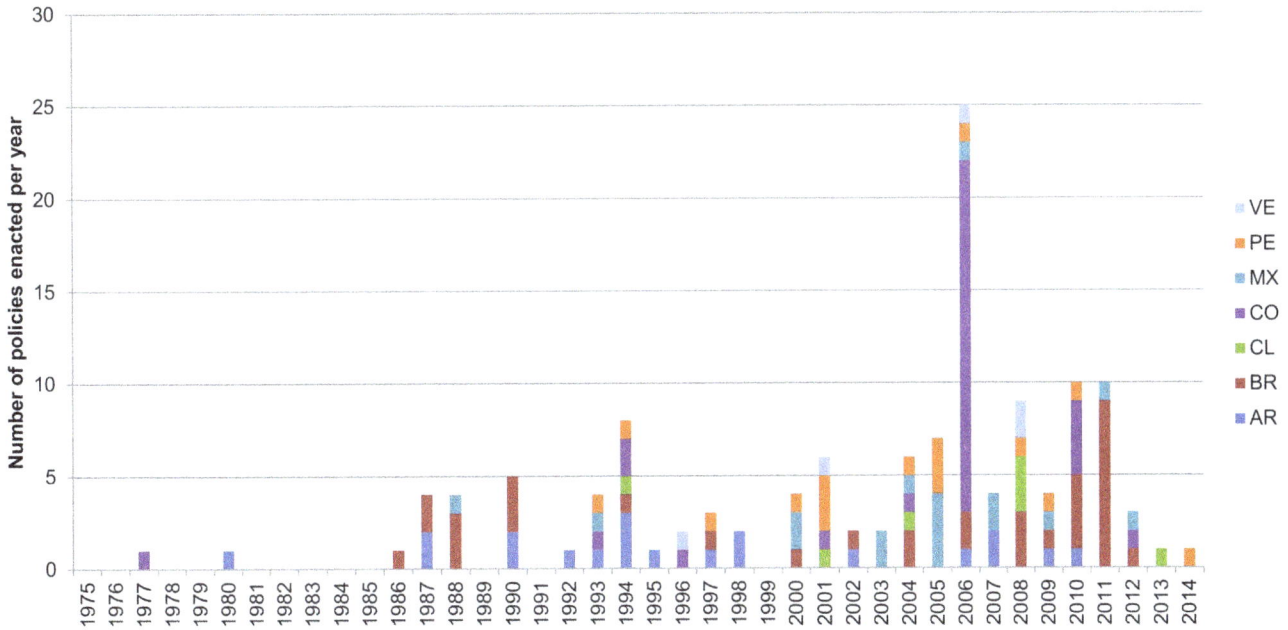

Figure 2. Timeline of the policies included in the study. The graphic represents the number of policies related to the environmental licensing system enacted per year on each of the studied countries. Revoked policies have not been included. Countries' names have been abbreviated to the codes set by ISO 3166.

of mitigation (See Table S3 for details). Apart from this, most provisions related to impact avoidance are found in habitat-specific or protected areas policies, which establish general thresholds for what can or cannot be done in certain habitats (such as wetlands) or in proximity to protected areas.

Minimization and Restoration: Similar to avoidance it is laws directed at the energy sector that includes the highest percentage of provisions regarding minimization and restoration. For most countries no provisions for minimization or restoration requirements are found in the other sectors. Many of the provisions that

refer to minimization or restoration in the general environmental licensing process occur in reference to habitat-specific documents. Most commonly, those policies set a list of environmental assets that shall be restored if negatively impacted. While a few laws make specific recommendations for certain projects it is typically in reference to how those activities shall be carried out or establish performance standards to be met (e.g. survival rates for reforestation activities). See Table S4 for details.

Offsets: Although most countries enable the use of offsets only Brazil, Colombia, Mexico and Peru explicitly require their

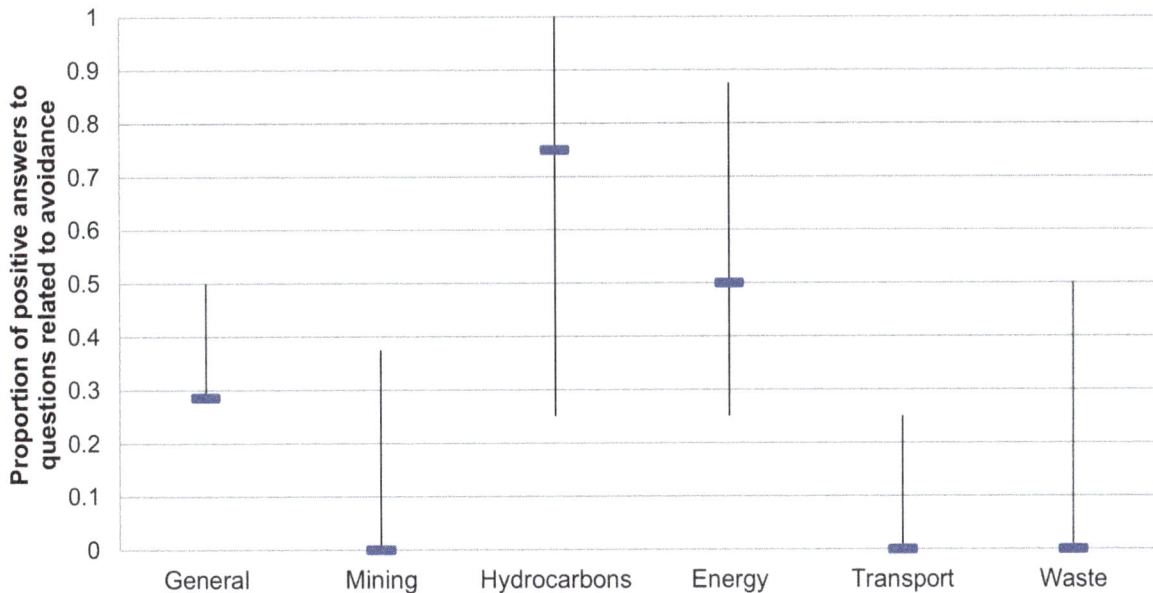

Figure 3. Median and standard-deviation of avoidance provisions in current sector-level policies.

implementation for specific impacts. In Chile some basic provisions are established in the national EIA regulation (see Decree 40 of 2013), although more specific guidance is being developed by the Ministry of the Environment. Detailed guidance is provided by Brazil, Colombia, Mexico and Peru all of which have specific regulations regarding offsets. These countries all include regulations that are already implemented or in the case of Peru are about to be passed into law. None of these regulations is sector-specific. While Brazilian and Mexican policies are aimed at impacts to specific natural assets, Colombia and Peru have a broader scope. For more details see section on offsets below.

Monitoring: While all countries require the use of EIA and many have requirements that emphasize the use of offsets few have explicit language requiring monitoring of development impacts and mitigation activities. Some countries add provisions specific to particular sectors explicitly requiring post-project monitoring, but such information is lacking in most general-scoped EIA policies. Several of the documents that make provisions for monitoring require specific activities to be included in the plan (schedule, indicators, human resources, etc.). Some policies state when the monitoring activities should be performed (e.g. construction and closure phases), but only three documents were found to set the duration of monitoring activities, which ranged from 3 to 10 years after the completion of the project or implementation of the mitigation activities (See Table S5 for details).

Offset Specific Review

Overview by country. Brazilian and Mexican schemes are the first for which specific offset policies were enacted and in turn include a relatively high number of policy documents (especially in the Brazilian case). In marked contrast the Colombian and Peruvian frameworks are recent, and have few policy related documents (See Table 1 and Figure 4). Here we consider the aspects of country-level offset policies highlighting aspects that promote conservation outcomes. Table 3 summarizes the results that are described below with more detail.

In Brazil all projects subject to EIA can utilize offsets and those EIAs shown to cause negative impacts on protected areas must implement offsets according to the scheme set by Law 9985. The effective implementation of the offsets can be carried out either by the developer (Normative Instruction 20 of 2011, article 11) or by the agency responsible for managing the protected area (ICMBio [Chico Mendes Institute for Biodiversity Conservation] in the case of Federal Protected Areas). Offsets will always be aimed at supporting conservation units of the National System of Protected Areas (SNUC, by its Brazilian acronym; Law 9985). During the time this system has been operational, it has generated over US$200 million to be invested in protected areas (Gustavo Pinheiro, personal communication).

In Colombia offsets are required for all projects subject to EIA that cause significant impacts on terrestrial ecosystems (Resolution 1517 of 2012, second article). The developer of the project is responsible for implementing the offsets, although the location is decided by the National Environmental License Authority (ANLA) in accordance with the provisions set in the regulation [35]. The newly enacted framework provides guidance for offset design and includes a series of rules developed for selecting offset sites that meet the conservation needs of potentially impacted biological targets (i.e. size, condition, landscape context) as well as rules for impacts to offset ratio determinations based on a structured and transparent approach [36]. Offsets can either benefit the National System of Protected Areas (SINAP) or be independent of it [35,36].

In Mexico offsets are always required for land-use change in forest areas (Ley General de Desarrollo Forestal Sustentable 2003). The agent responsible for offset implementation is the National Forest Commission (CONAFOR by its Spanish acronym) (Reglamento de la Ley General de Desarrollo Forestal Sustentable 2005), which decides the allocation of offset funds in projects implemented by different entities (agrarian communities, land owners, public administrations, research and education institutions and NGOs among others). There are no requirements for

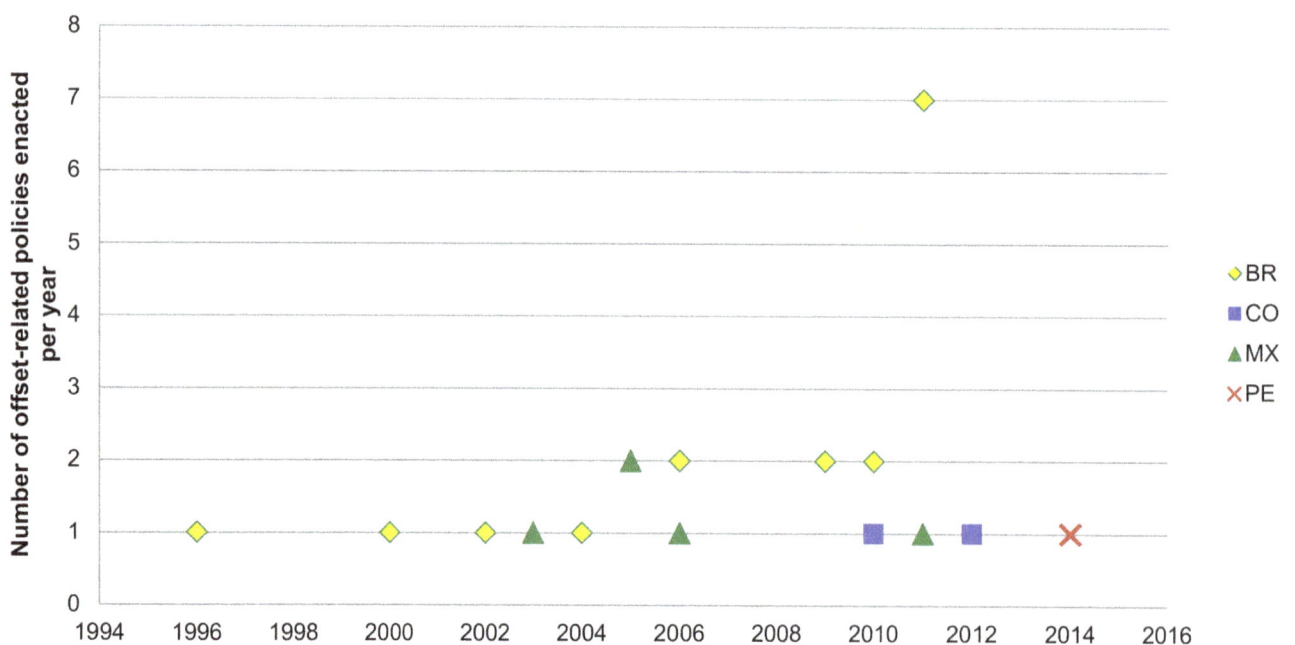

Figure 4. Number of policies related to each country's offset framework issued per year. Includes both current and revoked policies. Countries' names have been abbreviated to the codes set by ISO 3166.

Table 3. Summary of results of the review of offset frameworks by country.

	Brazil	Colombia	Mexico	Peru
Offset goal	Balancing project impact on protected areas with equivalent gains on the SNUC	Biodiversity no-net-loss	Balancing land-use change of forests with equivalent forest gains	Biodiversity no-net-loss or net-gain
Offset currency	Finance-based	Area	Forest area and restoration cost	Area
Equivalence	Does not prioritize in-kind	In-kind	Since the money goes into a fund, equivalency is supposed but not monitored	In-kind
Offset timing	Payment shall be made within 10 days from the date the ToC are signed. Direct implementation by the developer shall be done within 120 days from that date (deadline can be extended)			When environmental license is approved
Time lag				Allows for the use of CBs to reduce losses due to time lags
Offsets longevity	Considered permanent, as they benefit the SNUC	For the length of the project		For the duration of impacts
Uncertainty				Allows for the use of CBs to reduce uncertainty
Thresholds		Sets exclusion areas		
Additionality				Requires demonstrable gains
Link to landscape-level conservation goals	Linked to the SNUC	Coordinates with country's conservation portfolio		Foresees coordination with national conservation priorities
Offset monitoring		Requires comparing results against base line		
Transparency	Offset projects and license applications shall be made public	A public register of offset places will be set		A public register of offset places will be set

SNUC: National System of Protected Areas in Brazil; ToC: Terms of Commitment; CBs: conservation banks.

integrating offset activities into broader conservation priorities, and payment to the Mexican Forest Fund is the only tool enabled for developers to comply with the legal requirements regarding offsets (Ley General de Desarrollo Forestal Sustentable 2003). In 2013, approximately US$ 30 million have been allocated in reforestation projects related to this offset scheme [37].

In Peru most projects subject to EIA would be covered under the new law, although this is subject to the discretion of the Ministry of Environment and EIAs can be exempt from inclusion. The proposed law establishes that the developer be responsible for implementing the required offsets. Offsets are not required to be integrated into existing conservation priorities. While the law enables the developer to directly implement offsets, it also makes provisions for the creation of conservation banks.

Overview by Offset Criteria. Our review of the key offset criteria (Offset Goal, Offset Currency, Equivalence, Offset Timing, Time lag, Offset longevity, Uncertainty, Thresholds, Additionality, Linking offsets to Landscape-level conservation goals, Monitoring) suggests that relative to the idealized form of the regulations there are both situations when criteria appear to conform and many opportunities where regulations can be improved.

Offset Goal: not all the reviewed frameworks explicitly state the objective of compensatory mitigation, and only Peru and Colombia set no-net-loss and net-gain of biodiversity as goals for offsets. To ensure they meet these goals it will be necessary to include a framework to adjust impact to offsets ratios (see "Offset currency" subsection below).

Offset Currency: acreage seems to be the most common currency for calculating the equivalence between impacts and offsets. None of the reviewed Latin-American frameworks incorporates ecological function (e.g. carbon storage, water purification) of either the impacted sites or of the offset sites as part of the valuation process. In the case of Colombia, a set of acreage ratios (from 1:4 to 1:10) has been developed according to the national significance of the impacted ecosystems [36]. Regulatory guidance on this issue has yet to be developed for Peru. While not driven by a goal of no-net-loss in Mexico, the Agreement of 2005 establishes a set of acreage ratios (from 1:1.3 to 1:6) that are calculated according to eight criteria: ecosystem type, degree of conservation, presence of endangered or threatened species, affected ecosystem services, proximity to protected areas, project characteristics (= how its design affects the area), degree to which soil and vegetation resources are affected, and benefits the project will bring to the area (environmental, or social). These ratios are used to calculate the payment that the developer must make to the Forest Fund (Agreement of 2011).

Equivalence: in-kind offsets are explicitly prioritized in the Colombian and Peruvian frameworks, and each country uses a different method for calculating the equivalency between impacts and offsets. In the case of Colombia, offsets are required to match impacted ecosystems [36]. The new law in Peru does not include a detailed calculation system but defines a list of variables (e.g. type of habitat impacted, priority areas for conservation and ecosystem services,) that will have to be considered when selecting area to be used for offsets(see Annex II of the Law). The Brazilian approach includes calculations of how much money developers shall put towards offsets depending on the significance of the impacts resulting from the development activities (Decree 4340 of 2002, Decree 6848 of 2009), but there is no way of assessing the equivalence between impacted assets and offset measures.

Offset Timing: provisions regarding when offsets are implemented are not present in all the reviewed frameworks, and when they are they tend to be somewhat ambiguous. In Brazil, time limits for proving offset implementation or payment are conditioned to the signed terms of commitment. But that date is not clear so there is no way of knowing if offsets are implemented before or after the project impacts occur. In Peru offsets shall be implemented when environmental license is approved, which probably means before the project impacts occur, although this is not completely clear.

Time lag: none of the reviewed Latin-American frameworks includes provisions regarding time lags, although in the case of Peru the implementation of a national network of conservation banks, where offsets credits are generated before impacts are incurred, could help address this issue.

Offsets longevity: both Colombia and Peru clearly state the minimum duration of offsets. In Colombia offsets should last at least for the length of the project, while in Peru they are required to match the duration of the impacts. Although the topic is not specifically addressed in Brazil, benefits can be considered permanent since they benefit the National System of Protected Areas.

Uncertainty: none of the reviewed Latin-American offset frameworks includes provisions regarding uncertainty, although in the case of Peru the implementation of a national network of conservation banks would help address this issue.

Thresholds: only Colombia has set clear limits as to what can be offset.

Additionality: only Peru explicitly requires demonstrable gains that ensure that the offsetting process results in additional conservation outcomes.

Linking offsets to landscape-level conservation goals: Colombia is the country that clearly establishes the link between offsets and broader conservation plans, although Brazil also requires measures to benefit the Protected Areas System. Peru includes some broad guidance on this issue, although specifics have not been set yet, probably because they will be developed in forthcoming regulatory guidance.

Monitoring: provisions regarding post-implementation monitoring of offsets are scarce, and only Colombia requires results to be compared against the area ecological baseline. In Mexico, the CONAFOR is responsible for supervising offset project implementation and that they meet the agreed terms. None of the frameworks establishes specific time requirements for monitoring.

Discussion

Previous studies by Tanaka in 2010 [38] and The Biodiversity Consultancy in 2013 [39] identified 56 countries in the world that have or are developing national legislation or policies around offsets. In Latin America five territories required the use of offsets (Brazil, Colombia, French Guiana, Mexico and Paraguay), nine enabled the use of this tool (Argentina, Bolivia, Costa Rica, Cuba, Guyana, Honduras, Nicaragua, Panama, Uruguay and Venezuela), and three were developing policies related to offsets (Bahamas, Chile, Belize and Peru).

Our review of environmental licensing and offset policy frameworks in Argentina, Brazil, Chile, Colombia, Mexico, Peru and Venezuela shows that these systems have been evolving recently (see Figure 2), although not yet adequate in all cases. We have found significant variation on how EIAs are utilized, the importance of adhering to the mitigation hierarchy, offset goals and approaches for addressing key challenges to implementing offsets. Despite this divergence most countries have a sound foundation from which to develop policy for biodiversity offsets, but several issues require further guidance, including how best to: (1) ensure conformance with the mitigation hierarchy; (2) identify the most environmentally preferable offsets within a landscape context; (3) determine appropriate mitigation replacement ratios to ensure that biodiversity losses and gains are equivalent; and (4) ensure appropriate time and effort is given to monitor offsets performance.

EIA frameworks

The ability of an offset framework to deliver conservation outcomes for biodiversity depends heavily on the existence of a strong Environmental Impact Assessment process. This is because the EIA process is key to ensure that all significant impacts to biodiversity are accounted for and balanced through the application of the mitigation hierarchy [40]. When the EIA process is weak or lacking, offsets may fail to deliver potential value. Our review revealed several important flaws in the EIA process. For example, some of the surveyed countries do not require all EIAs to consider indirect impacts. Indirect impacts are impacts on the environment, which are not a direct result of the project, often produced away from or as a result of a complex pathway. In the case of building a new road, for example, they not only include environmental pressure exerted by the road itself (impacts on vegetation, wildlife and the physical environment etc.), but also the land occupied by producers of road construction materials e.g. mining operations providing the road base materials. These impacts are generally off-site, and may even occur a great distance away from the direct impacts of development. But failure to consider indirect impacts underestimates environmental impacts and can obviously undermine any attempt to achieve a goal of no-net-loss [41]. Most countries also fail to incorporate the mitigation hierarchy as part of the EIA process. Grounding decisions squarely in the mitigation hierarchy will ensure that offset usage conforms to necessary conservation outcomes [42,43]. Only the EIA regulations of Chile, Colombia and Mexico properly reference the complete mitigation hierarchy (avoid, minimize, restore, offset) although none of them explicitly requires adherence to it.

Mitigation hierarchy

Offset frameworks clearly need to emphasize the importance of the mitigation hierarchy—avoiding and minimizing/restoring impacts before proceeding to compensatory mitigation—without reference to the hierarchy in the EIA process there is little opportunity to ensure projects conform to it. Most guidance tends to focus on avoiding impacts to "difficult-to-replace" and "high significance" resources, but ultimately provides wide discretion to regulatory authorities on decisions about when to avoid, minimize, or offset [33,42,44]. Our review has shown that while advancing quite detailed offset policies, countries do not seem to have strong requirements regarding impact avoidance. Avoidance requirements found in environmental licensing policies were not very strong according to our survey (see Figure 3), and it has only been found to be explicitly required in two of the offset frameworks reviewed (Colombia and Peru). Several authors have suggested that if not implemented according to the mitigation hierarchy and a set of standards, the expanded use of biodiversity offsets could provide a "license to trash", allowing development in areas where impacts should have been avoided or more effectively minimized [33,45,46]. We propose that the strengthening of avoidance requirements in mitigation frameworks will require the inclusion of explicit statements requiring adherence to the mitigation hierarchy and prioritization of avoidance measures in policies related to the environmental licensing process (both general and sectorial). Additionally, offset policies should also address this issue from the perspective of avoiding impacts on elements that cannot be replaced or for impacts that are themselves irreversible [32,34,47]. Such guidance should focus on common species as well as rare and species at risk of imminent extinction as proposed by Regnery et al. [48]. Guidance should also incorporate science-based criteria, irreplaceability and vulnerability, examined through a systematic conservation planning framework as put forward by Kiesecker et al. [42]. Latin American countries are not alone in their lack of strong policy and regulation related to avoidance. There is broad agreement among scholars, scientists, policymakers, and regulators that in most mitigation frameworks the first and most important step in the mitigation hierarchy, avoidance, is ignored more often than it is implemented [33,44].

The No Net Loss goal

Offsets are intended as the last option for addressing environmental impacts of development after efforts have been undertaken to minimize impacts on-site through application of the other steps of the mitigation hierarchy: avoid, minimize, restore [6]. They seek to ensure that inevitable negative environmental impacts of development are balanced by environmental gains, with the overall aim of achieving a net neutral or positive outcome [33,49]. As a goal, no-net-loss or net-positive-impact provides a benchmark against which the scope and effectiveness of mitigation actions can be measured. Without a goal, mitigation is simply a collection of actions; there is no clear basis for assessing which actions are more important to take (to achieve what?) or how much is enough. Impact and offset accounting will matter greatly in evaluating a project's progress toward its goal. It is worth noting that no-net-loss accounting is not an entirely new frontier: the principles underpinning mitigation accounting are similar to those developed for greenhouse gas emissions accounting (see for example "net positive climate impacts" [50]). The goal of a mitigation framework should be the first thing to be clearly set by the policies that regulate it [51]. Our results show that most mitigation policies do not define their environmental goals, and only in the cases of Colombia and Peru are these goals clearly defined in their offset policies. This lack of information about policy goals has also been noted for other countries and other environmental regulations [52]. This is a major subject to be addressed in future policy development. Only when goals are clearly defined can mitigation measures be properly designed, and progress evaluated.

Offsets timing and habitat banks

While there remain many offset accounting challenges that need to be addressed e.g. timing and permanence of offsets, significant progress is being made driven by science and practice [53–55]. One of the most effective ways of avoiding these problems is to implement offsets in advance so that they deliver conservation benefits before the impacts occur. However, provisions regarding when offsets shall be implemented are not present in many of the reviewed frameworks, and when they are, they are not clearly stated. Adding a clear requirement for implementing offsets in advance of project impacts should be a priority for future policy updates in all the countries reviewed. However, impact prediction may not be accurate, and offsets that were implemented in advance may have to be adjusted as real impacts are evaluated in the field. Adaptive management will play a key role in this regard (see subsection about monitoring). From the business perspective, delivering offset benefits before impacts occur may be impractical under some circumstances, as they require long time to be fully established. Business objectives are also subject to change as markets fluctuate making detailed development plans challenging to assess proactively. But where offsets are implemented after project work begins it will be important to minimize losses due to time lags. Sometimes the use of multipliers (e.g. increasing the size of the offset) has been proposed to balance the losses due to time lags [56,57]. However, recent research suggests that this approach does not guarantee against the shortages triggered by temporal delays that can threaten the achievement of meaningful offset gains [54,55]. Habitat banks (also called 'biodiversity banks', 'conservation banks' or, in the US, 'mitigation banks') may help reduce uncertainty and the need to consider time lags because they provide the opportunity to implement anticipated offsets: by the time a credit is bought the offset activities it accounts for have long been implemented. Habitat banks also provide advantages to on-site and small parcel mitigation. By consolidating necessary services to create, maintain, and monitor, habitat banks are able to provide services at a lower cost [51,58]. Because habitat bank credits are created prior to impacts, purchasing credits from a habitat bank decreases permitting time [59]. The cost of achieving a certain level of performance and duration is often lower than other offsets options and regulatory burden and risk is passed from developer to habitat bank. We propose that habitat banking can help implement offsets and provide positive conservation outcomes that may not have been achieved otherwise [60]. For example, buying habitat banking credits is sometimes the only feasible offsetting option for small companies which have no capacity to carry out offset projects by themselves. More importantly, habitat banks aggregate multiple offset activities into few, larger projects, which are more likely to deliver conservation outcomes [61]. Such aggregation would probably be harder to achieve through other means. However, of all the reviewed countries only the new offset law in Peru allows for the use of habitat banking (called 'conservation banking'). Incorporation of this tool into existing mitigation frameworks may improve the implementation of offsets and gains for conservation.

Landscape scale

Historically mitigation has occurred primarily in a reactive fashion at small spatial scales on a site-by-site basis but there is general consensus among research and practitioners that mitigation should be a more comprehensive approach that considers whole systems, anticipates impacts, and recommends effective actions to keep our natural systems healthy [42]. Integrating mitigation at a landscape scale moves beyond a project-by-project approach to one that can support a dynamic vision consistent with broader conservation goals. A landscape vision is essential because it ensures that the biologically and ecologically important features remain essential throughout the process. Without this vision, the sight of the overarching conservation targets is lost, establishing priorities becomes difficult, and limited resources may be squandered. In this sense, the Colombian and Peruvian offset frameworks are progressive, as they have been developed from a landscape conservation perspective. These frameworks also require offsets for impacts to all natural ecological systems. Compare this to the use of offsets in the United States, where offsets are typically only used to address impacts to wetlands and for threatened species. These new frameworks in Colombia and Peru can serve as an example to be followed by future offset policies not only in Latin America but globally.

Moving forward, we hope that offset frameworks develop guidelines that prompt practitioners to think strategically about offset site selection, and to develop practical guidelines for how to select offset sites. Site selection for offsets should be an exercise in landscape ecology. Using quantitative site selection tools [62,63], or blending this process as part of landscape level conservation plans, to provide a transparent, flexible and rule-based approach towards guiding site selection. Moreover, if political pressures constrain practitioners to a particular political extent, quantitative site selection tools will allow them to assess if meeting goals are possible given those constraints [49]. When it comes to offsets, failure to systematically select suitable sites could reduce the potential benefits for conservation.

Monitoring

Post-implementation monitoring should be a key component of every mitigation framework. Monitoring is a way of ensuring compliance with policy requirements, evaluating the achievement of the mitigation goals, and getting feedback on the effectiveness of the activities implemented [34,64]. It is also the primary driver of adaptive management, a necessary procedure for getting long-term conservation outcomes [33,46,65]. In some way, it is also an essential component for transparency of the process, since the public does not only need to know which activities are proposed and how mitigation funds are allocated (information that many of the reviewed countries already provide for offsets), but also if and how such actions are carried out. However, the lack of post-implementation monitoring is a common problem in mitigation and conservation projects in general [44,66]. Even when follow-up programs are required, they are often required for a short period, and because of the short temporal scale problems with offset implementation frequently go undetected [54]. Many of the countries in our survey lack provisions that guide the monitoring of impacts and mitigation measures, and the few cases that do require monitoring typically require short monitoring periods. Lack of enforcement of environmental policies related to offsets is a common problem [58,67]. The requirement of solid monitoring processes is the first step to address these issues and will need to be key component of any mitigation policies if they are to promote sustainable development.

Conclusion

Our results indicated that all the surveyed countries have national-level Environmental Impact Assessment laws or regulations and most enable the use of offsets but only Brazil, Colombia, Mexico and Peru explicitly require their implementation. While several countries may have quite detailed offset policies, most countries do not seem to have strong requirements regarding impact avoidance which could undermine the use of offsets. While the most recent frameworks (those from Colombia and Peru) show more adherence to the theoretical recommendations we outlined there are still some principles that have not been included in most country level frameworks. In some cases, this may be due to the lack of scientific agreement on how to address certain issues in practice. To ensure that the use of offsets advances biodiversity conservation going forward it will be necessary to develop further guidance on how best to: (1) ensure conformance with the mitigation hierarchy; (2) identify the most environmentally preferable offsets within a landscape context; (3) determine appropriate mitigation replacement ratios; and (4) ensure appropriate time and effort is given to monitor offsets performance. Despite these shortcomings most countries have a strong foundation from which to develop policy for biodiversity offsets.

In addition to these issues the Business and Biodiversity Offsets Program, by far the largest multi-stakeholder effort to examine biodiversity offsets, stresses the importance to ensure that offsets involve stakeholder participation, the fair and equitable distribution of offsets benefits and use of traditional knowledge in offset design. While we agree these are important issues they were not included in our analysis given our focus on the theoretical scientific issues involved in offset design.

Although policies and regulatory guidance alone will not deliver conservation value without regulatory oversight and implementation capacity. The effectiveness of an offset program demands a responsible administrative entity with firm requirements for adequate oversight, performance accountability, and process transparency and fairness. Achieving these objectives requires several administrative functions, including: 1) communication and maintenance of standards and protocols; 2) application of standards to individual projects to analyze impacts and determine needs for mitigation; 3) coordination and oversight of mitigation planning to target mitigation funding toward projects with high conservation return on investment; 4) oversight of mitigation funds to ensure appropriate fiduciary management and impartial allocation; 5) a process that utilizes monitoring and provides a mechanism to adjust activities based on monitoring results; and 6) procedures for sanctions against failure to achieve legal requirements to make sure that laws are effectively implemented. An independent third-party entity that oversees these functions will be essential.

Acknowledgments

We thank Linda Krueger, Jessica Wilkinson, Lila Gil, Luis Alberto Gonzales, Oscar Castillo, Jose Yunis, Gustavo Iglesias, Gustavo Pinheiro, Karen Oliveira, Laurel Mayer, Mark Weisshaar, Jacquelin Gutierrez, David Hults, Ed Bloom, Mauricio Trejo, Francisco Solis, and Felipe Osorio for their technical support in different stages of the project. We thank Irene Burgués (Conservation Strategy Fund) and eLaw (particularly Jennifer Glasson) for providing useful resources for the study, Kei Sochi for translation of Japanese documents, and TNC's Development by Design team in Fort Collins for helpful discussions.

Author Contributions

Conceived and designed the experiments: AV AB JK. Performed the experiments: AV. Analyzed the data: AV AB JK. Wrote the paper: AV JK. Approved final version for submission: AV AB JK.

References

1. Alternative Latin Investor (2011) Infrastructure Investment Latin America 2011. Available: http://www.alternativelatininvestor.com/assets/Infrastructure-Investment.pdf. Accessed 2014 Aug 5.

2. Barbero JA (2012) Infrastructure in the Development of Latin America. Bogotá, Colombia. Available: http://publicaciones.caf.com/media/33151/ideal_ingles_feb8.pdf Accessed 2014 Aug 5.

3. Morgan RK (2012) Environmental impact assessment: the state of the art. Impact Assess Proj Apprais 30: 5–14. doi:10.1080/14615517.2012.661557.

4. Lawrence DP (2003) Environmental Impact Assessment: Practical Solutions to Recurrent Problems. Hoboken, New Jersey: Wiley-Interscience.

5. Pritchard D (1993) Towards sustainability in the planning process: the role of EIA. Ecos - A Rev Conserv 14: 3–15.

6. Council on Environmental Quality (2000) Protection of the environment (under the National Environmental Policy Act). Washington, D.C.

7. Business and Biodiversity Offsets Programme (BBOP) (2013) To No Net Loss and Beyond: an overview of the Business and Biodiversity Offsets Programme (BBOP). Washington, D.C.

8. Villarroya A, Puig J (2013) A proposal to improve ecological compensation practice in road and railway projects in Spain. Environ Impact Assess Rev 42: 87–94. doi:10.1016/j.eiar.2012.11.002.

9. Madsen B, Carroll N, Kandy D, Bennett G (2011) 2011 Update: State of Biodiversity Markets. Washington, D.C.

10. International Finance Corporation (2012) Performance Standards on Environmental and Social Sustainability. Available: http://www.ifc.org/wps/wcm/connect/115482804a0255db96fbffd1a5d13d27/PS_English_2012_Full-Document.pdf?MOD=AJPERES. Accessed 2014 Aug 5.

11. European Bank for Reconstruction and Development (2008) Environmental and Social Policy. Available: http://www.ebrd.com/downloads/research/policies/2008policy.pdf. Accessed 2014 Aug 5.

12. Equator Principles (2013) The Equator Principles. Available: http://www.equator-principles.com/resources/equator_principles_III.pdf. Accessed 2014 Aug 5.

13. Rio Tinto (2004) Rio Tinto's Biodiversity Strategy. London, UK & Melbourne, Australia.

14. Barrick (2010) Barrick. Building Value in Everything We Do: Annual Report 2010. Toronto, Canada.

15. Teck Resources Limited (2013) Teck's Biodiversity Strategy. Vancouver, Canada. Available: http://www.tecksustainability.com/sites/base/pages/our-strategy/biodiversity. Accessed 2014 Aug 5.

16. Wood C (2003) Environmental impact assessment: a comparative review. Harlow: Pearson-Prentice Hall.

17. República de Chile (1980) Constitución Política de la República de Chile. Available: http://www.leychile.cl/Navegar?idNorma=242302. Accessed 2014 Aug 5.

18. República de Colombia (1991) Constitución Política de Colombia. Available: http://www.constitucioncolombia.com/indice.php. Accessed 2014 Aug 5.

19. Congreso Constituyente Democrático del Perú (1993) Constitución Política del Perú. Available: http://www.tc.gob.pe/legconperu/constitucion.html. Accessed 2014 Aug 5.

20. Asamblea Constituyente de Venezuela (1999) Constitución de la República Bolivariana de Venezuela. Available: http://www.tsj.gov.ve/legislacion/constitucion1999.htm. Accessed 2014 Aug 5.

21. Congreso Constituyente de los Estados Unidos Mexicanos (1917) Constitución Política de los Estados Unidos Mexicanos con reformas hasta 2014. Available: http://www.diputados.gob.mx/LeyesBiblio/ref/cpeum.htm. Accessed 2014 Aug 5.

22. Secretaría de Desarrollo Urbano y Ecología de México (1988) Ley general del equilibrio ecológico y la protección al ambiente. Available: http://biblioteca.semarnat.gob.mx/janium/Documentos/Ciga/agenda/DOFsr/148.pdf. Accessed 2014 Aug 5.

23. Pinto-Ferreira L (1989) Comentários à Constituição Brasileira. Vol.1. Sao Paulo: Saraiva.

24. McElfish Jr JM (1995) Minimal Stringency: Abdication of State Innovation. Environ Law Report 25: 10003.

25. Richardson N, Gottlieb M, Krupnick A, Wiseman H (2013) The State of State Shale Gas Regulation. Available: http://www.rff.org/rff/documents/RFF-Rpt-StateofStateRegs_Report.pdf. Accessed 2014 Aug 5.

26. Doswald N, Barcellos-Harris M, Jones M, Pilla E, Mulder I (2012) Biodiversity offsets: voluntary and compliance regimes. A review of existing schemes, initiatives and guidance for financial institutions. Cambridge, UK.

27. Glasson J, Salvador N (2000) EIA in Brazil: a procedures–practice gap. A comparative study with reference to the European Union, and especially the UK. Environ Impact Assess Rev 20: 191–225. doi:10.1016/S0195-9255(99)00043-8.

28. Espinoza G, Alzina V (2001) Review of Environmental Impact Assessment in Selected Countries of Latin America and the Caribbean. Methodology, Results, and Trends. Espinoza G, Alzina V, editors Santiago de Chile: Inter-American Development Bank (IDB) - Center for Development Studies.

29. Astorga A (2006) Estudio Comparativo de los Sistemas de Evaluación de Impacto Ambiental en Centroamérica. San Jose, Costa Rica. Available: http://ceur.usac.edu.gt/Biocombustibles/27_Estudio_comparativo_de_Sistemas_de_Evaluacion_de_Impacto_Ambiental_en_CA.pdf. Accessed 2014 Aug 5.

30. Sanchez-Triana E, Enriquez S (2007) A Comparative Analysis of Environmental Impact Analysis Systems in Latin America. Annual Conference of the International Association for Impact Assessment. p. 100.

31. Saborio-Coze A, Flores-Nava A (2009) Review of environmental impact assessment and monitoring of aquaculture in Latin America. Rome.

32. Business and Biodiversity Offsets Programme (BBOP) (2012) Standard on Biodiversity Offsets. Washington, D.C. Available: http://www.forest-trends.org/documents/files/doc_3078.pdf. Accessed 2014 Aug 5.

33. McKenney BA, Kiesecker JM (2010) Policy development for biodiversity offsets: a review of offset frameworks. Environ Manage 45: 165–176. doi:10.1007/s00267-009-9396-3.

34. Bull JW, Suttle KB, Gordon A, Singh NJ, Milner-Gulland EJ (2013) Biodiversity offsets in theory and practice. Oryx 47: 369–380. doi:10.1017/S003060531200172X.

35. Saenz S, Walschburger T, González JC, León J, McKenney B, et al. (2013) Development by Design in Colombia: Making Mitigation Decisions Consistent with Conservation Outcomes. PLoS One 8: e81831. doi:10.1371/journal.pone.0081831.

36. Saenz S, Walschburger T, González J, León J, McKenney B, et al. (2013) A Framework for Implementing and Valuing Biodiversity Offsets in Colombia: A Landscape Scale Perspective. Sustainability 5: 4961–4987. doi:10.3390/su5124961.

37. CONAFOR (2013) Programa de Compensación Ambiental por Cambio de Uso del Suelo en terrenos Forestales. Available: http://www.conafor.gob.mx/portal/index.php/tramites-y-servicios/apoyos-2013. Accessed 2013 Dec 20.

38. Tanaka A, Ohtaguro S (2010) Biodiversity offsets that enable strategic ecological restorations - Current situation of institutionalizing biodiversity offset in various countries and its implications to Japan [in Japanese]. City Plan Mag 59: 18–25.

39. The Biodiversity Consultancy (2013) Government policies on biodiversity offsets. Available: http://www.thebiodiversityconsultancy.com/wp-content/uploads/2013/07/Government-policies-on-biodiversity-offsets1.pdf. Accessed 2014 Aug 4.

40. Jesus J de (2013) Mitigation in Impact Assessment. IAIA FasTips No.6. Available: http://iaia.org/PublicDocuments/special-publications/Fastips_620Mitigation.pdf. Accessed 2014 Aug 5.

41. European Commission (1999) Guidelines for the Assessment of Indirect and Cumulative Impacts as well as Impact Interactions. Brussels. Available: http://ec.europa.eu/environment/eia/eia-studies-and-reports/pdf/guidel.pdf. Accessed 2014 Aug 5.

42. Kiesecker JM, Copeland H, Pocewicz A, McKenney B (2010) Development by design: blending landscape-level planning with the mitigation hierarchy. Front Ecol Environ 8: 261–266. doi:10.1890/090005.

43. Kiesecker JM, Sochi K, Heiner M, McKenney B, Evans JS, et al. (2013) Development by Design: Using a Revisionist History to Guide a Sustainable Future. In: Levin SA, editor. Encyclopedia of Biodiversity. Waltham, MA, U.S.A.: Academic Press. pp. 495–507.

44. Clare S, Krogman N, Foote L, Lemphers N (2011) Where is the avoidance in the implementation of wetland law and policy? Wetl Ecol Manag 19: 165–182. doi:10.1007/s11273-011-9209-3.

45. Walker S, Brower AL, Stephens RTT, Lee WG (2009) Why bartering biodiversity fails. Conserv Lett 2: 149–157. doi:10.1111/j.1755-263X.2009.00061.x.

46. Quétier F, Lavorel S (2011) Assessing ecological equivalence in biodiversity offset schemes: Key issues and solutions. Biol Conserv 144: 2991–2999. doi:10.1016/j.biocon.2011.09.002.

47. Pilgrim JD, Brownlie S, Ekstrom JMM, Gardner TA, von Hase A, et al. (2013) A process for assessing the offsetability of biodiversity impacts. Conserv Lett 6: 376–384. doi:10.1111/conl.12002.

48. Regnery B, Couvet D, Kerbiriou C (2013) Offsets and Conservation of the Species of the EU Habitats and Birds Directives. Conserv Biol 27: 1335–1343. doi:10.1111/cobi.12123.

49. Kiesecker JM, Copeland H, Pocewicz A, Nibbelink N, McKenney B, et al. (2009) A Framework for Implementing Biodiversity Offsets: Selecting Sites and Determining Scale. Bioscience 59: 77–84. doi:10.1525/bio.2009.59.1.11.

50. Climate Community and Biodiversity Alliance (2008) CCB Standards. Available: http://www.climate-standards.org/ccb-standards/. Accessed 2013 Dec 12.

51. Bekessy SA, Wintle BA, Lindenmayer DB, Mccarthy MA, Colyvan M, et al. (2010) The biodiversity bank cannot be a lending bank. Conserv Lett 3: 151–158. doi:10.1111/j.1755-263X.2010.00110.x.

52. Brownlie S, Botha M (2009) Biodiversity offsets: adding to the conservation estate, or "no net loss"? Impact Assess Proj Apprais 27: 227–231. doi:10.3152/146155109X465968.

53. Moilanen A, van Teeffelen AJA, Ben-Haim Y, Ferrier S (2009) How Much Compensation is Enough? A Framework for Incorporating Uncertainty and Time Discounting When Calculating Offset Ratios for Impacted Habitat. Restor Ecol 17: 470–478. doi:10.1111/j.1526-100X.2008.00382.x.

54. Maron M, Hobbs RJ, Moilanen A, Matthews JW, Christie K, et al. (2012) Faustian bargains? Restoration realities in the context of biodiversity offset policies. Biol Conserv 155: 141–148. doi:10.1016/j.biocon.2012.06.003.

55. Gardner TA, Von Hase A, Brownlie S, Ekstrom JMM, Pilgrim JD, et al. (2013) Biodiversity Offsets and the Challenge of Achieving No Net Loss. Conserv Biol 27: 1254–1264. doi:10.1111/cobi.12118.

56. Business and Biodiversity Offsets Programme (BBOP) (2012) Guidance Notes to the Standard on Biodiversity Offsets. Washington, D.C. Available: http://www.forest-trends.org/documents/files/doc_3099.pdf. Accessed 2014 Aug 5.

57. Overton JM, Stephens RTT, Ferrier S (2013) Net present biodiversity value and the design of biodiversity offsets. Ambio 42: 100–110. doi:10.1007/s13280-012-0342-x.

58. Norton DA (2009) Biodiversity offsets: two New Zealand case studies and an assessment framework. Environ Manage 43: 698–706. doi:10.1007/s00267-008-9192-5.

59. Latimer W, Hill D (2007) Mitigation banking: Securing no net loss to biodiversity? A UK perspective. Plan Pract Res 22: 155–175. doi:10.1080/02697450701584337.

60. Wende W, Herberg A, Herzberg A (2005) Mitigation banking and compensation pools: improving the effectiveness of impact mitigation regulation in project planning procedures. Impact Assess Proj Apprais 23: 101–111. doi:10.3152/147154605781765652.

61. Dickie I, Tucker G (2010) The use of market-based instruments for biodiversity protection - the case of habitat banking. Technical report. Available: http://www.ieep.eu/assets/472/eftec_habitat_banking_technical_report.pdf. Accessed 2014 Aug 5.

62. Arponen A, Kondelin H, Moilanen A (2007) Area-based refinement for selection of reserve sites with the benefit function approach. Conserv Biol 21: 527–533.

63. Moilanen A (2013) Planning impact avoidance and biodiversity offsetting using software for spatial conservation prioritisation. Wildl Res 40: 153. doi:10.1071/WR12083.

64. Hayes N, Morrison-Saunders A (2007) Effectiveness of environmental offsets in environmental impact assessment: practitioner perspectives from Western Australia. Impact Assess Proj Apprais 25: 209–218. doi:10.3152/146155107X227126.

65. Hilderbrand R, Watts A, Randle A (2005) The myths of restoration ecology. Ecol Soc 10: 19.

66. Ferraro PJ, Pattanayak SK (2006) Money for Nothing? A Call for Empirical Evaluation of Biodiversity Conservation Investments. PLoS Biol 4: e105. doi:10.1371/journal.pbio.0040105.

67. Burgin S (2008) BioBanking: an environmental scientist's view of the role of biodiversity banking offsets in conservation. Biodivers Conserv 17: 807–816. doi:10.1007/s10531-008-9319-2.

68. Morris RKA, Alonso I, Jefferson RG, Kirby KJ (2006) The creation of compensatory habitat—Can it secure sustainable development? J Nat Conserv 14: 106–116. doi:10.1016/j.jnc.2006.01.003.

8

Resource-Mediated Indirect Effects of Grassland Management on Arthropod Diversity

Nadja K. Simons[1]*, **Martin M. Gossner**[1], **Thomas M. Lewinsohn**[2], **Steffen Boch**[3], **Markus Lange**[4], **Jörg Müller**[5], **Esther Pašalić**[1], **Stephanie A. Socher**[3], **Manfred Türke**[1], **Markus Fischer**[3,6], **Wolfgang W. Weisser**[1]

1 Terrestrial Ecology Research Group, Department of Ecology and Ecosystem Management, School of Life Sciences Weihenstephan, Technische Universität München, Freising, Germany, **2** Department of Animal Biology, Institute of Biology, University of Campinas, Campinas, Sao Paulo, Brazil, **3** Institute of Plant Sciences, University of Bern, Bern, Switzerland, **4** Max-Planck-Institute for Biogeochemistry, Jena, Germany, **5** Institute of Biochemistry and Biology, University of Potsdam, Potsdam, Germany, **6** Biodiversity and Climate Research Centre, Senckenberg Gesellschaft für Naturforschung, Frankfurt/Main, Germany

Abstract

Intensive land use is a driving force for biodiversity decline in many ecosystems. In semi-natural grasslands, land-use activities such as mowing, grazing and fertilization affect the diversity of plants and arthropods, but the combined effects of different drivers and the chain of effects are largely unknown. In this study we used structural equation modelling to analyse how the arthropod communities in managed grasslands respond to land use and whether these responses are mediated through changes in resource diversity or resource quantity (biomass). Plants were considered resources for herbivores which themselves were considered resources for predators. Plant and arthropod (herbivores and predators) communities were sampled on 141 meadows, pastures and mown pastures within three regions in Germany in 2008 and 2009. Increasing land-use intensity generally increased plant biomass and decreased plant diversity, mainly through increasing fertilization. Herbivore diversity decreased together with plant diversity but showed no response to changes in plant biomass. Hence, land-use effects on herbivore diversity were mediated through resource diversity rather than quantity. Land-use effects on predator diversity were mediated by both herbivore diversity (resource diversity) and herbivore quantity (herbivore biomass), but indirect effects through resource quantity were stronger. Our findings highlight the importance of assessing both direct and indirect effects of land-use intensity and mode on different trophic levels. In addition to the overall effects, there were subtle differences between the different regions, pointing to the importance of regional land-use specificities. Our study underlines the commonly observed strong effect of grassland land use on biodiversity. It also highlights that mechanistic approaches help us to understand how different land-use modes affect biodiversity.

Editor: Christian Rixen, WSL Institute for Snow and Avalanche Research SLF, Switzerland

Funding: The work was funded by the DFG Priority Program 1374 "Infrastructure-Biodiversity-Exploratories" (DFG-WE 3081/21-1.). www.dfg.de/spp/en. Additional funds for exchange visits to analyze data were provided by Unicamp (PRP/Faepex) for TML and MMG and by DAAD (Project TUMBRA) for NKH. www.toek.wzw.tum.de/index.php?id=117. This work was supported by the German Research Foundation (DFG) and the Technische Universität München within the funding programme Open Access Publishing. The funders had no role in study design, data collection and analysis, decision to publish, or preparation of the manuscript.

Competing Interests: The authors have declared that no competing interests exist.

* Email: nadja.simons@tum.de

Introduction

Negative effects of intensive grassland land use on biodiversity have been found for many taxa including plants [1], herbivorous and carnivorous arthropods [2–4], pollinating insects [5] and birds [6]. Despite the growing consensus that intensive land use has generally negative effects on biodiversity [7], the particular mechanisms that lead to a decrease in biodiversity are often not fully understood [8], because land use consists of various modes that can have opposing or additive effects on biodiversity. In semi-natural grasslands important land-use modes are mowing, grazing and fertilization. Several observational and experimental studies found decreasing diversities of plants and arthropods with increasing frequency of mowing events e.g. [4,9–11], with increasing fertilization intensity e.g. [12,13] or with increasing

grazing intensity [14–16]. Whereas effects of land-use modes on plants are often direct, e.g. when mowing hinders seed set of late-flowering plants, effects on higher trophic levels such as insect herbivores or carnivores may be either direct or mediated by changes in the plant community.

Several hypotheses have been proposed to describe the effects of differently diverse plant communities on the diversity and abundance of herbivores and predators. The 'Resource Heterogeneity Hypothesis' (RHH) predicts that more diverse resources provide more niches for a greater number of specialized species at higher trophic levels [17–20], i.e. herbivore diversity is promoted by increased plant diversity and predator diversity increases in response to herbivore diversity – a positive trophic cascade. In contrast to the RHH, the 'More Individuals Hypothesis' (MIH) proposes that diversity of consumers increases when resource quantity increases, i.e. plant biomass for herbivores and herbivore biomass for predators [21]. According to the MIH, this positive effect of resource abundance (or biomass) on consumer diversity is mediated by an increase in total consumer abundance. Borer et al. [22] studied plant diversity effects on arthropod diversity in experimental plant communities and found that more diverse plant communities hosted more diverse herbivore communities, but that this effect was mediated by higher overall plant and herbivore biomass. While the RHH assumes a direct link between resource and consumer diversity, the MIH assumes an indirect link through resource and consumer abundance, i.e. resource abundance increases consumer abundance which in turn increases consumer diversity. Thus, arthropod biomass and diversity may be differently affected by increasing land-use intensity depending on the mechanistic relationships between land use and the herbivore and carnivore communities.

We combined detailed information on grassland land-use modes (fertilization, grazing and mowing) with biomass and species richness data of plants and arthropods to analyse effects of land-use intensity on the arthropod community. Although experimental short-term manipulation of land use can elucidate immediate effects of the land-use modes (such as mowing), it is not clear whether these mechanisms operate similarly when different land-use practices are combined in agricultural grasslands and under long term conditions. This can only be assessed by studying grasslands which have been used as meadows or pastures continuously for several years or decades. In our study, we build on the work of Socher et al. [1] who tested for direct and indirect effects of grassland land use on plant diversity and biomass. We studied effects of land use on the arthropod community including both direct and indirect (via the plant community) chains of effects. The study system includes grasslands which have been managed for at least 20 years and comprise a wide range of land-use intensities. By including three different regions in Germany and by sampling in two consecutive years, we were able to consider the generality of observed patterns.

Based on the two hypotheses mentioned above and previous knowledge on mechanisms of grassland land use, we defined two alternative models. In the first model ('Resource Heterogeneity Model') we tested whether land-use intensity affects herbivore and predator diversity via changes in the diversity of their respective resources (RHH). According to the RHH we expected positive effects of plant diversity on herbivore diversity and of herbivore diversity on predator diversity (Figure 1 A). In the second model ('Resource Abundance Model'), we tested whether effects of land-use intensity affect arthropods through changes in resource quantity (MIH) and added herbivore and predator biomass to the model (Figure 1 B). According to the MIH, we expected plant biomass to have a positive effect on herbivore diversity through herbivore biomass and positive effects of herbivore biomass on predator diversity, respectively.

Within both models (Figure 1 A & B), we expected mowing to decrease plant diversity (by a loss of disturbance-sensitive species) and to decrease plant biomass (through mechanical disturbances during the growing period). Grazing was expected to increase plant diversity (increasing number of niches for plants or preventing competitive exclusion) and decrease plant biomass (recurrent disturbance of plant growth). Another possible plant species response to grazing or mowing is overcompensating for tissue loss leading to the opposite expectation (i.e. increase of plant biomass following grazing or mowing). However, overcompensation has usually been demonstrated for particular species, not at the community level e.g. [23,24], where a decrease of plant biomass is more likely. Fertilization was expected to decrease plant

diversity (dominance of fast-growing species) and increase plant biomass (increased nutrient input). Based on the findings of Socher et al. [1] we expected plant biomass to be negatively correlated with plant diversity.

In our study we asked the following questions: 1) How do land-use modes, singly or in combination, affect arthropod diversity? 2) How are effects of land-use intensity on herbivore diversity mediated by the responses of plants to land use (do they follow the 'Resource Heterogeneity Hypothesis' or the 'More Individuals Hypothesis')? 3) Are the responses of predators to land use governed by the same mechanisms as responses of herbivores?

Materials and Methods

Study sites and land use

The Biodiversity Exploratory research program (www. biodiversity-exploratories.de) was established in 2006 within three regions in Germany: (1) Schorfheide-Chorin (SCH) in north-east Germany (3–140 m a.s.l., 53°02′ N 13°83′ E, annual mean precipitation 500–600 mm, mean temperature 8–8.5°C). (2) Hainich-Dün (HAI) in central Germany (285–550 m a.s.l., 51°20′ N 10°41′ E, 500–800 mm, 6.5–8°C). (3) Swabian Alb (ALB) in south-west Germany (460–860 m a.s.l., 48°43′ N 9°37′ E, 700–1000 mm, 6–7°C). Within each region, 50 plots of 50 m × 50 m size serve as basis for surveys of biodiversity or ecosystem processes. Those plots were chosen from a total of 500 candidate plots in each region on which initial vegetation and land-use surveys were conducted. The 50 plots per region cover the whole regional gradient of land-use modes and intensity [25].

Land-use modes on the studied grasslands include mowing with different frequency (meadows), grazing by different kinds of livestock (cattle and sheep pastures) or both mowing and grazing (mown pastures). Grasslands are either unfertilized or fertilized with different amounts of fertilizer. During the study years all grassland plots continued to be managed by farmers in the same way as the surrounding grasslands. Grassland land use was assessed each year (since 2006) by standardized interviews with farmers and land-owners. From these interviews, we derived information on fertilization, grazing and mowing: Fertilization intensity was calculated as the total amount of nitrogen applied per hectare and year, in the form of organic or chemical fertilizer. For grazing, information on livestock units and the duration of grazing periods were combined as a measure of grazing intensity. Mowing intensity was calculated as the number of mowing events per year.

For our analyses we used land-use information from the two years prior to sampling and the sampling year. Mowing intensity included mowing events in the sampling year only up to the sampling event (e.g. for samples taken in June, mowing events later than June in the sampling year were not included). As mowing not only has long-term but also short-term effects on arthropods (e.g. Humbert et al. [26] showed that mowing with machinery leads to high mortality in Orthoptera), we included the number of days between the arthropod sampling day and the last mowing event prior to the respective sampling as an additional land-use variable ('Time after mowing'). As two arthropod samplings were conducted per year (see below), we used the mean number of days for each year. For a detailed description of land-use intensity calculations and an overview of land-use information see Appendix S1 as well as Table S1 & Table S2.

Plant and arthropod sampling and plot selection

Plant diversity and plant biomass on the plots were assessed between mid-May and mid-June in 2008 & 2009 using vegetation surveys and aboveground biomass harvests following the methods

A Resource Heterogeneity Model

B Resource Abundance Model

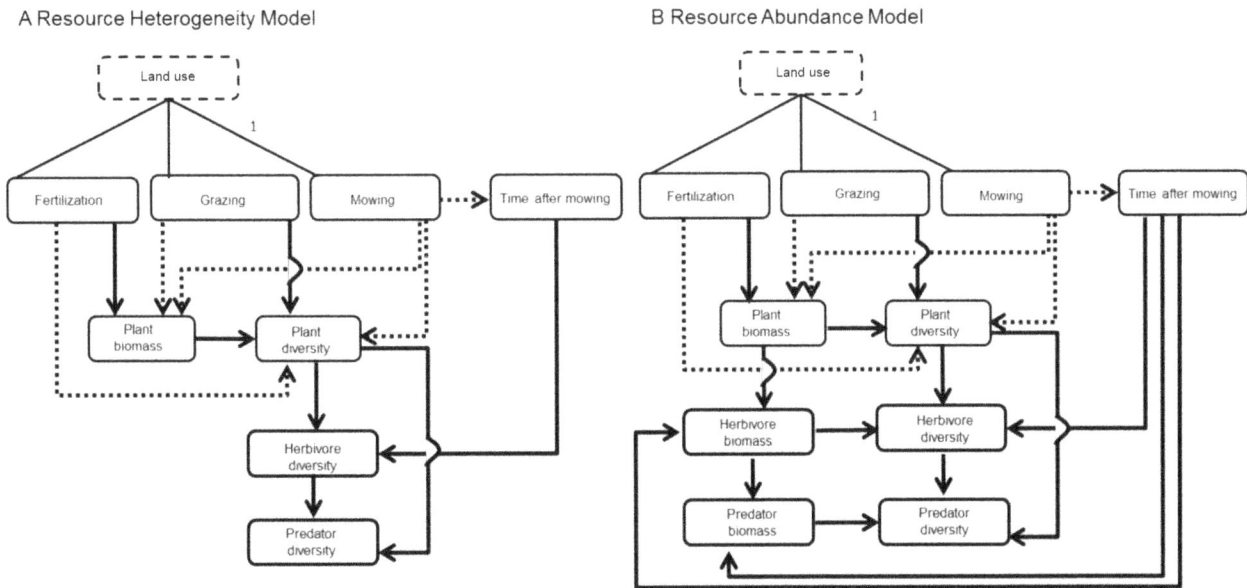

Figure 1. Concepts for structural equation models. 'Land use' is a latent (i.e. unmeasured) variable influencing the management components. The path coefficient for the correlation between 'Land use' and cutting frequency was a priori set to 1. The arrows indicate expected causal effects between measured variables; solid arrows represent expected positive effects and dotted arrows expected negative effects. For more detailed information on model definition and variable calculation see the Method section and Appendix S1.

described by Socher et al. [1]: On a 4 m × 4 m subplot in each plot all vascular plant species were identified in the field following the nomenclature of Wisskirchen & Häupler [27], categorized into functional groups (grasses, herbs, legumes) and their cover (%) was estimated. As vegetation was generally relatively uniform across the plot, we consider our subplot sample as representative for the 50 m × 50 m plot. Plant biomass was clipped at a height of 2 to 3 cm above ground and dried for 48 hours at 80°C before weighing. Occasionally occurring shrubs and litter were excluded before drying. In 2008, biomass samples were taken in two 25 cm × 50 cm subplots next to the vegetation survey plot. In 2009, biomass was harvested on eight subplots of 50 cm × 50 cm size adjacent to one side of the vegetation survey plot to better reflect small-scale variability. In both years, total plant biomass [g/m²] was recorded as the mean from the subplots. In mown and grazed plots, the subplots were fenced until biomass harvest to obtain productivity estimates unaffected by current year grazing or mowing. For the analyses we used plant species richness per plot as measure for plant diversity.

We sampled arthropods by sweep-netting with a total of 60 double-sweeps (one double-sweep is defined as moving the net from the left to the right and back perpendicular to the walking direction) along three plot borders (150 m in total) in June and August 2008 & 2009 during periods of dry weather conditions. The two sampling months were chosen because they cover the main activity period of arthropods and were found to represent the variation in diversity among plots equally well as sampling in five months (tested with a subset of plots in 2009, data not shown). We chose sweep-netting over suction sampling because of logistic difficulties with suction sampling on our large number of plots and because suction sampling was found to underrepresent Heteroptera (Gossner et al., unpublished data from our plots and [28]).

Sampling was conducted within seven days or fewer per region and within two weeks in all regions. Arthropods were preserved in 70% ethanol, sorted by taxonomic group and identified to species level by taxonomic experts. We focused on Araneae, Hemiptera

(Cicadomorpha, Fulgoromorpha and Heteroptera), Coleoptera and Orthoptera because of their numerical dominance in the studied grasslands. Only adult individuals were considered for the analysis, because identification of juveniles is often difficult. Samples from both months within one year were pooled for analysis. Because sampling effort was standardized there was no need to adjust for differences in the number of sampled individuals. Accordingly, we used species richness as diversity measure.

Due to external influences, such as aggressive cows preventing us from entering several plots, the number of plots with a complete dataset (plant and arthropod data) differed somewhat between years: in 2008 we considered 124 plots and in 2009 we considered 141 plots (see Table 1 for the number of plots per region). Of those plots, 117 plots were considered in both years.

Ethics Statement

Field work permits were issued by the responsible state environmental offices of Baden-Württemberg, Thüringen and Brandenburg (according to § 72 BbgNatSchG, i.e. nature conservation law of Brandenburg). Out of the 150 initially sampled plots, 17 are part of grassland areas that are protected under the Habitats Directive (FFH), 58 are protected as LSG (landscape conservation site), 17 are protected as NSG (nature conservation site) and 5 are situated within a former military training area. Except the 5 plots which are situated in the former military training area and 6 plots that are situated on state property, all other grasslands are privately owned by the farmers. One individual of *Phytoecia cylindrica* (Coleoptera, Linnaeus, 1758), which is protected under BArtSchV (Federal Protection of Species Order) was sampled in 2008.

Classification of trophic guilds and arthropod biomass assessment

All identified arthropod species were assigned to one of three trophic guilds (herbivores, predators and decomposers) based on

Table 1. Mean ± standard error for the sampled variables within each of the three regions.

	Vascular plants		Herbivores			Predators		
	No. species	Biomass (g)	No. species	Abundance	Biomass (g)	No. species	Abundance	Biomass (g)
2008								
Swabian Alb N=43	33.3±10.3	598.1±302.9	33.3±10.6	628.6±570.9	15.3±14.3	5.4±3.5	10.6±8.3	0.4±0.4
Hainich-Dün N=38	25.9±9.6	592.6±228.8	31.4±12.3	307.4±205.0	9.6±8.6	5.6±2.8	15.8±10.0	0.5±0.5
Schorfheide-Chorin N=43	16.0±3.7	752.0±281.7	26.7±7.7	281.1±145.0	8.8±4.9	5.1±2.7	13.2±9.5	0.3±0.3
F-value$_{2,121}$	46.72	4.56	4.64	11.89	5.09	0.29	3.25	0.91
P	<0.001	0.012	0.011	<0.001	0.008	0.752	0.042	0.404
Means over all regions	25.0±2.2	648.0±11.0	30.0±1.9	410.0±19.6	11.2±3.1	5.0±1.3	13.0±2.6	0.4±0.02
2009								
Swabian Alb N=47	32.1±11.2	272.1±123.1	20.5±5.6	180.9±144.5	6.7±4.8	3.8±2.0	4.6±2.8	0.3±0.3
Hainich-Dün N=48	33.6±12.1	299.7±128.7	16.6±7.1	73.0±93.4	2.8±4.4	4.3±2.5	8.5±7.2	0.2±0.2
Schorfheide-Chorin N=46	19.8±4.2	331.2±130.5	22.1±8.0	163.2±105.1	4.7±4.0	5.2±2.9	9.8±7.2	0.3±0.6
F-value$_{2,140}$	27.25	2.49	7.648	11.76	9.25	4.1	9.0	0.5
P	<0.001	0.086	<0.001	<0.001	<0.001	0.019	<0.001	0.61
Means over all regions	29.0±2.2	301.0±7	20.0±1.6	139.0±10.6	4.7±2.2	4.0±1.2	8.0±2.3	0.3±0.7

N = number of plots sampled in the respective year. Arthropod biomass was estimated from the species' mean length (see text). Means were tested for differences between regions with ANOVA. Degrees of freedom are indicated with the F-value$_{(df\ Region,\ df\ Residuals)}$. Abundances are expresses as number of adult individuals whereas biomass is given in grams. Species lists of herbivores and predators can be found in Appendix S5 and Appendix S6, original datasets are available from the Dryad Digital Repository:

their known main food resource as adults (Appendix S2). Because decomposers made up less than 2% of individuals in 90% of our samples, they were not included in the analysis. Arthropod biomass per plot was estimated by applying the general power function developed by Rogers et al. [29] to each sampled species:

$$biomass[mg] = 0.305 * L^{2.62}$$

Where L is the mean length (in mm) of a species, derived from the same sources as given for the identification of feeding guilds (Appendix S2). The contribution of a species to total biomass was then calculated as: species-specific biomass × abundance of the species. Biomass for each plot was calculated for herbivores and predators separately by summing over all species. We used biomass rather than abundances as measure for resource abundance to be consistent with plant assessments.

Statistical analysis

To compare means of the response variables among regions, we used ANOVA (aov) with 'Region' as explanatory variable in R [30].

For structural equation modelling, there are, in principle three main approaches [31]: 1) Strictly confirmatory, where a single model is defined and tested against the data. 2) Alternative modelling, where similar models are tested and compared by model selection for best fit. 3) Model generating, where one tentative initial model is simplified until the model fit cannot be improved. We used a combination of approaches 1 and 3 by defining two models based on the 'Resource Heterogeneity' (RHH) and 'More Individuals' (MIH) hypotheses (confirmatory approach, Figure 1) but using step-wise deletion of paths within these models when the model structure did not fit the data (model generating). The second step was included to assess whether the discrepancies between model and data structure were due to erroneous assumptions about interactions among variables (in this case model fit would be improved by step-wise deletion) or due to missing (i.e. not measured) variables (in that case model fit would not be improved).

Grassland land use was included in the models by the variables fertilization, grazing and mowing as well as 'Time after mowing' (as described in the first section of Materials and Methods). In addition, we added a latent (i.e. unmeasured) variable 'Land use' to describe the combined effect of land use. This latent variable accounts for correlations between the different land-use modes. As latent variables have no underlying data which the model algorithm can use to calculate its variance, either the latent's variance or one regression weight between the latent and one of the connected variables has to be fixed to 1. We chose to fix the regression weight between the latent variable and mowing frequency because mowing is a good descriptor of land-use intensity and it is correlated with both fertilization and grazing intensity [32]. The two other regression weights (between 'Land use' and grazing or fertilization) and the latent's variance can then be estimated by the model algorithm. To test for a non-linear effect of grazing intensity, we tried a quadratic transformation of grazing intensity, but because it did not improve model fit we kept the linear relationship. As the correlations between the land-use modes varied between regions, the model sometimes did not converge; in these cases, we included correlations between mowing and fertilization and between mowing and grazing instead of the latent variable. Except for the additional variables and pathways that were added in the 'Resource Abundance Model', all pathways

were expected to be identical in both models (Fig. 1 A & B), as motivated in the Introduction.

We used the software package 'sem' in R for structural equation modelling [30] which fits models using Maximum Likelihood based on observed and expected covariance matrices. The model fit is estimated as the overall model p-value which indicates if the two covariance matrices are significantly different from each other (p<0.05, bad model fit) or not (p>0.05, good model fit [31], p.128f). A step-wise selection procedure was applied by hand using the corrected Akaike's Information Criterion (AICc) for models without appropriate fit (model p-value <0.05). If a second measure of model quality, the Goodness-of-fit (GoF) index, was above 0.75 for the final model, models without adequate fit were still used for comparison among regions.

To calculate the total effect of 'Land use' on the plant and arthropod variables, the standardized path coefficients within each possible pathway between 'Land use' and the respective variable were multiplied and all resulting products were summed (e.g. coefficients from 'Land use' to fertilization and from fertilization to plant biomass were multiplied and added to the products of pathways including grazing and mowing). Total effects among plant and arthropod variables were calculated accordingly. Data was transformed as necessary (further details in Appendix S3).

Results

Over all plots we recorded 271 vascular plant species in 2008 and 281 vascular plant species in 2009; 54,660 herbivore individuals from 392 species were sampled in 2008 and 19,507 herbivores from 335 species in 2009; 1,823 predator individuals from 162 species were sampled in 2008 and 1,075 predators from 154 species in 2009. The average plant biomass per plot was 648 g/m^2 in 2008 and 301 g/m^2 in 2009, mean estimated herbivore biomass per plot was 11.2 g and 4.7 g (2008 and 2009 respectively) and mean estimated predator biomass per plot was 0.4 g and 0.3 g (including 1 and 5 plots with no predators in 2008 and 2009 respectively). The lower plant and herbivore biomass in 2009 compared to 2008 can be attributed to a shorter time-span between the onset of the vegetation period and sampling (as calculated for 'Time after mowing', see Appendix S1). Increasing plant biomass was correlated with a decrease in plant species diversity (see below) and with an increase in relative cover of grass species (compared to herbaceous species) ($R^2_{adj} = 0.269$ in 2008 and $R^2_{adj} = 0.081$ in 2009, $P<0.001$ in both years and all regions; Spearman's correlation).

In 2008 and 2009, plant and herbivore species richness as well as herbivore abundance and herbivore biomass differed significantly between regions and were highest in ALB (Table 1). Plant biomass also differed significantly between regions in 2008 and was highest in SCH while in 2009 it was similar among regions (Table 1). Predator species richness differed between regions only in 2009 (being highest in SCH) and was very similar among regions in 2008. Predator abundances differed between regions both in 2008 and 2009 and were lowest in ALB. Predator biomass did not differ between regions in both years (Table 1).

Resource Heterogeneity Model

The Resource Heterogeneity Model fitted the 2008 and 2009 data well for each region (Figure 2). Except for the omission of the latent variable 'Land use' in three cases, no further path-selection procedure was applied. Despite this omission, correlations among the land-use modes were consistent between years and regions, where fertilization was positively correlated with 'Land use' (or

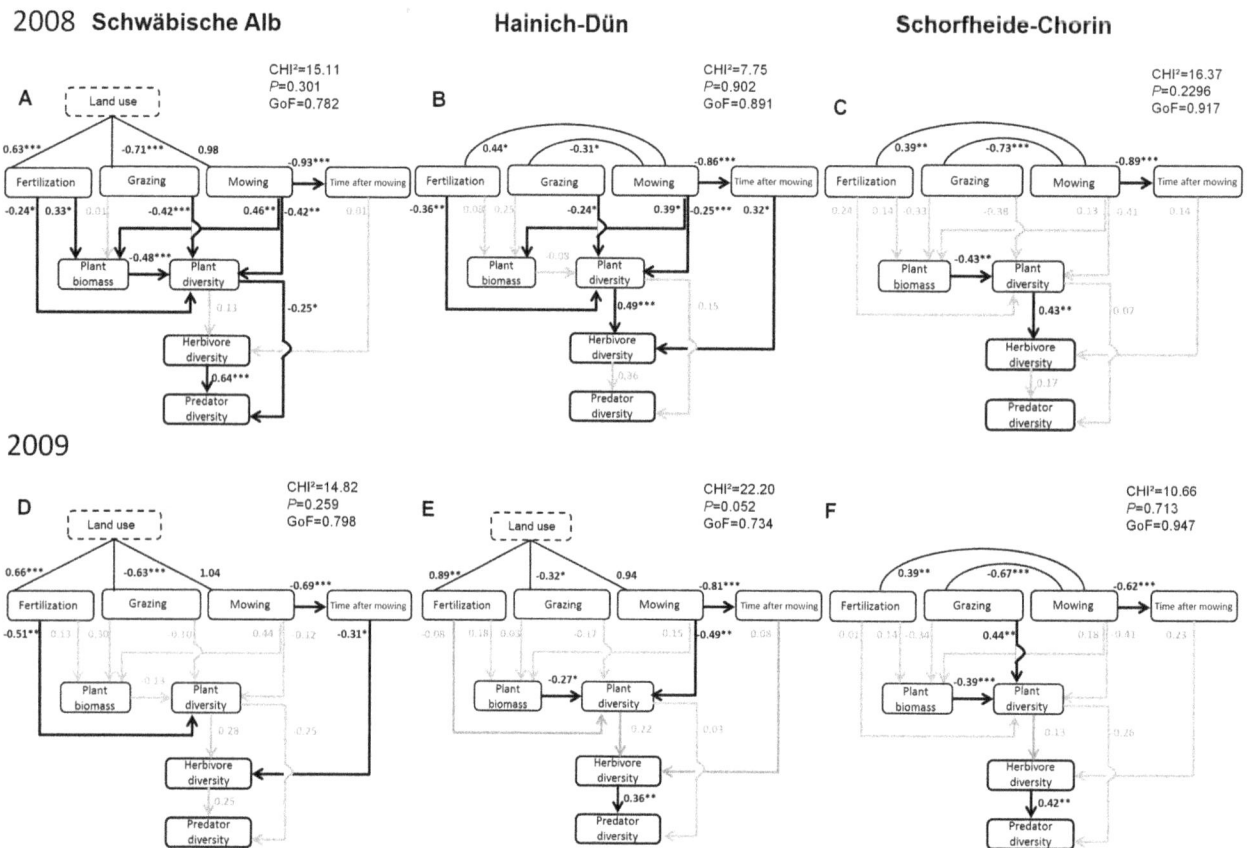

Figure 2. Standardized regression weights and significance levels from the Resource Heterogeneity Model. Models are shown for 2008 (A, B, C) and 2009 (D, E, F) for the Swabian Alb (A, D), Hainich-Dün (B, E) and Schorfheide-Chorin (C, F). Black solid lines and numbers indicate significant paths, grey arrows and numbers indicate non-significant paths. Correlation graphs for plant and arthropod measures can be found in Appendix S4. GoF= Goodness of fit. Significance levels: P<0.05: */P<0.01: **/P<0.001: ***. Total effects in Tables 2 & 3 were calculated using the standardized regression weights, e.g. plant diversity effects on predator diversity in Swabian Alb 2008 were calculated as 0.13 (coefficient from plant diversity to herbivore diversity)*0.64 (herbivore diversity to predator diversity) *−0.25 (plant diversity to predator diversity) = −0.17.

mowing) and grazing was negatively correlated with 'Land use' (or mowing).

Fertilization generally had the predicted negative effect on plant diversity and a positive effect on plant biomass, although effects were significant only in ALB (for diversity in 2008 & 2009 and for biomass only in 2008) and HAI (for diversity in 2008). We found positive but not significant effects of grazing intensity on plant biomass (Figure 2) but grazing intensity consistently reduced plant diversity (except for a significant increase in SCH in 2009). Mowing frequency consistently increased plant biomass and decreased plant diversity (although only significant in ALB and HAI and mostly in 2008). Differences in effect direction were found in SCH (positive effect of fertilization on plant diversity, and negative effects of grazing and mowing on plant biomass) but all of those effects were not significant.

As expected, plant biomass was always negatively related to plant diversity even though the paths for HAI 2008 and ALB 2009 were not significant. Effects of plant diversity on herbivore diversity were positive in all regions and years, as predicted by the RHH (although only significant in HAI and SCH). Similarly, we found a positive effect of herbivore diversity on predator diversity in all regions and years (here it was significant in 3 of 6 cases).

When summing the direct and indirect effects of 'Land use' on the target variables, we found total effects to be positive for plant biomass and negative for plant diversity (Table 2). The only exception was SCH, due to the strong influence of grazing (negative on plant biomass in 2008 and positive on plant diversity in 2009). The total effect of 'Land use' on herbivore diversity was negative in all regions and both years, whereas total effects on predator diversity were generally very weak (<0.1). The total effects of plant diversity on herbivore diversity were all positive as they are identical to the direct effects in this model. Total effects of plant diversity on predator diversity were, however, generally weak. Overall, 'Land use' decreased plant diversity directly and indirectly through an increase in plant biomass, which led to a decrease in herbivore diversity and a following decrease in predator diversity.

Resource Abundance Model

The Resource Abundance Model fitted the 2008 data adequately well for HAI and SCH, but not for ALB, even after applying path-selection procedures. For 2009, we found adequate model fits for all three regions after applying path-selection to HAI (Figure 3). Compared with the Resource Heterogeneity Model, the relationships among land-use modes remained the same, except that the latent variable could be kept in the model for HAI in 2008 (because the model converged in the first run). The direct effects of the land-use modes on plant diversity and biomass changed slightly in magnitude but not in direction compared with

Table 2. Standardized total effects derived from the Resource Heterogeneity Model for both years.

		Land use (mode) effects on				Plant diversity effects on	
		Plant biomass	Plant diversity	Herbivore diversity	Predator diversity	Herbivore diversity	Predator diversity
Swabian Alb	2008	0.65	−0.59	−0.09	0.09	0.13	−0.17
	2009	0.35	−0.44	0.10	0.03	0.28	0.06
Hainich-Dün	2008					0.49	0.32
Fertilization	2008	0.08	−0.37	−0.18	−0.06		
Grazing	2008	0.25	−0.26	−0.13	−0.05		
Mowing	2008	0.39	−0.56	−0.54	−0.19		
	2009	0.28	−0.56	−0.19	−0.08	0.22	0.11
Schorfheide-Chorin	2008					0.43	0.15
Fertilization	2008	0.41	0.18	0.10	0.01		
Grazing	2008	−0.33	−0.24	−0.10	0.0		
Mowing	2008	0.17	0.01	−0.01	0.0		
	2009					0.13	−0.20
Fertilization	2009	0.14	−0.06	0.0	0.02		
Grazing	2009	−0.03	0.46	0.06	0.02		
Mowing	2009	−0.19	−0.10	−0.16	−0.09		

Total effects were calculated by multiplying the standardized path coefficients on the single pathways between two variables and summing up those values for all possible pathways. Standardized total effects can range between −1 and 1. Effects are shown from the first row on the second row. For a description on how total effects were calculated see last paragraph in Material and Methods and refer to legend of Figure 2 for an example. Original datasets are available from the Dryad Digital Repository.

the first model and between years. The change in magnitude is due
to the fact that effects were standardized (i.e. values are relative to
each other) which is influenced by the change in number of
variables included in the model. As in the first model, plant
biomass was negatively related to plant diversity (although not
significant in HAI 2008) and plant diversity had a positive effect on
herbivore diversity (although only significant in ALB & HAI 2008
and SCH 2009; Figure 3).

We did not find a significant direct effect of plant biomass on
herbivore biomass as predicted by the MIH in any region or year.
As the relationship between plant and herbivore biomass may be
weakened if there are shifts in the relative abundances of
differently sized herbivore species, we also tested the Resource
Abundance Model using arthropod abundances instead of biomass
but again did not find any effect of plant biomass on herbivore
abundances, neither in 2008 or 2009 (Figure S1: Standardized
regression weights and significance levels from the resource
abundance model including arthropod abundances). Herbivore
biomass had a significant positive effect on herbivore diversity in
all regions and both years. Herbivore biomass also had a positive
effect on predator biomass, significant in three out of six cases
(ALB 2008, 2009, SCH 2009). Predator biomass itself had a
significant positive effect on predator diversity in all regions and
both years.

Total effects of 'Land use' on plant and arthropod diversity as
well as on plant biomass were identical in direction and similar in

effect strength to the Resource Heterogeneity Model for both
years, including the reverse effects in SCH due to grazing. Total
effects of plant diversity on herbivore diversity were positive, but
total effects of plant biomass on herbivore diversity were absent or
very weak. Total effects of herbivore biomass on predator diversity
were positive (Table 3).

In summary, 'Land use' increased plant biomass, but this effect
was not leading to an increase in herbivore biomass. Nevertheless,
increasing herbivore biomass generally increased predator biomass
which in turn increased predator diversity.

Discussion

We tested whether effects of land-use intensity on the diversity
of arthropod herbivores and predators are mediated by changes in
the diversity or biomass of their respective resources, i.e. plants
and herbivores. Although the strength of effects varied between
regions and years, we found negative effects of land-use intensity
on the diversity of both arthropod groups which were mediated by
different pathways. For herbivores, the results were consistent with
the 'Resource Heterogeneity Hypothesis', as we found significant
effects of plant diversity on herbivore diversity with no evidence for
effects mediated by plant biomass. For predators, results were
more consistent with the 'More Individuals Hypothesis', i.e.
predator diversity and biomass were more strongly affected

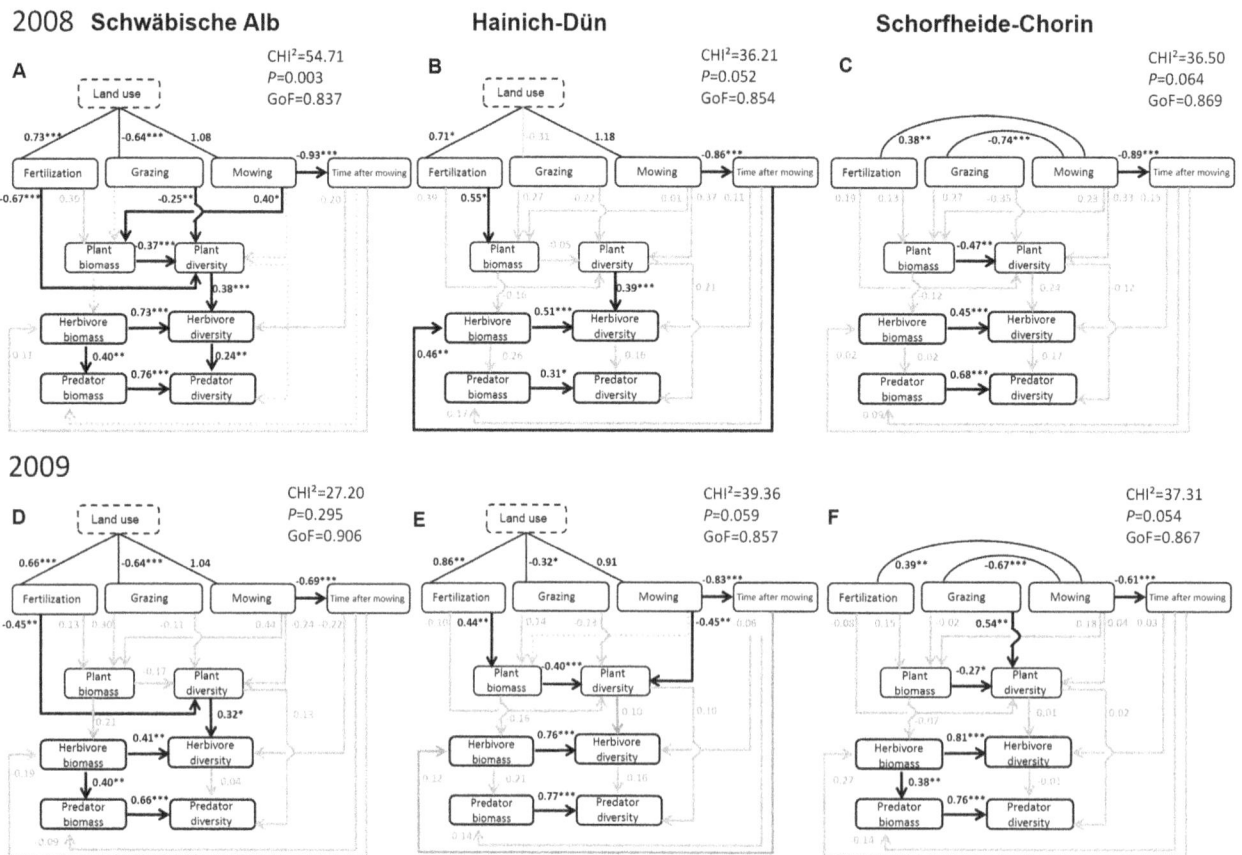

Figure 3. Standardized regression weights and significance levels from the Resource Abundance Model. Models are shown for 2008 (A, B, C) and 2009 (D, E, F) for the Swabian Alb (A, D), Hainich-Dün (B, E) and Schorfheide-Chorin (C, F). Black solid lines and numbers indicate significant paths, grey arrows and numbers indicate non-significant paths. Grey dotted paths were excluded during the step-wise selection procedure. Correlation graphs for plant and arthropod measures can be found in Appendix S4. GoF = Goodness of fit. Significance levels: P<0.05: */P<0.01: **/ P<0.001: ***.

Table 3. Standardized total effects derived from the Resource Abundance Model for both years.

		Land use (model) effects on				Plant diversity effects on		Plant biomass effects on	Herbivore biomass effects on
		Plant biomass	Plant diversity	Herbivore diversity	Predator diversity	Herbivore diversity	Predator diversity	Herbivore diversity	Predator diversity
Swabian Alb	2008	0.65	−0.56	−0.09	−0.06	0.38	0.09	−0.14	0.48
	2009	0.05	−0.48	0.07	0.02	0.32	0.15	0.04	0.28
Hainich-Dün	2008	0.84	−0.71	−0.70	−0.36	0.39	0.27	−0.10	0.16
	2009	−0.37	−0.43	−0.53	−0.38	0.77	0.52	0.03	0.66
Schorfheide-Chorin	2008					0.24	−0.08	0.17	0.09
Fertilization	2008	−0.28	0.32	0.09	−0.02				
Grazing	2008	0.23	−0.46	−0.12	0.03				
Mowing	2008	0.13	−0.39	−0.25	0.06				
	2009					0.01	0.03	−0.06	0.28
Fertilization	2009	−0.02	−0.08	0.0	0.0				
Grazing	2009	−0.18	0.59	0.01	0.02				
Mowing	2009	0.15	0.0	0.01	0.02				

Total effects were calculated by multiplying the standardized path coefficients on the single pathways between two variables and summing up those values for all possible pathways. Standardized total effects can range between −1 and 1. Effects are shown from the first row on second row. For an example on how effects were calculated see legend of Figure 2.

indirectly by changes in herbivore biomass than by direct effects of herbivore diversity.

Total land-use effects on arthropods and differences between regions

The negative total effect of land-use intensity on herbivore and predator diversity is consistent with previous reports on the consequences of grassland land use on arthropods [2–4], but the effects of the individual land-use modes (especially grazing) differed between regions and years and were only partly consistent with expectations. For instance, moderate grazing was found to have a positive effect on arthropod diversity in several studies [33] but we found only weak total effects of grazing intensity on arthropod diversity (<0.2). A wider range of grazing intensities in our study system compared to other studies might explain the absence of a positive effect in our case; e.g. Dennis et al. [34] found higher abundance and diversity of arthropods under moderate grazing intensity of sheep compared to high grazing intensity. The different types of livestock in our study system likely changed effects of grazing as well because herbivore diversity in 2008 was significantly lower on cattle-grazed plots compared to sheep-grazed plots and mixed grazing by cattle and horses had a significant positive effect on predator diversity compared with cattle or sheep grazing (Figure S2: Effects of livestock type on plant and insect species richness in 2008). Differences in the grazing gradients between regions and changes in grazing practices between years can also explain the change from a negative effect of grazing to a positive effect of grazing on plant diversity in SCH.

The different correlations among the land-use modes in the different regions are, in fact, a striking result of our study. Whereas all modes were significantly correlated with the latent variable 'Land use' in ALB, fertilization and grazing were sometimes (HAI) or always (SCH) independent of each other in the other regions. The discrepancies were probably caused by the range of land-use options realized in the different regions. In ALB, where most of the grasslands are managed by small farming enterprises or farming families, we find extensively grazed, unfertilized plots, e.g. sheep pastures on nutrient-poor hillsides, as well as intensively grazed, fertilized plots. Thus, grazing and fertilization are closely linked in this region. In the other two regions, plots with low grazing intensity sometimes receive low fertilizer input (for instance, in organic farming practices) but in other cases they are highly fertilized mown pastures that are only grazed at the end of the plant growth period. This weakened the correlation between grazing and fertilization.

Effects of land-use intensity on plant biomass and diversity

Socher et al. [1] extensively discussed the effects of fertilization, grazing and mowing on plant biomass and diversity using the same plots as the present study, thus here we only summarize the main points: Increasing fertilization intensity generally decreased plant diversity and increased biomass as found in many preceding studies. Negative effects of high mowing frequency on plant diversity support previous findings from different types of grasslands and various regions [35]. The negative effect of grazing on plant diversity appears to contrast the general finding that grazing increases plant diversity [36]. However, most previous studies compared grazed with ungrazed sites and thus found that grazing increased plant diversity via increased sward heterogeneity [11,37]. In our case, grazing ranged from no grazing to very high grazing intensities, which could result in a non-linear or hump-shaped effect of grazing on plant diversity. As including a non-

linear effect of grazing did not improve our model (see Method section) it seems likely that the negative effect of very high grazing intensities is exceeding the positive effect moderate grazing has compared to non-grazed sites.

Effects of changes in the plant community on herbivores

Our results showed clear evidence that land-use effects on herbivore diversity are mediated by plant diversity as predicted by the 'Resource Heterogeneity Hypothesis'. The fact that we did not find effects on herbivore diversity mediated by plant biomass is in contrast to findings from experimental plant communities [22], which showed that arthropod diversity was only indirectly affected by plant diversity through increased plant biomass. This indirect effect on arthropod diversity was additionally mediated by arthropod biomass (measured as biovolume) and therefore followed the 'More Individuals Hypothesis'[22]. The differences between results from our study in managed grasslands and results from experimental plant communities may be due to the absence of correlations between the proportion of particular plant functional groups and plant diversity in biodiversity experiment, as both variables are manipulated similarly. In contrast, the grasslands in our study system showed an increasing cover proportion of grasses with decreasing plant diversity. As found by Haddad et al. [38] the presence of grass species (which are productive but have low nutritional quality for herbivores) led to a decrease in total insect abundance by 25% even though total plant biomass increased. Only when grass species were absent and all plants were of higher nutritional quality, insect abundance was best explained by plant biomass. Hence, the higher cover of grass species on grasslands with low plant diversity in our study system led to a higher plant biomass but at the same time could not sustain higher herbivore biomass possibly because the overall nutritional value for the arthropods did not increase together with plant biomass.

Effects of changes in the herbivore community on predators

We found direct and indirect effects between herbivore and predator diversity, indicating mechanisms in accordance with both the 'Resource Heterogeneity' and the 'More Individuals Hypothesis'. One example would be the ALB 2008 where both direct and indirect effects were significant and strong. This indicates a complementary role of both mechanisms which might be the result of different predator groups reacting to either one of the mechanisms. In our study, the total effect of herbivore biomass on predator diversity was stronger in five out of six cases than the direct effect of herbivore diversity (compare Figure 3 and Table 3). This agrees with results from a plant diversity experiment, where effects of plants on herbivores were consistent with the 'Resource Heterogeneity Hypothesis' but effects of herbivores on predators were more in agreement with the 'More Individuals Hypothesis' [12]. Further research is needed to understand how predator diversity is affected by land use, as herbivore biomass was not affected by any of the included factors. This is relevant for sustainable land use of grasslands in agricultural dominated landscapes, because high predator abundance and diversity enhances biocontrol potential e.g. [39].

Where to go from here

We tested the effect of land-use intensity on important grassland herb-dwelling arthropods over a wide geographic range. To achieve standardized sampling in the vegetation layer across a large number of plots, sweep-net sampling was used which is a suitable method to representatively sample important herbivores (e.g. Heteroptera) as well as predators (e.g. Araneae) among arthropods (e.g. [40–42]). Nevertheless, it is a less well-performing method to sample other functional groups such as pollinators (butterflies and bees) [40] or ground-dwelling species [43]. Additional methods (such as suction sampling, pitfall traps or pan traps) might therefore be advised if a study's focus is not only on herb-dwelling species. Disentangling the differences which might apply to different functional groups within herbivores and predators (e.g. sucking vs. chewing herbivores or predators with different hunting strategies) will further increase our understanding of land-use effects. One promising approach was recently proposed by Lavorel et al. [44] who included producer and consumer traits in structural equation models to understand how land use affects ecosystem services through changes in the trait composition of the groups which provide the services.

Conclusions

Our results emphasize the importance of studying indirect effects of land-use intensity on the arthropod community, as they showed that herbivores and predators respond to changes of different aspects of their resources. We confirm that herbivore diversity is responding positively to higher plant diversity in grasslands, whereas herbivore biomass matters more than diversity for how predator diversity is affected by land use. By including different regions we showed on the one hand that the negative effects of high fertilization intensity and high mowing frequency on arthropod diversity are consistent over large scales; but on the other hand the variability of land-use traditions clearly indicates that findings cannot be easily extended to a wider geographical context. Our results thus not only emphasize the importance of land use for biodiversity changes, but also the need for more differentiated approaches to disentangle how different land-use modes have different effects on biodiversity, and how chains of effects differ for different aspects of biodiversity.

Supporting Information

Figure S1 Standardized regression weights and significance levels from the resource abundance model including arthropod abundances. Models are shown after step-wise deletion of non-significant paths. Black solid lines and numbers indicate significant paths; grey arrows indicate non-significant paths. Grey, dotted paths were excluded during the step-wise selection procedure. Significance level: $p<0.05$: */$p<0.01$: **/$p<0.001$: ***.

Figure S2 Effects of livestock type on plant and insect species richness in 2008. Means per plot and standard errors are shown. Horizontal lines indicate significant differences based on Tukey's HSD test. Significance levels: $p<0.05$: */$p<0.01$: **/$p<0.001$: ***.

References

1. Socher SA, Prati D, Boch S, Müller J, Klaus VH, et al. (2012) Direct and productivity-mediated indirect effects of fertilization, mowing and grazing on grassland species richness. Journal of Ecology 100: 1391–1399.
2. Bell JR, Wheater CP, Cullen WR (2001) The implications of grassland and heathland management for the conservation of spider communities: a review. Journal of Zoology 255: 377–387.
3. Di Giulio M, Edwards PJ, Meister E (2001) Enhancing insect diversity in agricultural grasslands: the roles of management and landscape structure. Journal of Applied Ecology 38: 310–319.
4. Nickel H, Hildebrandt J (2003) Auchenorrhyncha communities as indicators of disturbance in grasslands (Insecta, Hemiptera) - a case study from the Elbe flood plains (northern Germany). Agriculture Ecosystems & Environment 98: 183–199.

Table S1 Mean and range of land-use activities in the three regions and during the years considered for the analysis with samplings from 2008.

Table S2 Mean and range of land-use activities in the three regions and during the years considered for the analysis with samplings from 2009.

Appendix S1 Assessment and calculation of land-use information.

Appendix S2 Classification of trophic guilds.

Appendix S3 Structural equation model setup and path-selection procedure.

Appendix S4 Bivariate correlations between model variables.

Appendix S5 List of arthropod species sampled in 2008

Appendix S6 List of arthropod species sampled in 2009.

Acknowledgments

Special thanks go to Leonardo Ré Jorge, who gave valuable comments during the discussion on the model structure. We thank Luis Sikora for his help with sweep-netting the grasslands and Roland Achtziger, Theo Blick, Boris Büche, Michael-Andreas Fritze, Günter Köhler, Frank Köhler, Franz Schmolke and Thomas Wagner for identifying the arthropods and providing data on trophic guilds and body size. A database on Orthoptera was kindly provided by Frank Dziock. We thank the managers of the three exploratories, Swen Renner, Sonja Gockel, Kerstin Wiesner and Martin Gorke for their work in maintaining the plot and project infrastructure; Andreas Hemp and Uta Schumacher for insuring successful field work in the Schorfheide; Simone Pfeiffer and Christiane Fischer giving support through the central office, Michael Owonibi for managing the central data base and Eduard Linsenmair, Dominik Hessenmöller, Jens Nieschulze, Ingo Schöning, François Buscot, Ernst-Detlef Schulze and the late Elisabeth Kalko for their role in setting up the Biodiversity Exploratories project.

Author Contributions

Conceived and designed the experiments: MF MMG WWW. Performed the experiments: SB ML JM EP SAS MT. Analyzed the data: NKS MMG TML. Contributed to the writing of the manuscript: NKS MMG TML SB ML JM EP SAS MT MF WWW.

5. Weiner CN, Werner M, Linsenmair KE, Bluthgen N (2011) Land use intensity in grasslands: Changes in biodiversity, species composition and specialisation in flower visitor networks. Basic and Applied Ecology 12: 292–299.
6. Chamberlain DE, Fuller RJ, Bunce RGH, Duckworth JC, Shrubb M (2000) Changes in the abundance of farmland birds in relation to the timing of agricultural intensification in England and Wales. Journal of Applied Ecology 37: 771–788.
7. Allan E, Bossdorf O, Dormann CF, Prati D, Gossner MM, et al. (2014) Interannual variation in land-use intensity enhances grassland multidiversity. Proceedings of the National Academy of Sciences of the United States of America 111: 308–313.

8. Littlewood NA, Stewart AJA, Woodcock BA (2012) Science into practice – how can fundamental science contribute to better management of grasslands for invertebrates? Insect Conservation and Diversity 5: 1–8.

9. Marini L, Fontana P, Scotton M, Klimek S (2008) Vascular plant and Orthoptera diversity in relation to grassland management and landscape composition in the European Alps. Journal of Applied Ecology 45: 361–370.

10. Morris MG, Lakhani KH (1979) Responses of grassland invertebrates to management by cutting. 1. Species-diversity of Hemiptera. Journal of Applied Ecology 16: 77–98.

11. Woodcock B, Potts S, Tscheulin T, Pilgrim E, Ramsey A, et al. (2009) Responses of invertebrate trophic level, feeding guild and body size to the management of improved grassland field margins. Journal of Applied Ecology 46: 920–929.

12. Haddad NM, Crutsinger GM, Gross K, Haarstad J, Knops JMH, et al. (2009) Plant species loss decreases arthropod diversity and shifts trophic structure. Ecology Letters 12: 1029–1039.

13. van den Berg LJL, Vergeer P, Rich TCG, Smart SM, Guest D, et al. (2011) Direct and indirect effects of nitrogen deposition on species composition change in calcareous grasslands. Global Change Biology 17: 1871–1883.

14. Ryder C, Moran J, Mc Donnell R, Gormally M (2005) Conservation implications of grazing practices on the plant and dipteran communities of a turlough in Co. Mayo, Ireland. Biodiversity and Conservation 14: 187–204.

15. Sjodin NE, Bengtsson J, Ekbom B (2008) The influence of grazing intensity and landscape composition on the diversity and abundance of flower-visiting insects. Journal of Applied Ecology 45: 763–772.

16. Watkinson AR, Ormerod SJ (2001) Grasslands, grazing and biodiversity: editors' introduction. Journal of Applied Ecology 38: 233–237.

17. Hutchinson GE (1959) Homage to Santa Rosalia or Why are there so many kinds of animals? American Naturalist 93: 145–159.

18. Lewinsohn TM, Roslin T (2008) Four ways towards tropical herbivore megadiversity. Ecology Letters 11: 398–416.

19. Southwood TRE, Brown VK, Reader PM (1979) The relationships of plant and insect diversities in succession. Biological Journal of the Linnean Society 12: 327–348.

20. Strong DR, Jr., Lawton JH, Southwood TRE (1984) Insects on Plants: Community Patterns and Mechanisms. Cambridge, MA: Harvard University Press.

21. Srivastava DS, Lawton JH (1998) Why more productive sites have more species: An experimental test of theory using tree-hole communities. American Naturalist 152: 510–529.

22. Borer ET, Seabloom EW, Tilman D, Novotny V (2012) Plant diversity controls arthropod biomass and temporal stability. Ecology Letters 15: 1457–1464.

23. Andreasen C, Hansen CH, Moller C, Kjaer-Pedersen NK (2002) Regrowth of weed species after cutting. Weed Technology 16: 873–879.

24. Becklin KM, Kirkpatrick HE (2006) Compensation through rosette formation: the response of scarlet gilia (Ipomopsis aggregata: Polemoniaceae) to mammalian herbivory. Canadian Journal of Botany-Revue Canadienne De Botanique 84: 1298–1303.

25. Fischer M, Bossdorf O, Gockel S, Hansel F, Hemp A, et al. (2010) Implementing large-scale and long-term functional biodiversity research: The Biodiversity Exploratories. Basic and Applied Ecology 11: 473–485.

26. Humbert JY, Ghazoul J, Walter T (2009) Meadow harvesting techniques and their impacts on field fauna. Agriculture Ecosystems & Environment 130: 1–8.

27. Wisskirchen R, Haeupler H (1998) Standardliste der Farn- und Blütenpflanzen Deutschlands. Stuttgart (Hohenheim): Eugen Ulmer.

28. Brook A, Woodcock B, Sinka M, Vanbergen A (2008) Experimental verification of suction sampler capture efficiency in grasslands of differing vegetation height and structure. Journal of Applied Ecology 45: 1357–1363.

29. Rogers LE, Hinds WT, Buschbom RL (1976) A general weight versus length relationshop for insects. Annals of the Entomological Society of America 69: 387–389.

30. R Core Team (2013) R: A language and environment for statistical computing. 3.0.2 ed. Vienna, Austria: R Foundation for Statistical Computing.

31. Grace JB (2006) Structural Equation Modeling and Natural Systems. Cambridge, UK: Cambridge University Press.

32. Blüthgen N, Dormann CF, Prati D, Klaus VH, Kleinebecker T, et al. (2012) A quantitative index of land-use intensity in grasslands: Integrating mowing, grazing and fertilization. Basic and Applied Ecology 13: 207–220.

33. Scohier A, Dumont B (2012) How do sheep affect plant communities and arthropod populations in temperate grasslands? Animal 6: 1129–1138.

34. Dennis P, Skartveit J, McCracken DI, Pakeman RJ, Beaton K, et al. (2008) The effects of livestock grazing on foliar arthropods associated with bird diet in upland grasslands of Scotland. Journal of Applied Ecology 45: 279–287.

35. Hopkins A, Wilkins RJ (2006) Temperate grassland: key developments in the last century and future perspectives. Journal of Agricultural Science 144: 503–523.

36. Marion B, Bonis A, Bouzille JB (2010) How much does grazing-induced heterogeneity impact plant diversity in wet grasslands? Ecoscience 17: 229–239.

37. Rook AJ, Dumont B, Isselstein J, Osoro K, WallisDeVries MF, et al. (2004) Matching type of livestock to desired biodiversity outcomes in pastures - a review. Biological Conservation 119: 137–150.

38. Haddad NM, Tilman D, Haarstad J, Ritchie M, Knops JMH (2001) Contrasting effects of plant richness and composition on insect communities: A field experiment. American Naturalist 158: 17–35.

39. Geiger F, Bengtsson J, Berendse F, Weisser WW, Emmerson M, et al. (2010) Persistent negative effects of pesticides on biodiversity and biological control potential on European farmland. Basic and Applied Ecology 11: 97–105.

40. Buffington ML, Redak RA (1998) A comparison of vacuum sampling versus sweep-netting for arthropod biodiversity measurements in California coastal sage scrub. Journal of Insect Conservation 2: 99–106.

41. Doxon E, Davis C, Fuhlendorf S (2011) Comparison of two methods for sampling invertebrates: vacuum and sweep-net sampling. Journal of Field Ornithology 82: 60–67.

42. Spafford R, Lortie C (2013) Sweeping beauty: is grassland arthropod community composition effectively estimated by sweep netting? Ecology and Evolution 3: 3347–3358.

43. Standen V (2000) The adequacy of collecting techniques for estimating species richness of grassland invertebrates. Journal of Applied Ecology 37: 884–893.

44. Lavorel S, Storkey J, Bardgett R, Bello F, Berg M, et al. (2013) A novel framework for linking functional diversity of plants with other trophic levels for the quantification of ecosystem services. Journal of Vegetation Science 24: 942–948.

9

Asexual Propagation of Sea Anemones That Host Anemonefishes: Implications for the Marine Ornamental Aquarium Trade and Restocking Programs

Anna Scott*, Jannah M. Hardefeldt, Karina C. Hall

National Marine Science Centre and Marine Ecology Research Centre, School of Environment, Science and Engineering, Southern Cross University, Coffs Harbour, New South Wales, Australia

Abstract

Anemonefishes and their host sea anemones form an iconic symbiotic association in reef environments, and are highly sought after in the marine aquarium trade. This study examines asexual propagation as a method for culturing a geographically widespread and commonly traded species of host sea anemone, *Entacmaea quadricolor*. Two experiments were done: the first to establish whether size or colour morph influenced survival after cutting into halves or quarters; and the second to see whether feeding was needed to maximise survival and growth after cutting. Survival rates were high in both experiments, with 89.3 and 93.8% of the anemones cut in half, and 62.5 and 80.4% cut in quarters surviving in experiments 1 and 2, respectively. Anemones that were cut in half were larger in size, and healed and grew quicker than those cut in quarters. However, even though survival was lower when the individuals were cut in quarters, this treatment produced the greatest number of anemones. Feeding increased oral disc diameter growth and reduced wet weight loss, but did not significantly influence pedal disc diameter. Given that the anemones took up to 56 d to form an off-centre mouth, it is highly likely that feeding may have produced greater effect if the experiment was run for longer. This low technology method of propagation could be used to produce individuals throughout the year and the anemones could then be used to supply the aquarium trade or restock depleted habitats, thus supporting biodiversity conservation in coral reef areas.

Editor: Wan-Xi Yang, Zhejiang University, China

Funding: Funding for this research was provided by the Associate Deans Research Support Scheme and the Marine Ecology Research Centre, Southern Cross University. The funders had no role in study design, data collection and analysis, decision to publish, or preparation of the manuscript.

Competing Interests: The authors have declared that no competing interests exist.

* Email: anna.scott@scu.edu.au

Introduction

The trade of marine ornamentals for aquariums is rapidly expanding, causing concerns about the sustainability and environmental impacts of the industry [1–3]. Collection often occurs on coral reefs in developing island nations, where it can be difficult to implement effective management [2,4,5]. The fishery is highly selective, and harvesting from limited areas has led to localised depletions or extinctions of target species in some cases [6–8].

Sea anemones that host symbiotic anemonefishes are highly sought after by aquarists [2,9]. An alteration in the population dynamics of either partner generally effects the other due to the obligate nature of the symbiosis [6,8,10]. These host anemones represent high-value species for collectors (e.g. in the Philippines the price paid to fisherman for anemones can be up to 13 times the amount for anemonefishes), which means they are often preferentially harvested [8]. The development of reliable and cost-effective methods for culturing anemones could facilitate the supply of animals for the aquarium trade, or the restocking of reefs that have already been impacted by natural or anthropogenic disturbances [11]. In doing so, it would create a new and ecologically sustainable industry, and thus support biodiversity conservation in coral reef areas [12,13].

Anemones could be cultured using sexual reproduction, which has been studied in two (*Entacmaea quadricolor* and *Heteractis crispa*) of the ten species of sea anemones that host anemonefishes [14–17]. Both of these anemones have separate sexes, and broadcast spawn gametes for external fertilization and development during predictable annual spawning periods [14,15,17,18]. The larvae settle onto a variety of surfaces, where metamorphosis and development occurs [16]. Alternatively, it may be possible to propagate anemones using asexual reproduction. *Entacmaea quadricolor* and *Heteractis magnifica* can reproduce asexually using longitudinal fission [19], where the anemone divides by stretching in opposite directions, thinning the tissue and tearing perpendicular to the axis of stretch [20,21]. Porat and Chadwick-Furman [22] cut six *E. quadricolor* in half to generate individuals for a split-pair laboratory experiment with anemonefishes, and found that the majority survived (67%). Likewise, anecdotal evidence from aquarium hobbyists suggests that a variety of anemone species might be able to be propagated by cutting them in half.

This study aimed to determine whether the most geographically widespread host anemone, *E. quadricolor*, could be cultured using fragmentation. We investigated: 1) whether anemones could survive being cut in halves or quarters, and which cut type would produce the maximum number of anemones; 2) if survival was influenced by initial size or colour morph; and 3) whether feeding was needed to maximise survival and growth. Given that this species naturally undergoes longitudinal fission [19], and there is some published [22] and anecdotal evidence to suggest these anemones can survive being fragmented, it was hypothesized that: anemones would survive being cut; that survival would be influenced by size (with larger individuals faring better than smaller individuals) but not colour morph; and that feeding would increase survival and accelerate growth.

Materials and Methods

Ethics statement

This work was done under the conditions specified by NSW Fisheries Permit P02/0025-4.0.

Experiment 1 – survival

Twenty-four *E. quadricolor* were collected on SCUBA from 15–18 m depth on 14 March 2012 at North Solitary Island, Solitary Islands Marine Park, New South Wales, Australia (29°55′S, 153°23′E). Anemones were gently removed from the substratum by hand, placed into a mesh bag, and transferred into a 70-L white polyvinyl chloride (PVC) tub on the boat. They were transported to the National Marine Science Centre, Coffs Harbour and maintained in a shaded rectangular 3 000-L outdoor tank with flow-through (10 L min^{-1}) coarse-sand filtered (~30 μm) seawater for 12 d before the start of the experiment. Seawater was sourced from Charlesworth Bay (30°15′S, 153°8′E) through a gravel filter-box system that was located 150 m offshore and at a depth of approximately 3 m. Half of the anemones had a purple column and brown tentacles with white tips (morph 1); while the other half had a red column and brown tentacles with a white ring below green tips (morph 2). It is likely that morph 1 was male, and morph 2 was female, as Scott and Harrison [17] found that the colour of the column and tentacles generally corresponded to a particular sex at the collection location.

The experiment was done in two 3 000 L tanks that were located outdoors to provide natural photoperiod and light. Shade cloth covers were placed over the tanks to ensure that light induced bleaching did not occur (light levels did not exceed 400 μmol photons m^{-2} s^{-1}). Twenty-four white PVC tubs (40 L, 29.5×39 cm wide, 30 cm deep) were evenly distributed between the tanks. Each tub had four circular outlets (2.5 cm diameter) located 3.5 cm below the top that were covered with 1 mm square mesh; a rock (approximately 15×10×6 cm) for habitat; and was supplied with flow-through seawater (sourced as described above but supplied at 0.5 L min^{-1}) to maintain ambient seawater temperature (20.5–24.5°C). Temperature was logged every 15 min in each of the tanks using a Thermochron iButton temperature logger (Maxim, USA).

The anemones were weighed on a top-pan balance after gentle squeezing and blotting with absorbent paper to remove as much water as possible (wet weight 61–274 g). Anemones were then randomly allocated to tubs and acclimatised for 3 d before the start of the experiment. Eight individuals (four of each colour morph) were randomly assigned to each of the following treatments: control (uncut), cut in half, or cut in quarters.

At the start of the experiment, the oral disc diameter (ODD) of each anemone was measured along the long and short axes to the nearest mm with vernier callipers and averaged (ODD 109–235 mm). Anemones were then removed from the tubs and fragmented with a 165 mm steel chefs knife. This was done on a plastic cutting board, whilst wearing latex gloves. Tentacles were

moved toward the outer edge of the oral disc, and cuts were made through the anemones mouth (along a random plane to the directional mouth). The resulting fragments were immediately placed back into their respective tubs, with the pedal disc directed downwards. The knife and cutting board were cleaned between cutting each anemone. Control anemones were also removed from the tubs; however they were simply placed back into the tubs without cutting. If the water in the tubs became cloudy after cutting a ~80% water exchange was done to rapidly clear the water.

The number of alive fragments (i.e. survival) was recorded at 6 h, daily for the first week, and then on day 9, 11, 14, 21, 28, and 35. Anemones were fed small pieces of prawn on a weekly basis, starting 7 d after cutting. Tubs were cleaned every 3 d to remove boluses or mucus produced by the anemones and residual food, and to prevent algal build up on the tubs. This was done using a scouring pad and siphon hose.

Experiment 2 – survival, effect of feeding on size, and visual observations of recovery

The methodology for experiment 2 was similar to that described above, except for the following departures. Forty-two *E. quadricolor* of four colour morphs: i) red column, brown tentacles with white ring below green tips (n = 18); ii) purple column, brown tentacles with white tips (n = 13); iii) light brown column, brown tentacles white ring below brown tips (n = 7); and iv) pink column, brown tentacles with white tips (n = 4), were collected from 12–18 m depth on the 5 October 2012. The anemones were measured before cutting (ODD 98–215 mm, pedal disc diameter [PDD, which was measured using the same methods described for ODD] 45–108 mm, and wet weight 32–171 g). The linear measurements of ODD and PDD were correlated (correlation coefficient r = 0.77), but each only showed weak correlation with wet weight (r = 0.4 for both).

Anemones were randomly allocated directly into 42 separate tubs rather than held temporarily in a holding tank. The tubs were located in three 1 200 L tanks supplied with flow-through seawater (sourced the same way as described in experiment 1, but then filtered to 10 μm with Filtaflo sediment filters and supplied at 0.6 L min-1), which was heated in two 3 000 L header tanks using Aquahort heater/chiller units to maintain a similar temperature range (21–24°C) to that in experiment 1. Temperature was logged every 3 h in twelve tubs (four in each tank).

Fourteen individuals were randomly assigned to each of the treatments: control (no cut), cut in half, or cut in quarters. Half of the anemones from each of these treatments were fed weekly. This feeding frequency was selected as the growth rates of juvenile *Heteractis cripsa* fed once or three-times weekly have been shown to be equivalent [11]. Pieces of prawn flesh were placed onto the oral disc until each anemone was satiated, and consequently the amount of food received by each individual was proportional to its size. The remaining anemones were not provided with any additional heterotrophic food. Although individuals were randomly allocated to the various treatments, the treatment types were divided among the three experimental tanks.

The number of alive fragments (i.e. survival) and their recovery (cut healing status, and whether or not their oral disc was expanded and their pedal disc was attached to the tub) were recorded at 6 h, daily for the first week, and then on day 9, 11, 14, 21, 28, 35, 42, 56, 70, 84 and 105. Throughout the experiment ODD, PDD, and wet weight were used as indicators of anemone size. Measurements of ODD and PDD were first taken one week after cutting to allow the anemones time to re-expand, and were repeated on day 28, 42, 56, 70, 84 and 105. Fragments were weighed immediately after cutting, and then on d 28, 56, 84, and 105. Tubs were cleaned every 2–3 d.

Data analysis

Survival – experiments 1 and 2. Survival differences among the different treatments were investigated using the Fisher's exact test. Variations in survival over time among the different treatments were analysed for each experiment using nonparametric Kaplan-Meier (K-M) product-limit analyses. The census times at which fragments were last observed alive were used as the time of death, so estimates of survival times are slightly conservative. As non-parametric survival analyses cannot explicitly incorporate information from replication, survival curves were estimated for the total number of fragments in each treatment combination pooled across replicate tubs.

Data from experiment 1, were further analysed with a parametric survival analysis (Cox proportional hazards model), to determine whether pre-fragmented size or morph had a significant effect on survival over time. Pre-fragmented sizes were categorised as either small (<100 g) and large (>100 g) for analyses. Survival data were analysed using the 'survival' library in R (version 2.11.1, http://www.r-project.org).

Growth – experiment 2. After cutting, there were uneven numbers of anemones in each tub (one control, two halves, or four quarters). In tubs containing multiple fragments it was not possible to individually identify each fragment between measurements, so growth data (i.e. wet weight, ODD, and PDD) were averaged per tub for all cut anemones and tubs were used as replicate units for each treatment combination. Actual size measurements were used for all statistical analyses rather than percentage change to alleviate complications associated with negative growth and percentages for statistical computations. To account for the influence of initial size on subsequent growth, the first size measurement of each fragment (immediately after cutting for wet weight and on day 7 for ODD and PDD) was used as a covariate in all analyses. All growth variables were log-transformed to meet the assumptions of normality and homoskedacticity.

To assess the effects of cut types (no cut, half, and quarter), feeding (fed and unfed), and tanks (1, 2, and 3) on growth during the experiment, generalized linear models (GLM) were fitted with the 'lme4' package in R, with parameters estimated by restricted maximum likelihood (REML). Baseline models included the design/random variables of initial fragment size as a fixed covariate, and tank and days as random factors. Days were nested within tubs to account for the repeated measures experimental design. The treatment factors of cut types, feeding and days, and associated interaction terms were fitted in various combinations of fixed factors along with the design/random variables in separate GLM, and compared by maximum likelihood comparisons of the change in deviance from the baseline model using the Akaike Information Criterion (AIC).

Results

Survival

In experiment 1, all of the eight controls survived, along with 15 of the 16 halves (93.8%) (Figure 1). In contrast, significant mortality occurred in the quarters, with 20 of the 32 fragments (62.5%) surviving (Figures 1 and 2a, Fisher's, $\chi^2 = 10.6$, df = 2, p = 0.005). Nevertheless, anemones cut into quarters still produced the greatest number of anemones at the end of the experiment (20 quarters vs 15 halves). Among the quarters, fragments from morph 1 (75%) had greater survival than those from morph 2 (50%), but this result was not significant (Figure 2b, K-M, Wald = 1.7, df = 1,

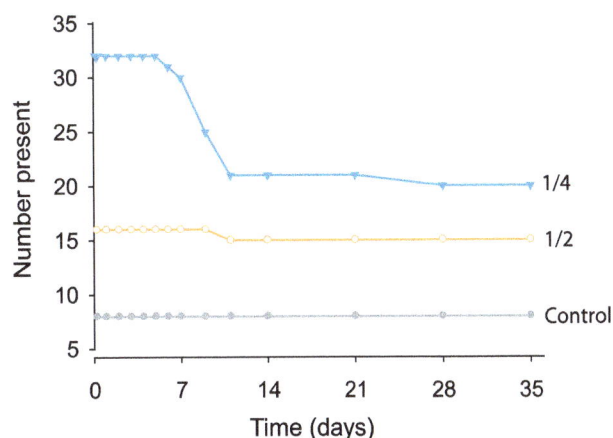

Figure 1. Number of *Entacmaea quadricolor* present in each treatment over time (in days) during experiment 1.

p = 0.197). Likewise, although more of the quarters from large anemones survived (75%), compared with those from small anemones (50%), the difference was not significant (Figure 2c, K-M, Wald = 2.3, df = 1, p = 0.131).

In experiment 2, no significant mortality was recorded among either halves or quarters relative to controls (Figure 3, Fisher's, $\chi^2 = 4.2$, df = 2, p = 0.124). All controls survived, and 89.3% and 80.4% of halves and quarters survived, respectively. Thus, the anemones cut in quarters produced almost double the number of anemones (n = 45 surviving fragments) than those cut in half (n = 25). Given the low mortality in this experiment, none of factors analysed (cut type, feeding or tanks) had a significant effect on the survival of fragments over time (Figure 4, K-M, p>0.05 in all cases).

Effects of feeding on size

Larger fragments resulted from cutting larger anemones for a given cut type. This initial fragment size had a significant influence on final fragment size as measured by all three variables, with anemones of a larger initial size remaining larger throughout the experiment (GLM; p<0.001 in all cases).

Although the ODD of the sea anemones showed no consistent positive or negative change over time across all treatments during the experiment (non-significant day effect, GLM, χ^2 deviance = 10.0, df = 5, p = 0.08), it did vary significantly among cut types and feeding treatments (Figure 5a–c). The ODD of quarters increased less during the experiment than that of halves or controls (significant cut type effect, GLM, χ^2 deviance = 9.9, df = 2, p = 0.007), even after the effect of initial fragment size was accounted for. The ODD of the fed anemones grew larger than unfed anemones (significant feeding effect, GLM, χ^2 deviance = 9.5, df = 1, p = 0.002), particularly among controls (Figure 5a) and to a lesser extent among halves later in the experiment (Figure 5b).

In comparison, feeding had no influence on the PDD of the anemones (non-significant feeding effect, GLM, χ^2 deviance = 2.6, df = 1, p = 0.108). Nevertheless, PDD generally increased throughout the experiment for all treatment groups (significant day effect, GLM, χ^2 deviance = 21.0, df = 5, p<0.001). Cut type also influenced PDD growth, with the controls and halves growing more than the quarters (significant cut type effect, GLM, χ^2 deviance = 12.5, df = 2, p = 0.002).

Although all of the anemones gradually decreased in wet weight, including the controls (significant day effect, GLM, χ^2

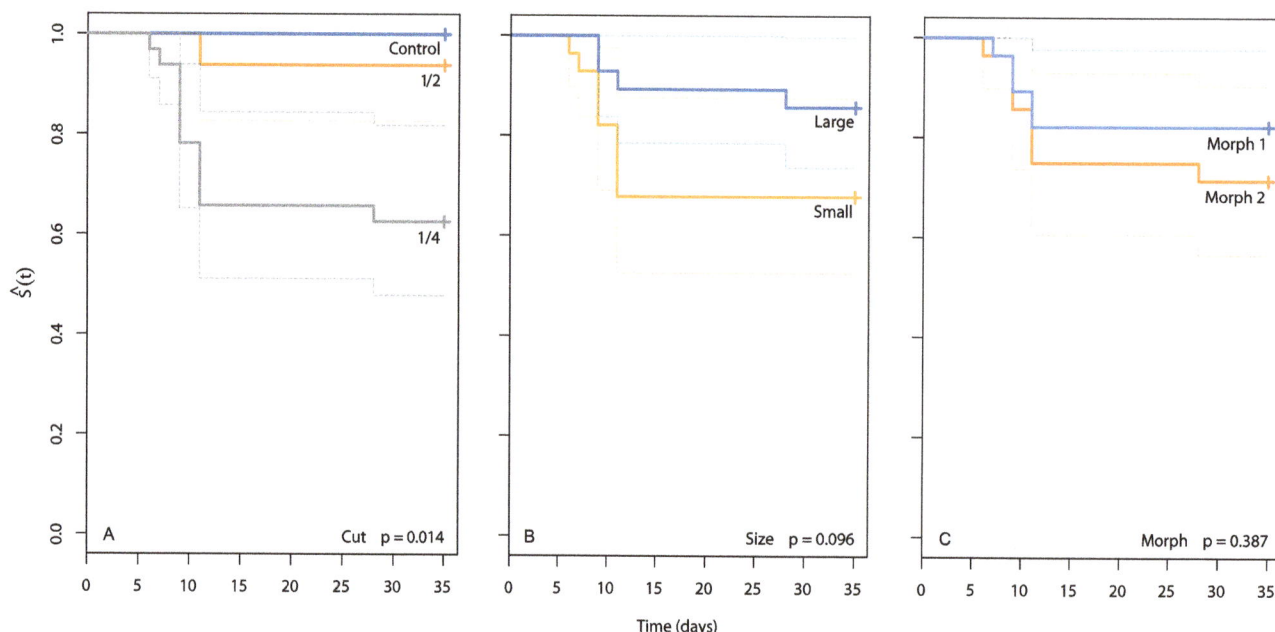

Figure 2. Kaplan-Meier survival [Ŝ(t)] plots for *Entacmaea quadricolor* **among different: (a) cut types, (b) sizes, and (c) morphs during experiment 1.** Dashed lines represent confidence intervals.

deviance $= 68.6$, df $= 3$, p<0.001), those that received food showed reduced wet weight loss in comparison to those that were unfed (significant feeding effect, GLM; χ^2 deviance $= 5.3$, df$= 1$, p $= 0.022$). There was an interaction between these two factors, with feeding having a greater influence towards the end of the experiment than at the start (significant feeding $\times$ day effect, GLM, χ^2 deviance $= 9.5$, df$= 3$, p $= 0.023$).

Visual observations

The majority of individuals in all of the treatments had attached to the bottom of the tubs 6 h after the start of experiment 2 (93% of the controls, 96% of the halves, and 89% of the quarters).

Likewise, most of the anemones that weren't cut or were cut in half had re-expanded their oral disc and tentacles (93% and 73%, respectively); however only 32% of the quarters had re-expanded. Most anemones still had an open cut at this time (Figure 6a). While in others, the column had started to curl inwards towards the region of the cut (36% of the anemones cut in halves and 7% of those cut in quarters), or the column had completely covered the cut (39% of the anemones cut in half, Figure 6b). After 28 d, scar tissue covered the region where the cut was made in 96% of the cut individuals (Figure 6c).

All of the anemones accepted food 7 d after the start of the experiment (anemones that had been fragmented moved the food

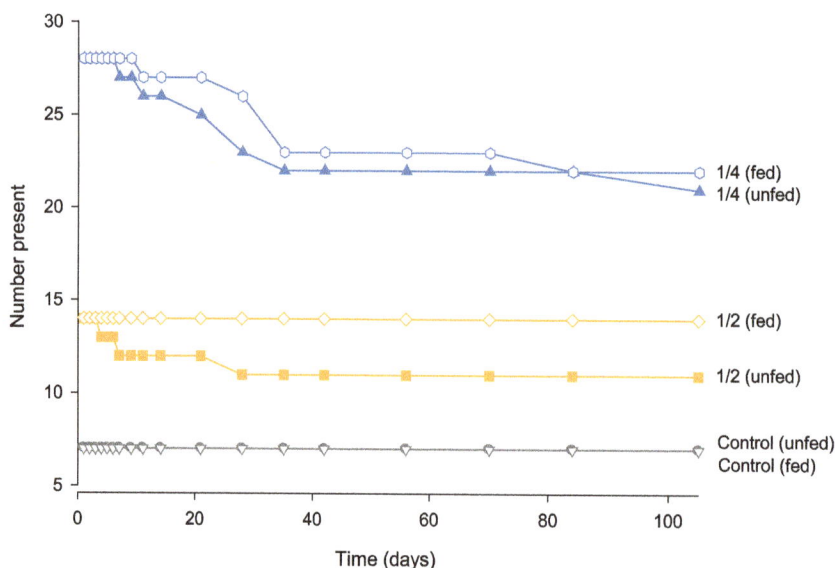

Figure 3. Number of *Entacmaea quadricolor* **present in each treatment over time (in days) during experiment 2.**

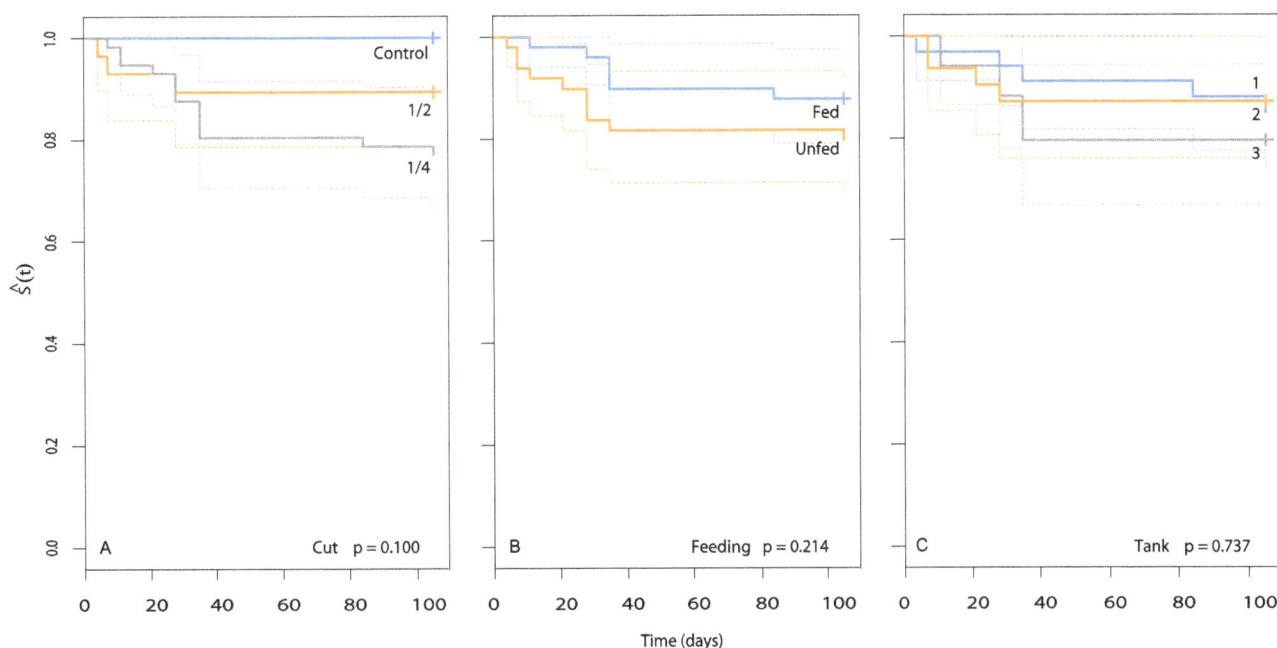

Figure 4. Kaplan-Meier survival [Ŝ(t)] plots for *Entacmaea quadricolor* among different: (a) cut types, (b) feeding treatments, and (c) tanks during experiment 2. Dashed lines represent confidence intervals.

towards the area of the cut). By day 56, off-centre mouths were clearly visible in all of the cut individuals (Figure 6d), except for one of the quarters, and by day 84, 68% of the anemones which were cut in half had mouths that were almost centrally located within the oral disc. This level of regeneration was not found for any of the anemones cut in quarters at any stage of the experiment.

Asexual reproduction did not occur naturally during the experiment; however, sexual reproduction did occur. Eggs were found in two tubs on the 16 January (19 nights after the full moon). One of these anemones had been cut in half, and the other had been cut in quarters 99 d earlier. It is not known if any males spawned, as sperm would have been rapidly removed from the tubs due to the flow-through seawater.

Discussion

E. quadricolor can be propagated asexually by cutting them in half or quarters. The survival rates of the fragments from the sea anemones that were cut in half were greater than has previously been reported (i.e. 89.3 and 93.8% for experiments 1 and 2 respectively vs. 67% in Porat and Chadwick-Furman [22]). Cutting the anemones in quarters produced more individuals; however mortality rates were comparatively higher, the individuals produced were smaller, and also tended to recover and grow more slowly. The latter finding contrasts other growth experiments with intact anemones, where small anemones often tend to grow faster than larger ones [23,24]. This difference likely arises from the smaller fragments needing more time to heal and grow a mouth. Irrespective of whether anemones are cut in halves or quarters, the methods detailed provide a low technology solution for culturing this species that would allow for the year round production of individuals for the aquarium trade. Furthermore, a major advantage of this technique for industry would be that once procured, desirable colour morphs that attract higher prices could be cloned thereby increasing profitability.

Feeding increased ODD growth and reduced wet weight loss, whereas it did not significantly influence PDD growth. Ammonia supplements or the presence of anemonefish (which produce wastes that can be used by their host) have similarly been shown to reduce size loss in *E. quadricolor* [25,26]. It is possible that more frequent feeding may have been necessary to optimise growth during our study. For example, Chomsky et al. [24] found that *Actinia equina* needed to be fed twice a week for significant PDD growth. However, given that the size difference of fed and unfed anemones was greatest in individuals that were not cut, the similarity between feeding treatments in the cut anemones may have been due to the time it took for them to grow a fully functioning mouth (56 d). It is therefore likely that running the experiment for longer would have resulted in a greater feeding effect. Regardless of the treatment, all of the anemones would have received some nutritional benefits from their *Symbiodinium*, microscopic particulate matter, or dissolved organic matter that would have provided energy and potentially influenced size [26–28].

Precisely measuring size change in sea anemones using non-destructive methods is not simple given their ability to expand, contract, and store variable amounts of water in the body [29,30]. In this study, we found that wet weight showed greater variation and less correlation with the linear size measurements of ODD and PDD. Wet weight has been used in a number of studies on sea anemone growth (for example [11,23,24]), however these were generally on smaller anemones. Although we did try to standardise the amount of water contained within the anemones it is likely that some variation among individuals was amplified by their larger body size. Despite this, a significant positive influence of feeding on wet weight was still detected.

It is important to recognise that the findings of this study relating to survival might not apply to all sea anemone species. In anemones that use fission as a reproductive mode, wound healing and regeneration occur by rapid cell proliferation at the wound site [31]. Species that do not naturally asexually reproduce may

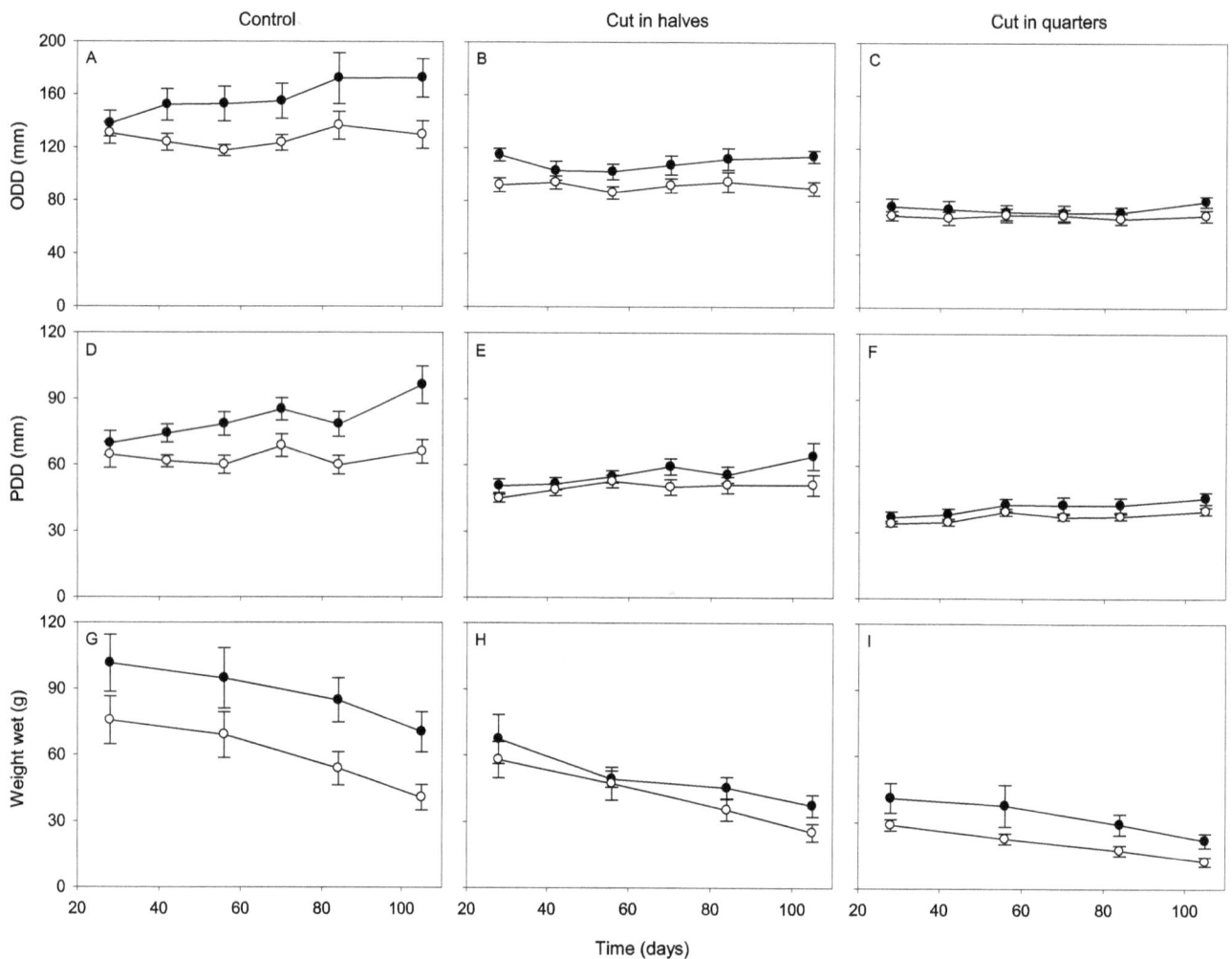

Figure 5. Change in size of *Entacmaea quadricolor* over time (in days) during experiment 2. Estimated means (±SE) of fitted (a–c) ODD, (d–f) PDD, and (g–i) wet weight from generalized linear models.

not have this capacity, and therefore may not be able to recover from cutting. Another important ethical consideration is whether cutting the anemones may cause them pain. Cnidarians possess a basic nervous system comprised of a diffuse nerve net with sensory neuron agglomerations at key structures [32]. Column stimulation produces nerve impulses that cause a closure reflex, which might appear to be a nociceptive response; however, nociceptors in Cnidaria only respond to mechanical stimuli and not heat stimuli and without a central processing unit they are unlikely to experience pain [32].

The practical application of the technique developed during this study would allow the creation of a new industry, which could have economic as well as environmental benefits. Collecting pressures on wild populations could therefore be reduced by supplying the aquarium trade with captive-bred anemones. Fragmentation is already used to supply a small percentage of hard and soft corals (which are closely related to sea anemones) for the trade [2,33,34]. The study species has been propagated by natural asexual reproduction for research purposes [Chadwick and Delbeek personal communication] and cutting has been used by aquarium hobbyists. Although marine ornamental aquaculture is still in its infancy, it is thought that the propagation of animals for the trade will become more important as further restrictions are

placed on wild collection and consumers become more aware of the potential adverse impacts of these activities [12,13]. The potential for diversification and growth in this sector is promising due to high product value and the benefits to biodiversity conservation; however viability is highly dependent on the price at which individuals can be produced [2,13].

Because aquarium collecting provides employment in rural low-income coastal areas that have otherwise limited resources and economic options, it would be preferable if aquaculture occurred in the areas that are currently exporting anemones [2,5,12]. Given the simplicity of the technique developed during this study it could easily be used without personnel needing a large amount of training, thus allowing livelihoods to be maintained in a more sustainable manner. This may have flow-on benefits such as building the technical capacity of communities and fostering awareness for better stewardship of resources [35]. Furthermore, it may be possible to apply this technique in the field (either in cages or on the reef flat), which would negate the need for captive breeding facilities that are expensive to set up and maintain.

Asexual propagation could also be used for restoration programs and could potentially occur *in situ*. The restocking of areas would help restore the breeding populations needed to help replenish areas that have already been denuded by human or

Figure 6. Recovery of *Entacmaea quadricolor* **fragments after cutting: (a) open cut, (b) column tissue covering the cut, (c) column coloured scar tissue, and (d) off-centre mouth.**

natural disturbances [36]; and therefore have subsequent benefits for anemonefishes that cannot survive in the field without their hosts [10]. Given that this technique would produce individuals with limited genetic variability, the consequences of this would need to be considered [37], along with the causative factors that necessitated restocking in the first place. Management initiatives, such as better regulation and enforcement, may also be needed to address the issues that caused the reductions [36].

The findings of this study demonstrate a feasible solution for reducing aquarium collecting pressures that are impacting host sea anemone abundance in some areas of their distribution [8,38], and could also allow for the restoration of areas already impacted by human or natural disturbances. Culturing these species would offer economic and environmental benefits by providing an alternative source of these high-value species, which provide essential microhabitat for obligate symbiotic anemonefishes.

Acknowledgments

We would like to thank R. Edgar and S. Dalton for help in the field; T. Piddocke, K. Cowden, D. Eggeling and K. Kaposi for help in the laboratory; and N. Chadwick for comments on the manuscript.

Author Contributions

Conceived and designed the experiments: AS KH. Performed the experiments: JH AS KH. Analyzed the data: KH. Contributed reagents/materials/analysis tools: AS KH JH. Wrote the paper: AS KH.

References

1. Whitehead MJ, Gilmore E, Eager P, McMinnity P, Craik W, et al. (1986)Aquarium fishes and their collection in the Great Barrier Reef region. Townsville: GBRMPA. 1–39 p.
2. Wabnitz C, Taylor M, Green E, Razak T (2003) From ocean to aquarium: the global trade in marine ornamental species. Cambridge, UK: UNEP-WCMC.
3. Shuman CS, Hodgson G, Ambrose RF (2004) Managing the marine aquarium trade: is eco-certification the answer? Environ Conserv 31: 339–348.
4. Wood E (2001) Global advances in conservation and management of marine ornamental resources. Aquarium Sciences and Conservation 3: 65–77.
5. Wood EM (2001) Collection of coral reef fish for aquaria: global trade, conservation issues and management strategies. UK: Marine Conservation Society. 80 pp.
6. Edwards AJ, Shepherd AD (1992) Environmental implications of aquarium-fish collection in the Maldives, with proposals for regulation. Environ Conserv 19: 61–72.
7. Gasparini J, Floeter S, Ferreira C, Sazima I (2005) Marine ornamental trade in Brazil. Biodiversity and Conservation 14: 2883–2899.
8. Shuman CS, Hodgson G, Ambrose RF (2005) Population impacts of collecting sea anemones and anemonefish for the marine aquarium trade in the Philippines. Coral Reefs 24: 564–573.
9. Olivotto I, Planas M, Simões N, Holt GJ, Avella MA, et al. (2011) Advances in breeding and rearing marine ornamentals. Journal of the World Aquaculture Society 42: 135–166.
10. Fautin DG, Allen GR (1997) Anemonefishes and their host sea anemones: a guide for aquarists and divers. Perth: Western Australian Museum.

11. Scott A (2012) Effects of feeding on the growth rates of captive-bred *Heteractis crispa*: a popular marine ornamental for aquariums. Bull Mar Sci 88: 81–87.
12. Tlusty M (2002) The benefits and risks of aquacultural production for the aquarium trade. Aquaculture 205: 203–219.
13. Moorhead JA, Zeng C (2010) Development of captive breeding techniques for marine ornamental fish: a review. Reviews in Fisheries Science 18: 315–343.
14. Scott A, Harrison P (2007) Broadcast spawning of two species of sea anemone, *Entacmaea quadricolor* and *Heteractis crispa*, that host anemonefish. Invertebr Reprod Dev 50: 163–171.
15. Scott A, Harrison PL (2007) Embryonic and larval development of the host sea anemones *Entacmaea quadricolor* and *Heteractis crispa*. Biol Bull 213: 110–121.
16. Scott A, Harrison PL (2008) Larval settlement and juvenile development of sea anemones that provide habitat for anemonefish. Mar Biol 154: 833–839.
17. Scott A, Harrison PL (2009) Gametogenic and reproductive cycles of the sea anemone, *Entacmaea quadricolor*. Mar Biol 156: 1659–1671.
18. Scott A, Harrison PL (2005) Synchronous spawning of host sea anemones. Coral Reefs 24: 208.
19. Dunn DF (1981) The clownfish sea anemones: Stichodactylidae (Coelenterata: Actiniaria) and other sea anemones symbiotic with pomacentrid fishes. Trans Am Philos Soc 71: 1–115.
20. Stephenson TA (1929) On methods of reproduction as specific characters. J Mar Biol Assoc UK 16: 131–172.
21. Mire P (1998) Evidence for stretch-regulation of fission in a sea anemone. J Exp Zool 282: 344–359.

22. Porat D, Chadwick-Furman NE (2005) Effects of anemonefish on giant sea anemones: ammonium uptake, zooxanthella content and tissue regeneration. Mar Freshwat Behav Physiol 38: 43–51.

23. Tsuchida CB, Potts DC (1994a) The effects of illumination, food and symbionts on growth of the sea anemone *Anthopleura elegantissima* (Brandt, 1835). 1. Ramet Growth. J Exp Mar Biol Ecol 183: 227–242.

24. Chomsky O, Kamenir Y, Hyams M, Dubinsky Z, Chadwick-Furman NE (2004) Effects of feeding regime on growth rate in the Mediterranean sea anemone *Actinia equina* (Linnaeus). J Exp Mar Biol Ecol 299: 217–229.

25. Porat D, Chadwick-Furman NE (2004) Effects of anemonefish on giant sea anemones: expansion behavior, growth, and survival. Hydrobiologia 530/531: 513–520.

26. Roopin M, Chadwick NE (2009) Benefits to host sea anemones from ammonia contributions of resident anemonefish. J Exp Mar Biol Ecol 370: 27–34.

27. Fitt WK, Pardy RL (1981) Effects of starvation, and light and dark on the energy metabolism of symbiotic and aposymbiotic sea anemone, *Anthopleura elegantissima*. Mar Biol 61: 199–205.

28. Muller-Parker G, Davy SK (2001) Temperate and tropical algal sea anemone symbiosis. Invertebr Biol 120: 104–123.

29. Hirose Y (1985) Habitat, distribution and abundance of coral reef sea-anemones (Actiniidea and Stichodactylidae) in Sesoko Island, Okinawa, with notes on expansion and contraction behavior. Galaxea 4: 113–127.

30. Leal MC, Nunes C, Engrola S, Dinis MT, Calado R (2012) Optimization of monoclonal production of the glass anemone *Aiptasia pallida* (Agassiz in Verrill, 1864). Aquaculture 354: 91–96.

31. Passamaneck YJ, Martindale MQ (2012) Cell proliferation is necessary for the regeneration of oral structures in the anthozoan cnidarian *Nematostella vectensis*. BMC Dev Biol 12: 34.

32. Smith ESJ, Lewin GR (2009) Nociceptors: a phylogenetic view. J Comp Physiol A 195: 1089–1106.

33. Borneman EH, Lowrie J (2001) Advances in captive husbandry and propagation: An easily utilized reef replenishment means from the private sector? B Mar Sci 69: 897–913.

34. Olivotto I, Planas M, Simões N, Holt GJ, Avella MA, et al. (2011) Advances in breeding and rearing marine ornamentals. J World Aquacult Soc 42: 135–166.

35. Purcell SW (2012) Principles and science of stocking marine areas with sea cucumbers In: Hair CA, Pickering TD, Mills DJ, editors. Asia–Pacific Tropical Sea Cucumber Aquaculture: ACIAR Proceedings, 136. Australian Centre for International Agricultural Research, Canberra. pp.92–103.

36. Bell JD, Bartley DM, Lorenzen K, Loneragan NR (2006) Restocking and stock enhancement of coastal fisheries: potential, problems and progress. Fish Res 80: 1–8.

37. Baums IB (2008) A restoration genetics guide for coral reef conservation. Mol Ecol 17: 2796–2811.

38. Edwards AJ, Shepherd AD (1992) Environmental implications of aquarium-fish collection in the Maldives, with proposals for regulation. Environ Conserv 19: 61–72.

10

Genetic Variability and Population Structure of *Disanthus cercidifolius* subsp. *longipes* (Hamamelidaceae) Based on AFLP Analysis

Yi Yu[1], Qiang Fan[1], Rujiang Shen[1], Wei Guo[3], Jianhua Jin[1], Dafang Cui[2]*, Wenbo Liao[1]*

1 Guangdong Key Laboratory of Plant Resources and Key Laboratory of Biodiversity Dynamics and Conservation of Guangdong Higher Education Institutes, School of Life Sciences, Sun Yat-Sen University, Guangzhou, China, **2** College of Forestry, South China Agriculture University, Guangzhou, China, **3** Department of Horticulture and Landscape Architecture, Zhongkai University of Agriculture and Engineering, Guangzhou, China

Abstract

Disanthus cercidifolius subsp. *longipes* is an endangered species in China. Genetic diversity and structure analysis of this species was investigated using amplified fragments length polymorphism (AFLP) fingerprinting. Nei's gene diversity ranged from 0.1290 to 0.1394. The AMOVA indicated that 75.06% of variation was distributed within populations, while the between-group component 5.04% was smaller than the between populations-within-group component 19.90%. Significant genetic differentiation was detected between populations. Genetic and geographical distances were not correlated. PCA and genetic structure analysis showed that populations from East China were together with those of the Nanling Range. These patterns of genetic diversity and levels of genetic variation may be the result of *D. c.* subsp. *longipes* restricted to several isolated habitats and "excess flowers production, but little fruit set". It is necessary to protect all existing populations of *D. c.* subsp. *longipes* in order to preserve as much genetic variation as possible.

Editor: Ting Wang, Wuhan Botanical Garden, Chinese Academy of Sciences, Wuhan, China

Funding: Support for this study was provided by a grant by the National Natural Science Foundation of China (Project 31170202) to WL, a grant by the National Natural Science Foundation of China (Project 30670141) to WL, a grant by the National Natural Science Foundation of China (Project 31100159) to QF, a grant by the National Natural Science Foundation of China (Project 40672017) to JJ. The funders had no role in study design, data collection and analysis, decision to publish, or preparation of the manuscript.

Competing Interests: The authors have declared that no competing interests exist.

* Email: lsslwb@mail.sysu.edu.cn (WL); cuidf@scau.edu.cn (DC)

Introduction

Disanthus Maxim. (Hamamelidaceae) is a monotypic genus endemic to China and Japan [1]. *Disanthus cercidifolius* subsp. *longipes* is the only subspecies in *Disanthus* which is distributed in China (southern Zhejiang, central and northwestern Jiangxi, and southern Hunan), while its sister, *D. c.* subsp. *cercidifolius*, is endemic to Japan [2]. *D. c.* subsp. *longipes* was first reported by Cheng [3] in 1938, and then revised by Chang [4] in 1948. In a systematic study of *Disanthus* [2], the author believed that the Chinese species of *Disanthus* was a sister population of *D. c.* subsp. *cercidifolius*, which is distributed in warm and humid forests of the Cathayan Land, and also was a Tertiary relic. *D. c.* subsp. *longipes* was characterized by its morphology and preference for humid, acid soils and shady habitats. The inflorescences are paired, axillary; Capitula are 2-flowers, purple; Leaves are heart shaped, green then turning to purple, orange, and red. It is usually a small tree, 2–3 m high, occasionally reaching heights of 6–8 m in forests when growing along streams. Because of severe habitat fragmentation that caused population decline, *Disanthus* was listed in the 1992 Red List of Endangered Plant Species of China [5], the 1994 IUCN Red List of Threatened Species (www.iucnredlist.org), the Key Wild Plants under State Protection [6] and the 2004 China Species Red List as bring a species at high risk of extinction in the wild [7].

Although study of *Disanthus* has attracted many investigators [8–22], little attention has been paid to its genetic analysis and population structure, except Xiao [23]. Xiao investigated the genetic diversity of *D. c.* subsp. *longipes* based on nine allozyme loci and found a higher genetic variation within populations as well as significantly lower variation among populations. However, sampling of Xiao's study was limited to part of the distribution area of *D. c.* subsp. *longipes* only. So a comprehensive study of the populations' genetic structure at different geographic scales is still needed.

Preserving the genetic diversity of endangered species is one of the primary goals in conservation planning. Because survival and evolution of species depended on the maintenance of sufficient genetic variability within and among populations to accommodate new selection pressures caused by environmental changes [24,25]. For endemic endangered species, intraspecific variation is a prerequisite for any adaptive changes or evolution in the future, and have profound implications for species conservation [26,27]. The knowledge of the levels and patterns of genetic diversity is important for designing conservation strategies for threatened and endangered species [28,29]. So identifying variations with molecular markers has provided the abundant information concerning genetic diversity in plant species [30–36]. Amplified fragment length polymorphism (AFLP) [37] is a PCR-based technique which has been successfully applied to the identification and estimation of molecular genetic diversity and population structure [32–34,38–41]. This technique can generate information on multiple loci in a single assay without prior sequence knowledge, and [42]. Using AFLP markers, we would know (1) the degree of genetic diversity within and among populations; (2) which factors might explain genetic variation; and (3) how to apply this information to develop recommendations for management of this endangered species.

Materials and Methods

Ethics Statement

Field studies were approved by Hunan Provincial Bureau of Forestry for collection in Xinning County (1XN), the Dupanglin National Nature Reserve (2DP), the Mangshan National Nature Reserve (3MS), and approved by Guangdong Provincial Bureau of Forestry for collection in Nanling National Nature Reserve (4NL), and approved by Zhejiang Provincial Bureau of Forestry for collection in QianJiang Source National Forestry Park (5QJ) and Zhulong town, Longquan City (6ZL), and approved by Jiangxi Provincial Bureau of Forestry for collection in Guanshan National Nature Reserve (7GS), and Mount Sanqingshan National Park (8SQ).

Specimen collection

The specimens were collected from eight populations from April to June of 2008. All populations but one grew in evergreen and deciduous broad-leaved mixed forests (often lived with species such as in genus *Cyclobalanopsis* and *Sorbus, etc.*) at altitudes of 450–1200 m. The exception (3MS) was the one growing in bamboo forests in the Mangshan Mountains of Yizhang. For molecular analysis, 10–11 individuals per population were sampled. The locations and information of populations are provide in Table 1 and Fig. 1. Before DNA extraction, all the dried leaves were

preserved in silica gel [43]. All the voucher specimens are deposited at the Herbarium of Sun Yat-sen University (SYS).

DNA extraction and AFLP reactions

Genomic DNA was extracted from silica gel-dried leaves using the cetyl trimethylammonium bromide (CTAB) method [44]. The extracted DNA was dissolved in 100 µL of Tris-hydrochloride (TE) buffer [10 mmol/1 Tris-HCl (pH 8.0), 1 mmol/EDTA (pH 8.0)] and used as a template for the polymerase chain reaction (PCR).

AFLP reactions were performed following the method reported by [37] with the following modifications. The restriction digest and ligation steps were done as separate reactions. For the digestion, approximately 500 ng of genomic DNA was incubated at 37°C (EcoRI) or 65°C (MseI) for 2 h in a 20 µL volume reaction containing 10× H Buffer (TOYOBO, Shanghai) and 10 U restriction enzymes EcoRI or MseI. For the ligation, 20 µL of a ligation mix consisting of 10× T4 DNA Ligase (TOYOBO, Shanghai), 1 µL EcoRI-adapter, 1 µL MseI-adapter, and 2 U T4 DNA Ligase was added to the sample and kept at 22°C for 3 h. After ligation, the samples were diluted 10-fold with sterile deionized water (dH$_2$O). A pre-selective polymerase chain reaction (PCR), using PTC-200 thermocycler (MJ research, Waltham, MA) was done using primer pairs with a single selective nucleotide extension. The reaction mix (total volume 20 µL) consisted of 4 µl template DNA from the restriction/ligation step, 1 µL primer (EcoRI/MseI), and 15 µL AFLP Core Mix (13.8 µL dH$_2$O, 1.6 µL MgCl$_2$, 1.6 µL dNTPs (2.5 mM), 1 U Taq DNA polymerase, and 10× H buffer). After an initial incubation at 94°C for 2 min, 20 cycles at 94°C for 20 s, 56°C for 30 s, and 72°C for 2 min, with a final extension at 60°C for 30 min, were performed. The PCR products of the amplification reaction were diluted 10-fold with dH$_2$O and used as a template for the selective amplification using two AFLP primers, each containing three selected nucleotides.

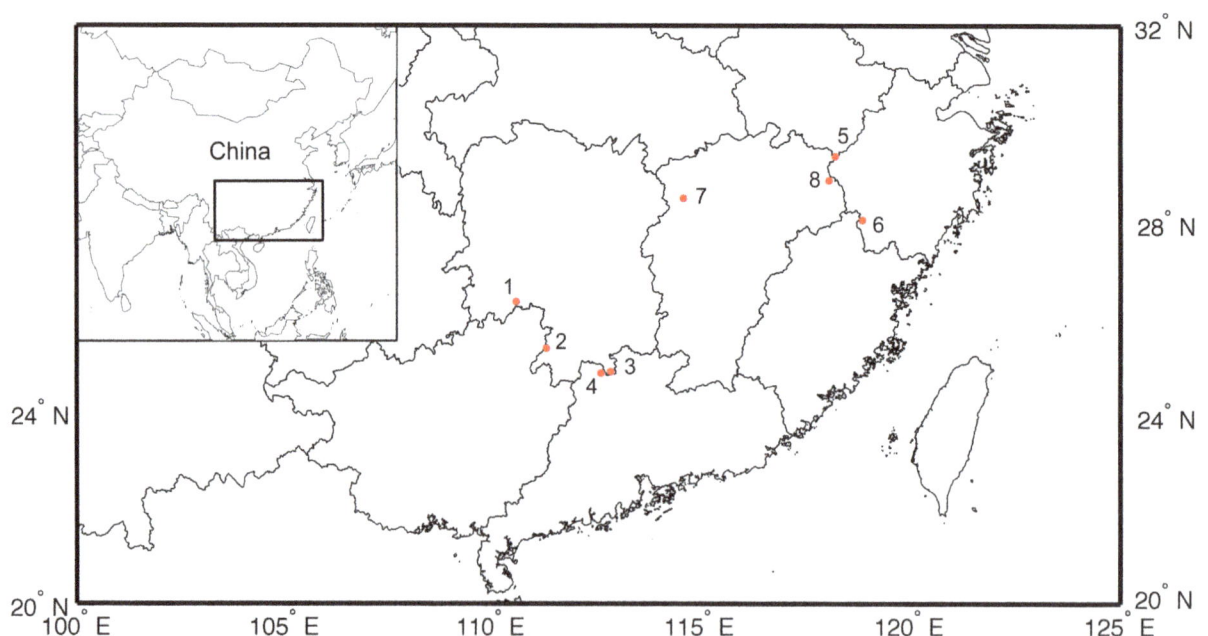

Figure 1. Location of eight populations in two groups sampled in a study of genetic diversity of *Disanthus cercidifolius* subsp. *longipes*. Populations are represented by black dots and located as Table 1. Note: 1:1XN, 2:2DP, 3:3MS, 4:4NL, 5:5QJ, 6:6ZL, 7:7GS, 8:8SQ.

Table 1. Information of location of populations.

ID	GPS co-ordinates	Location of populations	Location type	Altitude (m)	Collection voucher
1XN	26°25'08"N, 110°36'31"E	Xinning county, Hunan Province	mountain slope on the edge of forests	690–740	SHEN Rujiang and GUO Wei
2DP	25°27'43"N, 111°20'20"E	Dupanglin National Nature Reserve, Hunan Province	in bushes by stream	700–850	SHEN Rujiang and GUO Wei
3MS	24°58'12"N, 112°53'16"E	Mangshan National Nature Reserve, Hunan Province	in bamboo forests by stream	680–923	SHEN Rujiang and GUO Wei
4NL	24°56'18"N, 112°39'40"E	Nanling Naitional Nature Reserve, Guangdong Province	mountain slope	680–700	SHEN Rujiang
5QJ	29°23'59"N, 118°12'58"E	Qianjiang Source National Forestry Park, Zhejiang Province	mountain slope	650–700	SHEN Rujiang and GUO Wei
6ZL	28°06'28"N, 118°51'27"E	Zhulong town, Longquan City, Zhejiang Province	mountain slope by streams	1040	SHEN Rujiang and GUO Wei
7GS	28°33'22"N, 114°35'42"E	Guanshan National Nature Reserve, Jiangxi Province	mountain slope	560–600	CHEN Lin
8SQ	28°54'57"N, 118°03'52"E	Mount Sanqingshan National Park, Jiangxi Province	mountain slope in forests	620–1260	Observation team of Sun Yat-sen University

Nine primer combinations labeled with fluorescent 6-carboxy fluorescein (6-FAM) were probed for selective amplification, and only primer combinations with the greatest number of bands (EcoRI/MseI: AAC/CTG; ACA/CTG and ACT/CAT) were selected. Selective amplification was done using the following touchdown PCR conditions: 94°C for 2 min first, then 10 cycles at 94°C for 20 s, and 66°C for 30 s, with a 1°C decrease per cycle then extension at 72°C for 2 min; followed by 20 cycles at 94°C for 20 s, 56°C for 30 s, and 72°C for 2 min. After amplification, 3 μL of the samples were diluted 3-fold with sterile deionized water (dH$_2$O), and mixed with 10 μL formamide and 0.2 μL Size standard-600 (Beckman Coulter, Fullerton, CA). The mix was used for sequence analysis.

Raw data were collected on a CEQ8000 Sequencer (Beckman Coulter). AFLP products were resolved using a Beckman Coulter CEQ8000 genetic analyzer. Semi-automated fragment analysis was performed using the fragment analysis software of the CEQ8000. The chromatograms of fragment peaks were scored as present (1) or absent (0), and a binary qualitative data matrix was constructed. A total of 82 individuals were run twice with all primer combinations.

Data analysis

As a measure of population diversity, the binary data matrix was input to POPGENE version 1.32 [45], assuming Hardy–Weinberg equilibrium. The following indices were used to quantify the amount of genetic diversity within each population examined: the number of AFLP fragments (Frag$_{tot}$), the percentage of polymorphic fragments (Frag$_{poly}$), Nei's (1973) gene diversity (H), and Shannon's information index (I). In addition, the number of unique fragments (Frag$_{uni}$) and DWs, frequency-down-weighted marker values [46], were calculated as measures of divergence. For each population, the number of occurrences of each AFLP marker in that population was divided by the number of occurrences of that particular marker in the total dataset, then these values were summed up [46,47]. The value of DW was expected to be high in long-term isolated populations where rare markers should accumulate due to mutations, whereas newly established populations were expected to exhibit low values [46]. To even out the unequal sample sizes, DWs were calculated using five randomly chosen individuals. The value of DW was calculated by AFLPdat [48], which is based on R version 2.9.0 (www.r-project.org).

Nei's genetic distance [49] of D. c. subsp. longipes populations were calculated using the software POPGENE version 1.32 [45], then cluster analysis was performed based on the unweighted pair group method with arithmetic averaging (UPGMA) [50] using NTSYS pc version 2.1e [51]. In addition, based on the genetic distance index devised by [52], the UPGMA dendrogram of individuals was drawn using TREECON version 1.3b [53]. The robustness of the branches was estimated by 1000 bootstrap replicates.

ARLEQUIN version 3.0 [54] was used to perform an analysis of molecular variance (AMOVA) [55] to assess the hierarchical genetic structure among populations and within populations. The AFLPdat program [48], based on R version 2.9.0 (www.r-project.org), was used to convert the AFLP data matrix to the ARLEQUIN input format. The AMOVA was performed by partitioning genetic variation among and within populations regardless of their geographic distribution. In this study, the traditional F-statistics [56] cannot be used in dominant marker AFLP, therefore Φ-statistics [55] replace F-statistics. The significance of Φ values was tested by 1000 permutations. Φ-statistics are computed from a matrix of Euclidean squared distances between every pair of individuals [40]. Two models of the eight populations were tested to investigate regional relationships. Firstly, we treat all populations as a single group to obtain a value for Φ_{st} as an overall measure of population divergence (a two-level analysis), and then we divided the populations into two groups, the East China region and the Nanling Range (three-level hierarchical analyses).

A distance matrix of Φ_{st} between every pair of populations was calculated in ARLEQUIN as a measure of interpopulation genetic differentiation, from which 1000 bootstrapped replicate matrices were then computed, so gene flow (Nm) based on Φ_{st} [56] could be calculated. Isolation-by-distance was investigated by computing the correlation between geographic distance and genetic distance (Φ_{st}) between every pair of populations and applying the Mantel test, using NTSYSpc version 2.1e [51]. The Mantel z-statistic value was tested non-parametrically by creating a null distribution of z using 1000 random permutations and comparing the observed z value [40].

Table 2. Number of loci evaluated for each of three AFLP primer combinations utilized in assays of 82 individuals of *Disanthus cercidifolius* subsp. *longipes* from eight populations.

Primer combination	Number of loci
EcoRI-AAC/MseI-CTG	144
EcoRI-ACA/MseI-CTG	152
EcoRI-ACT/MseI-CAT	157
Total	453

The AFLP data were also subjected to a principal components analysis (PCA), which may help reveal unexpected relationships among a large number of variables, reducing them to two or three new uncorrelated variables so they retain most of the original information [57]. We chose Jaccard's similarity coefficient [50] to calculate the eigenvalue and eigenvector. The standardized data were projected onto the eigenvectors of the correlation matrix and represented in a two-dimensional scatter plot [57]. Plots of samples in relation to the first three principal components were constructed with populations designated as either populations of the East China region or populations of the Nanling Range. The data from the PCA was analyzed using the computer program NTSYSpc version 2.1e [51]. A two-dimensional representation of genetic relationships among *D. c.* subsp. *longipes* genotypes was carried out using SPSS version 16.0 [58].

STRUCTURE version 2.3.1 [59] was used to investigate structure at the individual level. In this study, we inferred that each individual of *D. c.* subsp. *longipes* comes purely from one of the populations. It was applied using a "no admixture" model, 100 000 burn-in period, and 50 000 MCMC replicates after burn-in. The approach requires that the number of clusters K be predefined, and the analysis then assigns the individuals to clusters probabilistically [40]. We performed ten runs for each value of K (1 to 10). To determine the K value, we used both the LnP(D) value and Evanno's ΔK [60]. LnP(D) is the log likelihood of the observed genotype distribution in K clusters and can be output by STRUCTURE simulation [59]. Evanno's ΔK took consideration of the variance of LnP(D) among repeated runs and usually can indicate the ideal K. The suggested Δk = M(|L(k+1)-2L(k) +L(k-1)|)/S[L(k)], where L(k) represents the kth LnP(D), M is the mean of 10 runs, and S their standard deviation[60,61]. The output uses color coding to show the assignments of individuals in each population to the clusters.

Population structure was also investigated by HICKORY version 1.1 [62], in order to assess the importance of inbreeding in the data and the assumption of Hardy-Weinberg equilibrium. HICKORY makes it possible to evaluate departures from Hardy-Weinberg equilibrium in dominant as well as co-dominant markers [40]. The program AFLPdat [48] was used to convert the AFLP data matrix to the HICKORY input format. The deviance information criterion (*DIC*) criteria, *Dbar*, *Dhat*, and *pD*, for assessing the importance of inbreeding were computed using the default values: Burn-in 5000; sample 100 000; thin 20 [62].

Results

Population genetic diversity

In this study, three of the nine AFLP primer combinations were used (Table 2). The number of fragments for each primer combination (with percentage of polymorphisms within parenthesis), were: EcoRI-AAC/MseI-CTG: 144 (98.61%), EcoRI-ACA/

MseI-CTG: 152 (99.34%), and EcoRI-ACT/MseI-CAT: 157 (96.18%). The length of the fragments varied from 64 bp to 560 bp. The three AFLP primer combinations produced a total of 453 fragments in 82 individuals, of which 444 (98.01%) were polymorphic. The total number of fragments per population (Frag$_{tot}$) varied between 198 (7GS) and 227 (5QJ). The percentage of polymorphism (Frag$_{poly}$) across the eight populations ranged from 43.71% (7GS) to 50.11% (5QJ). The number of fragments that only occur in one population (Frag$_{uni}$) varied between 6 (7GS) and 16 (8SQ). The DW ranged from 56.85 (5QJ) to 111.19 (1XN), mean 76.88, and SD = 21.11. The diversity within a population (H) ranged from 0.1290 (2DP) to 0.1394 (5QJ) (Table 3).

AMOVA

The two-level AMOVA in ARLEQUIN gave a Φ_{st} value of 0.2328 (P<0.001), with of 23.28% variation among populations and 76.72% within populations (Table 4). The three-level hierarchical AMOVA analyses of the two-group models shows that three-quarters of the variation (75.06%) was concentrated within populations, while the between group component (5.04%) was less than the between populations-within-group component (19.90%). All three Φ values were significant based on a 1000 permutation test (Table 4).

Pairwise genetic distance and gene flow

The pairwise genetic distance (Φst) and gene flow (*Nm*) matrix were used to establish the level of genetic divergence among the populations (Table 5). Estimates of pairwise genetic distance using AFLP date ranged from 0.1364 (P<0.001) for the most related populations (4NL and 6ZL), to 0.3428 (P<0.001) in the most divergent populations (3MS and 8SQ). The gene flow (*Nm*) among populations ranged from 0.4792 to 1.5830. Except for one population in Nanling National Nature Reserve (4NL), most of populations' Φst were above 0.2, and the corresponding *Nm* were below 1. These results indicate that the genetic distance of *D. c.* subsp. *longipes* is not close, yet nor dependent on geographic distance.

Relationship between geographic distance and genetic distance (The Mantel tests)

A Mantel test showed no correlation between the geographic distance and the Φst genetic distance (r = 0.38917, P = 0.9920, Fig. 2). This result implies that *D. c.* subsp. *longipes* does not demonstrate a historical pattern of isolation-by-distance.

Principal components analysis

A PCA of the AFLP-based distance data was performed to examine relationships among the populations of *D. c.* subsp. *longipes*. The PCA showed that populations from East China and the Nanling Range clearly do not cluster separately (Fig. 3). The

Table 3. Region, population (ID), sample size (N), total number of fragments per population (Frag$_{tot}$), percentage of polymorphic fragments (Frag$_{poly}$), number of fragments that only occur in one population (Frag$_{uni}$), frequency-down-weighted marker values (DW), Nei's (1973) gene diversity (H), Sannon's index (I) sampled in a study of *Disanthus cercidifolius* subsp. *longipes* (H. T. Chang) K. Y. Pan populations.

Region	ID*	N	Frag$_{tot}$	Frag$_{poly}$	Frag$_{uni}$	DW	H	I
Nanling range	1XN	10	216	47.68	12	111.19	0.1357±0.1779	0.2106±0.2570
	2DP	11	203	44.81	11	56.96	0.1290±0.1790	0.1991±0.2580
	3MS	10	204	45.03	10	93.96	0.1373±0.1839	0.2100±0.2652
	4NL	10	215	47.46	9	58.09	0.1383±0.1710	0.2136±0.2608
East China region	5QJ	11	227	50.11	10	56.85	0.1394±0.1778	0.2172±0.2564
	6ZL	10	209	46.14	12	77.45	0.1393±0.1822	0.2141±0.2631
	7GS	10	198	43.71	6	64.85	0.1353±0.1852	0.2062±0.2662
	8SQ	10	223	49.23	16	95.68	0.1373±0.1775	0.2138±0.2560
All	-	82	453	98.01	-	-	0.1781±0.1633	0.2918±0.2223

Table 4. Analysis of molecular variance (AMOVA) in *Disanthus cercidifolius* subsp. *longipes* for 82 individuals from eight populations.

Model	Source of variation	df	Sum of squares	Variance component	Percentage variance %	Φ-statistics	P
Two levels	among populations	7	1055.95	11.14	23.28	$\Phi_{st}=0.23283$	P<0.001
	within populations	74	2716.05	36.7	76.72		
	Total	81	3772	47.84			
three levels	between groups	1	237.66	2.46	5.04	$\Phi_{ct}=0.05040$	P<0.05
	between populations-within-groups	6	818.29	9.73	19.90	$\Phi_{sc}=0.20955$	P<0.001
	within populations	74	2716.05	36.70	75.06	$\Phi_{st}=0.24939$	P<0.001
	Total	81	3772	48.89			

Computed with ARLEQUIN version 3.0. The *P* values represent the probability of obtaining an equal or more extreme value by chance, estimated from 1000 permutations.

Table 5. Pairwise genetic distance (Φst, lower diagonal, $P<0.001$) and gene flow (Nm, upper diagonal) between eight populations of *Disanthus cercidifolius* subsp. *longipes* based on AFLP data.

	1XN	2DP	3MS	4NL	5QJ	6ZL	7GS	8SQ
1XN	-	1.4673	0.7122	1.1283	0.7637	0.7610	0.8506	0.6140
2DP	0.1456	-	0.8055	1.1586	0.6833	0.7813	0.7398	0.5624
3MS	0.2598	0.2369	-	1.4639	0.6874	0.7389	0.8037	0.4792
4NL	0.1814	0.1775	0.1459	-	1.0727	1.5830	1.2885	0.6644
5QJ	0.2466	0.2679	0.2667	0.1890	-	1.1441	0.7897	1.3148
6ZL	0.2473	0.2424	0.2528	0.1364	0.1793	-	1.1783	0.6917
7GS	0.2272	0.2526	0.2373	0.1625	0.2405	0.1750	-	0.5466
8SQ	0.2893	0.3077	0.3428	0.2734	0.1598	0.2655	0.3138	-

Significance levels are based on 1000 iterations and indicate that the probability that random Φst values are higher than the observed values. $Nm = (1 - Fst)/4*Fst$ [56].

natural divide is separated into three clusters: cluster 1 (1XN, 2DP), cluster 2 (5QJ, 8SQ), and cluster 3, a mixture that includes some individuals from cluster 1 or cluster 2. The first and second principal component axes, PC1 and PC2, accounted for 25.46% and 25.17% of the total variation, respectively.

Hierarchical and cluster analysis

Nei's genetic distances [49] between populations showed two main clusters of populations (Fig. 4). The cluster that comprises 5QJ and 8SQ is clearly differentiated from the others. Populations 1XN and 2DP are another cluster, and population 4NL and 6ZL are grouped with 7GS, although the UPGMA dendrogram of the 82 individuals based on Nei and Li's genetic distance is not clearly grouped (Fig. 5). Although most individuals in the same population cluster together, some individuals in different populations (except 1XN and 3MS) also are clustered together.

In the genetic structure analysis (STRUCTURE), the highest estimate of the likelihood of the data, given the number of clusters chosen, was $k = 10$ (ten clusters). We can't get a clear knee in Lnp(D) (Fig. 6). Evanno's ΔK took consideration of the variance of LnP(D) among repeated runs and usually can indicate the best k. See the Fig. 6, we can find that when the $k = 6$ and we get the highestΔK value. So the best $k = 6$. The diagram (Fig. 7) showing assignment of individuals to the clusters revealed a clinal structure to the data. In the six-cluster model ($k = 6$), gene pool 1 (including 5QJ, 8SQ) and gene pool 2 (including 1XN, 2DP) are restricted with a few from another pool, only being distributed in the southwest (gene pool 1) and northeast (gene pool 2). Gene pool 3 (including 3MS, 4NL), while Gene pool 4, pool 5 and pool 6, consist of 6ZL and 7GS. These 4 pools have the most widespread pattern.

The HICKORY results showed that there is inbreeding in populations of *D. c.* subsp. *longipes* (Table 6), because the *DIC* and *Dbar* parameter were lower in "*Full model*" than in other models. This pattern of results allows the "*Full model*" to be considered best, according to the HICKORY manual [62].

Discussion

Genetic diversity in endemic and endangered species

Accurate estimates of genetic diversity are useful for optimizing sampling strategies aiming at the conservation and management of genetic resources [26,63,64]. According to Hamrick and Godt (1989), there are strong associations between geographic range and genetic diversity [65]. Allozyme analyses concluded that endemic and geographically limited plant species generally possess less genetic variation, due to genetic drift and restricted gene flow [66–68]. However, in our study, the percentage of polymorphic fragments (Frag$_{poly}$) and Shannon's information index (I) are higher than those of endemic species based on allozyme [65]. This may account for the new technique can generate more genetic diversity information. Historical events have also been shown to be responsible for variations in genetic diversity [68]. Numerous allozyme studies and an increasing number of cpDNA and mtDNA studies now provide substantial evidence that putative refugee plant populations harbor higher levels of genetic diversity relative to their likely descendant populations [69]. *Disanthus cercidifolius* subsp. *longipes* is a Tertiary relic, which is distributed in warm and humid forests of the Cathayan Land in the Tertiary [2,70]. Lots of present endemics, several of which inhabit Pleistocene refugia during the Quaternary glacial period, were able to maintain higher levels of diversity because of population stability during the glacial cycle. *Disanthus cercidifolius* subsp. *longipes* may be one of them and preserved mutation during the

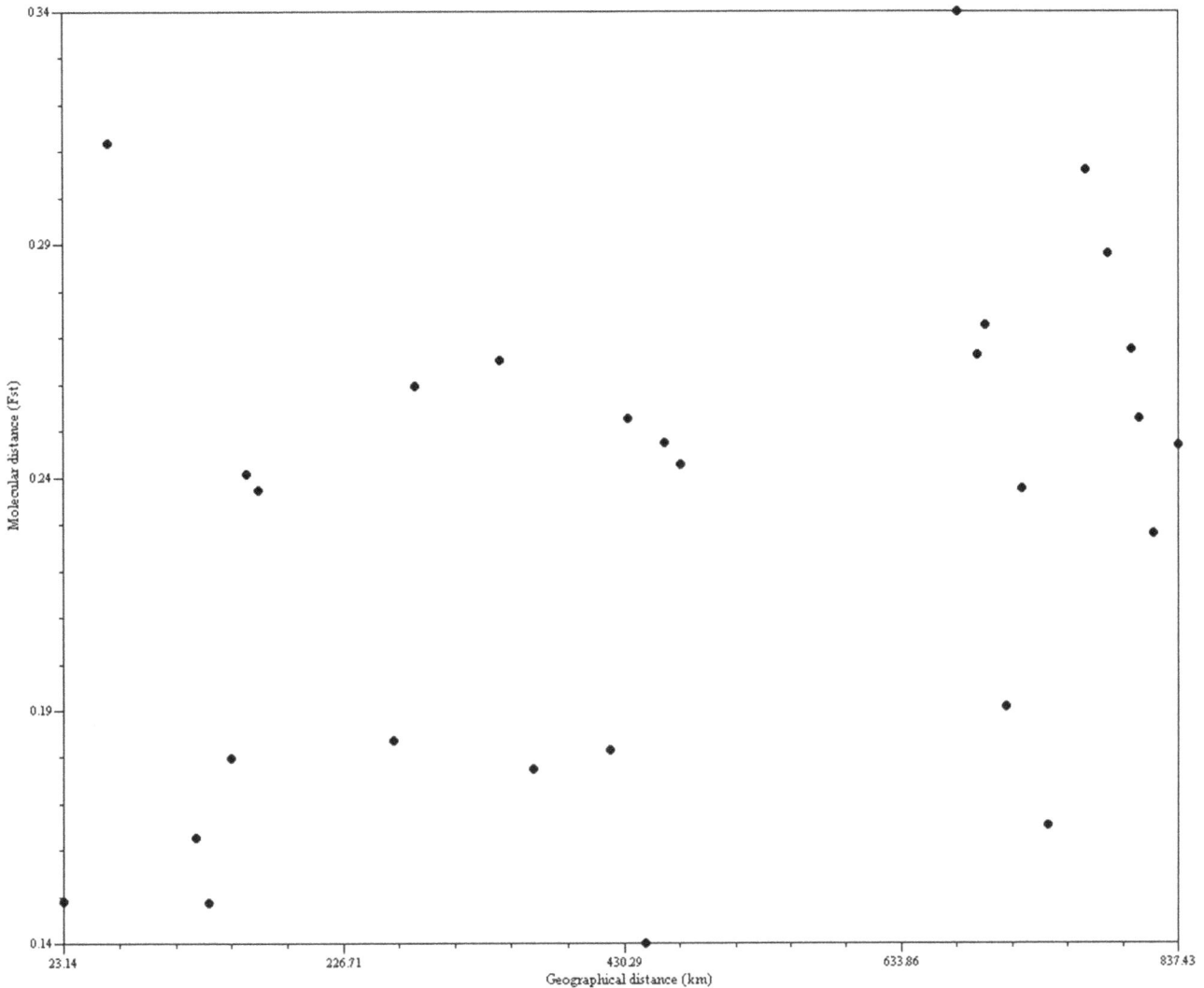

Figure 2. *Disanthus cercidifolius* **subsp.** *longipes,* **bivariate plot showing no correlation between matrices of pair-wise geographic (km) and genetic distance (Φ*st*) in eight populations, comprising 82 individuals.** Computed with NTSYSpc ver. 2.1e.

relatively long glacial period. So it shows high genetic diversity [69,71]. Compared by AFLP, RAPD, and ISSR markers, the genetic diversity indices in this study revealed an intermediate level of genetic diversity of *D. c.* subsp. *longipes* compared to other endangered species [34,41,66,72,73].

Results of studies of genetic diversity within populations show that all populations were on the similar level, except for the Dupanglin National Nature Reserve (2DP) which is shown to be lower than the others. The lack of genetic diversity resulting from the total existing individual in population 2DP is limited (fewer than 100 individuals). In practice, a larger population often has higher genetic variation [72,74].

Genetic variation and gene flow

In this study AMOVA showed that the largest portion (76.72%) of genetic variance is contributed by genetic variation within populations (Φ_{st} value 0.23283) and only a small portion (23.28%) is due to differences among groups. Some other studies of endemic and endangered species show a similar pattern [34,41,66,72,73]. Moderate or high diversity and low population partitioning in rare plants have previously been attributed to a number of factors

[66,75–77], including insufficient length of time for genetic diversity to be reduced following a natural reduction in population size and isolation; adaptation of the genetic system to small population conditions; recent fragmentation (human disturbance) of a once-continuous range, i.e., genetic system; and extensive gene flow due to the combination of bird pollination and high outcrossing rates [66,78]. The direct estimate of gene flow (*Nm*) based on Φ_{st} 0.23283 was 0.82374, which means that the number of migrants per generation is lower than one. The present distribution of *D. c.* subsp. *longipes* is restricted to several isolated habitats [5,7]. However, fossil and palynological evidences suggested that, *Disanthus* was wide spread in warm and humid forests in the Tertiary [2,70]. Accordingly, a reasonable hypothesis is that the modern range of *Disanthus* species was the result of population fragmentation and contraction after the Quaternary glacial cycles. Considering that *D. c.* subsp. *longipes* has poor success in pollination, "excess flowers production, but little fruit set" [79], and fragmented habitat, inbreeding, which generally leads to decreased fitness, or inbreeding depression, of a population, some of the above causes result in the endangered status of *D. c.* subsp. *longipes*.

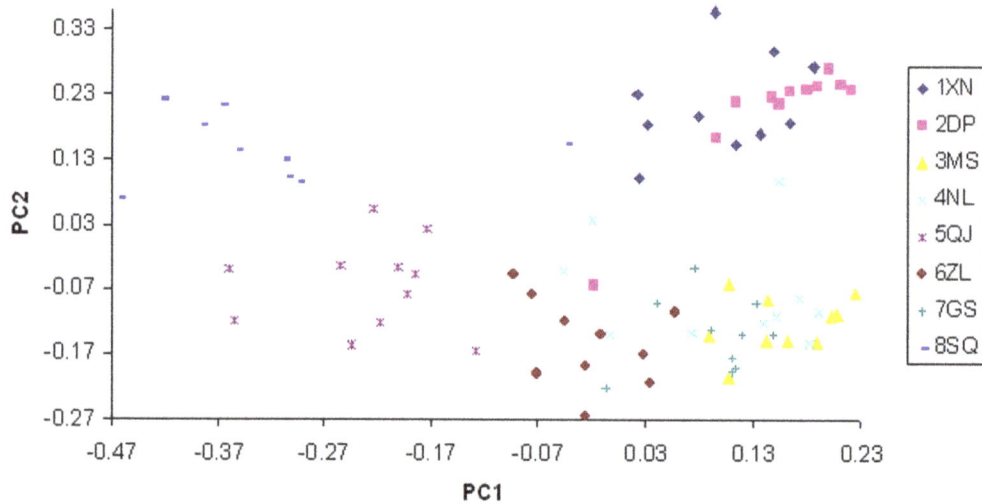

Figure 3. Principal coordinates analysis of 82 individuals from eight populations, based on dissimilarity matrix (Jaccard's coefficient). Accessions are plotted according to the values of first (x axis) and the second (y axis) components and with different symbols according to population. Principal coordinate axes shown (pc1 and pc2) represent 25.46% and 25.17% of respective variance in the dissimilarity matrix.

Genetic structure at different hierarchical levels

Although AFLP markers are dominant, they provide no information on heterozygote frequencies, and our investigation provides no direct information on the reproductive strategy of *D. c.* subsp. *longipes*. In general, outcrossing and long-lived seed plants maintain the most genetic variation within populations, while predominantly selfing, short-lived species harbor compara-tively higher variation among populations [65,80]. According to data based primarily on allozyme analysis, $\Phi_{st} = 0.2$ for outcrossing species and 0.5 for inbreeders [65]. The overall Φ_{st} value 0.23283 and the gene flow value for *D. c.* subsp. *longipes* are similar to other species that outcross [81–84]. Permutation tests of the fixation index ($\Phi_{st} = 0.23$) indicated significant genetic structuring. The populations of *D. c.* subsp. *longipes* spread from east to west (Fig. 1). And in the field investigation of our study, we

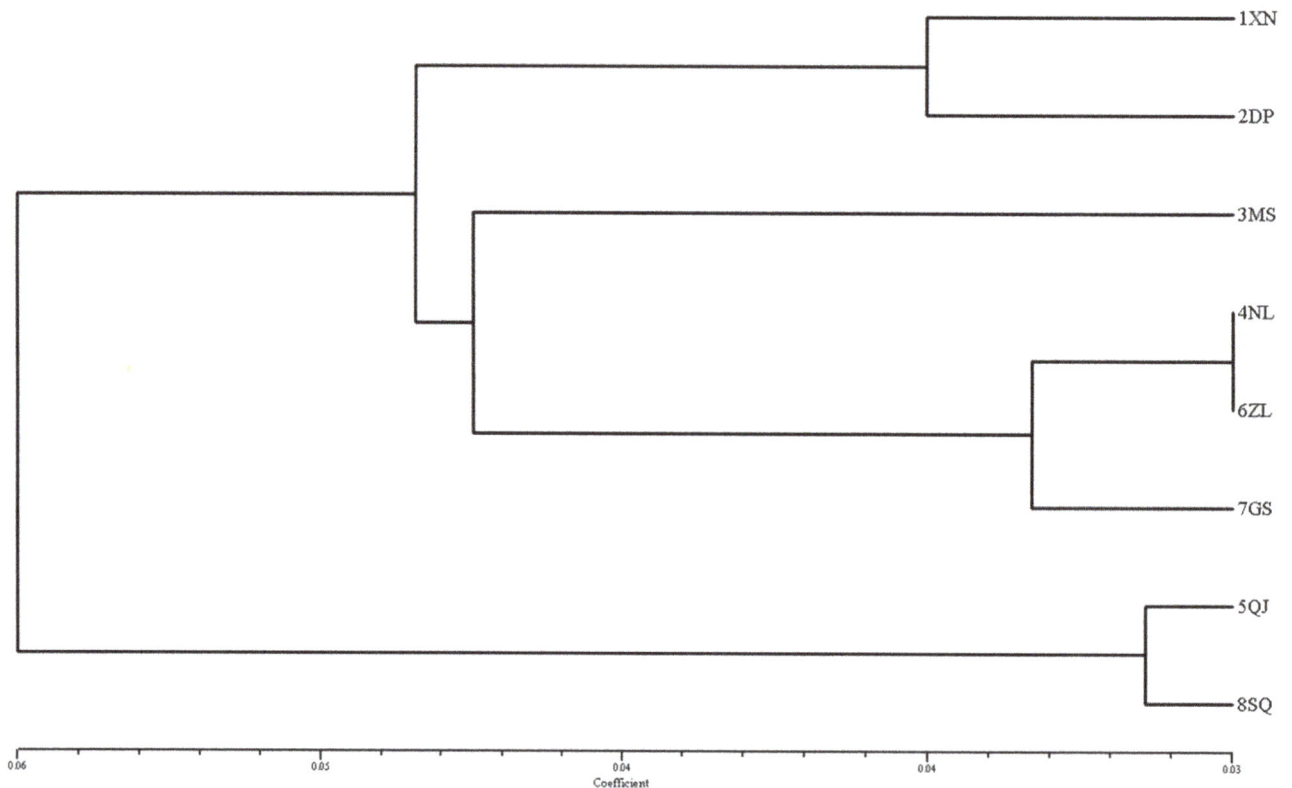

Figure 4. UPGMA dendrogram based on genetic distance. Note: A = 1XN, B = 2DP, E = 3MS, G = 4NL, H = 5QJ, I = 6ZL, J = 7GS, K = 8SQ.

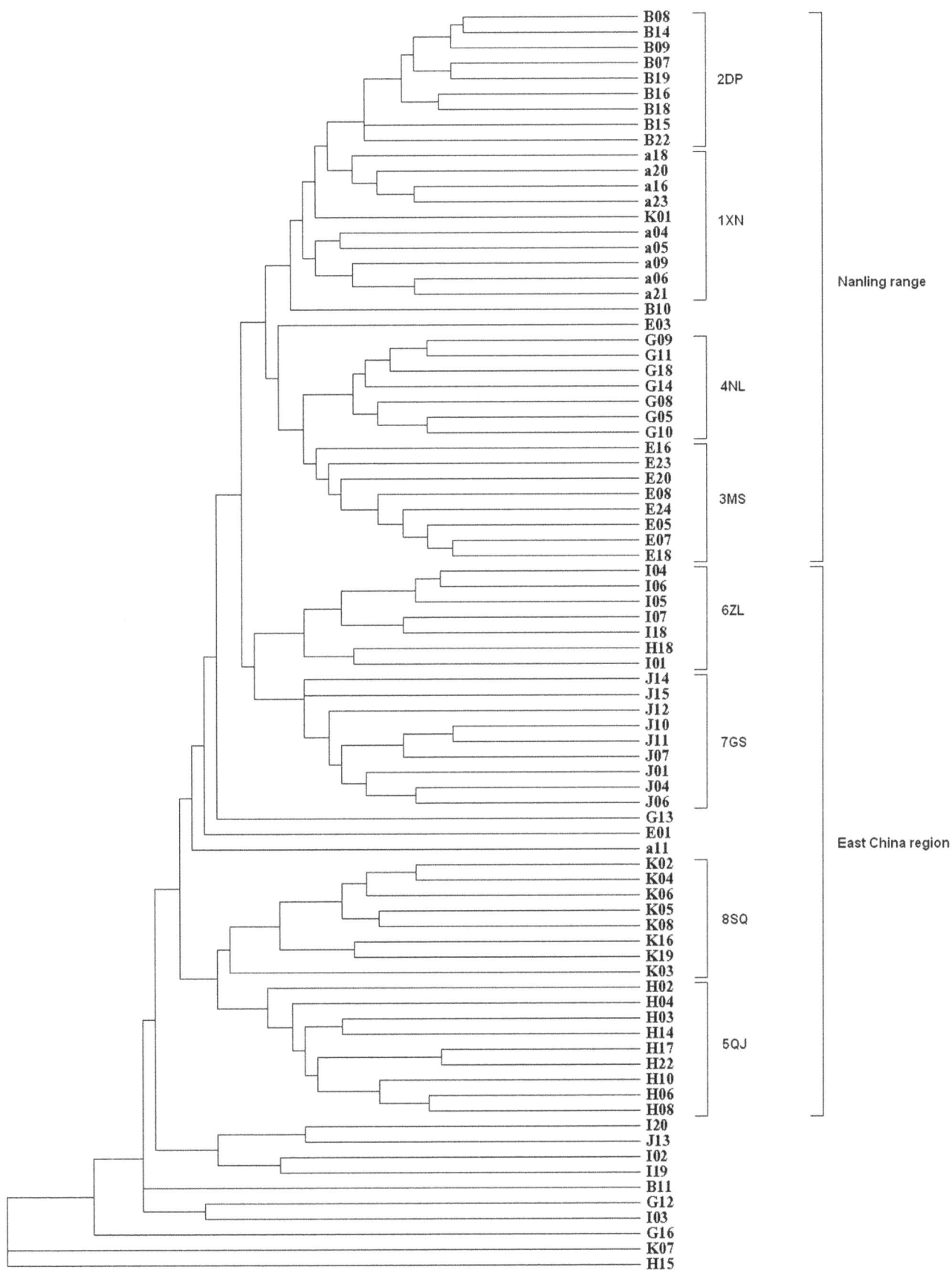

Figure 5. The UPGMA dendrogram based on Nei & Li's genetic distance.

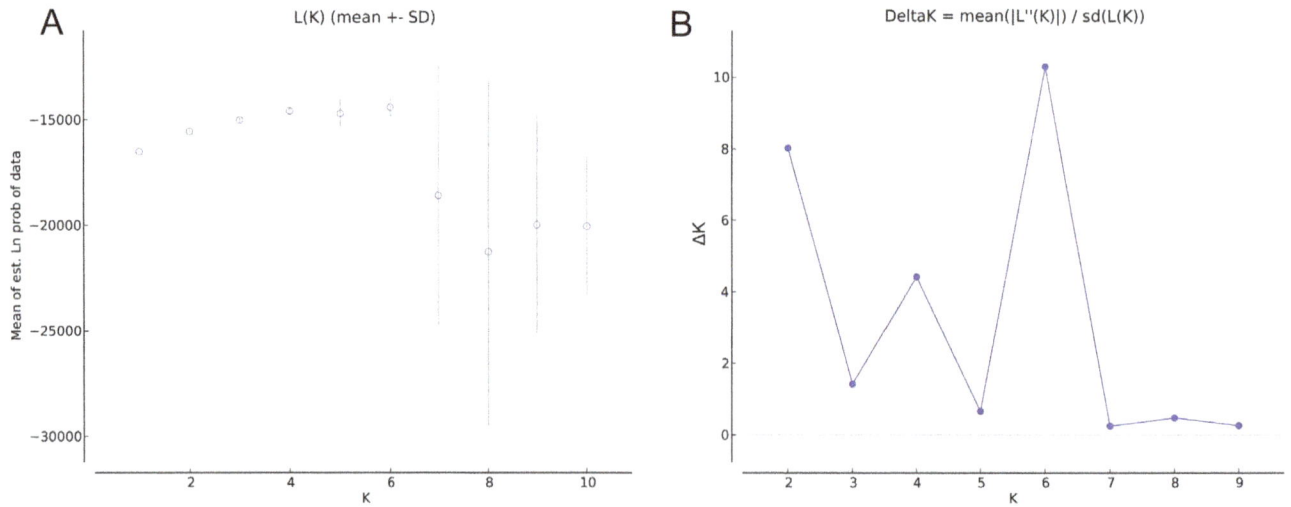

Figure 6. The mean LnP(D) and ΔK over 10 repeats of STRUCTRUE simulations. A. Mean lnP(D) value with k = 1–10. B. ΔK value with k = 2–9.

found that *D. c.* subsp. *longipes* is one of the dominant species in its habitat.

HICKORY software analysis suggests that *D. c.* subsp. *longipes* populations are not genetically differentiated. There is inbreeding in populations of *D. c.* subsp. *longipes* (Table 6), because the *DIC* and *Dbar* parameter were lower in "*Full model*" than that in other models [62]. While inbreeding generally leads to decreased fitness (inbreeding depression) of a population, it also can be advantageous, allowing plants to adapt to disadvantageous conditions [85–88]. *D. c.* subsp. *longipes* originally had an outcrossing system, but under some conditions (e.g., poor efficiency in wind or insect pollination), it would adopt inbreeding or mixed systems to reduce the risk of reproductive failure. This result is consistent with Xiao's conclusions [89] from his study of the reproductive ecology of *D. c.* subsp. *longipes*. If we use co-dominant markers, it most likely would provide an affirmative result and better understanding of the processes.

Hierarchical cluster analysis (Fig. 5) revealed that most individuals of a specific population were grouped together, despite the mixing-in of some individuals from different populations. Most geographically close populations tended to cluster together. The DW values show that the two close populations are clearly divergent, that one is long-term isolated (DW value higher) and the other is newly established (DW value lower). The DW value of 7GS is close to average, but its site is isolated. It might be that there once had been other populations between 6ZL and 7GS. In fact, there is a population of *D. c.* subsp. *longipes* in the Junfeng Mountains, Jiangxi Province, according to Pan's data [2], but we have not obtained any specimens for analysis since 2008. Moreover, although *D. c.* subsp. *longipes* in the Jinggang Mountains, Jiangxi Province, might have been transplanted from Guanshan National Nature Reserve based on the data [90], we found a large population area of *D. c.* subsp. *longipes* in the Jinggang Mountains in October 2009, evidence that the present distribution of *D. c.* subsp. *longipes* appears to be relic from a once extensive range and population.

Genetic structure analysis of the individual samples using STRUCTURE shows that six gene pools are represented in the data, and each gene pool is relatively independent except a few mixtures from other clusters. 1XN and 2DP are the first cluster separated from the other populations; 5QJ and 8SQ form the distinct cluster 2 from the others; 3MS and 4NL also cluster with each other and become distinct from rest populations. These three clusters show high correlation with geographic distribution. And the rest three clusters mixed in 6ZL and 7GS. There is a close relationship between 6ZL and 7GS. It shows that 6ZL is related to 7GS rather than to 5QJ or 8SQ, although 6ZL is close to 5QJ and 8SQ in geographic location. It seems likely that there were some populations between current populations 6ZL and 7GS, or gene flow could have been accomplished by pollen or seed dispersal.

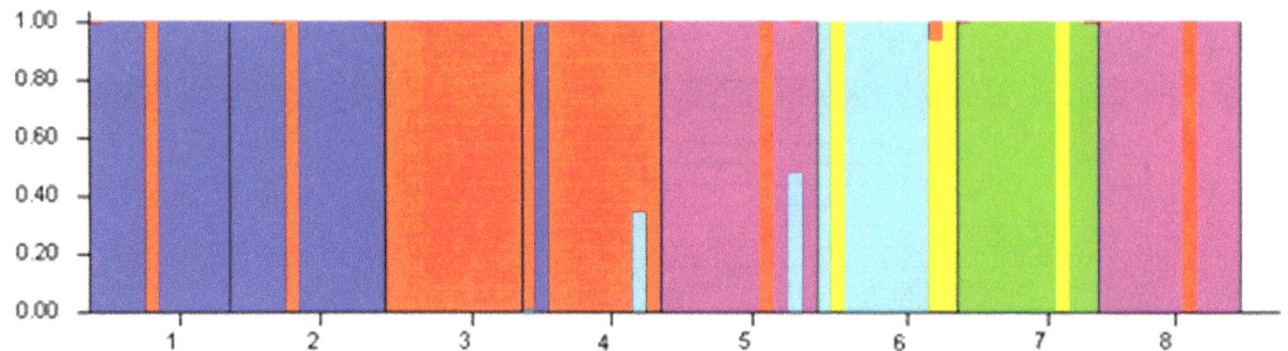

Figure 7. Diagram of results from STRUCTURE(k = 6).

Table 6. Genetic structure analysis using HICKORY ver. 1.1 [62].

Model	Parameter				
	Dbar	Dhat	pD	DIC	f
Full	6750.93	5395.03	1355.9	8106.82	0.980572
f = 0	6780.4	5350.93	1429.47	8209.87	
Theta = 0	12092.5	11716.4	376.128	12468.6	0.994715
f free	6844.98	5363.43	1481.55	8326.53	0.500214

Default values for computations were used as follows: Burn-in 5000; sample 100 000; thin 20.

The results of PCA based on AFLP data indicate a different result, which the eight populations form a triangle that has 1XN and 2DP at one angle, 5QJ and 8SQ at a second angle, but 3MS and 4NL mixed with 6ZL and 7GS in the third angle. The reasons for divergence of the East China region and the Nanling Range are not clear.

Guide for conservation measures

In the face of the species extinction generated by human beings [91,92], the urgent need for the conservation of biodiversity is global. In fact, many endemic species became endangered by the loss or fragmentation of habitats and change in natural conditions [26,29,64,93,94]. *D. c.* subsp. *longipes*, with its narrow archipelago-like distribution, habitat fragmentation, and reproductive-physiologic barriers, is typical species that can be easily affected. Although *D. c.* subsp. *longipes* is endangered in China, it has not yet been given high priority for protection. The original aim of this work is to provide some insight into the population biology of *D. c.* subsp. *longipes* in order to guide conservation measures. The guides which can be suggested by our study listed below. Firstly, Most of the genetic variation is located within populations, so we need to protect all the existing populations of *D. c.* subsp. *longipes* in order to preserve as much genetic variation as possible. Secondly, we should promote the further research and technology development to enable better reproduction of *Disanthus*. Finally, it is necessary to enforce the governmental law to forbid people stealing or purchasing wild *Disanthus*.

Conclusions

The results of the AFLP on populations of *Disanthus cercidifolius* subsp. *longipes* show a pattern of high within-population diversity and low among-population divergence. Although the distribution of *D. c.* subsp. *longipes* could be grouped as the East China region and the Nanling Range, the population divergence is clear. The estimates of genetic differentiation and gene flow suggest that the species primarily outcrosses, but can resort to mixed reproductive strategies. These patterns of genetic diversity and levels of genetic variation may be the result of *D. c.* subsp. *longipes* restricted to several isolated habitats and "excess flowers production, but little fruit set". The status of relic and their low reproductive success results in *D. c.* subsp. *longipes* being endangered.

Acknowledgments

We are grateful to the anonymous reviewers and the editor for their critical review; their comments substantially contributed to the revision and improvement of this work. We thank Dr. Sufang Chen for her valuable advice on writing.

Author Contributions

Conceived and designed the experiments: YY RS WL. Performed the experiments: YY RS. Analyzed the data: YY WG QF. Contributed reagents/materials/analysis tools: JJ DC. Wrote the paper: YY WL DC.

References

1. Mabberley DJ (1997) The Plant-book: A Portable Dictionary of the Vascular Plants. Cambridge University Press.
2. Pan K, Lu A, Wen J (1991) A systematic study on the genus *Disanthus* Maxim. (Hamamelidaceae). Cathaya 3: 1–28.
3. Cheng WC (1938) Observations on Mong-shan (Hunan). Science(Sci Soc China) 22: 400.
4. Chang HT (1948) Additions to the Hamamelidaceus flora of China. Sunyatsenia 7: 70.
5. Fu L (1992) China Plant Red Data Book: Rare and Endangered Plants. Beijing: Science Press.
6. State Forestry Administration and Ministry of Agriculture (1999) Key Wild Plants under State Protection. Beijing.
7. Wang S, Xie Y (2004) China Species Red List. Beijing: Higher Education Press.
8. Zavada M, Dilcher D (1986) Comparative pollen morphology and its relationship to phylogeny of pollen in the Hamamelidae. Ann Missouri bot Gard 73: 348–381.
9. Endress PK (1989) A suprageneric taxonomical classification of the hamamelidaceae. Taxon 38: 371–376.
10. Pan KY, Lu AM, Wen J (1991) A systematic study on the genus *Disanthus* Maxim. (Hamamelidaceae). Cathaya 3: 1–28.
11. Li J (1997) Systematics of the Hamamelidaceae based on morphological and molecular evidence. Ph.D. Thesis, University of New Hampshire.
12. Li J, Bogle AL, Donoghue MJ (1999) Phylogenetic relationships in the Hamamelidoideae inferred from sequences of trn non-coding regions of chloroplast DNA. Harvard Papers in Botany 4: 343–356.
13. Shi XH, Xu BM, Li NL, Sun YT (2002) Preliminary study on dormancy and germination of *Disanthus cercidifolius* Maxim var. *longipes* HT Chang seeds. Seed 6: 5–7.
14. Xiao YA, He P, Deng HP, Li XH (2002) Numerical analysis of population morphological differentiation of *Disanthus cercidiifolius* Maxim. var. *longipes* in Jinggangshan. Journal of Wuhan Botanical Research 20: 365–370.
15. Li K, Tang X (2003) The community characteristics and species diversity of the *Disanthus cercidiifolius* var. *longipes* shrubland in the Guanshan nature reserve of Jiangxi Province. Journal of Nanjing Forestry University 27: 73–75.
16. Li X, Xiao Y, Hu W, Zeng J (2005) Isozymes analysis of genetic differentiation of the endangered plant *Disanthus cercidifolius* var. *longipes* HT Chang (Hamamelidaceae). Journal of Jianggangshan University (Natural Sciences) 26: 34–38.
17. Huang S, Fang Y, Tan X, Yan J, Fang S (2007) Effects of different concentrations of NAA on cutting regeneration of *Disanthus cercidifolius* var. *longipes*. Journal of Plant Resources and Environment 16: 74.
18. Gao P, Yang A, Yao X, Huang H (2009) Isolation and characterization of nine polymorphic microsatellite loci in the endangered shrub *Disanthus cercidifolius* var. *longipes* (Hamamelidaceae). Molecular Ecology Resources 9: 1047–1049.
19. Bogle A, Philbrick C (1980) A generic atlas of hamamelidaceous pollens. Contributions to the Gray Herbarium, Harvard University 210: 29–103.
20. Chang HT (1979) Hamaelidaceae. Flora Reipublicae Popularis Sinicae. Beijing: Science Press. pp.36–116.
21. Zhang ZY, Chang HT, Endress PK (2003) *Disanthus* Maximowicz. In: Z. Wu and P. H. Raven, editors. Flora of China. Beijing: Science Press and St. Louis: Missouri Botanical Garden Press.

22. Endress PK (1993) Hamamelidae. In: K. Kubitzki, J. G. Rohwer and V. Bittrich, editors. The Families and Fenera of Vascular Plants. Berlin: Springer-Verlag.

23. Xiao YA (2001) The study on population adaptability and genetic diversity of the endangered species *Disanthus cercidifolius* Maxim. var. *longipes* HT Chang. M.Sc. Thesis, Southwest China Normal University.

24. Soule M, Simberloff D (1986) What do genetics and ecology tell us about the design of nature reserves? Biological Conservation 35: 19–40.

25. Barrett S, Kohn J (1991) Genetic and evolutionary consequences of small population size in plants: implications for conservation. Genetics and conservation of rare plants: 3–30.

26. Ge S, Hong D, Wang H, Liu Z, Zhang C (1998) Population genetic structure and conservation of an endangered conifer, *Cathaya argyrophylla* (Pinaceae). International Journal of Plant Sciences 159: 351–357.

27. Millar C, LiBBY W (1991) Strategies for conserving clinal, ecotypic, and disjunct population diversity in widespread species. Genetics and conservation of rare plants: 149–170.

28. Francisco-Ortega J, Santos-Guerra A, Kim S, Crawford D (2000) Plant genetic diversity in the Canary Islands: a conservation perspective. American Journal of Botany 87: 909.

29. Qiu Y, Li J, Liu H, Chen Y, Fu C (2006) Population structure and genetic diversity of *Dysosma versipellis* (Berberidaceae), a rare endemic from China. Biochemical Systematics and Ecology 34: 745–752.

30. Song J, Murdoch J, Gardiner SE, Young A, Jameson PE, et al. (2008) Molecular markers and a sequence deletion in intron 2 of the putative partial homologue of LEAFY reveal geographical structure to genetic diversity in the acutely threatened legume genus *Clianthus*. Biological Conservation 141: 2041–2053.

31. Perez-Collazos E, Segarra-Moragues JG, Catalan P (2008) Two approaches for the selection of Relevant Genetic Units for Conservation in the narrow European endemic steppe plant *Boleum asperum* (Brassicaceae). Biological Journal of the Linnean Society 94: 341–354.

32. Yang J, Qian ZQ, Liu ZL, Li S, Sun GL, et al. (2007) Genetic diversity and geographical differentiation of *Dipteronia* Oliv. (Aceraceae) endemic to China as revealed by AFLP analysis. Biochemical Systematics and Ecology 35: 593–599.

33. Prohens J, Anderson GJ, Herraiz FJ, Bernardello G, Santos-Guerra A, et al. (2007) Genetic diversity and conservation of two endangered eggplant relatives (*Solanum vespertilio* Aiton and *Solanum lidii* Sunding) endemic to the Canary Islands. Genetic Resources and Crop Evolution 54: 451–464.

34. Jian SG, Zhong Y, Liu N, Gao ZZ, Wei Q, et al. (2006) Genetic variation in the endangered endemic species *Cycas fairylakea* (Cycadaceae) in China and implications for conservation. Biodiversity and Conservation 15: 1681–1694.

35. Pfosser M, Jakubowsky G, Schluter PM, Fer T, Kato H, et al. (2005) Evolution of *Dystaenia takesimana* (Apiaceae), endemic to Ullung Island, Korea. Plant Systematics and Evolution 256: 159–170.

36. Strand AE, Leebens-Mack J, Milligan BG (1997) Nuclear DNA-based markers for plant evolutionary biology. Molecular Ecology 6: 113–118.

37. Vos P, Hogers R, Bleeker M, Reijans M, van de Lee T, et al. (1995) AFLP: a new concept for DNA fingerprinting. Nucleic Acids Res 23: 4407–4414.

38. Armstrong TTJ, De Lange PJ (2005) Conservation genetics of *Hebe speciosa* (Plantaginaceae) an endangered New Zealand shrub. Botanical Journal of the Linnean Society 149: 229–239.

39. Kim SC, Lee C, Santos-Guerra A (2005) Genetic analysis and conservation of the endangered Canary Island woody sow-thistle, *Sonchus gandogeri* (Asteraceae). Journal of Plant Research 118: 147–153.

40. Andrade IM, Mayo SJ, Van den Berg C, Fay MF, Chester M, et al. (2007) A preliminary study of genetic variation in populations of *Monstera adansonii* var. *klotzschiana* (Araceae) from north-east brazil, estimated with AFLP molecular markers. Annals of Botany 100: 1143–1154.

41. Tang SQ, Dai WJ, Li MS, Zhang Y, Geng YP, et al. (2008) Genetic diversity of relictual and endangered plant *Abies ziyuanensis* (Pinaceae) revealed by AFLP and SSR markers. Genetica 133: 21–30.

42. Wolfe A, Liston A (1998) Contributions of PCR-based methods to plant systematics and evolutionary biology. Molecular systematics of plants II: 43–86.

43. Chase MW, Hills HH (1991) Silica gel: an ideal material for field preservation of leaf samples for DNA studies. Taxon 40: 215–220.

44. Doyle JJ, Doyle JL (1987) A rapid DNA isolation procedure for small quantities of fresh leaf tissue. Phytochemical bulletin 19: 11–15.

45. Yeh FC, Yang RC, Boyle TBJ, Ye ZH, Mao JX (1997) POPGENE, the user-friendly shareware for population genetic analysis. Molecular Biology and Biotechnology Centre, University of Alberta, Canada.

46. Schönswetter P, Tribsch A (2005) Vicariance and dispersal in the alpine perennial *Bupleurum stellatum* L.(Apiaceae). Taxon: 725–732.

47. Ortiz M, Tremetsberger K, Talavera S, Stuessy T, García-Castaño JL (2007) Population structure of *Hypochaeris salzmanniana* DC.(Asteraceae), an endemic species to the Atlantic coast on both sides of the Strait of Gibraltar, in relation to Quaternary sea level changes. Mol Ecol 16: 541–552.

48. Ehrich D (2006) AFLPdat: a collection of R functions for convenient handling of AFLP data. Mol Ecol Notes 6: 603–604.

49. Nei M (1978) Estimation of average heterozygosity and genetic distance from a small number of individuals. Genetics 89: 583–590.

50. Sneath PHA, Sokal RR (1973). Numerical Taxonomy, 2nd edn. San Francisco: Freeman.

51. Rohlf FJ (2000) NTSYS-pc: numerical taxonomy and multivariate analysis system, version 2.1. New York: Exeter Software.

52. Nei M, Li WH (1979) Mathematical model for studying genetic variation in terms of restriction endonucleases. Proc Natl Acad Sci USA 76: 5269–5273.

53. Van de Peer Y, De Wachter Y (1994) TREECON for Windows: a software package for the construction and drawing of phylogenetic trees for the Microsoft Windows environment. Comp Applic Biosci 10: 569–570.

54. Excoffier L, Laval G, Schneider S (2005) Arlequin (version 3.0): An integrated software package for population genetics data analysis. Evolutionary Bioinformatics 1: 47–50.

55. Excoffier L, Smouse P, Quattro J (1992) Analysis of molecular mariance inferred from metric distances among DNA haplotypes: Application to human mitochondrial DNA restriction data. Genetics 131: 479–491.

56. Wright S (1950) Genetical structure of populations. Nature 166: 247–249.

57. Nie Z-L, Wen J, SUN H (2009) AFLP Analysis of *Phryma* (Phrymaceae) Disjunct between Eastern Asia and Eastern North America. Acta Botanica Yunnanica 31: 289–295.

58. Inc S (2007) SPSS 16.0 for Windows. SPSS Inc, Chicago, IL.

59. Pritchard JK, Stephens M, Donnelly P (2000) Inference of population structure using multilocus genotype data. Genetics 155: 945–959.

60. Evanno G, Regnaut S, Goudet J (2005) Detecting the number of clusters of individuals using the software STRUCTURE: a simulation study. Mol Ecol 14: 2611–2620.

61. Zhang DL, Zhang HL, Wang MX, Sun JL, Qi YW, et al. (2009) Genetic structure and differentiation of *Oryza sativa* L. in China revealed by microsatellites. Theor Appl Genet 119: 1105–1117.

62. Holsinger K, Lewis P (2003) HICKORY: a package for analysis of population genetic data V. 1.0. University of Connecticut, Storrs, USA.

63. Cardoso M, Provan J, Powell W, Ferreira P, De Oliveira D (1998) High genetic differentiation among remnant populations of the endangered *Caesalpinia echinata* Lam.(Leguminosae-Caesalpinioideae). Molecular Ecology 7: 601–608.

64. Bouza N, Caujapé-Castells J, González-Pérez M, Batista F, Sosa P (2002) Population structure and genetic diversity of two endangered endemic species of the Canarian laurel forest: *Dorycnium spectabile* (Fabaceae) and *Isoplexis chalcantha* (Scrophulariaceae). International Journal of Plant Sciences 163: 619–630.

65. Hamrick J, Godt M (1989) Allozyme diversity in plant species. In: A. H. D. Brown, M. T. A. Clegg, L. Kahler and B. S. Weir, editors. Plant Population Genetics, Breeding and Genetic Resources. Sunderland, Mass: Sinauer. pp.43–63.

66. Ge XJ, Yu Y, Zhao NX, Chen HS, Qi WQ (2003) Genetic variation in the endangered Inner Mongolia endemic shrub *Tetraena mongolica* Maxim.(Zygophyllaceae). Biological Conservation 111: 427–434.

67. Hamrick JL, Godt MJW (1996): Conservation genetics of endemic plant species. In Conservation genetics: case histories from nature. In: Avise JC, Hamrick JL, Editors. New York: Chapman and Hall. pp.281–304.

68. Karron J (1991) Patterns of genetic variation and breeding systems in rare plant species. Genetics and Conservation of Rare Plants: 87–98.

69. Ge XJ, Yu Y, Zhao NX, Chen HS, Qi WQ (2003) Genetic variation in the endangered Inner Mongolia endemic shrub *Tetraena mongolica* Maxim. (Zygophyllaceae). Biol Conserv 111: 427–434.

70. Wu Z, Lu A, Tang Y, Chen Z, Li DZ (2003) The families and gerera of angiosperms in China. Beijing: Science Press.

71. Lewis PO, Crawford DJ (1995) Pleistocene Refugium Endemics Exhibit Greater Allozymic Diversity than Widespread Congeners in the Genus *Polygonella* (Polygonaceae). American Journal of Botany 82: 141–149.

72. Kwon J, Morden C (2002) Population genetic structure of two rare tree species (*Colubrina oppositifolia* and *Alphitonia ponderosa*, Rhamnaceae) from Hawaiian dry and mesic forests using random amplified polymorphic DNA markers. Mol Ecol 11: 991–1001.

73. Lacerda DR, Acedo MDP, Filho JPL, Lovato MB (2001) Genetic diversity and structure of natural populations of *Plathymenia reticulata* (Mimosoideae), a tropical tree from the Brazilian Cerrado. Molecular Ecology 10: 1143–1152.

74. Jian S, Zhong Y, Liu N, Gao Z, Wei Q, et al. (2006) Genetic variation in the endangered endemic species *Cycas fairylakea* (Cycadaceae) in China and implications for conservation. Biodiversity & Conservation 15: 1681–1694.

75. Schaal B, Hayworth D, Olsen K, Rauscher J, Smith W (1998) Phylogeographic studies in plants: problems and prospects. Mol Ecol 7: 465–474.

76. Slatkin M (1987) Gene flow and the geographic structure of natural populations. Science 236: 787.

77. Zawko G, Krauss S, Dixon K, Sivasithamparam K (2001) Conservation genetics of the rare and endangered *Leucopogon obtectus* (Ericaceae). Mol Ecol 10: 2389–2396.

78. Maguire T, Sedgley M (1997) Genetic diversity in *Banksia* and *Dryandra* (Proteaceae) with emphasis on Banksia cuneata, a rare and endangered species. Heredity 79: 394–401.

79. Xiao YA, Zeng JJ, Li XH, Hu WH, He P (2006) Pollen and resource limitations to lifetime seed prodution in a wild population of the endangered plant *Disanthus cercidifolius* Maxim. var. *longipes* H. T. Chang (Hamamelidaceae). Acta Ecologica Sinica 26: 496–502.

80. Huang Y, Zhang C, Li D (2009) Low genetic diversity and high genetic differentiation in the critically endangered *Omphalogramma souliei* (Primulaceae): implications for its conservation. Journal of Systematics and Evolution 47: 103–109.

81. Kothera L, Richards C, Carney S (2007) Genetic diversity and structure in the rare Colorado endemic plant *Physaria bellii* Mulligan (Brassicaceae). Conservation Genetics 8: 1043–1050.

82. Jacquemyn H, Honnay O, Galbusera P, Roldan-Ruiz I (2004) Genetic structure of the forest herb Primula elatior in a changing landscape. Molecular Ecology 13: 211–219.

83. Juan A, Crespo M, Cowan R, Lexer C, Fay M (2004) Patterns of variability and gene flow in *Medicago citrina*, an endangered endemic of islands in the western Mediterranean, as revealed by amplified fragment length polymorphism (AFLP). Mol Ecol 13: 2679–2690.

84. Morjan CL, Rieseberg LH (2004) How species evolve collectively: implications of gene flow and selection for the spread of advantageous alleles. Mol Ecol 13: 1341–1356.

85. Allard R, Jain S, Workman P (1968) The genetics of inbreeding populations. Adv Genet 14: 55–131.

86. Antonovics J (1968) Evolution in closely adjacent plant populations. V. Evolution of self-fertility. Heredity 23: 219–238.

87. Jain S (1976) The evolution of inbreeding in plants. Annual Review of Ecology and Systematics 7: 469–495.

88. Lloyd D (1980) Demographic factors and mating patterns in angiosperms.

89. Xiao YA (2005) Studies on Reproductive Ecology and photosynthetic adaptability of the Endangered Plant *Disanthus cercidifolius* var.*longipes* HT Chang. Ph.D. Thesis, Southwest China Normal University.

90. Liu R (1999) *Disanthus cercidifolius* var. *longipes*. Plants 4: 7.

91. Vitousek P, Mooney H, Lubchenco J, Melillo J (1997) Human domination of Earth's ecosystems. Science 277: 494.

92. Chapin F, Zavaleta E, Eviner V, Naylor R, Vitousek P, et al. (2000) Consequences of changing biodiversity. Nature 405: 234–242.

93. Palacios C, Gonzalez-Candelas F (1997) Lack of genetic variability in the rare and endangered *Limonium cavanillesii* (Plumbaginaceae) using RAPD markers. Molecular Ecology 6: 671–675.

94. Qiu Y, Luo Y, Comes H, Ouyang Z, Fu C (2007) Population genetic diversity and structure of *Dipteronia dyerana* (Sapindaceae), a rare endemic from Yunnan Province, China, with implications for conservation. Taxon 56: 427–437.

11

Local Scale Comparisons of Biodiversity as a Test for Global Protected Area Ecological Performance

Bernard W. T. Coetzee[1,2]*, **Kevin J. Gaston**[3], **Steven L. Chown**[2]

1 Centre for Invasion Biology, Department of Botany and Zoology, Stellenbosch University, Stellenbosch, Western Cape, South Africa, **2** School of Biological Sciences, Monash University, Melbourne, Victoria, Australia, **3** Environment and Sustainability Institute, University of Exeter, Penryn, Cornwall, United Kingdom

Abstract

Terrestrial protected areas (PAs) are cornerstones of global biodiversity conservation. Their efficacy in terms of maintaining biodiversity is, however, much debated. Studies to date have been unable to provide a general answer as to PA conservation efficacy because of their typically restricted geographic and/or taxonomic focus, or qualitative approaches focusing on proxies for biodiversity, such as deforestation. Given the rarity of historical data to enable comparisons of biodiversity before/after PA establishment, many smaller scale studies over the past 30 years have directly compared biodiversity inside PAs to that of surrounding areas, which provides one measure of PA ecological performance. Here we use a meta-analysis of such studies (N = 86) to test if PAs contain higher biodiversity values than surrounding areas, and so assess their contribution to determining PA efficacy. We find that PAs generally have higher abundances of individual species, higher assemblage abundances, and higher species richness values compared with alternative land uses. Local scale studies in combination thus show that PAs retain more biodiversity than alternative land use areas. Nonetheless, much variation is present in the effect sizes, which underscores the context-specificity of PA efficacy.

Editor: Robert Guralnick, University of Colorado, United States of America

Funding: This work was funded by the National Research Foundation of South Africa, Stellenbosch University, and by a Harry Crossley grant to BWTC. The funders had no role in study design, data collection and analysis, decision to publish, or preparation of the manuscript.

Competing Interests: The authors have declared that no competing interests exist.

* Email: bwtcoetzee@gmail.com

Introduction

Nearly 12% of the world's terrestrial surface is now classified as some form of protected area (PA) [1]. Indeed, the designation and maintenance of PAs are considered key global strategies to address the growing extinction crisis [1]. The unprotected world has been so transformed by human activity that it can now be characterized more readily by a set of human biomes than by the classic biogeographic regions [2]. Therefore, affording an area protection, a long-standing and current centrepiece of conservation strategy [3], appears to be an effective means of conserving its biodiversity features. Conservation scientists have rightly been concerned, however, that simple assumptions of positive ecological performance may be misleading [4–9]. Studies have recognized that effective PA management is key to biodiversity protection [7], and demonstrated that PA designation achieves good conservation return on investment at a relatively low cost [10]. Evidence exists, however, that in many cases PA systems are inefficiently planned to maximize benefits to biodiversity often owing to their spatial location [4,6], and worrying declines in biodiversity even within PAs in particular regions have been identified [5,8].

Much interest has focused on determining PA effectiveness in terms of preventing landscape cover changes (e.g. [11–13]), but these assessments serve only as a proxy for PA performance, as the measures used cannot necessarily capture the implications of land use change for biodiversity features. Where the latter are investigated, outcomes are typically available for specific areas, such as the tropics [8], or particular taxa, such as birds [14] or mammals [5]. Given that negative pressures on biodiversity and evidence for population declines are global in extent [15,16], the overall significance of terrestrial PAs for maintaining biodiversity values thus remains unclear. Protected area policy demonstrating their efficacy should ideally be evidence-based, that is, informed by rigorously established objective scientific evidence, as should be the case for conservation policy generally [17]. However, such evidence is not as well developed as it should be [18], despite urgent calls for so doing both in the scientific [18,19] and policy [3] arenas.

Protected area ecological performance would best be assessed by determining for every established PA what the overall biodiversity status is compared with what would have happened in the absence of protection. Plainly such a comparison cannot readily be achieved. One experimentally tractable alternative is the assessment of biodiversity before and after land cover change, but such studies are extremely rare (although see [20] for a notable exception). The scarcity of data to enable comparisons before/after PA establishment almost invariably necessitates comparisons of each PA with some other area that is unprotected, but similar in all but this designation. As a consequence, assessments of PA performance are typically restricted to small spatial scales and particular taxa (e.g. [14,21]), but there is no clear indication of the generality of their often-contrasting outcomes [18,21,22]. The biodiversity response when comparing a PA with some other area that is unprotected can vary widely, with different studies finding both higher and/or lower biodiversity values across areas [8,20–22]. Results from such studies suggest that PA ecological performance is context-specific and can be influenced by several local factors [8,18,22]. As a consequence, the generality of PA efficacy in maintaining biodiversity across regions remains unclear.

Here we use local scale studies comparing biodiversity between PAs and surrounding alternative land use areas to test if PAs contain higher biodiversity values than surrounding areas, using a meta-analysis. Specifically, our aim is to assess the ecological performance of terrestrial PAs, compared with areas in close proximity that are not protected, thus outside PAs, using three key biodiversity attributes: the abundances of individual species (hereafter 'species abundances'), assemblage abundances (summed across species) and assemblage species richness. These are key measures of biodiversity [23]. We also use an information theoretic approach with candidate explanatory variables to explore reasons for the variation in effect size. We then consider sources of bias in interpreting results, and highlight the benefits and shortcomings of our approach in determining PA efficacy.

Methods

Literature search

Our search and data extraction protocol follows best practice guidelines in conducting meta-analysis (see Appendix S1). We used keyword searches in Web of Science, Scopus, and Google Scholar for relevant papers published from 1975–2011, and their references, and included those reporting pairwise comparisons of biodiversity measurements either inside and outside protected areas (PAs) or between areas within PAs (details follow). The initial search string was: (bird* OR mammal* OR reptile* OR amphibia* OR arthropod* OR insect* OR fish* OR plant* OR vegetation*) AND ("protected area" OR "protected areas" OR "national park" OR "national parks" OR "reserve" OR "reserves" OR "game reserve" OR "game reserves") AND (effective* OR inside* OR performance* OR assessment* OR evaluation* OR estimate* OR comparison* OR contrast*) AND (outside* OR adjacent* OR neighbour* OR adjoining* OR bordering* OR near*). We also used unstructured and opportunistic literature searches with sections of the initial string (particularly to identify studies on different taxonomic groups), expert knowledge of available data sets, and results of a recent multi-database systematic review on protected area efficacy [9]. Grey literature (informally published written material [such as reports, theses and books]) was targeted but exceedingly rare, as found by others [9].

Data capture

From suitable papers we retained studies that measured biodiversity responses inside PAs, and outside PAs, in a replicated study design. These data represent three response variables: (i) species abundances (the abundances of individual species), (ii) abundances per assemblage, (iii) species richness per assemblage, following [24]. Species abundance represents indices of abundance; for example counts, density, capture frequency, occupancy estimates and biomass, for a single species both inside and outside PAs, but only in cases where taxonomy was resolved to the species level (species included are listed in the Dataset S1; N = 243). Assemblage abundance represents indices of abundance; for example counts, density, capture frequency, occupancy estimates and biomass, from cases where abundance was reported across species assemblages or could be calculated across sampling sites. We included these data as estimated by the original authors, but ensured that authors followed a replicated study design, and identical calculations of such indices both inside and outside PAs, and so we consider this response variable as an additional indicator of abundance across species groups (following [24]). Species richness included, for example, observed/estimated/rarefied richness, species density and genera/family richness, from cases where richness was reported or could be calculated across sampling sites. The use of these three variables as estimates of biodiversity was constrained by the approaches adopted by the studies we examined. Although biodiversity can be measured in a variety of ways, these three measures are commonly used as effective measures of biodiversity [23].

Most data we used included mean, standard deviation and sample sizes both inside and outside PAs. Where standard deviation was not reported (20%; 300/1484), we calculated it from either standard error or confidence intervals using imputation methods [25]. Studies with no suitable variance measures were omitted. Where the species could not be correctly identified to species level, data were omitted from the species abundance analysis, but included in the assemblage abundance analyses in cases where these values could be summed across groups of unidentified species. Invasive and domesticated species were omitted as they are not considered here to be of conservation interest. Observations on presence or absence of species were omitted as effect size cannot be calculated. Where data were reported across mixed habitats and/or vegetation types, we only took those from matched pairs, i.e. in areas where the same habitat type was reported. To avoid inadequate treatment of fragmentation effects, data from habitat fragments, typically forest fragments, were excluded following [24]. We did not find studies comparing biodiversity features before and after PA establishment. Few studies considered census-area effects [26] directly, but as individual studies had similar (often identical) sampling designs inside and outside the PAs and in consequence comparable sampling areas, we consider the potential confounding effects of sampling area negligible. Data reporting diversity indices were omitted to avoid pseudoreplication [24], as they are secondary (or derived) measures of species richness and/or abundance, and data on demographics or community structure were omitted because the direction of the expected response was not straightforward to interpret [24,27]. WebPlotDigitizer v2.4 [28] was used to capture data from figures, which is considered a robust technique [29].

We found 861 pairwise observations inside and outside PAs from 86 studies distributed amongst 32 countries and 57 PAs that met our criteria (Figure 1; all studies used in the meta-analysis are in Dataset S1; Figure S1 in File S1). During the search, we also discovered comparisons within PAs only, that typically included a pristine baseline site (as judged by the authors) in the PA and an anthropogenically disturbed area also inside the PA (disturbances such as logging, clearing or hunting pressure). To determine if PA designation may offset negative anthropogenic influences, such as land transformation, we compared this portion of the dataset to the comparisons made inside and outside PAs. If the effect sizes of comparisons between areas within PAs are lower than for comparisons made inside and outside PAs, we can infer that PAs offset negative anthropogenic influences, such as land transformation, to a greater degree than no PA designation. We identified an additional 623 such pairwise comparisons from 41 studies between sites within PAs only.

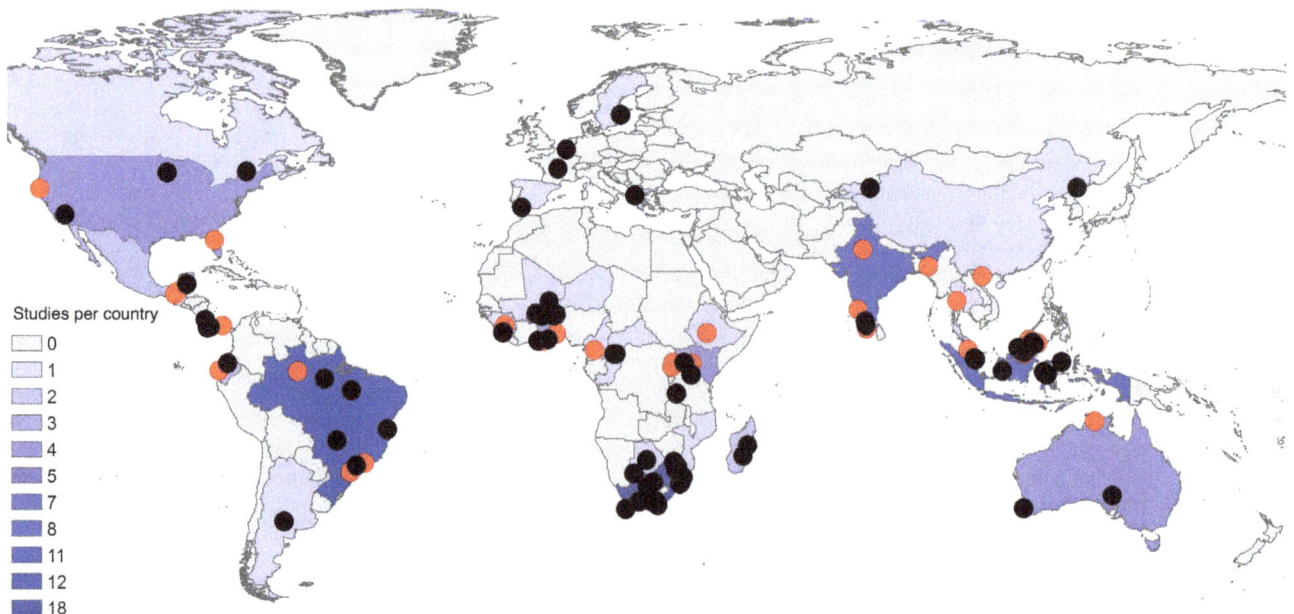

Figure 1. Map of the study sites by the centroid coordinates of protected areas for inside-outside pairwise comparisons (black dots; n = 71) and inside only comparisons (red dots; n = 32). Both categories include data where studies reported across clusters of protected areas.

Data analysis

To estimate effect size, we calculated the Hedges g metric for pairwise comparisons. This is the weighted average of the mean standardized difference (based on pooled variance measures). The metric is the most commonly used, and preferred, to compare pairs of means where variance is available, and unlike others is insensitive to unequal sampling variances in paired groups [27,30]. It is defined as

$$g = \frac{x_{inside} - x_{outside}}{SD_{pooled}}$$

where

$$SD_{pooled} = \sqrt{\left(\frac{(n_{inside} - 1)SD^2_{inside} + (n_{outside} - 1)SD^2_{outside}}{n_{inside} + n_{outside} - 2}\right)}$$

Since Hedges g is a biased estimator of population effect size [27,30], we used the commonly applied conversion factor J to compute the bias-corrected Hedge's g^* metric [27], or $g^* = gJ$, where

$$J = 1 - \frac{3}{4(n_{inside} + n_{outside} - 2) - 1}$$

We then calculated effect sizes using a random-effects model that weights individual comparisons by the inverse of within-study variance plus between study variance, following [24,27,30], with a maximum likelihood variance estimator. Random effects models encompass the variance both between-studies and within-studies, and as such, they are the most appropriate for the majority of ecological meta-analyses because they account for variation in

study-specific effects [27]. All analyses were conducted in the "metafor" package [31] in R [32].

We set the direction of the sign of the effect size as positive when the biodiversity value for PAs was greater than outside PAs, implying that PAs contain higher biodiversity values [33]. We selected a single effect size measure that could incorporate variance, which was standardized across response variables. We calculated effect sizes for pairwise comparisons across data inside and outside PAs (N = 861), and also for comparisons inside PAs only (N = 623). We then calculated effect sizes for the three response variables. For inside and outside comparisons only we further calculated effect sizes for (i) five major taxonomic groups (mammals, birds, herptiles [reptiles and amphibians combined], arthropods, plants), (ii) continents (excluding Antarctica with no data), and (iii) International Union for Conservation of Nature's Protected Area Management Categories (a globally recognised PA categorisation system [34]). The categories are primarily based on their management objectives, in which categories 1–4 reflect stricter goals for biodiversity conservation [1 being the strictest], and 5–6 generally allow extractive use via exploitation of biodiversity features [34], see full definitions in Appendix S2). Finally, (v) we calculated effect sizes for the status of species on the IUCN Red List (a global inventory of the threat status of species according to predetermined criteria [35]).

We calculated two commonly used metrics to characterize heterogeneity between pair-wise comparisons as employed in the "metafor" package [31] in R [32]: the Q-statistic and the I^2 value. Total heterogeneity in effect size can be tested with a Q-statistic where a significant value indicates that the estimated effect size is more heterogeneous than expected by chance [27]. The total heterogeneity of the study, Q_{TOTAL} or Q_T can be calculated as

$$Q_T = \sum_{k=1}^{K} W_k (\hat{\theta}_k - \hat{\mu})^2$$

The I^2 value is presented as the total percentage of heterogeneity that can be attributed to between-study variance [27]. The metric quantifies the heterogeneity by comparing the calculated Q_T to its expected value under the assumption of homogeneity [27], as

$$I^2 = Max \left[100 \times \frac{Q_T - (K-1)}{Q_T}, 0 \right]$$

We conducted three commonly used tests for detecting bias in meta-analysis in the "metafor" package [31] in R [32]: funnel plot, cumulative meta-analysis and Orwin's fail safe N [27,30,31]. A funnel plot graphs effect size against standard error, and assumes that studies with the largest sample sizes will have lower standard error, and so will be near the average effect size, while studies with smaller sample sizes will be spread on both sides of the average effect size. Variation from this assumption can indicate bias, although the source of such bias may be unclear from a funnel plot [27]. However, positive asymmetry is typically taken to mean publication bias, in that those studies with positive effects are submitted and/or accepted for publication with a greater frequency then those with negative effects [27]. A cumulative meta-analysis sorts all pair-wise comparisons by precision, thus starting with the studies with the largest standard error, after which the comparison with the next largest standard error is added and the effect size is recalculated, and so continues iteratively [27]. The resulting graph enables inspection of the development of the observed effect size with the addition of more precise data. Orwin's fails safe N is a metric of the number of studies averaging null results that would have to be added to the observed outcomes to reduce the average effect size to half the observed effect size. All tests for publication bias in this study showed that it was negligible [27] (Table S1; Figure S2 & S3 in File S1).

To address the potential spatial pseudoreplication in the dataset that could arise from multiple responses reported within studies [22], PAs, countries or species, we recalculated effect sizes after sampling one pairwise comparison only per study, PA, country or species, respectively. This resampling was repeated 10 000 times for each of these four parameters and the estimated mean and 95% confidence intervals compared with the overall effect size, which was conducted in R [32].

We used an information theoretic approach [36] to assess the influence of a candidate set of models and variables to explain the variation in effect size, where data were available for all variables. Models tested the influence of (i) pre-planned subgroups in the meta-analysis (variables: response variable, taxonomic group, PA IUCN Category), (ii) design, location and structural attributes of the PAs (variables: continent, latitude, longitude, PA area in km^2, and PA establishment date, using [34]) and (iii) influence of socio-economic conditions of the countries in which PAs are located (variables: World Governance Index [37], Gross Domestic Product (GDP), Country Population size and the Gini coefficient of income inequality [38]). We followed an exhaustive search approach, which entails fitting all possible model formulations, with a Generalized Linear Model (GLM; [39]). We assumed a Gaussian distribution with a log link function, which was identified as the appropriate family and link function by visual inspection of quantile-quantile plots, using the "glmulti" package [40], in R [32]. We selected the highest ranked model based on the lowest Akaike Information Criterion [39,40] value. Furthermore, to address possible pseudoreplication, one pairwise comparison per study was selected at random and the respective GLM model refitted as above. We selected the highest ranked model based on

the lowest Akaike Information Criterion value [39,40], and repeated this procedure 1000 times, to calculate the proportionally highest ranked model for each candidate dataset.

At least some of the variation in effect size may also be accounted for by the scale over which studies were conducted. To test the influence of distance between PA boundaries, and the maximum distance between pair-wise comparisons, a best GLM model by exhaustive fit was performed as above [40]. These independent variables were (i) the maximum distance to protected area boundary within studies, and (ii) the maximum distance between pair wise comparisons within studies, meaning, within each study, the maximum distance between sampling points assigned to all points in that study. Since the data on the distance between comparisons were only available for a reduced subset of pair-wise comparisons (N = 569), we performed a separate GLM as above with only the distance variables. We also estimated Pearson's product moment correlation coefficient between Hedges g* metric values for pair-wise comparisons and the two distance variables in R [32].

Results

The mean effect size using the random effects model, which provides an indication of the general trend across all 861 comparisons, was 0.444 (95% confidence intervals 0.324–0.564; Table 1). Substantial variation was present in the direction and size of effects in response variables for different pairwise comparisons ($I^2 > 87.9\%$; Q-statistic significant at <0.01; Table 1). However, when fitting the random effects model, PAs had higher species abundances (Figure 2A; N = 330), assemblage abundances (Figure 2A; N = 297) and assemblage species richness (Figure 2A; N = 234) than land use areas outside PAs (see Table 1).

Effect sizes for PAs with no IUCN category designation were lower than those with a designation, but remained positive and overlapped with the overall effect size and so we included them here (Table 1). Studies that reported across clusters of PAs rather than individual PAs remained positive and were thus included in the overall assessment of effect size (Table 1). When resampling effect sizes to account for pseudoreplication, they remained positive and overlapped with the overall effect size for both inside-outside PAs and inside PA only comparisons, but were less positive for species responses (Table S2). The variance of these resampled effect sizes also increased, but we note that effect size precision increases with additional data (Figure S3 in File S1). Thus, the present results overall can be considered robust to pseudoreplication.

Although variable, the mean effect sizes confirm that on average PAs contain significantly higher numbers of species and more individuals for mammals, birds, herptiles and arthropods, but the effect is non-significant for plants (Figure 2B; Table 1). Small mammals showed a smaller effect size for species abundance (< 1 kg; N = 25; 0.042; CI: −0.236–0.320) than did large mammals (>1 kg; N = 114; 0.372; CI: 0.131–0.613). These results suggest that while most species benefit from PA establishment, a suite of them, typically plants, fare better outside PAs in typically anthropogenically transformed habitat.

Protected area efficacy by continent generally showed positive effect sizes, apart from South America (strongly negative and only just non-significant) and Australia (also non-significant; Figure 2C). We note that sample sizes for Europe and Australia are low.

Improved biodiversity outcomes with an increase in IUCN management category would seem an obvious *a priori* outcome, but such a simple relationship was not clear (Figure 2D). IUCN

Table 1. Effect sizes (ES), lower bound (lbCI) and upper bound (ubCI) confidence intervals and sample sizes (N), Tau square (I^2), Q statistic (Q) with its p-value (Qp) for at different subgroup designations.

Description	Subgroup	ES	lbCI	ubCI	N	I^2	Q	Qp
Overall	Inside Outside PAs	0.444	0.324	0.564	861	94.9	8620.1	<0.01
Overall	Comparisons inside PAs only	0.172	0.083	0.261	623	46.8	1439.9	<0.01
Overall	PAs with IUCN designation	0.621	0.488	0.755	496	92.9	3622.3	<0.01
Overall	PAs with no IUCN designation	0.161	−0.050	0.372	365	95.9	4910.5	<0.01
Overall	Clusters of PAs	0.560	0.400	0.719	189	86.1	980.9	<0.01
Overall	Unique PAs identified	0.413	0.264	0.562	672	95.8	7638.9	<0.01
Variable	Species abundance	0.517	0.382	0.652	330	87.9	1869.6	<0.01
Variable	Assemblage abundance	0.349	0.083	0.615	297	97.6	4997.7	<0.01
Variable	Assemblage species richness	0.457	0.238	0.676	234	93.2	1666.9	<0.01
Variable	Species abundance (inside)	−0.086	−0.235	0.063	295	50.2	721.4	<0.01
Variable	Assemblage abundance (inside)	0.165	0.044	0.286	152	9.9	270.8	<0.01
Variable	Assemblage species richness (inside)	0.529	0.364	0.694	176	51.7	418.5	<0.01
Taxon	Mammals	0.179	0.010	0.344	216	88.2	1129.6	<0.01
Taxon	Birds	0.657	0.410	0.910	332	97.2	5415.6	<0.01
Taxon	Herptiles	0.487	0.159	0.816	55	89.6	425.4	<0.01
Taxon	Arthropods	0.654	0.421	0.882	152	64.1	406.1	<0.01
Taxon	Plants	0.043	−0.214	0.301	106	95.4	1167.8	<0.01
Continent	Africa	0.450	0.319	0.581	422	92.9	3320.0	<0.01
Continent	Asia	0.986	0.769	1.203	185	71.2	571.4	<0.01
Continent	South America	−0.441	−0.883	0.002	152	98.2	3950.6	<0.01
Continent	North America	0.587	0.340	0.835	78	84.0	387.2	<0.01
Continent	Australasia	0.056	−0.160	0.273	18	4.2	16.9	<0.01
Continent	Europe	2.544	0.362	4.726	6	97.1	133.3	<0.01
IUCN Cat.	IUCN 1	0.316	0.038	0.594	35	0.0	32.1	0.560
IUCN Cat.	IUCN 2	0.754	0.581	0.926	367	94.8	3204.3	<0.01
IUCN Cat.	IUCN 3	0.646	−0.473	1.764	1	NA	0.0	0.999
IUCN Cat.	IUCN 4	−0.099	−0.222	0.025	50	42.1	104.5	<0.01
IUCN Cat.	IUCN 5	1.371	0.838	1.904	27	83.8	112.6	<0.01
IUCN Cat.	IUCN 6	−0.044	−0.382	0.295	16	0.0	6.1	0.978

Table 1. Cont.

Description	Subgroup	ES	lbCI	ubCI	N	I^2	Q	Qp
IUCN Cat.	No IUCN Designation	0.161	-0.050	0.372	365	95.9	4910.5	<0.01
Red List	Not Evaluated	0.596	0.234	0.957	62	91.0	402.8	<0.01
Red List	Data deficient	-0.479	-1.170	0.212	6	85.2	32.4	<0.01
Red List	Least Concern	0.460	0.291	0.628	168	83.4	744.8	<0.01
Red List	Near Threatened	1.027	0.262	1.791	27	95.2	323.8	<0.01
Red List	Vulnerable	0.814	0.333	1.294	28	89.2	195.4	<0.01
Red List	Endangered	0.557	0.266	0.848	22	60.9	50.1	<0.01
Red List	Critically Endangered	-0.119	-0.548	0.311	6	66.3	14.5	0.013
	Small mammals	0.042	-0.236	0.320	25	3.4	21.0	0.637
	Large mammals	0.372	0.131	0.613	114	93.1	796.6	<0.01

PA = Protected area. Cat = Category.

Category 2 PAs, and to a lesser extent Category 1 PAs, had a high positive effect size. However, although the sample sizes are low, so did IUCN Category 5 PAs that allow much extractive use within their borders. PAs with no IUCN designation, and those of categories 4 and 6 had no significant effect and few data are available on Category 3 PAs.

Species listed as Least Concern, Near Threatened, Vulnerable or Endangered by the IUCN generally had greater abundances inside than outside PAs (Figure 3; Table 1). A small sample size of only six observations from one species (*Gorilla gorilla*) that is Critically Endangered was non-significant. However, at least for species abundance, few data on those species of greatest conservation concern, as measured by their Red List status, are available (Figure 3). Indeed, published studies comparing the abundances of highly threatened species both inside and outside PAs are rare, and many such species do not occur in PAs [18,41].

The overall effect size from pairwise comparisons inside PAs only (0.172; 95% confidence intervals: 0.083–0.261) was lower than that of the inside and outside comparisons only (their 95% confidence intervals did not overlap). When fitting the random effects model, pristine areas in PAs only, did not have significantly higher species abundances (−0.086; CI: −0.235–0.063; n = 295), but did have significantly higher assemblage abundances (0.165; CI: 0.044–0.286; n = 152) and significantly higher assemblage species richness (0.529; CI: 0.364–0.694; n = 176) than anthropogenically disturbed areas.

The variation explained by the fitted explanatory models was low, with the meta-analytical and socio-economic models each accounting for about 5% and 7%, respectively, of the variation in effect size (Table S3). By contrast, the PA-model accounted for 25% of the variation (Table S3). For all candidate explanatory variables there were multiple competing best-fit models (Table S4). Distance among comparison sites explained only *c*. 1% of the variation in effect size for studies included in our meta-analysis (Table S5). Indeed, despite being significant the relationship between effect size and the greatest distance between comparison sites was weak (Pearson's r = 0.146; p<0.001) as was the relationship with distance to PA boundary (r = 0.085; p<0.05).

Discussion

An initial assessment of global PA efficacy should be to determine if differences exist in biodiversity between PAs and unprotected land in a direction demonstrating higher biodiversity values in the former. Most of the studies included here did indeed find higher species abundances, assemblage abundance, and species richness inside PAs compared to areas outside them. This pattern holds across taxonomic groups (although non-significant for plants) and continents (although non-significant for South America and Australia).

What mechanisms underlie the non-significant effects for plants and for South America and Australia are not entirely obvious. Clearly, habitat change has major effects on plants, with many studies documenting replacement of particular species and changes in habitat structure [2,24,42,43]. Likewise there has been a growing focus on the scarcity of large old trees [44]. The plant studies analysed here included 53 pairwise comparisons of species richness, 41 of assemblage abundance and 12 of species abundances. All of the 12 species abundance studies were concerned with trees. This balance of investigations to date may account for the lack of an effect, and may rather reflect the globally emerging fingerprint for biodiversity loss under global change, where community turnover, but not necessarily species richness change is observed [43]. In the case of Australia, the small

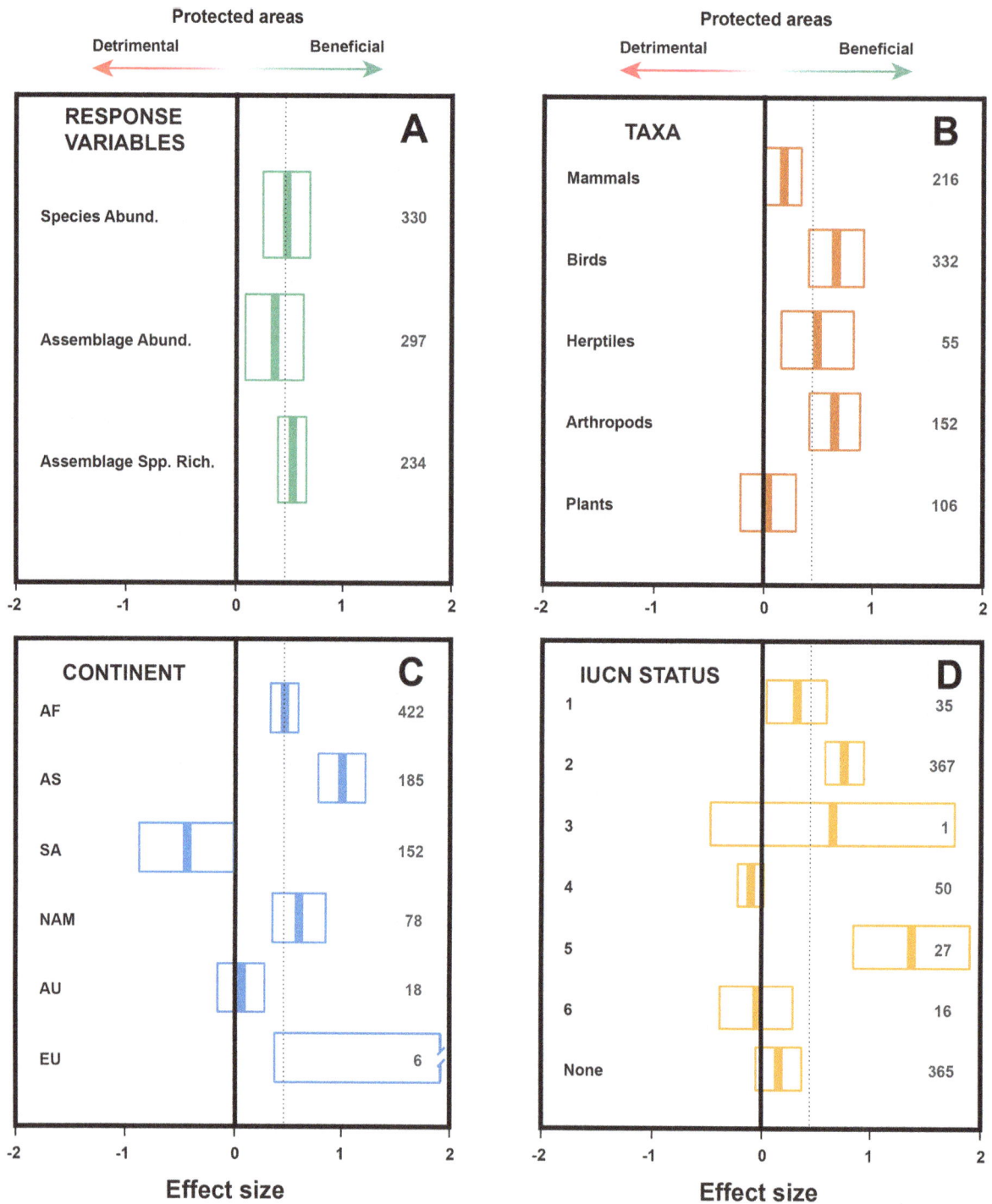

Figure 2. Effect sizes and 95% confidence intervals of response variables (A), taxa (B), continent (C) and the Protected Areas IUCN category (D). Positive boxplot values indicate a net positive impact of protected areas (PAs) on biodiversity. Sample sizes are in grey, the vertical black lines show a zero effect size, while the dashed lines show the overall effect size of 0.444. The effect size for the truncated bar (Europe; Panel C) with large variance due to low sample sizes is 2.54 CI: 0.36–4.73. Abund = Abundance; Spp. Rich. = Species richness; AF = Africa; AS = Asia; SA = South America; NAM = North America; AU = Australasia; EU = Europe. IUCN categories are detailed in Appendix S2.

sample size may be driving the absence of an effect size. Alternatively, most data from Australia (84%; 16/19) came from only one study that documented the recovery of small mammal populations after their isolation from invasive predators [45], and so it is not clear to what extent this outcome reflects general patterns on the continent. For South America the situation is also

difficult to explain. The taxon with the most data for South America was the mammals (53%; 81/152), many of them large (> 1 kg; 54%; N = 44), and their increase outside PAs contributes to the non-significant outcome. One explanation may be hunting pressure. Carrilo et al. [46] found that increasing hunting pressure inside a PA diminished large mammal populations in Costa Rica.

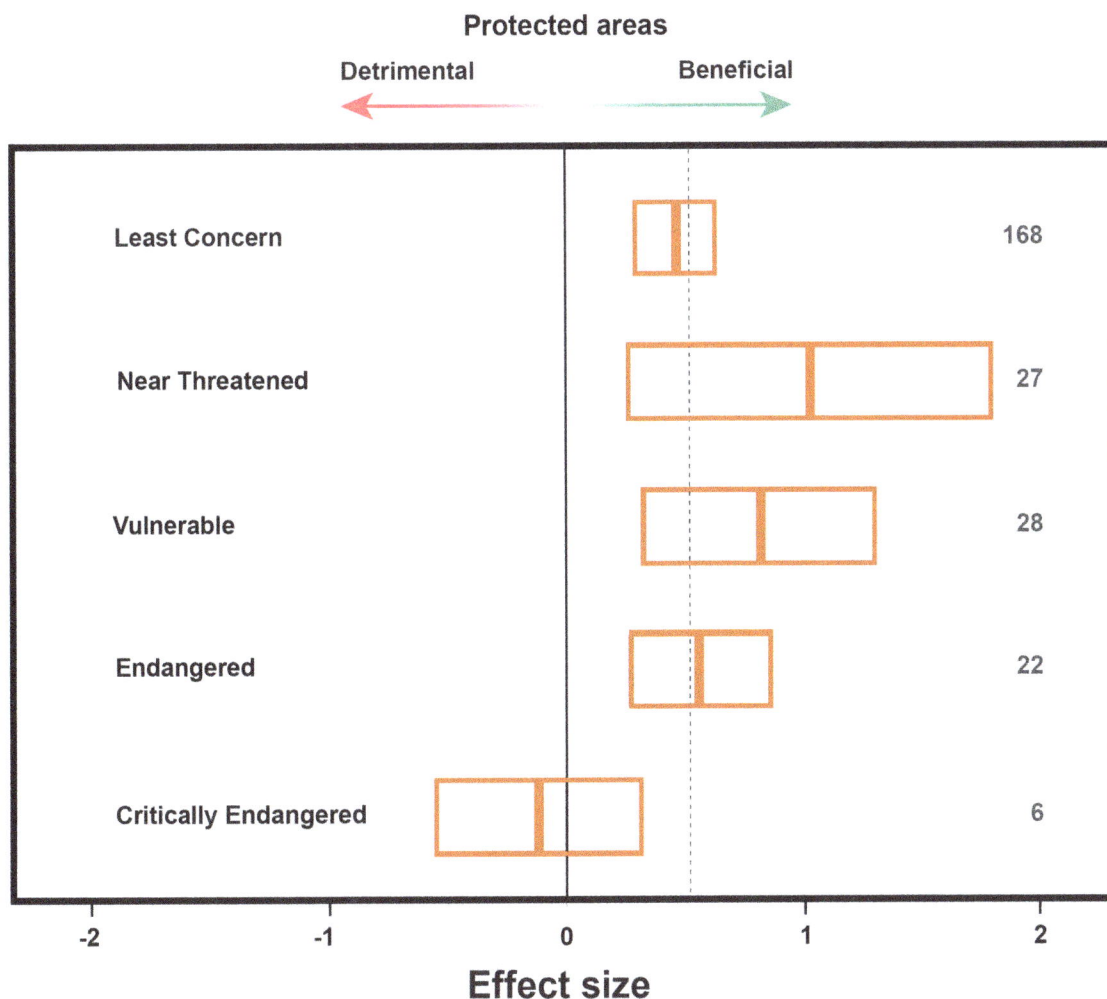

Figure 3. Effect sizes and 95% confidence intervals for species abundance responses by their IUCN Red List status. Positive boxplot values indicate a net positive impact of protected areas on species abundance. Note that multiple responses may be reported for one species (unique species = 168, total cases = 251) and taxonomic uncertainties, Not Evaluated and Data Deficient species are excluded (n = 79). The vertical black line shows a zero effect size while the dashed line indicates the overall effect size for species abundance responses (0.517). Taxa included mammals, birds, herptiles and plants.

In contrast to our general finding, Negroes et al. [47] found that private forest reserves in Brazil were responsible for conserving medium to large-sized vertebrates, more so than PAs. While primary forest is globally irreplaceable for conserving biodiversity [24], land use areas under low extraction, or regenerating forests, seem to contribute to a degree to an integrated landscape level conservation strategy, which may be particularly true in South America [48]. Furthermore, in a comparable meta-analysis, Gibson et al. [24] found that primary forests in South America retain more biodiversity than transformed forests, but since they did not focus on PAs exclusively, they had a larger sample size (N = 909). This finding emphasises that greater sample sizes could increase a positive effect size signal between transformed and more natural areas, a finding that our data corroborate (Figure S3 in File S1).

Despite effect sizes generally being positive, PA efficacy clearly varies considerably amongst PAs, species and local contexts, as demonstrated here and in region-specific studies [5,8,17,21]. Determining the mechanisms driving the pattern of higher species abundances, assemblage abundance, and species richness inside compared to outside PAs remains challenging. Using GLMs we sought to explore a range of proximate factors that might explain this variation. We included explanatory variables such as spatial structure, socio-economic conditions, and structural attributes of PAs, which have been shown elsewhere to have effects on biodiversity values [4,6,49–55]. These assessments provided some insight, but did not account for much of the variation in effect sizes found here. Since comparisons inside and outside PAs in our database were mainly made at the local scale, the geographic context explained little of the observed effect size variation. In consequence, the extent of site matching, which may play a role in increasing estimates of PA efficacy outcomes in some cases [56], may be less important in our database comparing biodiversity features themselves, rather than proxies such as deforestation (but see [56]). Socio-economic factors influence conservation outcomes in some regions [50–52], but here also fared relatively poorly at explaining variation. Likewise structural variables such as PA size and location explained little of the variation. Some of these variables are important drivers of biodiversity variation generally, such as area and latitudinal position (mostly via energy availability) [57,58,59]. However, they have much less of an influence *per se* on differences in diversity inside and outside PAs. Such an outcome suggests that PA efficacy itself is invariate, at least with regard to these variables. In other words the largest effect is of the PAs themselves, as we found here. Our meta-analysis of studies which compare pristine and transformed areas within PAs themselves bears out this suggestion. The pristine areas typically have higher biodiversity values than the transformed areas. Moreover, the effect size here is weaker than the effect size found when comparisons are between sites inside and outside PAs. Together with evidence that pristine forests retain biodiversity features to a greater degree than transformed areas [24], these outcomes suggest that PAs must offset negative anthropogenic influences to a greater degree than no PA designation.

Nonetheless, additional variation may be attributable to other factors such as characteristics of the organisms themselves. For mammals at least, we were able to show that larger species in particular are reduced outside PAs, in keeping with other evidence that smaller mammals are typically better able to tolerate conditions outside PAs [8,21,53,60]. Others have also highlighted the contrasting responses in biodiversity documented by assessments among PAs, but comparing different taxonomic groups [8,21]. Interspecific variability in population responses to landscape change is well-known (e.g. [61]), and some taxa obviously fare better in transformed landscapes, even within PAs. Whether this is the case for species that have a high threat status is more difficult to discern given that information for such species is so scarce [41]. Unfortunately, the scope of our data did not enable us to pursue in more detail possible factors underlying effect size variation among continents and other taxa, largely because of the risk of misleading subgroup effects [25,27]. A range of factors could explain the variation we found given substantial life history differences among these taxa (e.g. dispersal distances, life span, migration propensity, tolerance of disturbance [62–64]), and considerable differences in the evolutionary history of species on continents [59,65].

Similarly, the relationship between the designated management status of a PA and effective conservation of its biodiversity features also seems to be unclear. This outcome provides further evidence for calls that the IUCN PA management categories should be reassessed to reflect biodiversity outcomes rather than management objectives (see [66]). Thus, challenges for conservation science include determining which mechanisms drive positive PA efficacy, under which conditions PAs fail species with differential responses to threats, and concomitantly, IUCN threat designations, and how management categories and biodiversity outcomes can best be aligned.

In addition to these proximate mechanisms that might influence effect size, several ultimate mechanisms may also be important. These include (i) the observation that PAs are non-randomly placed, typically biased towards areas of inaccessibility, which in itself would reduce threatening anthropogenic processes [54]; (ii) the persistence of existing differences in abundance and richness between the areas at the time of PA designations, as a result of the choice of location [18]; (iii) that lower levels of threatening processes, such as habitat alteration or exploitation, have prevailed inside PAs than elsewhere; (iv) that active management of PAs has maintained or increased abundance and richness relative to outside PAs [18]; and/or (v) leakage effects, where threatening processes are displaced from PAs to surrounding areas [55].

Given that data on biodiversity condition before and after establishment of PAs is so rare, studies have understandably had to focus on some comparison of PAs to other areas that may embody likely outcomes if the area had not received PA designation. However, the effect size as calculated here can inherently not explicitly consider the counterfactual (the biodiversity outcome

that would have occurred if there had been no PA designation, see [18,56,67]). Site matching approaches attempt to address this bias arising when observable biophysical and socioeconomic factors may affect biodiversity in addition to PA designation [56,67]. While these studies may detect a weaker signal for the influence of PA designation itself, they still typically focus on proxies of biodiversity to determine performance outcomes, such as deforestation offsets. Although the irreplaceability of primary forests for biodiversity has been established [24], measures of deforestation cannot characterize changes in species richness and abundance itself. As a consequence, the vast majority of published studies on biodiversity have taken a broader view of the counterfactual, and made what for the investigators seemed *a priori* sensible comparisons between PAs and other areas that they held to embody likely outcomes if the former had not received designation. However, due to the way the original studies were designed, the dataset developed here cannot be analysed with site matching approaches. Thus some concern might remain that the observed effects have been influenced by any one or many of the mechanisms above. Nonetheless, the ultimate mechanisms that drive patterns of higher biodiversity retention within PAs are clearly far from universal, and their geographic distribution and intensity is poorly known [18,67]. Therefore, a comprehensive assessment of PA effectiveness can benefit from assessments that consider the net outcomes of both observed effects (as is the case here) in addition to those approaches that can better quantify assessments of effects of bias (i.e. site matching approaches [56] or experimental designs [20]).

In conclusion, despite the ultimate mechanisms underlying our findings not being firmly established, and much variation in effect sizes across regions and among taxa remaining unexplained, a signal is clear: PAs have positive biodiversity values compared with alternative land uses. In consequence, our results, together with emerging qualitative evidence [9], suggest that in general PA establishment itself may confer a net benefit to biodiversity. Thus, our approach provides a quantitative demonstration of the value of PAs as an effective strategy for conserving biodiversity. In other words, studies undertaken at a local scale to date clearly indicate that an ecological foundation exists from which the economic, political and social benefits of PAs are being realized [68]. This outcome provides evidence in support of the value of the Aichi Biodiversity Target 11 of the Convention on Biological Diversity's (CBD) new Strategic Plan for Biodiversity [69], the nationally agreed goals to fulfil signatory countries commitments under the CBD [69]. Target 11 calls for the expansion and effective management of PAs. The outcomes of our analyses show why this Target is worth achieving.

Supporting Information

File S1 Contains the following files: **Table S1.** Orwin's fail safe N is 1238 to reach an overall effect size of 0.222. **Table S2.** Effect sizes determined by resampling one pairwise comparison per unit of study, per species, per country and per protected area (PA), to assess the potential spatial pseudoreplication in our dataset arising from multiple responses. **Table S3.** Best GLM models by exhaustive fit for the Meta Analysis model, Protected Areas (PA) model and Socio-Economic model. **Table S4.** Proportion of five highest ranked models for Meta Analysis model, PA model and the Socio-Economic model. **Table S5.** Best GLM model by exhaustive fit for two variables, the maximum distance to protected area boundary within studies, and the maximum distance between pair wise comparisons within studies. **Figure S1.** PRISMA flow diagram, depicting the flow of information through different phases of the search process conducted. **Figure S2.** Funnel plot of effect size standard error plotted against effect size for all inside-outside pairwise comparisons. **Figure S3.** Cumulative meta-analysis of the dataset sorted by precision, with effect sizes and 95% confidence intervals (n = 861). **Appendix S1.** PRISMA (Preferred Reporting Items for Systematic Reviews and Meta-Analyses) checklist. **Appendix S2.** Detailed descriptions of IUCN protected area management categories.

Dataset S1 Complete dataset and references included in the meta-analysis.

Acknowledgments

Comments by Melodie A. McGeoch, Richard A. Fuller and anonymous reviewers improved the manuscript. This paper is dedicated to Cor J.S. Coetzee.

Author Contributions

Conceived and designed the experiments: BWTC SLC. Performed the experiments: BWTC. Analyzed the data: BWTC. Contributed reagents/materials/analysis tools: KJG. Contributed to the writing of the manuscript: BWTC KJG SLC.

References

1. Jenkins CN, Joppa L (2009) Expansion of the global terrestrial protected area system. Biol Conserv 142: 2166–2174.
2. Ellis EC, Ramankutty N (2008) Putting people in the map: anthropogenic biomes of the world. Front Ecol Environ 6: 439–447.
3. Bertzky B, Corrigan C, Kemsey J, Kenney S, Ravilious C, et al. (2012) Protected Planet Report 2012: Tracking progress towards global targets for protected areas. UNEP-WCMC, Cambridge. Available: http://www.unep-wcmc.org/ppr2012_903.html. Accessed February 2014.
4. Rodrigues ASL, Andelman SJ, Bakarr MI, Boitani L, Brooks TM, et al. (2004) Effectiveness of the global protected area network in representing species diversity. Nature 428: 640–643.
5. Craigie ID, Baillie JEM, Balmford A, Carbone C, Collen B, et al. (2010) Large mammal population declines in Africa's protected areas. Biol Conserv 143: 2221–2228.
6. Fuller RA, McDonald-Madden E, Wilson KA, Carwardine J, Grantham HS, et al. (2010) Replacing underperforming protected areas achieves better conservation outcomes. Nature 466: 365–367.
7. Leverington F, Hockings M, Paveses H, Costa K, Courrau J (2010) Management effectiveness evaluation in protected areas - a global study. The University of Queensland, Australia. Available: http://www.wdpa.org/me/PDF/global_study_2nd_edition.pdf. Accessed February 2014.
8. Laurance WF, Useche DC, Rendeiro J, Kalka M, Bradshaw CJA, et al. (2012) Averting biodiversity collapse in tropical forest protected areas. Nature 489: 290–294.
9. Geldmann J, Barnes M, Coad L, Craigie ID, Hockings M, et al. (2013) Effectiveness of terrestrial protected areas in reducing habitat loss and population declines. Biol Conserv 61: 230–238.
10. Balmford A, Bruner A, Cooper P, Costanza R, Farber S, et al. (2002) Economic reasons for conserving wild nature. Science 297: 950–953.
11. Defries R, Hansen A, Newton AC, Hansen MC (2005) Increasing isolation of protected areas in tropical forests over the past twenty years. Ecol Appl 15: 19–26.
12. Joppa LN, Loarie SR, Pimm SL (2008) On the protection of "protected areas." Proc Natl Acad Sci USA 105: 6673–6678.
13. Nagendra H (2008) Do parks work? Impact of protected areas on land cover clearing. Ambio 37: 330–337.
14. Greve M, Chown SL, van Rensburg BJ, Dallimer M, Gaston KJ (2011) The ecological effectiveness of protected areas: a case study for South African birds. Anim Cons 14: 295–305.
15. Butchart SHM, Walpole M, Collen B, van Strien A, Scharlemann JPW, et al. (2010) Global biodiversity: indicators of recent declines. Science 328: 1164–1168.
16. Chown SL, Lee JE, Hughes KA, Barnes J, Barrett PJ, et al. (2012) Challenges to the future conservation of the Antarctic. Science 337: 158–159.
17. Sutherland WJ, Pullin AS, Dolman PM, Knight TM (2004) The need for evidence-based conservation. Trends Ecol Evol 19: 4–7.
18. Gaston KJ, Jackson SF, Cantú-Salazar G, Cruz-Piñón G (2008) The ecological performance of protected areas. Annu Rev Ecol Syst 39: 93–113.
19. Ferraro PJ, Pattanayak SK (2006) Money for nothing? A call for empirical evaluation of biodiversity conservation investments. PLoS Biol 4: e105.
20. Laurance WF, Camargo JLC, Luizão RCC, Laurance SG, Pimm SL, et al. (2011) The fate of Amazonian forest fragments: a 32-year investigation. Biol Cons 144: 56–67.
21. Gardner TA, Caro T, Fitzherbert EB, Banda T, Lalbhai P (2007) Conservation value of multiple-use areas in East Africa. Cons Biol 21: 1516–2155.
22. Caro TM, Gardner TA, Stoner C, Fitzherbert E, Davenport TRB (2009) Assessing the effectiveness of protected areas: paradoxes call for pluralism in evaluating conservation performance. Divers Distrib 15: 178–182.
23. Magurran AE, McGill BJ (2011) Biological diversity: Frontiers in measurement and assessment. Oxford University Press, Oxford. 345 p.
24. Gibson L, Lee TM, Koh LP, Brook BW, Gardner TA (2011) Primary forests are irreplaceable for sustaining tropical biodiversity. Nature 478: 378–381.
25. Higgins JPT, Green S (2011) Cochrane handbook for systematic reviews of interventions. Version 5.0.2 [updated September 2009]. Available: http://www.cochrane.org/sites/default/files/uploads/Handbook4.2.6Sep2006.pdf. Accessed February 2014.
26. Gaston KJ, Blackburn T, Gregory RD (1999) Does variation in census area confound density comparisons? J Appl Ecol 36: 191–204.
27. Koricheva J, Gurevitch J, Mengersen K (2013) Handbook of meta-analysis in ecology and evolution. Princeton University Press, New Jersey, 498 p.
28. Rohatgi A (2014) WebPlotDigitizer v2.4. Available: http://arohatgi.info/WebPlotDigitizer. Accessed February 2014.
29. Schmid CH, Stewart GB, Rothstein HR, Lajeunesse MJ, Gurevitch J (2013) Software for statistical meta-analyis. In: Eds Koricheva J, Gurevitch J, Mengersen K (2013) Handbook of meta-analysis in ecology and evolution. Princeton University Press, New Jersey, 498 p.
30. Borenstein M, Hedges LV, Higgins JPT, Rothstein HR (2009) Introduction to meta-analysis. Wiley-Blackwell, UK. 450 p.
31. Viechtbauer W (2012) Conducting meta-analyses in R with the metafor package. J Stat Soft 3: 1–48.
32. R Foundation (2014) R: A language and environment for statistical computing. Available: http://www.R-project.org. Accessed February 2014.
33. Gaston KJ, Charman K, Jackson SF, Armsworth PR, Bonn A, et al. (2006) The ecological effectiveness of protected areas: the United Kingdom. Biol Conserv 132: 76–87.
34. IUCN UNEP (2014) The world database on protected areas. Available: http://www.protectedplanet.net. Accessed February 2014.
35. IUCN (2014) The IUCN Red List of threatened species. Available: www.iucnredlist.org. Accessed February 2014.
36. Burnham K, Anderson DR (2002) Model selection and multimodel inference: a practical information-theoretic approach. Springer-Verlag, New York. 520 p.
37. World Governance Index (2014) Available: www.govindicators.org. Accessed February 2014.
38. CIA (2012). The World Factbook Available: https://www.cia.gov/library/publications/the-world-factbook/index.html. Accessed February 2014.
39. Bolker BM (2008) Ecological models and data in R. Princeton University Press, New Jersey. 396 p.
40. Calcagno V, Mazancourt C (2010) glmulti: An R package for easy automated model selection with (Generalized) Linear Models. J Stat Soft 34: 1–29.
41. Ricketts TH, Dinerstein E, Boucher T, Brooks TM, Butchart SHM, et al. (2005). Pinpointing and preventing imminent extinctions. Proc Natl Acad Sci USA 102: 18497–18501.
42. Foley JA, Defries R, Asner GP, Barford C, Bonan G, et al. (2005) Global consequences of land use. Science 309: 570–574.
43. Dornelas M, Gotelli NJ, McGill B, Shimadzu H, Moyes F, et al. (2014) Assemblage time series reveal biodiversity change but not systematic loss. Science 344: 296–299.
44. Lindenmayer D, Laurance W, Franklin J (2012) Global decline in large old trees. Science 338: 1305–1306.
45. Moseby KE, Hill BM, Read JL (2009) Arid recovery - a comparison of reptile and small mammal populations inside and outside a large rabbit, cat and fox-proof exclosure in arid South Australia. Austral Ecol 34: 156–169.
46. Carrillo E, Wong G, Cuarón AD (2000) Monitoring mammal populations in Costa Rican protected areas under different hunting restrictions. Conserv Biol 14: 1580–1591.
47. Negrões N, Revilla E, Fonseca C, Soares AMVM, Jácomo ATA, et al. (2011) Private forest reserves can aid in preserving the community of medium and large-sized vertebrates in the Amazon arc of deforestation. Biodivers Conserv 20: 505–518.
48. Barlow J, Gardner TA, Araujo IS, Ávila-Pires TC, Bonaldo AB, et al. (2007) Quantifying the biodiversity value of tropical primary, secondary, and plantation forests. Proc Natl Acad Sci USA 104: 18555–18560.
49. Thomson JR, Weiblan G, Thomson BA, Alfaro S, Legendre P (1996) Untangling multiple factors in spatial distributions: lilies, gophers and rocks. Ecology 77: 1698–1715.
50. Reyers B, van Jaarsveld AS, McGeoch MA, James AN (1998) National biodiversity risk assessment: a composite multivariate and index approach. Biodivers Conserv 7: 945–965.

51. Veech JA (2003) Incorporating socioeconomic factors into the analysis of biodiversity hotspots. Appl Geog 23: 73–88.

52. Balmford A, Moore JL, Brooks TM, Burgess N, Hansen LA, et al. (2001) Conservation conflicts across Africa. Science 291: 2616–2619.

53. Woodroffe R, Ginsberg JA (1998) Edge effects and the extinction of populations inside protected areas. Science 280: 2126–2128.

54. Joppa LN, Pfaff A (2009) High and far: biases in the location of protected areas. PLoS ONE 4: e8273.

55. Ewers RM, Rodrigues ASL (2008) Estimates of reserve effectiveness are confounded by leakage. Trends Ecol Evol 23: 113–116.

56. Andam KS, Ferraro PJ, Pfaff A, Sanchez-Azofeifa GA, Robalino JA (2008) Measuring the effectiveness of protected area networks in reducing deforestation. Proc Natl Acad Sci USA 105: 16089–16094.

57. Rosenzweig ML (1995) Species diversity in space and time. Cambridge University Press, Cambridge.

58. Gaston KJ (2000) Global patterns in biodiversity. Nature 405: 220–227.

59. Jetz W, Fine PVA (2012) Global gradients in vertebrate diversity predicted by historical area-productivity dynamics and contemporary environment. PLoS Biol 10: e1001292.

60. Peres CA (2004) Effects of subsistence hunting on vertebrate community structure in Amazonian forests. Conserv Biol 14: 240–253.

61. Henle K, Davies KF, Kleyer M, Margules CR, Settele J (2004) Predictors of species sensitivity to fragmentation. Biodivers Conserv 13: 207–251.

62. Nicolakakis N, Sol D, Lefebvre L (2003) Behavioural flexibility in birds predicts species richness, but not extinction risk. Anim Behav 65: 445–452.

63. Hays GC, Scott R (2013) Global patterns for upper ceilings on migration distance in sea turtles and comparisons with fish, birds and mammals. Func Ecol 27: 748–756.

64. Di Marco M, Buchanan GM, Szantoi Z, Holmgren M, Grottolo Marasini G, et al. (2014) Drivers of extinction risk in African mammals: the interplay of distribution state, human pressure, conservation response and species biology. Proc R Soc B 369: 20130198.

65. Hawkins BA, McCain CM, Davies TJ, Buckley LB, Anacker BL, et al. (2012) Different evolutionary histories underlie congruent species richness gradients of birds and mammals. J Biogeogr 39: 825–841.

66. Boitani L, Cowling RM, Dublin HT, Mace GM, Parrish J (2008) Change the IUCN protected area categories to reflect biodiversity outcomes. PLoS Biol 6: 436–438.

67. Joppa LN, Pfaff A (2010) Global protected area impacts. Proc R Soc B 278: 1633–1638.

68. Stolton S, Dudley N (2010) Arguments for protected areas: multiple benefits for conservation and use. Routledge, London. 296 p.

69. CBD (2012) Convention on Biological Diversity: Aichi Biodiversity Targets. Available: http://www.cbd.int/sp/targets/. Accessed February 2014.

12

Separating Macroecological Pattern and Process: Comparing Ecological, Economic, and Geological Systems

Benjamin Blonder[1]*, **Lindsey Sloat**[2], **Brian J. Enquist**[2,3], **Brian McGill**[4]

1 Sky School, University of Arizona, Tucson, Arizona, United States of America, **2** Department of Ecology and Evolutionary Biology, University of Arizona, Tucson, Arizona, United States of America, **3** Santa Fe Institute, Santa Fe, New Mexico, United States of America, **4** School of Biology and Ecology, University of Maine, Orono, Maine, United States of America

Abstract

Theories of biodiversity rest on several macroecological patterns describing the relationship between species abundance and diversity. A central problem is that all theories make similar predictions for these patterns despite disparate assumptions. A troubling implication is that these patterns may not reflect anything unique about organizational principles of biology or the functioning of ecological systems. To test this, we analyze five datasets from ecological, economic, and geological systems that describe the distribution of objects across categories in the United States. At the level of functional form ('first-order effects'), these patterns are not unique to ecological systems, indicating they may reveal little about biological process. However, we show that mechanism can be better revealed in the scale-dependency of first-order patterns ('second-order effects'). These results provide a roadmap for biodiversity theory to move beyond traditional patterns, and also suggest ways in which macroecological theory can constrain the dynamics of economic systems.

Editor: Rachata Muneepeerakul, Arizona State University, United States of America

Funding: BB was supported by a National Science Foundation pre-doctoral fellowship. LS was supported by the National Center for Ecological Analysis and Synthesis. BJE was supported by a National Science Foundation Advancing Theory in Biology grant. BJM was funded by a National Science Foundation Experimental Program to Stimulate Competitive Research grant EPS-0904155 at the University of Maine. The funders had no role in study design, data collection and analysis, decision to publish, or preparation of the manuscript.

Competing Interests: The authors have declared that no competing interests exist.

* Email: bblonder@gmail.com

Introduction

Decades of research have identified four central patterns that together describe the broad-scale organization of biological diversity [1]. These include the: abundance distribution of species [2]; the relationship between species richness and area [3,4]; the decrease in assemblage similarity with increasing distance [5]; and the spatial dispersion of individuals within species [6,7]. A central question is how local biotic and abiotic interactions and variation in rates of speciation and extinction influence these large-scale patterns of diversity [8]. Indeed, much of the ongoing debates within biodiversity science result from the fact that many different models have been proposed to explain these individual patterns [2,4,9–12]. More recently, several theories such as maximum entropy [13] and neutral theory [14–16] have claimed to be able to simultaneously predict these patterns [1].

A central problem for theories of biodiversity is that they all make similar predictions for these near-universal patterns despite beginning from disparate assumptions [17]. One potentially troubling implication for ecology is that these patterns may not reflect anything unique about organizational principles of biology or the functioning of ecological systems [11,18,19]. Instead, they may be a statistical inevitability for any complex system with a large number of variables influencing the system's dynamics [20–

22]. If non-ecological systems show similar patterns, then the fundamental validity of theories of biodiversity that invoke ecological mechanisms as an explanation would be challenged. Stronger tests of theory require alternative approaches.

There is an opportunity to identify a different set of patterns that arise from only ecological processes, and which can therefore distinguish between ecological and non-ecological systems [23,24]. We hypothesize that distinguishing biodiversity theories using empirical patterns is possible with second-order effects but not with first-order effects. We define first-order effects as a set of functions that describe macroecological patterns across scales, and second-order effects as the scale-dependent parameters of these functions. We specifically hypothesize that:

(1) Any system where objects are partitioned in categories (species) across space and many variables interact multiplicatively will be described by a common set of functions, i.e. first-order effects. These first-order effects can be predicted based on common assumptions of multiple unified biodiversity theories [1] or are statistically inevitable consequences of the Central Limit Theorem [20,21]. Thus, any system should be characterized by an approximate log-normal species-abundance distribution (SAD) [25,26], an approximate [3,27] power law [28] species-area relationship (SAR), a monotonically declining Jaccard similarity-

distance function [5], and a positive intraspecific clustering function at different distances (i.e. clumped at all scales; see Methods) [6,29]. Many biodiversity theories predict some or all of these patterns [1].

(2) Ecological and non-ecological patterns, however, can be separated by changes in these patterns change with scale. Thus, quantifying the scale-dependent parameters, i.e. second-order effects [7,30–32] provide a novel way to assess mechanism in macroecology. For example, ecological processes (e.g. dispersal limitation, speciation) will have different scale dependences depending on the system of interest. Spatial scale may affect the slope of the species-area relationship [27] as well as the statistical moments of the species abundance distribution [33]. Some current unified theories of biodiversity are beginning to incorporate scale-dependence and these second-order effects into their predictions [13,14] while for others the role of scale remains unclear [19,34].

We hypothesize that ecological and non-ecological systems can be distinguished based on several patterns (Table 1). Our approach is to: 1) establish baseline expectations for first-order effects based on different biodiversity theories; 2) identify the potentially scale-dependent parameters of first-order effects; 3) plot these parameters as a function of spatial scale; and 4) detect changes in these functions from the baseline expectation. For example, the decay of similarity with distance pattern is predicted by several theories to be a negative exponential function (1). The slope of this function on logarithmic scale should be scale-invariant (2). However, a plot of the local slope of empirical data (3) might show a peak at large distance scales (4) indicative of a second-order effect that can only be explained by additional mechanisms not incorporated into the original biodiversity theories.

Here we use this approach to provide general insights into the degree to which non-ecological systems can be explained by ecological theory. We compared first- and second-order effects across a broad set of ecological and non-ecological systems. We compiled five large datasets that each describe the abundance, location, and identity of objects in multiple categories (species) throughout the continental United States, encompassing two ecological systems (North American birds and trees), two economic systems (US Census county business patterns, and a commercial database of franchise locations for several hundred major corporations) and one geological system (USGS mineral resources database) (Table 2). These datasets were chosen because they are either complete censuses or are known to be well sampled, have very large number of objects and categories, occur over the same large region, and have high spatial resolution (Figure 1).

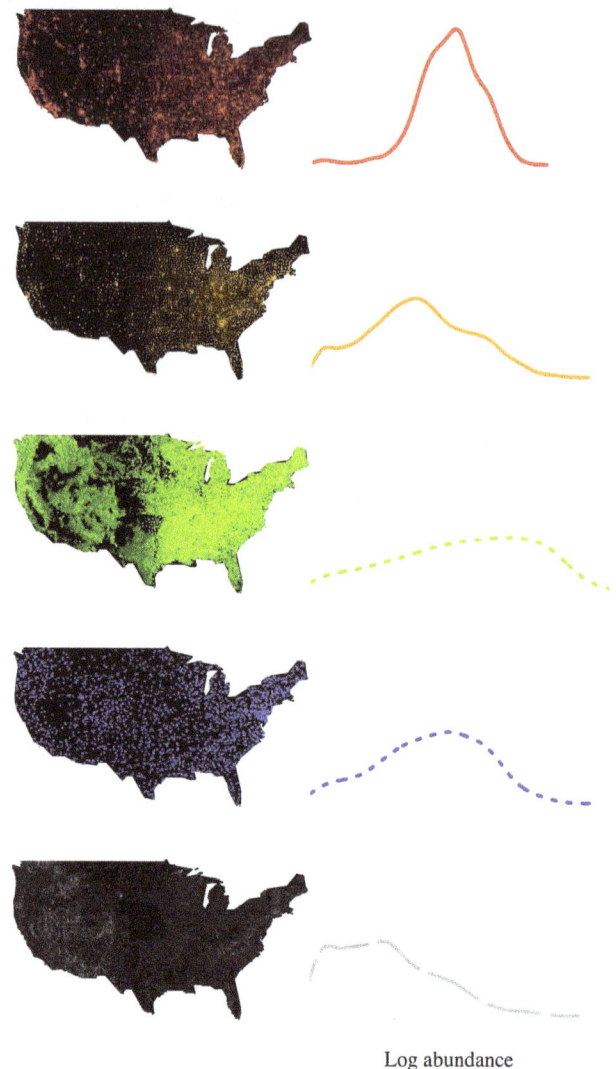

Log abundance

Figure 1. Distribution of objects across categories and space. Left column, site locations for each dataset (colored as described in Table 2). Site brightness is proportional to richness. Right column, relative abundance distribution for log-transformed abundance data at full scale (a first-order effect). All datasets are shown with the same axes.

Materials and Methods

Dataset assemblage

We generated community matrices for each of five datasets in which the ith row and jth column represented the abundance of

Table 1. First and second order effects.

Macroecological pattern	First-order effect	Second-order effect
Species abundance distribution	Log-normal (approximate)	Changes in mean, coefficient of variation, skewness, kurtosis at different area scales
Species area relationship	Power law (approximate)	Changes in local slope at different area scales
Similarity-distance relationship	Monotonic decreasing	Changes in local slope at different distance scales
Fraction of clumped species	Positive	Changes in local slope at different distance scales

First-order effects describe all datasets, while second-order effects may provide scale-dependent approaches for distinguishing datasets.

Table 2. Summary statistics for each economic, ecological, and geological dataset.

Dataset	Corporate locations	Industrial codes	Trees	Birds	Minerals
Type	Economy	Economy	Ecology	Ecology	Geology
Lines drawn as	solid red	solid orange	dotted green	dotted blue	dashed gray
# Species	455	3,777	384	584	746
# Individuals	660,935	7,628,863	11,887,262	1,640,449	587,571
# Sites	20,936	3,106	391,981	2,251	54,837
Most common five species	Subway, Shell, T-Mobile, McDonald's, BP	Offices of physicians (exc mental health), Independent artists, writers & performers, Offices of lawyers, Offices of dentists, Limited-service restaurants	Loblolly pine, Red maple, Sweetgum, Sugar maple, White oak	Red-winged blackbird, European starling, American robin, Mourning dove, American crow	Gold, Sand & gravel, Construction, Silver, Copper

Maps display higher species richness at each site in brighter colors. In all subsequent graphs we have used the line-coloring scheme shown here.

species j at site i. Each community matrix was augmented with the latitude and longitude of each site. Each dataset was clipped to sites contained within the continental United States. All datasets were transformed from their raw form to community matrices using R for data manipulation and MATLAB for GIS analyses. These datasets are available in Information S1.

Dataset 1: Corporate locations. We purchased a commercial dataset containing the street addresses and latitude/longitudes of all locations of several hundred major corporations (AggData, Inc.). The data represent a census obtained between 2008 and 2010. We used shapefiles for the United States Census zip code tabulation areas (ZCTAs) to assign these point occurrences into assemblages. These approximately 20,000 areas cover the entire United States. While ZCTAs have unique complex boundaries and variable areas, they each cover roughly equivalent population levels and are a good comparable assemblage unit for this study. We then determined the latitude and longitude of each assemblage as the centroid of each ZCTA.

Dataset 2: Industrial codes. We downloaded the United States Census County Business Patterns dataset, which counts the numbers of businesses of different size classes in each of the North American Industrial Classification System's (NAICS) nested categories, within each of the counties of every state of the United States. These data were valid for the 2007 census year. The data include some intentional inaccuracies (low-abundance data swapped between sites or abundances randomized) to comply with privacy laws, but these effects are small in magnitude and should not affect our analyses. We restricted our analysis to only the most specific (six-digit) level of NAICS classification in order to closely match between biological and business species. To further improve this correspondence, we also assumed that businesses that fell into different size classes (1–10, 10–100, 100–1000, 1000+ employees) within a given NAICS category represented different species. We also obtained shapefiles for county boundaries and determined the latitude and longitude of each assembly as the centroid of each county.

Dataset 3: Birds. We obtained data from the North American Breeding Bird Survey, which counts the abundance of the bird species observed along hundreds of multi-kilometer transect routes by multiple volunteer birders. We used data from

the 2007 counts. We treated each route as an assemblage and determined its latitude and longitude as the midpoint of the route.

Dataset 4: Trees. We obtained data from the United States Forest Service's Forest Inventory of America, which counts the abundances of several hundred species of trees at hundreds of thousands of plots across the United States. At each plot, we used data from its most recent census, which ranged from 1985–2008. Plot data were pre-corrected for variable plot size and only included live trees. Because of privacy laws, these data contain intentional inaccuracies (plots on private land have their coordinates fuzzed and their abundances swapped) that are small in magnitude and do not affect our analyses. We treated each plot as an assemblage and used the plot center for the assemblage latitude and longitude.

Dataset 5: Minerals. We downloaded data from the United States Geological Survey's Mineral Resource Data System, which describes the locations of metallic and nonmetallic minerals throughout the world. We pooled the abundances of commodities, ores, and gangue at each site, because we were interested in geological processes and did not wish to stratify the data by economic value. Because each site contained a very low number of minerals (typically representing the useful output of a single mine) we chose to generate an equal-area grid (1000×1000) covering the bounding box of the continental United States, and pooled mineral abundances for all sites falling within each grid cell. We then defined the assemblage latitude and longitude as the center of the grid cell.

Data analysis

Species-abundance distribution. We sampled the abundance distribution at 100 spatial scales that logarithmically spanned a range from $0.1°$ to $40°$. At each scale, we chose 500 random sites. We defined a small circle on the surface of the earth whose radius was determined by the current spatial scale and whose center was the location of the current site. We then intersected this circle with a polygon defining the boundary of the region of interest (here, the continental United States). We calculated the surface area of this new polygon (in km^2) using spherical geometry and the known radius of the earth to determine the effective area of the site. We then pooled abundances for all species at all sites enclosed within this polygon, applied a log transformation, and calculated the mean, coefficient of variation, skewness, and kurtosis of this distribution.

Species-area relationship. We determined the species-area relationship using an identical procedure as for the species-abundance distribution, but calculated species richness instead of abundance distribution moments within each polygon. We log-transformed both area and richness before analysis so that the local/global slope of the curve reflects the scale dependent/independent power-law scaling exponent.

Similarity-distance function. We sampled the similarity-distance function at 100 spatial scales that logarithmically spanned a range from 0.1° to 40°. At each scale, we chose 1000 random pairs of sites. We calculated the distance between assemblages (in km) as the minimum arc length along the surface of the earth joining the centers of these assemblages. Then, for each site within each pair of sites we generated small circles centered on the location of each site with a radius equal to the current spatial scale. We intersected each small circle with the boundary polygon of the region of interest (here, the continental United States). To obtain the assemblage area we used the sum of the areas of both polygons (in km^2) calculated by the same approach described for the abundance distribution. We also pooled all abundances within each polygon and calculated the Jaccard similarity (number of species in common divided by the number of species in either polygon). To simplify the display of information results were plotted only for assemblages whose summed area was in the 10–100 km^2 or the 1000 to 10000 km^2 bins.

Intraspecific clumping. We assessed the fraction of species that exhibited intraspecific clumping at distances ranging from 10 to 5000 km in 10 km intervals. An individual species was defined to be intraspecifically clumped at a given distance scale if its observed pairwise distance distribution exceeded the upper 95% quantile of 100 samples from a null pairwise distance distribution. We calculated the pairwise distance distribution as the vector of distances (accounting for the curvature of the earth, as defined for the similarity-distance function) between every pair of sites at which this species occurred. We determined the null pairwise distance distribution by counting the number of sites at which this species occurred, randomly assigning that many occurrences of this species to randomly chosen sites, and repeating the pairwise distance calculation. This method accounts for sites that are non-uniformly or non-randomly positioned, corrects distances for the curvature of the earth, and generates conclusions that are consistent with more established methods for detecting clumping (e.g. pair correlation/o-ring function [6]). However, this method is computationally much faster for large datasets, because distance and intersection calculations can be pre-computed a single time.

First-order effects. Statistics were based on the metrics calculated using the methods described in the previous section. For the species-abundance distribution, we used abundance data at the largest spatial scale. We fit several candidate distributions to the data (Pareto, power-bend [35], Poisson log-normal, log-series, Weibull) and identified the distribution with the lowest AIC. For the species-area relationship, we used the log-transformed area and richness values described in the previous section, then reported the slope of the model (i.e. the power law exponent). For the similarity-distance relationship, we constructed a linear model using the sampled similarities and log-transformed distances for 10^4 km^2 assemblages and reported the slope of the model. For these two analyses, all models were highly significant (p<0.05) but trivially so because of the very large degrees of freedom. For the clumping analysis, we reported the fraction of species that were significantly clumped at the 5000 km distance scale. Because of the

very large number of degrees of freedom in all these analyses (up to 49,998), the standard error for every coefficient was much smaller than the coefficient value. We therefore did not present these uncertainty estimates or p-values because these statistical differences were unlikely to conclusively reflect biological differences.

Data visualization. We chose to show central trends in the data using LOESS (locally smoothed regression) and to quantify variation using error envelopes representing each middle quartile of a local subset of the data.

Results

We first quantified first-order effects in each ecological, economic, and geological dataset neglecting the effect of spatial scale. Our analyses show that, when data are aggregating at a continental spatial scale, each dataset is characterized by the expected first order effects (see Methods for details). All species abundance distributions were best fit by a log-normal distribution (ΔAIC to the next-best distribution >30), except for the tree dataset for which a log-normal or Weibull distribution were both appropriate (ΔAIC = 4). All species-area relationships had log-log slopes ranging from 0.28 to 0.50. All distance decay relationships had log-linear slopes ranging from −0.17 to −0.72. Lastly, most datasets showed intraspecific clumping (7–52% of species significantly clumped). The only exception was for trees, but this dataset included many widely cultivated species. In contrast, we found that second-order effects can distinguish between these datasets, suggesting the operation of different processes structuring each system across a range of scales.

We found that for scales of up to ~10^5 km^2 the coefficient of variation for the species-abundance distribution (Figure 2) for skewness, and kurtosis was larger for the economic datasets than for the ecological or geological datasets. These differences indicate the existence of a process specific to these economic systems that generates higher fractions of rare species at local scales. For example, it could be the case that rapidly growing economies have both higher birth and death rates reflected in large numbers of new businesses. However, all datasets had lower positive skewness at large scales, consistent with dispersal limiting the spread of rare species across space.

Across all systems, species-area relationships displayed a range of slopes and curvature indicating scale-dependent processes of richness accumulation (Figure 3). Consistent with dispersal limitation and a transition to novel species pools, we found a small increase in slope at intermediate scales for the biological data, A more striking pattern is the decrease in slope at large scales of the economic data, consistent with convergence to similar species pools on both coasts.

The similarity-distance function showed a range of decay rates and minimum similarities when comparing datasets (Figure 4). Ecological datasets decayed faster than economic or geological datasets, and an ecological dataset (trees) had the lowest minimum similarity at large distances. Minimum bird similarity was higher than minimum tree similarity, presumably because of the high dispersal potential and large range size of many birds. Economic datasets showed an increase in similarity at very large distances, consistent with high similarity of species pools on both coasts. This change in similarity was weaker for the geological dataset. Changes in the similarity-distance function with assemblage area were also consistent with dispersal limitation being more important in biological systems. We found that in communities encompassing larger areas, ecological systems maintained their decay rates and reached comparable minimum similarities, but

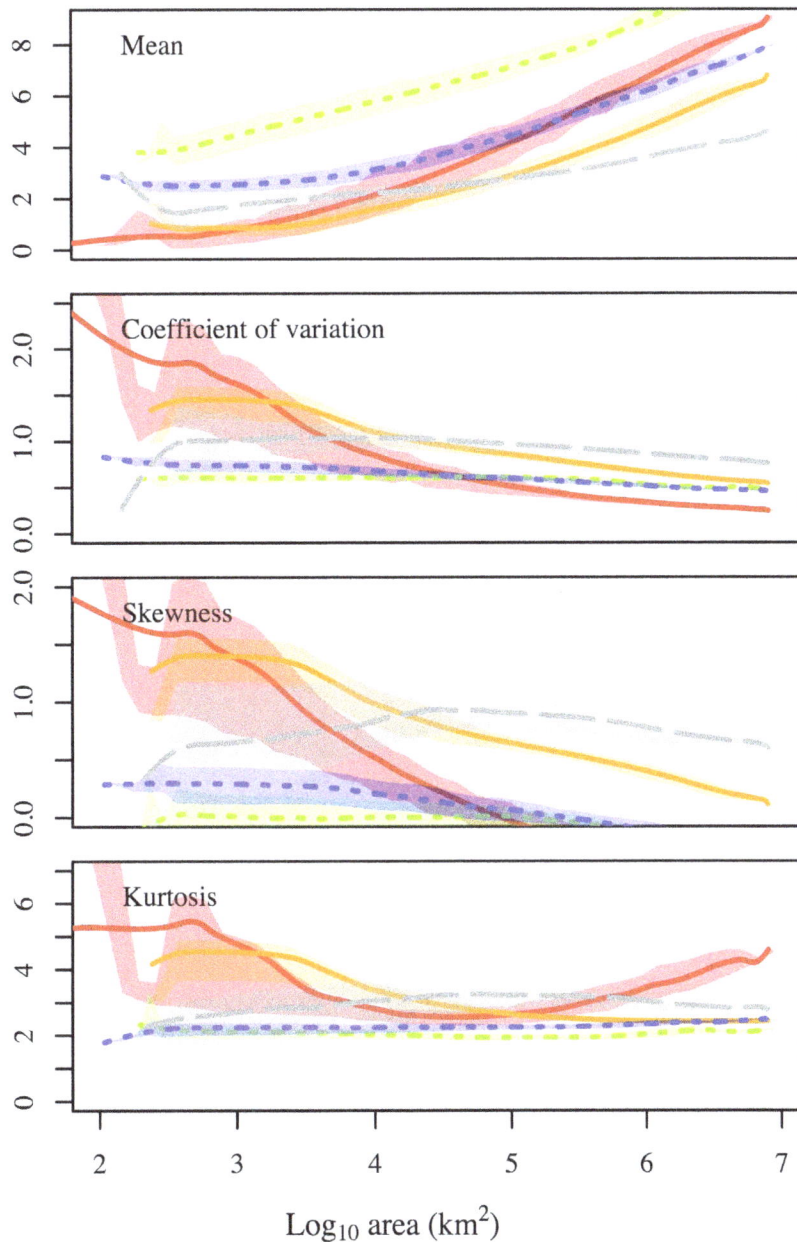

Figure 2. Central moments of the species-abundance distribution for log-transformed data. Line colors are described in Table 2.

that economic systems exhibited very limited decay and high minimum similarity.

The intraspecific clumping function showed more clumping at longer distances in ecological data sets compared to economic or geological datasets (Figure 5). The width of this leftmost part of the clumping function may provide insight into the average dispersal distance for species [6]. At intermediate distances, we found very low levels of clumping in all datasets, indicating that species distributions are spatially random at mid-continent scales. However, we also found increased clumping at whole-continent distances, especially in the economic datasets. This is broadly consistent with low dispersal limitation of businesses, and the high similarity of species pools on both coasts in economic systems.

Discussion

We have shown that spatial scaling can successfully separate four universal macroecological patterns. The consistency of first-order effects across ecological, economic, and geological systems [23,24] indicates that they provide little power to distinguish ecological mechanisms. We then identified second-order effects that were able to separate these five datasets – some of which are consistent with known ecological processes (e.g. dispersal limitation), and others not. Thus, future biodiversity theories should make simultaneous predictions for these scale-dependent second-order effects, in order to provide stronger tests and more separable predictions of theory. Our results indicate that theory especially should be able to make system-dependent predictions for the

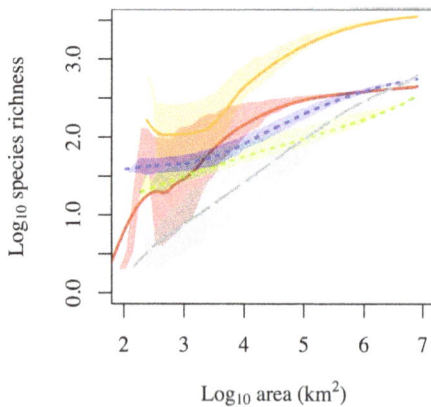

Figure 3. The species-area relationship distinguishes ecological datasets at large scales. Line colors are described in Table 2.

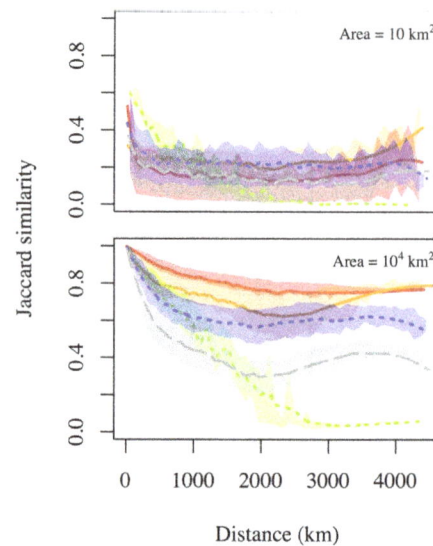

Figure 4. The decay in assemblage similarity with distance depends strongly on spatial scale. The rapidity of decrease and the minimum similarity are functions of dataset type and assemblage size. Line colors are described in Table 2.

spatial scaling of in skewness and kurtosis of the species abundance distribution, as well as of the intraspecific clumping function. Spatial scaling can become a powerful approach to distinguish mechanisms and guide the development and testing of more complex theories of biodiversity.

The general lack of dispersal limitation evident in economic systems is consistent with an 'everything is everywhere' perspective on economic diversity. For example, more than 80% of the major corporations remain the same in pairs of county-sized (10^4 km^2) communities at distances of up to 4500 km (Figure 4). Our results provide strong evidence for low beta diversity and high homogeneity of economic landscapes in the United States. Biodiversity theories will need to incorporate additional parameters to make scale-dependent predictions consistent with this finding.

Our results leave unresolved a potentially important zero-th order effect describing each system's state variables: the number of species and individuals found in each dataset. Although the value of these numbers set the scale of all first-order effects they may also ultimately constrain levels of variation in second-order effects. However, in all major theories of biodiversity, the number of individuals and the number of species are treated as free parameters [1]. Addressing the origin of the zeroth-order effect may provide as much insight as addressing the origin of second-order effects [36].

Our results question the importance of the species concept is to macroecological theory. All biodiversity theories and macroecological patterns are expressed in terms of species and individuals. For ecological systems these are natural and potentially preferred scales for understanding a system. However, there are many possible ways to partition objects in to categories for non-ecological systems, and it is unclear if any particular aggregation method should be preferred. For example, individuals businesses can be aggregated into NAICS codes, but the taxonomy and resolution of these codes is necessarily a human choice. Thus, macroecological theory may be applied best to biological species. However, many biodiversity theories are derived from very limited or no biological processes, suggesting that they should apply equally well to any partitioning of objects in to categories (e.g. taxon-invariance in the species-area relationship [37,38]). Therefore deviations from predictions, such as our second-order effects, should still reflect additional mechanisms.

We showed that our approach could be used for distinguishing different datasets or detecting situations where theory could be

modified to better accommodate empirical data. We do not intend the approach to be used for null-hypothesis significance testing, i.e. statistically rejecting the null hypothesis of no second-order effects. There are two reasons: first, the specific form of first-order effects may depend on the exact mathematical formulation of a biodiversity theory which limits our ability to derive general equations; and second, the form of second-order effects is likely to be more interesting than simply rejecting the ecological null hypothesis. Nevertheless, it should be possible to develop a mathematical formalism to infer second-order effects, a goal that may be useful for the developers of next-generation biodiversity theories.

The similarities in economies and ecosystems may indicate a set of shared processes and constraints whose elucidation will have fundamental or practical implications [24]. Ecological principles and theories have been used to understand economic phenomena like competition, wealth distributions and the growth of cities [39–

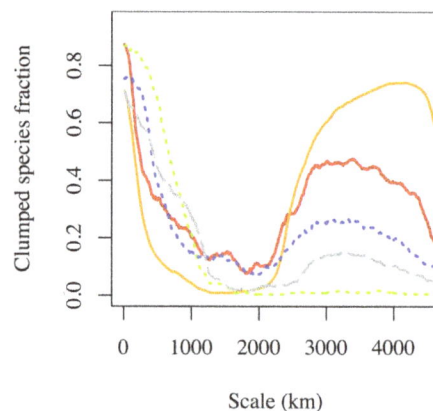

Figure 5. The fraction of species for which intra-specific clumping is consistently high at very small and very large scales. Among datasets, the clumping varies widely in magnitude with spatial scale. Line colors are described in Table 2.

41]. Because many first-order effects seem to occur regardless of system, extant macroecological theory may have practical consequences for other economic systems, e.g. financial networks [42]. For example, current work on up-scaling inexpensive local measurements of biodiversity for conservation purposes [27,43] may be relevant to economic reporting in developing regions, or in understanding the origin of other economic distributions [44]. Macroecological theory, because of its lack of intrinsic ecological mechanism, may also be applicable to many economic systems. In this way it may provide a more realistic understanding of limits to economic growth by identifying the first-order effects that provide universal and unavoidable constraints on economic systems, but also by identifying the zeroth- or second-order effects that may practically be modulated by policy shifts.

References

1. McGill B (2010) Towards a unification of unified theories of biodiversity. Ecol Lett 13: 627–642.
2. MacArthur R (1965) Patterns of species diversity. Biol Rev 40: 510–533.
3. Storch D, Keil P, Jetz W (2012) Universal species-area and endemics-area relationships at continental scales. Nature 488: 78–81.
4. Connor E, McCoy E (1979) The statistics and biology of the species-area relationship. Am Nat 113: 791–833.
5. Morlon H, Chuyong G, Condit R, Hubbell S, Kenfack D, et al. (2008) A general framework for the distance-decay of similarity in ecological communities. Ecol Lett 11: 904–917.
6. Condit R, Ashton P, Baker P, Bunyavejchewin S, Gunatilleke S, et al. (2000) Spatial patterns in the distribution of tropical tree species. Science 288: 1414–1418.
7. Plotkin J, Potts M, Leslie N, Manokaran N, LaFrankie J, et al. (2000) Species-area curves, spatial aggregation, and habitat specialization in tropical forests. J Theor Biol 207: 81–99.
8. Weir JT, Schluter D (2007) The Latitudinal Gradient in Recent Speciation and Extinction Rates of Birds and Mammals. Science 315: 1574–1576.
9. Gauch H, Whittaker R (1972) Coenocline simulation. Ecology 53: 446–451.
10. Preston F (1960) Time and space and the variation of species. Ecology 41: 612–627.
11. May R (1975) Patterns of species abundance and diversity. In: Cody M, Diamond J, editors. Ecology and evolution of communities. Cambridge, MA: Harvard University Press.
12. Sugihara G (1980) Minimal community structure - an explanation of species abundance patterns. Am Nat 116: 770–787.
13. Harte J, Zillio T, Conlisk E, Smith AB (2008) Maximum entropy and the state-variable approach to macroecology. Ecology 89: 2700–2711.
14. Chave J, Leigh E (2002) A spatially explicit neutral model of beta-diversity in tropical forests. Theor Popul Biol 62: 153–168.
15. Hubbell S (2001) A unified theory of biodiversity and biogeography: Princeton University Press.
16. Rosindell J, Cornell SJ (2009) Species-area curves, neutral models, and long-distance dispersal. Ecology 90: 1743–1750.
17. Chave J, Muller-Landau H, Levin S (2002) Comparing classical community models: theoretical consequences for patterns of diversity. Am Nat 159: 1–23.
18. MacArthur R (1969) Patterns of communities in the tropics. Biol J Linn Soc 1: 19–30.
19. Tilman D (2004) Niche tradeoffs, neutrality, and community structure: A stochastic theory of resource competition, invasion, and community assembly. Proc Natl Acad Sci USA 101: 10854–10861.
20. McGill B (2003) Strong and weak tests of macroecological theory. Oikos 102: 679–685.
21. McGill B, Nekola JC (2010) Mechanisms in macroecology: AWOL or purloined letter? Towards a pragmatic view of mechanism. Oikos 119: 591–603.
22. Allen AP, Li BL, Charnov EL (2001) Population fluctuations, power laws and mixtures of lognormal distributions. Ecol Lett 4: 1–3.
23. Gaston K, Blackburn T, Lawton J (1993) Comparing animals and automobiles - a vehicle for understanding body size and abundance relationships in species assemblages. Oikos 66: 172–179.
24. Nekola JC, Brown JH (2007) The wealth of species: ecological communities, complex systems and the legacy of Frank Preston. Ecol Lett 10: 188–196.
25. McGill B, Etienne RS, Gray JS, Alonso D, Anderson MJ, et al. (2007) Species abundance distributions: moving beyond single prediction theories to integration within an ecological framework. Ecol Lett 10: 995–1015.
26. MacArthur R (1960) On the relative abundance of species. Am Nat 94: 25–36.
27. Harte J, Smith AB, Storch D (2009) Biodiversity scales from plots to biomes with a universal species-area curve. Ecol Lett 12: 789–797.
28. Preston F (1962) The canonical distribution of commonness and rarity: part I. Ecology 43: 185–215.
29. McGill BJ (2011) Linking biodiversity patterns by autocorrelated random sampling. Am J Bot 98: 481.
30. Holyoak M, Loreau M (2006) Reconciling empirical ecology with neutral community models. Ecology 87: 1370–1377.
31. Martiny JBH, Eisen JA, Penn K, Allison SD, Horner-Devine MC (2011) Drivers of bacterial β-diversity depend on spatial scale. Proc Natl Acad Sci USA 108: 7850.
32. McGill B (2010) Matters of scale. Science 328: 575–576.
33. Rosindell J, Cornell SJ (2013) Universal scaling of species-abundance distributions across multiple scales. Oikos 122: 1101–1111.
34. McKane A, Alonso D, Sole R (2000) Mean-field stochastic theory for species-rich assembled communities. Phys Rev E 62: 8466–8484.
35. Pueyo S (2006) Diversity: between neutrality and structure. Oikos 112: 392–405.
36. Wiens J (2011) The causes of species richness patterns across space, time, and clades and the role of "ecological limits". Q Rev Biol 86: 75–96.
37. Harte J, Kitzes J, Newman EA, Rominger AJ (2013) Taxon categories and the universal species-area relationship. The American Naturalist 181: 282–287.
38. Storch D, Šizling A (2008) The concept of taxon invariance in ecology: do diversity patterns vary with changes in taxonomic resolution? Folia Geobot 43: 329–344.
39. Bettencourt L, Lobo J, Helbing D, Kühnert C, West G (2007) Growth, innovation, scaling, and the pace of life in cities. Proc Natl Acad Sci USA 104: 7301–7306.
40. Brown J, Burnside W, Davidson A, Delong J, Dunn W, et al. (2011) Energetic limits to economic growth. BioScience 61: 19–26.
41. Decker E, Kerkhoff A, Moses M (2007) Global patterns of city size distributions and their fundamental drivers. PLoS ONE 2: e934.
42. Haldane AG, May RM (2011) Systemic risk in banking ecosystems. Nature 469: 351–355.
43. He F, Gaston K (2000) Estimating species abundance from occurrence. Am Nat 156: 553–559.
44. Stanley M, Amaral L, Buldyrev S, Havlin S, Leschhorn H, et al. (1996) Scaling behaviour in the growth of companies. Nature 379: 804–806.

Acknowledgments

We thank the many people whose efforts have made these continental-scale datasets publicly available. The National Center for Ecological Analysis and Synthesis (NCEAS) provided computational resources.

Author Contributions

Conceived and designed the experiments: BB. Performed the experiments: BB LS. Analyzed the data: BB. Contributed reagents/materials/analysis tools: BE BM. Contributed to the writing of the manuscript: BB LS BE BM.

13

Spatial and Temporal Variations of Ecosystem Service Values in Relation to Land Use Pattern in the Loess Plateau of China at Town Scale

Xuan Fang[1], Guoan Tang[1]*, Bicheng Li[2], Ruiming Han[3]

1 Key Laboratory of Virtual Geographic Environment, Ministry of Education, School of Geography Science, Nanjing Normal University, Nanjing, China, **2** Research Center of Soil and Water Conservation and Ecological Environment, Chinese Academy of Sciences, Yangling, Shaanxi, China, **3** School of Geography Science, Nanjing Normal University, Nanjing, China

Abstract

Understanding the relationship between land use change and ecosystem service values (ESVs) is the key for improving ecosystem health and sustainability. This study estimated the spatial and temporal variations of ESVs at town scale in relation to land use change in the Loess Plateau which is characterized by its environmental vulnerability, then analyzed and discussed the relationship between ESVs and land use pattern. The result showed that ESVs increased with land use change from 1982 to 2008. The total ESVs increased by 16.17% from US$ 6.315 million at 1982 to US$ 7.336 million at 2002 before the start of the Grain to Green project, while increased significantly thereafter by 67.61% to US$ 11.275 million at 2008 along with the project progressed. Areas with high ESVs appeared mainly in the center and the east where largely distributing orchard and forestland, while those with low ESVs occurred mainly in the north and the south where largely distributing cropland. Correlation and regression analysis showed that land use pattern was significantly positively related with ESVs. The proportion of forestland had a positive effect on ESVs, however, that of cropland had a negative effect. Diversification, fragmentation and interspersion of landscape positively affected ESVs, while land use intensity showed a negative effect. It is concluded that continuing the Grain to Green project and encouraging diversified agriculture benefit to improve the ecosystem service.

Editor: Ricardo Bomfim Machado, University of Brasilia, Brazil

Funding: This study was sponsored by the Jiangsu Planned Projects for Postdoctoral Research Funds (No. 1401033C), the National Natural Science Foundation of China (No. 41401441), and the Priority Academic Program Development of Jiangsu Higher Education Institutions (PAPD) (No. 164320H101). The funders had no role in study design, data collection and analysis, decision to publish, or preparation of the manuscript.

Competing Interests: The authors have declared that no competing interests exist.

* Email: tangguoan@njnu.edu.cn

Introduction

Ecosystem contributes to human welfare by providing goods and services directly and indirectly [1–2]. With widely spreading of environmental problems, ecosystem service received increasing attention. Many studies showed human factors, such as urban sprawl [3,4,5], socioeconomic changes [6], agricultural policies [7,8], could affect natural or artificial ecosystems. Land use, an original and foundational human activity and represents the most substantial human alteration to systems on the planet of earth for long-term study [9], plays an important role in providing ecosystem services, including biodiversity, water filtration, retention of soil, etc. [10] Inappropriate land use may lead to significant degradation of local and regional ecological services [11]. Moreover, there were studies showed that ecosystem service trade-offs could successful apply to land use planning [12,13]. Understanding the relationship between ecosystem services and land use change is essential for maintaining a healthy ecosystem and getting sustainable services.

The growing body of literatures focused on how ecosystem service changes in response to land use change of different regions [14,15,16,17,18]. However, these studied focus on the impact of land use type on ecosystem service, while the spatial pattern of land that reflects ecological processed and functions [19] get less attention. Monitoring the characteristic of landscape patterns including area, shape, diversity, etc., is helpful to deeply understand the relationship between ecosystem service and land use change and then to provide complete references for land use planning.

The Loess Plateau is the area suffered from the most severe soil erosion in the world, and it is also a major agricultural production region in China [20]. Long-term poor land use has resulted in vegetation destruction and accelerated soil erosion [21]. To control soil erosion and restore the ecosystem, the Grain for Green project converting slope cropland to grassland or forestland was implemented in 1999 by the Chinese Government [22]. The land use on the plateau under the project has changed significantly. Studying the ecosystem service in relation to land use change before and after the Grain to Green project was crucial for ecosystem protection and agricultural sustainability for the area. Researchers have analyzed ecosystem service at different scales within the Loess Plateau [17,18,23]. However, town is a basic administrative area in China. Exploring the characteristic of ecosystem services change at town scale is of practical significance to provide operable land use planning.

Ecosystem service values (ESVs) is monetary assessment of ecosystem services. This paper examined the characteristics of ESVs at Hechuan town, a typical town in the hilly and gully region of the Loess Plateau. The objectives of this study were: 1) to analyze the changes in land use pattern from 1982 to 2008; 2) to access the spatial and temporal variation in ESVs in response to land use during this period; 3) to quantitively analysis the relationship between ESVs and land use pattern; and 4) to discuss how land use management is favorable for ecosystem service supply and the ecological and economic sustainable development.

Data and Methods

2.1 Ethics statement

No specific permits were required for the described studies, and the work did not involve any endangered or protected species.

2.2 Study area

The study area, Hechuan town (106°18′43″~106°32′16″E, 35°54′59″~36°06′05″N), is located in Guyuan city of the Ningxia Hui Autonomous Region of northwest China (Fig. 1), consisting 12 villages with 16,524 people. The reasons that Hechuan Town was chosen as the study area were, on the one hand, Hechuan town has the typical characteristics of Loess Plateau including the

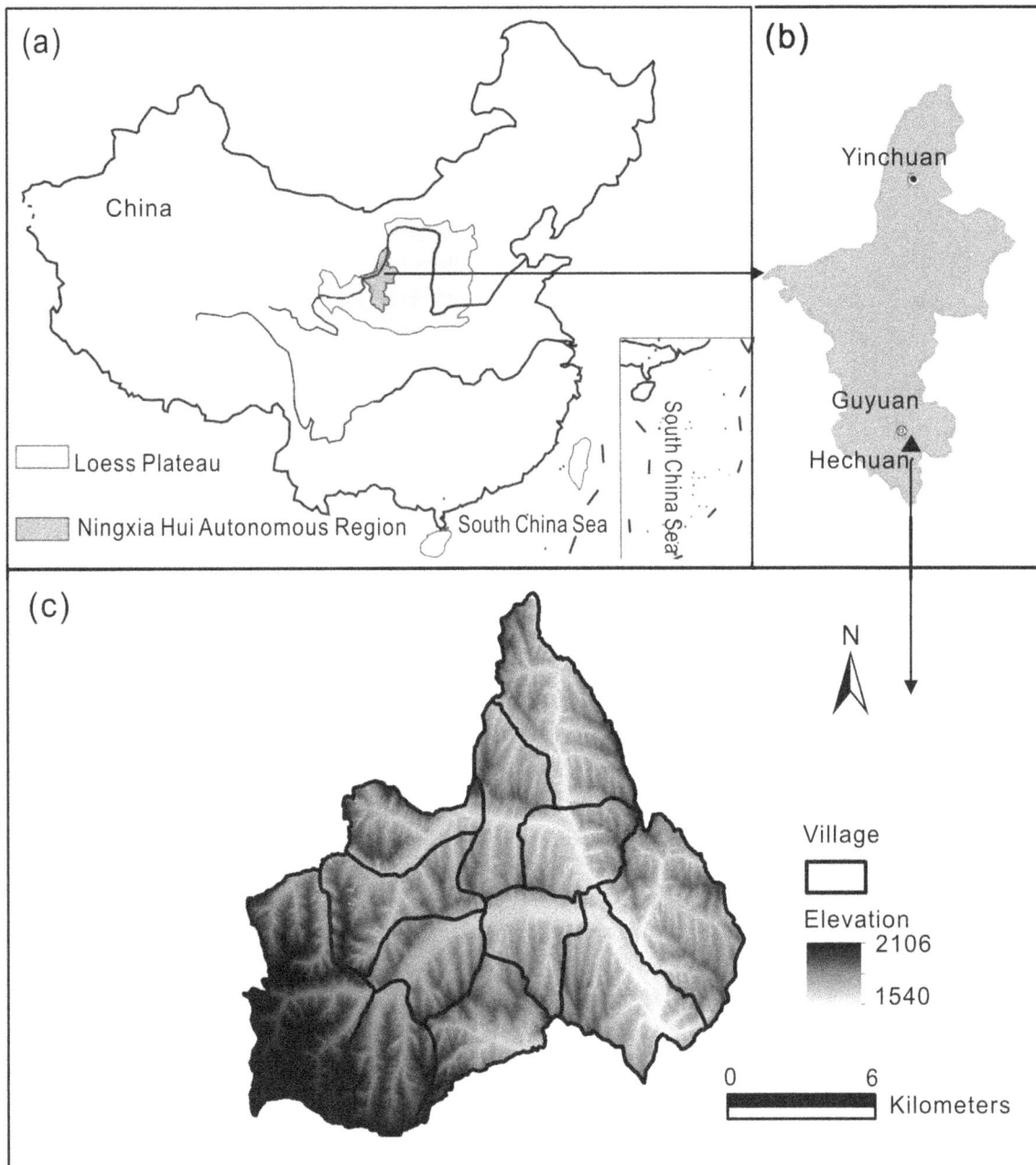

Figure 1. Location of the study area. Ningxia Province and the Loess Plateau, China (a), the location of Hechuan Town in the Loess area of Ningxia Province (b) and, the village boundary and the digital elevation model (DEM) map of Hechuan Town (c).

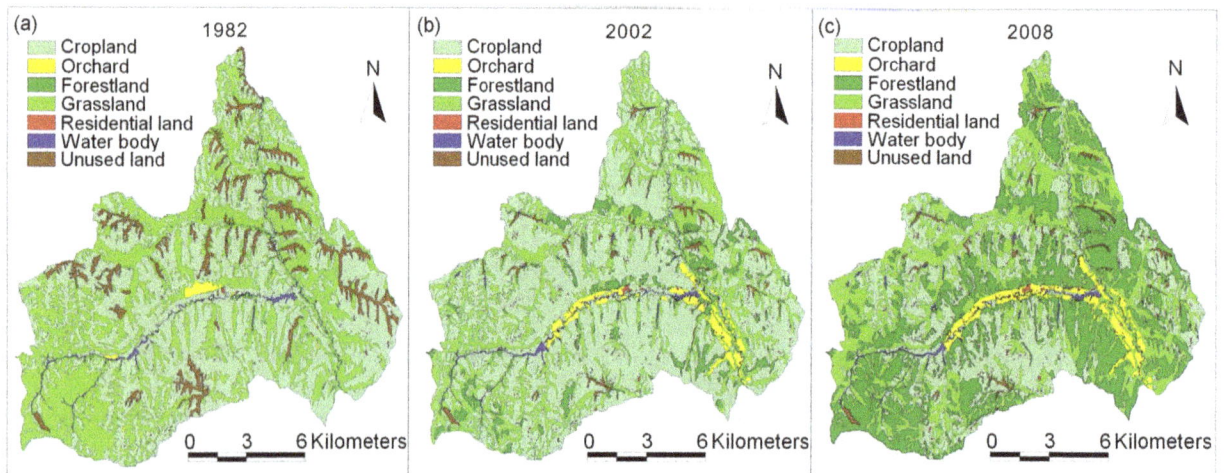

Figure 2. Land use maps of Hechuan town in 1982 (a), 2002 (b) and 2008 (c).

terrain of hill and gull, the fragile ecosystem and the backward economy; on the other hand, there was a long term ecological observation and experiment station in the study area, which facilitated the survey of land use and ecosystems. This town has an altitude ranging from 1540 to 2106 m, covering an area of 215.58 km^2. There exist the topographic differences in the town. The central area with river terrace stretches smoothly with a low elevation. The terrain in the northern area is fragmented while that of southern area is relatively simple. Hechuan town has a semi-arid continental temperate climate with the average annual temperature of 6.9°C and precipitation of 419 mm (1982–2002). Most of the annual precipitation is concentrated between June to September in the form of heavy storms that can cause severe soil erosion. The soil is composed of loessial soil and Dark loessial soils, which is erodible due to its weak cohesion and high infiltrability.

The ecosystem in Hechuan town is fragile with serious soil erosion and frequent natural disasters. Human disturbances of excessive land use, such as deforestation, overgrazing and over-reclamation further destructed the native natural grassland. Therefore, this area has long been in a vicious circle, endless cultivation and poverty. Since the early 1980s, a variety of comprehensive investigation of soil erosion was practiced by Chinese Academy of Sciences. Shanghuang watershed, located in the east of Hechuan town, was taken as a key test area. The

ecological restoration covering the whole town was started from implementing the Grain for Green project after 2002 (launched in 1999 by China government). Since then, abandoned cropland, shrubland (*Caragana korshinskii*, *Hippophae rhamnoides*) and artificial grassland (*Medicago sativa*) was generated, which made a significant change on landscape pattern and ecosystem components providing a variety of ecosystem services. Meanwhile, farming and grazing, the traditional way of living, had to be changed, and raising livestock, orchards, and migrant working diversified their incomes.

2.3 Data acquisition and preprocessing

Land use data was the key data for evaluating landscape pattern and ecosystem service. The land use data of 1982 was obtained by digitizing the land use patches from the 1:10,000 scale topographic maps of 1982, in which the information of land use types and its boundary are clearly shown. The 10 m resolution of remote sensing image could be considered to be corresponding with the scale of 1:50000 [24,25]. The land use data of 1982 acquired from 1:10000 topographic maps was therefore generalized to be at 1:50000 scale [26]. The land use data of 2002 and 2008 were respectively extracted from the 10 m resolution multispectral Spot-5 image of 2002 and 2008 by updating the land use patches of

Table 1. Equivalent weight factor of ecosystem service values (ESVs) per hectare of terrestrial ecosystem in China [30].

	Cropland	Forestland	Grass land	Water body	Barren land
Gas regulation	0.72	4.32	1.5	0.51	0.06
Climate regulation	0.97	4.07	1.56	2.06	0.13
Water supply	0.77	4.09	1.52	18.77	0.07
Soil formation and retention	1.47	4.02	2.24	0.41	0.17
Waste treatment	1.39	1.72	1.32	14.85	0.26
Biodiversity protection	1.02	4.51	1.87	3.43	0.40
Food production	1.00	0.33	0.43	0.53	0.02
Raw material	0.39	2.98	0.36	0.35	0.04
Recreation and culture	0.17	2.08	0.87	4.44	0.24
Total	7.90	28.12	11.67	45.35	1.39

Table 2. The ecosystem service values (ESVs) per hectare of different land use types in Hechuan town (US$·ha-1·yr-1).

	Cropland	Orchard	Forestland	Grass land	Water body	Unused land
Gas regulation	22.570	91.222	135.422	47.022	15.987	1.881
Climate regulation	30.407	88.244	127.585	48.902	64.576	4.075
Water supply	24.138	87.930	128.212	47.649	588.397	2.194
Soil formation and retention	46.081	98.118	126.018	70.219	12.853	5.329
Waste treatment	43.573	47.649	53.918	41.379	465.514	8.150
Biodiversity protection	31.975	99.999	141.378	58.620	107.523	12.539
Food production	31.348	11.912	10.345	13.480	16.614	0.627
Raw material	12.226	52.351	93.416	11.285	10.972	1.254
Recreation and culture	5.329	46.238	65.203	27.272	139.184	7.523
Total	247.647	623.663	881.498	365.828	1421.619	43.573

1982 one by one in visual interpretation method. The interpretation sign was established by understanding the Spot image characteristics and carrying out field surveys in order to further determine the relationship between the true ground and the image. The kappa accuracy index [27] was used to assess the accuracy of the interpretation. The stratified random sampling method was used to generate the reference points on the classified image for the accuracy test. These reference points were located in the field with a GPS with 5-m precision for ground truth. The total kappa indexes are all higher than 0.85, which are higher than the minimum acceptable (0.7) [28]. Considering the characteristic of the land use in study area and the interpretation level of the data and to facilitate the calculation of ESVs, the land use was classified into seven types: cropland, orchard, forestland, grassland, residential area, water area, and unused land (Fig. 2).

To acquire accurate area data of the land use for ESVs estimation and facilitate analyzing the spatial distribution of ESVs, the topographic maps and Spot images were transformed to the same projection and coordinate system (the Albers-Conical-Equal-Area projection system and Krasovsky 1940 coordinate system) before the extraction of land use data, and all acquired land use data were transformed to Arc-grid formats with the same grid size (10 m×10 m). The above data processing was completed using ERDAS and ArcGIS software.

2.4 Analysis on land use pattern

The transfer matrix analysis of land use was produced to understand how land use changed. Landscape metrics analysis was used for spatial pattern analysis of land use. Landscape metrics has been adopted widely; meanwhile, its abilities to indicate ecological process gained increasing attention [29,30,31]. Conceptual flaws in landscape pattern analysis, limitations inherent in landscape metrics and the improper use of pattern analysis may lead to the misuse of landscape metrics [32]. For better explanations and predictions of ecological phenomena from ecological pattern, the landscape metrics in this study was therefore selected by two steps. Firstly, the diversity, the fragmentation and the dominance of landscape were all considered, and then 34 metrics was selected, by understanding the knowledge of the landscape pattern and the ecological services indication of landscape metrics [33,34] and referring to the previous studies on landscape pattern [4,31,35,36]. Secondly, a correlation analysis for the 34 metrics was employed to ensure the low redundancy among landscape metrics. If the coefficient between two metrics was significant at 0.05 level, only one metric of them could be eventually selected.

Landscape-level metrics providing general landscape information and class-level metrics providing more specific information about variations at the local level and spatial patterns of land use classes [37] were used to monitor the characteristics of landscape pattern. The selected landscape-level metrics were patch density (PD), area-weighted mean shape index (SHAPE_AM), Interspersion and Justaposition Index (IJI), and Shannon's diversity index (SHDI). The selected class-level metrics were PD, the percentage of landscape (PLAND), SHAPE_AM and IJI. PD and SHAPE_AM could show the fragmentation of landscape. SHDI and PLAND reflect the dominance of some land use type and the diversity of landscape, respectively. IJI reflects whether the patches or classes are contiguous. Landscape metrics analysis was conducted with above metrics by FRAGSTATS 3.3, in which the eight-neighbor rule was used to derive the patch number. Besides these metrics, the land use intensity index (LUII) was also used to describe the landscape pattern. It was calculated by the following equation [31]:

$$LUII = \sum_{i=1}^{n} A_i \times C_i \qquad (1)$$

where $LUII$ is the land use intensity index, A_i is the percentage of for a give land use type i, and C_i is the coefficient value of intensity for a give land use type i, that is assigned 4 for build-ups, 3 for farmland and 2 for forest, orchard, grassland and water bodies, and 1 for unused land.

2.5 Estimation of ESVs

Costanza et al.'s model of ESVs estimation was adopted in this study [1,2]. The model classified ecosystem service into 17 types of service functions and estimated the ESVs by placing an economic value on different biomes [34]. For the defects of this model, such as overestimating the agriculture ESVs and underestimating the wetland ESVs, Xie et al. proposed refined coefficients for ESVs assessment both solving the above problem and making it apply to China [33,34]. Based on this model, the total ESVs in the study area was calculated using the following formulas:

$$ESV_k = \sum_f A_k VC_{kf} \qquad (2)$$

Table 3. Land use transition matrix from 1982 to 2002 and from 2002 to 2008 (%).

1982	2002								
	Cropland	Orchard	Forestland	Grassland	Residential land	Water body	Unused land	Total	Loss
Cropland	44.42	2.33	1.18	2.50	0.35	0.04	0.01	50.83	6.41
Orchard	0.17	0.05	0.26	0.09	0.00	0.00	0.00	0.57	0.53
Forestland	0.10	0.01	0.15	0.02	0.00	0.00	0.00	0.28	0.13
Grass land	12.63	0.39	4.74	22.15	0.01	0.08	0.00	40.01	17.86
Residential land	0.05	0.01	0.00	0.01	0.14	0.00	0.00	0.22	0.07
Water body	0.01	0.00	0.01	0.00	0.00	0.80	0.00	0.81	0.02
Unused land	1.39	0.00	0.16	3.98	0.00	0.07	1.66	7.27	5.61
Total	58.76	2.78	6.51	28.77	0.51	0.99	1.68	100.00	
Gain	14.35	2.73	6.36	6.62	0.37	0.19	0.01		

2002	2008								
	Cropland	Orchard	Forestland	Grass land	Residential land	Water body	Unused land	Total	Loss
Cropland	27.09	1.00	20.75	9.84	0.07	0.00	0.00	58.76	31.68
Orchard	0.00	2.75	0.01	0.00	0.01	0.00	0.00	2.78	0.03
Forestland	0.05	0.04	5.84	0.57	0.01	0.01	0.00	6.51	0.67
Grass land	0.12	0.00	7.17	21.42	0.02	0.04	0.00	28.77	7.35
Residential land	0.00	0.00	0.00	0.00	0.51	0.00	0.00	0.51	0.00
Water body	0.02	0.02	0.00	0.01	0.00	0.94	0.00	0.99	0.05
Unused land	0.00	0.00	0.27	0.08	0.00	0.00	1.33	1.68	0.35
Total	27.27	3.82	34.05	31.91	0.62	1.00	1.33	100.00	
Gain	0.18	1.07	28.21	10.50	0.11	0.05	0.00		

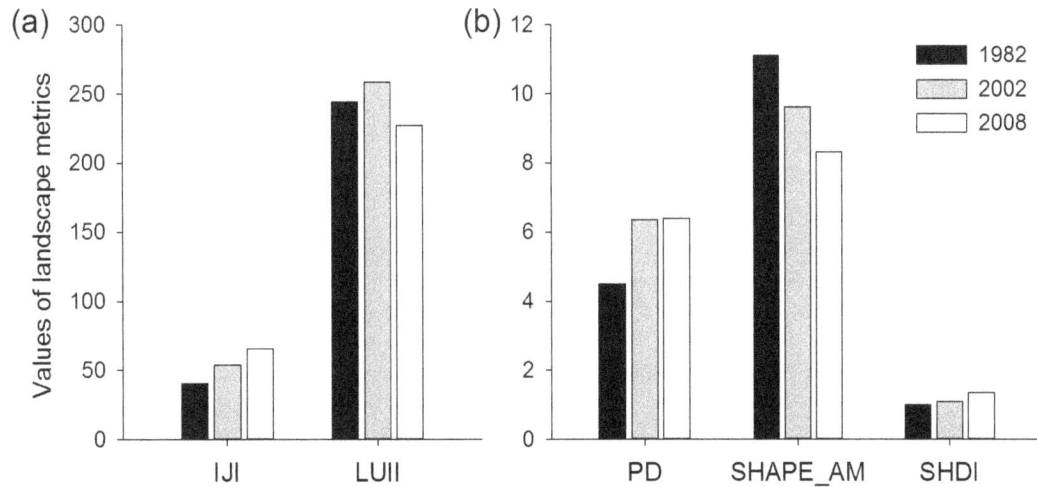

Figure 3. Landscape metrics at the landscape level in Hechuan Town in 1982, 2002 and 2008. IJI: Interspersion and Justaposition Index; LUII: land use intensity index; PD: patch density; SHAPE_AM: area-weighted mean shape index; SHDI: Shannon's diversity index.

$$ESV_f = \sum_k A_k VC_{kf} \qquad (3)$$

$$ESV = \sum_k \sum_f A_k VC_{kf} \qquad (4)$$

where ESV_k, ESV_f, ESV are the ESVs of land use type k, the ESVs of ecosystem service function type f, and the total ESVs respectively. A_k is the area (ha) for land use types. VC_{kf} is the value coefficient (US$·ha-1·yr-1) for land use type k and ecosystem service function type f, which is the key for ESVs estimating. Xie et al.'s model was used to determine VC_{kf}, which can be expressed as follows:

$$VC_{kf} = R_{kf} \times V_f \qquad (5)$$

where R_{kf} is the equivalent weight factor of ecosystem service, V_f is food production values of agriculture land per area per year.

The equivalent weight factor was presented for customizing Chinese terrestrial ecosystem based on Costanza et al.'s model by surveying 500 Chinese ecologists (Table 1) [34]. It is the ratio of the ESVs to the economic value of average natural food production provided by agricultural land per hectare per year. The factors of land use types in our study were basically assigned based on the nearest ecosystems in Xie et al.'s model. However, minor adjustments were made. The equivalent weight factor of orchard which was not put forward clearly in Xie et al.'s model was determined by the mean of grassland and forestland by referring some researches [5,18]. The factor of unused land equates to that of barren land, and that of residential land was determined to zero.

The value of food production service of agriculture land per area per year was considered to be 1/7 of the actual price of food production in Xie et al.'s model. With the average actual food production of cropland in Hechuan town from 1982 to 2008 of 901.77 kg/ha which was get from *Statistic yearbook of the Yuanzhou District, Guyuan City, Ningxia Hui Autonomous Region* and the average grain price of US$ 0.243 per kilogram (i.e. an equivalent of RMB Yuan 1.69 according to the average exchange

rate of 2008) in 2008, the value of food production service of cropland per area per year was calculated to be US$ 31.348 (i.e. an equivalent of RMB Yuan 217.713 according to the average exchange rate of 2008). ESVs of one unit area of each land use types were then assigned as shown in Table 2.

After the ESVs were calculated by above processing, a sensitivity analysis was conducted to test the land use type's representative for ecosystem types and the certainty of the coefficients value for ecosystem service. A coefficient of sensitivity (CS) was used to indicate the degree of sensitivity of ESVs to a coefficients value, calculated by the following formula [5]:

$$CS = \left| \frac{(ESV_j - ESV_i)/ESV_i}{(VC_{jk} - VC_{ik})/VC_{ik}} \right| \qquad (6)$$

where ESV_j an ESV_i are the total ESVs of the initial status j and the adjusted status i, and VC_{jk} and VC_{ik} are the initial and adjusted coefficients. A 50% adjustment in the coefficients was made in the study. The greater the CS responded to the adjustment, the more critical is the use of an accurate coefficient [38]. A CS lower than 1 indicates the ESVs is inelastic to the coefficient and the estimation of ESVS is reliable. Otherwise, a CS greater than 1 indicates the estimation of ESVs is sensitive to the coefficient.

2.6 Correlation and regression analysis

The data of ESVs and landscape metrics was used to analysis the relationship between ecosystem service and land use pattern change. Because the spatial variation of landscape pattern exist among 12 villages in Hechuan town, the land use data of the three years (1982, 2002 and 2008) for the 12 villages can be considered as representing different landscape pattern on a time-for-space perspective [39]. Therefore, there were totally 36 sample data. Correlation and regression was employed for the relationship analysis, in which Multiple stepwise regression was specifically chosen considering the multicollinearity among landscape metrics. The dependents were the nine categories and total ESVs, while the corresponding independents were the landscape-level and class-level landscape metrics.

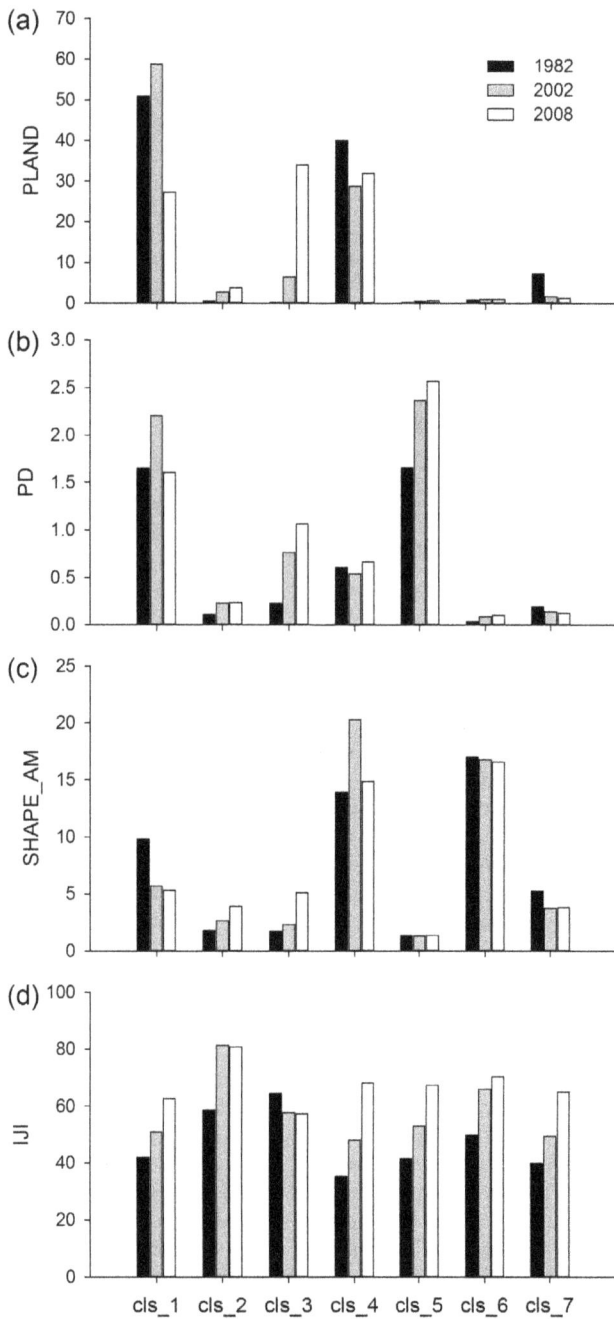

Figure 4. Landscape metrics at the class-level in Hechuan Town in 1982, 2002 and 2008. cls_1, cls_2, cls_3, cls_4, cls_5, cls_6, and cls_7 represent cropland, orchard, forestland, grassland, residential land, water body and unused land. PLAND: the percentage of landscape; PD: patch density; SHAPE_AM: area-weighted mean shape index; IJI: Interspersion and Justaposition Index.

Results

3.1 Changes of land use pattern

Table 3 showed the land use transition matrix. From 1982 to 2002, cropland as the dominant land use type increased from 50.83% to 58.76%. Grassland was the land use type with the largest change in area, decreasing from 40.01% to 28.77%. Orchard increased by 6.24% of total area, indicating the economic driving force of fruit trees on land use change. Forestland

Table 4. The change of ecosystem service values (ESVs) in Hechuan Town from 1982 to 2008.

		Cropland	Orchard	Forestland	Grass land	Water body	Unused land	Total
ESVs (10⁶ US$ yr⁻¹)	1982	2.714	0.077	0.051	3.155	0.249	0.068	6.315
	2002	3.137	0.374	1.237	2.269	0.304	0.016	7.336
	2008	1.456	0.514	6.470	2.517	0.305	0.013	11.275
Change of ESVs (10⁶ US$ yr⁻¹)	1982–2002	0.423	0.297	1.186	−0.887	0.054	−0.053	1.021
	2002–2008	−1.681	0.140	5.234	0.248	0.002	−0.003	3.939
	1982–2008	−1.258	0.437	6.419	−0.638	0.056	−0.056	4.960
Change of ESVS (%)	1982–2002	2.248	55.387	333.431	−4.045	3.145	−11.081	2.328
	2002–2008	−7.716	5.404	60.934	1.574	0.072	−2.992	7.731
	1982–2008	−6.674	81.578	1805.410	−2.913	3.232	−11.771	11.309
Average annual Change (%yr⁻¹)	1982–2002	0.112	2.769	16.672	−0.202	0.157	−0.554	0.117
	2002–2008	−1.286	0.901	10.155	0.262	0.012	−0.498	1.289
	1982–2008	−0.256	3.137	69.439	−0.112	0.124	−0.452	0.435

Table 5. Values of different ecosystem service functions in 1982, 2002, and 2008.

	1982			2002			2008		
	ESVs (10^6 US\$ yr^{-1})	%	Rank	ESVs (10^6 US\$ yr^{-1})	%	Rank	ESVs (10^6 US\$ yr^{-1})	%	Rank
Gas regulation	0.678	10.73	6	0.826	11.26	6	1.529	13.56	5
Climate regulation	0.791	12.53	5	0.936	12.75	5	1.540	13.65	4
Water supply	0.800	12.67	4	0.960	13.09	4	1.610	14.28	3
Soil formation and retention	1.141	18.06	1	1.260	17.17	1	1.764	15.65	1
Waste treatment	0.938	14.85	2	1.015	13.84	3	1.078	9.56	6
Biodiversity protection	0.915	14.49	3	1.054	14.37	2	1.738	15.42	2
Food production	0.466	7.38	7	0.506	6.90	7	0.367	3.25	9
Raw material	0.247	3.91	9	0.390	5.32	8	0.881	7.81	7
Recreation and culture	0.339	5.37	8	0.388	5.29	9	0.768	6.81	8
Total	6.315	100.00		7.336	100.00		11.275	100.00	

increased from 0.57% to 2.78%, reflecting that ecological restoration began to gain attention. From 2002 to 2008, cropland and forestland changed significantly, decreasing from 58.76% to 27.27% and increasing from 6.51% to 34.05% respectively. Land use structure was transferred from cropland dominated (58.76%) to cultivated land (27.27%), forestland (34.05%) and grassland (31.91%) relatively balanced distributed.

The most notable change of land use from 1982 to 2002 was the conversion from grassland to cropland and forestland with 12.63% and 4.74% of the total area respectively. The conversions from cropland (2.50%) and unused land (3.98%) to grassland were not adequate to compensate for the grass loss. From 2002 to 2008, the notable changes of land use were cropland to forestland, cropland to grassland, and grassland to forestland, with the rates of 20.75%, 9.84%, and 7.17% respectively. It was found that the conversion among land use types was more outstanding and concentrated than that before 2002, reflecting that the Grain for Green project as an ecological policy had great influence on land use change.

The results of landscape-level metric analysis were exhibited in Fig. 3. The significant increased PD from 1982 to 2002 reflected the landscape fragmentation. It was relative to the increase of patches on the land use types with intense human disturbance, such as cropland, residential land and artificial reservoir. Oppositely, the slight change of PD from 2002 to 2008 reflected that human disturbance became stable. The change of human disturbance was also demonstrated by the change of LUII which increased before 2002 and decreased after 2002. SHAPE_AM decreased in the study period, showing the landscape became more regular in shape. The increase of IJI suggested that the landscape became more contiguous and the ecological connectivity among land use types increased. SHDI increase obviously from 2002 to 2008, which related to that the land use structure became even.

Fig. 4 showed the change of class-level metrics. The PLAND of land use types indicated that cropland, forestland, and grassland had significantly influence on land use pattern. PD in orchard, forestland, and residential land increased obviously, attributing to the increasing area of these land use types and the fragmental terrain. SHAPE_AM showed that cropland and unused land became more regular in shape, while orchard and forestland more complicated. IJI increased generally in land use types. Orchard was the most contiguous with high IJI, which was relative to its concentrated distribution across the river terrace.

3.2 ESVs from 1982 to 2008

The ESVs of each land use type and the total ESVs was shown in Table 4. The total ESVs of Hechuan town was US\$ 6.315, US\$ 7.336 and US\$ 11.275 million in 1982, 2002 and 2008, respectively. From 1982 to 2002, the decline of ESVs caused by the decrease of grassland was offset by the increase of forestland, orchard and cropland, resulting that the total ESVs increased by US\$ 1.021 million. From 2002 to 2008, the total ESVs increased by US\$ 3.939 million, mainly due to the increase of forestland. The average annual change rate of total ESVs before and after 2002 was quite different, that is 0.81% and 8.95% respectively. It indicated the Grain to Green project implemented since 2002 had a significant effect on the ecosystem service. It was also shown from the value of ESVs produced by forestland occupying 57.39% of the total ESVs. Overall, the total ESVs increased US\$ 4.960 million during the study period, mainly due to the increase of ESVs by the increase of forestland and orchard far beyond the decrease of ESVs by the decrease of cropland and grassland. It was essentially because of the higher coefficient value of forestland and orchard than that of cropland and grassland.

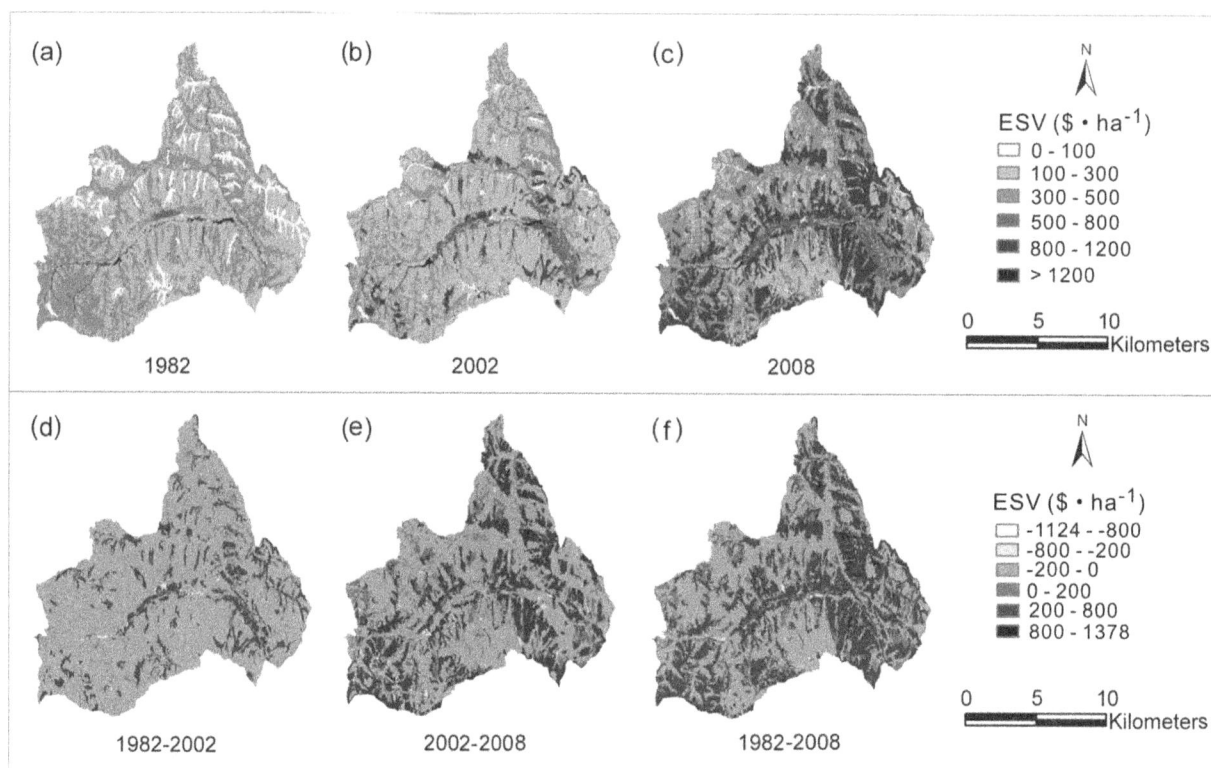

Figure 5. Spatial and temporal distribution of ecosystem service values (ESVs) in Hechuan Town from 1982 to 2008. The spatial distribution of ESVs in 1982 (a), 2002 (b) and 2008 (c), and the spatial-temporal changes of ESVs between time intervals from 1982 to 2002 (d), 2002 to 2008 (e) and 1982 to 2008 (f).

The ESVs of each ecosystem function type was shown in Table 5. Expect for food production, the values of ecosystem service functions increased especially after 2002. The decrease of food production was due to the great decline of cropland in the Grain to Green project. The ESVs proportion of each ecosystem function type to the total ESVs represented the contribution of each ecosystem function to the total ESVs. It was found that the functions of soil formation and retention, waste treatment, and food production were decline during 1982 to 2008, while other functions were improved. The rank of the contribution by each ecosystem service function was also estimated. It was basically stable except for relatively obvious decline in the rank of waste treatment and food production. In 2008, the rank order for each ecosystem service was as follows from high to low, soil formation and retention, biodiversity protection, water supply, climate regulation, gas regulation, waste treatment, raw material, recreation and culture, and food production. Soil formation and retention was the highest during the study period.

3.3 Spatial distribution of ESVs

Maps of ESVs in different periods (Fig. 5) showed the spatial distribution of ESVs of unit area in Hechuan town, directly reflecting the difference of ESVs among land use types. In 1982, the ESVs>4000 mostly appeared in the center of the town where river and river terrace located. It was because water body and orchard which intensely distributed in river terrace for its high water demand both had high ESVs. Therefore, due to the orchard increasing intensely and the forest increasing scatteredly, the increase of ESVs also mainly happened across the river terrace in 2002. Since 2008, the ESVs>4000 spread widely with the increase of forestland transformed from cropland. The lowest ESVs mostly

occurred in the gully where unused land was distributed in 1982. With vegetation recovery in the gully, the low ESVs happened from gully to terraced hillside where cropland with low ESVs was distributed in 2008. Fig. 5d–f showed the temporal change of ESVs spatial distribution. The change characteristic of 2002 to 2008 was adjacent to that during the total study period, reflecting that the change of ESVs mainly occurred after 2002, just after the Grain to Green project.

3.4 Relationship between ESVs and land use pattern

From the above analysis on the change of land use and ESVs in quantity and spatial distribution, we could infer there was some relationship between land use change and ecosystem service. To quantitively understand the relationship, the correlation analysis and regression analysis between ESVs and landscape pattern metrics was conducted.

Table 6 showed there existed significant correlations between ESVs and many landscape metrics (p<0.01), which explained that landscape pattern affected ESVs significantly. For example, the correlation coefficients between total ESVs and landscape metrics showed that there existed significantly positive relationship between SHDI (0.433), PLAND_3 (0.677), SHAPE_AM_3 (0.744), IJL_4 (0.513) and ESVs, and negative relationship between LUII (−0.634), PLAND_1 (−0.752) and ESVs. It reflected that the diversity and intensity of land use had important effects on total ESVs. It also reflected that cropland, forestland and grassland were the land use types which had significant effects on total ESVs. On quantity,the less the cropland and the more the forestland, the higher the total ESVs were. As to the landscape shape, the more regular the cropland and the more complex the forestland, the higher the total ESVs were. The higher the IJI of grassland, the

Table 6. Correlation coefficients between ecosystem service values (ESVs) and landscape pattern metrics.

	TESVs	ESVs_1	ESVs_2	ESVs_3	ESVs_4	ESVs_5	ESVs_6	ESVs_7	ESVs_8	ESVs_9
PD	0.035	0.497*	0.509*	0.539*	0.477*	0.516*	0.499*	-0.221	0.547	0.478*
SHAPE_AM	0.326	-0.216	-0.220	-0.188	-0.177	-0.088	-0.205	0.026	-0.290	-0.166
IJI	0.292	0.624*	0.635*	0.639*	0.597*	0.534*	0.621*	-0.293	0.687	0.586*
SHDI	0.433*	0.763*	0.764*	0.766*	0.741*	0.507*	0.765*	-0.636*	0.775*	0.765*
LUII	-0.634*	-0.681*	-0.658*	-0.618*	-0.675*	-0.113	-0.684*	0.977*	-0.599*	-0.734*
PLAND_1	-0.752*	-0.810*	-0.795*	-0.772*	-0.811*	-0.334	-0.815*	0.952*	-0.742*	-0.853*
PD_1	0.045	-0.055	-0.063	-0.056	-0.054	-0.113	-0.051	-0.134	-0.091	-0.022
SHAPE_AM_1	-0.476*	-0.369	-0.358	-0.330	-0.368	-0.063	-0.369	0.495	-0.328	-0.390
IJI_1	0.189	0.542*	0.552*	0.530*	0.510*	0.405	0.533*	-0.199	0.619	0.485*
PLAND_2	0.323	0.527*	0.541*	0.590*	0.520*	0.595*	0.534*	-0.246	0.558	0.525*
PD_2	0.420	0.457*	0.471*	0.515*	0.449	0.529*	0.463*	-0.193	0.491	0.452
SHAPE_AM_2	0.159	0.450	0.468*	0.541*	0.439	0.629*	0.460*	-0.149	0.493	0.452
IJI_2	0.207	0.392	0.409	0.483*	0.376	0.583*	0.402	-0.115	0.438	0.396
PLAND_3	0.677*	0.984*	0.983*	0.941*	0.975	0.558*	0.980*	-0.770*	0.988*	0.961*
PD_3	0.276	0.631*	0.637*	0.629	0.625	0.477*	0.629*	-0.383	0.653	0.606*
SHAPE_AM_3	0.744*	0.828*	0.827*	0.780*	0.820	0.449*	0.821*	-0.623*	0.836*	0.799*
IJI_3	0.040	0.231	0.241	0.291	0.208	0.356	0.236	-0.069	0.276	0.231
PLAND_4	0.224	-0.192	-0.212	-0.203	-0.159	-0.298	-0.181	-0.264	-0.311	-0.110
PD_4	-0.294	0.199	0.201	0.190	0.160	0.101	0.192	-0.086	0.257	0.171
SHAPE_AM_4	0.455	-0.061	-0.065	-0.049	-0.028	-0.029	-0.053	-0.086	-0.125	-0.022
IJI_4	0.513*	0.717*	0.719*	0.705*	0.697*	0.457*	0.715*	-0.539*	0.739*	0.701*
PLAND_5	-0.035	0.290	0.313	0.381	0.279	0.578*	0.297	0.081	0.357	0.271
PD_5	-0.047	0.244	0.269	0.322	0.244	0.548*	0.248	0.188	0.314	0.208
SHAPE_AM_5	0.118	0.081	0.082	0.053	0.072	-0.009	0.075	-0.004	0.105	0.055
IJI_5	0.307	0.461*	0.470*	0.525*	0.446*	0.512*	0.471*	-0.312	0.482	0.477*
PLAND_6	0.047	0.139	0.167	0.360	0.148	0.852	0.170	0.064	0.153	0.207
PD_6	-0.160	0.122	0.137	0.198	0.105	0.371	0.128	0.080	0.172	0.118
SHAPE_AM_6	0.378	0.088	0.086	0.140	0.080	0.141	0.099	-0.218	0.067	0.137
IJI_6	0.020	0.201	0.217	0.276	0.197	0.438	0.208	0.028	0.239	0.197
PLAND_7	-0.385	-0.447	-0.474*	-0.525*	-0.507*	-0.769*	-0.456*	-0.041	-0.442	-0.422
PD_7	-0.270	-0.105	-0.129	-0.182	-0.154	-0.494*	-0.114	-0.245	-0.108	-0.089
SHAPE_AM_7	-0.236	-0.313	-0.329	-0.434	-0.348	-0.650*	-0.333	0.150	-0.279	-0.358
IJI_7	0.210	0.416	0.408	0.287	0.402	-0.083	0.394	-0.254	0.443	0.341

TESVs: the total ecosystem service values (ESVs); ESVs_1: the ESVs of gas regulation; ESVs_2 climate regulation; ESVs_3: the ESVs of water supply; ESVs_4: the ESVs of soil formation and retention; ESVs_5: the ESVs of waste treatment, ESVs_6: the ESVs of biodiversity protection; ESVs_7 the ESVs of food production; ESVs_8: the ESVs of raw material; ESVs_9: the ESVs of recreation and culture.
PD: patch density; SHAPE_AM: area-weighted mean shape index; IJI: Interspersion and Justaposition Index; SHDI: Shannon's diversity index; LUII: land use intensity index; PLAND: percentage of landscape. The 1, 2, 3, 4, 5, 6, 7 after the above landscape metrics respects different landscape, that is cropland, orchard, forestland, grassland, residential land, water body and unused land, respectively.
*significant at 0.01 level.

Table 7. Regression analysis between ecosystem service values (ESVs) and landscape patterns (n = 36).

Dependent	Standardized coefficients regression	R^2	Sig.
Gas regulation	0.878×PLAND_3+0.166×PLAND_2-0.099×PLAND_1-0.068×IJI_1	0.990	*
Climate regulation	0.790×PLAND_3-0.197×PLAND_7-0.190×LUII+0.081×PLAND_2	0.998	*
Water supply	0.665×PLAND_3-0.317×PLAND_7-0.254×LUII+0.106×PLAND_2	0.955	*
Soil formation and retention	0.684×PLAND_3-0.301×PLAND_7-0.284×LUII+0.066×PLAND_2	0.998	*
Waste treatment	0.672×PLAND_6+0.365×PLAND_3-0.352×PLAND_7+0.051×PLAND_2+0.049×PLAND_5	0.993	*
Biodiversity protection	0.059×SHDI +0.861×PLAND_3-0.033×SHAPE_AM_3+0.133×PLAND_1	0.967	*
Food production	0.742×LUII-0.173×PLAND_3-0.052×SHDI+0.106×PLAND_1	0.991	*
Raw material	0.964×PLAND_3+0.091×SHDI+0.068×LUII	0.981	*
Recreation and culture	−0.747×PLAND_1+0.380×IJI_1	0.853	*
Total	-0.588×PLAND_1+0.569× SHAPE_AM_3-0.303×SHDI	0.709	*

*significant at 0.01 level.
PLAND_1: the percentage of cropland; PLAND_2: the percentage of orchard; PLAND_3: the percentage of forestland; PLAND_5: the percentage of residential land; PLAND_6: the percentage of water body; PLAND_7: the percentage of unused land; SHAPE_AM_3: the area-weighted mean shape index of forestland; IJI_1: the Interspersion and Justaposition Index of cropland; LUII: land use intensity index; SHDI: Shannon's diversity index.

higher the total ESVs were. This indicated that the connectivity of grassland was important for ecosystem service.

Correlation also occurred between ESVs of all the functions and landscape metrics (Table 6). However, the relationships between ESVs of different functions and landscape pattern were different. For example, the correlation between food production and landscape pattern was almost opposite from that between other ecosystem functions and landscape pattern. For example, PLAND_1 had a positive effect on food production; SHDI, PLAND_3, SHAPE_3, and IJI_4 had a negative effect on food production. It could infer that there were contradictions between food production and other ecosystem functions.

As shown in Table 7, the result of regression analysis further explained that the ESVs was correlated significantly with landscape pattern. The total ESVs could be predicted by PLAND on cropland, SHAPE_AM on grassland, and SHDI. ESVs of all kinds of ecosystem functions also could be explained by landscape metrics. These regression equations indicated that landscape-level metrics (such as SHDI and LUII) and class_level metrics (such as PLAND of forestland, orchard, and cropland, unused land, SHAPE of forestland, IJI of cropland) acted as predictors for categories of ecosystem services. Specifically, the proportion of forest (PLAND_3) accounted for almost all of the categories of ecosystem services.

Discussion

4.1 Reliability of ESVs

This study estimated ESVs by multiplying the area for each land use types by the corresponding value coefficients. As discussed in the previous researches, estimations using this method was coarse

with high variation and uncertainty for the following reasons, limitations on the economic evaluation [1], problems of double counting and scales [40,41,42], the complex, dynamic and nonlinear ecosystems [43], the imperfect matches of land use categories as proxies [38] and the accuracy of the ecosystem value coefficients [5]. This study also existed such uncertainty on ESVs estimation. For example, the value coefficient of orchard, determined by the average of forest and grassland, was an approximate estimation and need a further exploration. However, the estimation of temporal variation on ESVs was considered to be more reliable than that of cross-sectional analysis [5]. In addition, the sensitivity analysis of the estimated ESVs with 50% adjustment in the value coefficients was conducted. The result showed that the sensitivity coefficients of all land use categories were lower than 1 (Table 8), which suggested that despite of the above limitations, the estimated ESVs are reliable and useful for subsequent study.

4.2 Relationship between ESVs and landscape pattern

It is usually assumed that land use can affect the ecosystem service. Moreover, a few studies showed that there was a correlation between landscape pattern and ESVs [41,44]. This study signified this statement at town scale on the Loess Plateau. Land use configuration, land use intensity, landscape diversity, fragmentation and connectivity all affected ecosystem service.

The correlation analysis between ESVs and PLAND implied land use structure had significant impact on ecosystem service. Especially, the increase of forestland and the decrease of cropland played an important part in improving the ESVs in the past twenty years. It is closely related to the Grain to Green project comprehensively started in study area since 2002. In the project,

Table 8. The coefficient of sensitivity (CS) resulting from adjustment of ecosystem valuation coefficients.

	Cropland	Orchard	Forestland	Grass land	Water body	Unused land
1982	−0.430	−0.012	−0.008	−0.500	−0.039	−0.011
2002	−0.428	−0.052	−0.169	−0.309	−0.041	−0.002
2008	−0.129	−0.048	−0.574	−0.223	−0.027	−0.001

measures for optimizing land use structure were implemented, including restoring slope cropland into forest and grassland, banning grazing, transforming slopes into terraces, and building reservoirs, etc. Forestland and grassland increased by 423.19% (27.54% of the study area) and 10.93% (3.15% of the study area), and cropland decreased by 53.59%(31.49% of the study area) (Table 3). The increase of ESVs due to the increase of forestland occupied 46.28% of the total ESVs in 2008 (Table 4). The result of the correlation analysis between ESVs and PLAND reflected that vegetation recovery could strongly enhance ecosystem service, and it was coincident with many other studies on the Loess Plateau [17,18,23,45]. LUII, which also related to the proportion of land use types, implied the intensity of human activities. This study showed land use intensity had a negative effect on ecosystem service with negative correlation coefficients (-0.634) (Table 6). It was coincident with some studies on ESVs change under urbanization [5,31]. These studies showed that urbanization which means the increase of land use intensity led to considerable declines in ESVs.

Landscape diversity always presents high positive relevance with biodiversity [46]. Our results were coincident to previous statements given the positive relationships between SHDI and biodiversity conservation. However, there were studies reporting the negative relationships between them, in which the increase of SHDI was the result of rapid urban sprawl [31]. In our study, the increase of SHDI was because land use structure became more balanced, which was the result of the increase of forestland. In addition, landscape diversity could also promote agricultural production [47]. Our study disagreed with this statement, and showed that food production was weakened with landscape diversification. It was because that the increase of SHDI was the result of a larger number of conversion from cropland to forestland. Therefore, the relationship between landscape diversity and biodiversity conservation as well as food production should not be treat as the same but be understood considering the driving force of SHDI change.

Fragmentation could lead to declining habitat quality, lower wildlife survival, and limited movement of soil microorganisms [48], and subsequently cause the decrease of ecosystem service [30]. Our study disagreed with this statement. For example, PD of the total landscape, PD_Forest, PD_orchard and shape_ Forest revealed significantly positive impacts on most categories of ESVs (Table 6–7). The increase of PD and the decrease of connectivity of landscape were usually simultaneous, which is disagreed in our study (Fig. 3 and Fig. 4). The landscape became more contiguous as IJI shown. Table 6 showed the IJI had significantly positive impacts on ESVs. Especially, the increase of IJI of grassland promoted the total ESVs and all categories of ESVs. This maybe because the connectivity of landscape has contribution to habitat corridors [49] and forest production [50].

Based on the relationship between ESVs and landscape pattern, we could improve the ecosystem service by the adjustment of land use policy. On the one hand, continuing to implement the Grain to Green project is helpful for improving ESVS, because it could increase the vegetation coverage, decline the intensity of land use, and make cropland become regular by canceling the slope cropland. On the other hand, diversified agriculture gathering planing fruit trees, planting crops and breeding, which could promote the diversification of land use, should be encouraged to increase both ESVs and farmer's incomes.

Conclusion

ESVs at town scale in the Loess Plateau were estimated in Hechuan town of Ningxia Hui Autonomous Region from 1982 to 2008. It was concluded that ESVs varied with land use change. ESVs in 1982, 2002, and 2008 were US$ 6.315, US$ 7.336 and US$ 11.275 million respectively. Among all the land use types, forestland, grassland and cropland had important contribution (> 90%) on ESVs. The total ESVs increased slowly by 16.17% due to the decrease of grassland from 1982 to 2002, while the total ESVS increased significantly by 67.61% due to the increase of forestland from 2002 to 2008. Areas with high services level were mainly located in the center due to orchard and east due to forestland, while areas with low services level mainly located in the north and south sides due to cropland.

Land use pattern had a significant effect on ecosystem service in our study by analyzing and discussing the relationship between landscape pattern and ESVs. The proportion of forestland had a positive effect on ecosystem service while that of cropland had a negative effect on ESVs. The diversity and interspersion of landscape both had a positive effect on ESVs. Land use intensity which reflects the intensity of human activities had a negative effect on ESVs. Fragmentation had positive effect on ESVs, which was disagreed with the previous studies because the fragmentation in study area was related to the increased patch of such land use types as forestland, water body, orchard.

Based on the results of this study, it was conclude that land use pattern was important for ecosystem service. Therefore, we could improve the ecosystem service by the adjustment of land use policy. Continuing the Grain to Green project is reasonable and significant because it could increase the vegetation coverage and decline land use intensity. Diversified agriculture collecting planing fruit trees, growing food and breeding should be encouraged, because it could not only promote ecosystem service by increasing landscape diversification but also improve people's incomes.

Author Contributions

Conceived and designed the experiments: XF. Analyzed the data: XF. Contributed reagents/materials/analysis tools: GAT BCL. Contributed to the writing of the manuscript: XF GAT RMH.

References

1. Costanza R, Arge DR, Groot DR, Farber S, Grasso M, et al. (1997) The value of the world's ecosystem services and natural capital. Nature 387: 253−260.
2. Costanza R, Cumberland J, Daly H, Goodland R, Norgaard R (1997) An Introduction to ecological economics. Delray Beach Fla USA: St Lucie Press.
3. Kreuter UP, Harris HG, Matlock MD, Lacey RE (2001) Change in ESVs in the San Antonio area, Texas. Ecological Economics 39: 333−346.
4. Ronald CE, Yuji M (2013) Landscape pattern and ESV changes: Implications for environmental sustainability planning for the rapidly urbanizing summer capital of the Philippines. Landscape and Urban Planning 116: 60−72.
5. Li TH, Li WK, Qian ZH (2010) Variations in ESV in response to land use changes in Shenzhen. Ecological Economics 69: 1427−1435.
6. Cai YB, Zhang H, Pan WB, Chen YH, Wang XR (2013) Land use pattern, socio-economic development, and assessment of their impacts on ESV: study on

natural wetlands distribution area (NWDA) in Fuzhou city, southeastern China. Environ Monit Assess 185: 5111−5123.
7. Zaehle S, Bondeau A, Carter RT, Cramer W, Erhard M, et al. (2007) Projected changes in terrestrial carbon storage in europe under climate and land-use change, 1990–2100. Ecosystems 10: 380−401.
8. Eliska L, Jana F, Edward N, David V (2013) Past and future impacts of land use and climate change on agricultural ecosystem services in the Czech Republic. Land Use Policy, 33: 183−194.
9. Vitousek PM, Mooney HA, Lubchenco J, Melillo JM (1997) Human domination of earth's ecosystems. Science 277: 494−499.
10. Nasiri F, Huang GH (2007) Ecological viability assessment: A fuzzy multi-pleattribute analysis with respect to three classes of ordering techniques. Ecol Inform 2: 128–137.

11. Collin ML, Melloul AJ (2001) Combined land-use and environmental factors for sustainable groundwater management. Urban Water 3: 229–237.

12. Schmidta JP, Mooreb R, Alber M (2014) Integrating ecosystem services and local government finances into land use planning: A case study from coastal Georgia. Landscape and Urban Planning 122: 56–67.

13. Ernesto FV, Federico CF (2006) Land-use options for Del Plata Basin in South America: Tradeoffs analysis based on ecosystem service provision. Ecological Economics 57: 140–151.

14. Christine F, Susanne F, Anke W, Lars K, Franz M (2013) Assessment of the effects of forestland use strategies on the provision of ecosystem services at regional scale. Journal of Environmental Management 127: 96–116.

15. Ignacio P, Berta M, Pedro Z, David GDA, Carlos M (2014) Deliberative mapping of ecosystem services within and around Donana National Park (SW Spain) in relation to land use change. Reg Environ Change14: 237–251.

16. Mendoza-Gonzalez G, Martinez ML, Lithgow D, Perez-Maqueo O, Simonin P (2012) Land use change and its effects on the value of ecosystem services along the coast of the Gulf of Mexico. Ecological Economics 82: 23–32.

17. Su CH, Fu BJ (2013) Evolution of ecosystem services in the Chinese Loess Plateau under climatic and land use changes. Global and Planetary Change 101: 119–128.

18. Si J, Nasiri FZ, Han P, Li TH (2014) Variation in ESVs in response to land use changes in Zhifanggou watershed of Loess plateau: a comparative study. Environmental Systems Research 3: 2.

19. Turner MG, Gardner RH, O'Neill RV (2001) Landscape Ecology in theory and practice. New York: Springer-Verlag.

20. Ritsema CJ (2003) Introduction: soil erosion and participatory land use planning on the Loess Plateau in China. Catena 54: 1–5.

21. Fu BJ, Wang YF, Lu YH, He CS, Chen LD, et al. (2009) The effects of land-use combinations on soil erosion: a case study in the Loess Plateau of China. Progress in Physical Geography 33: 793–804.

22. Fu BJ, Chen DX, Qiu Y, Wang J, Meng QH (2002) Land Use Structure and Ecological Processes in the LoessHilly Area, China. Beijing: Commercial Press. (in Chinese).

23. Jing L, Zhiyuan R (2011) Variations in ESV in Response to Land use Changes in the Loess Plateau in Northern Shaanxi Province, China. Int. J. Environ. Res 5: 109–118.

24. Zhang TB, Tang JX, Liu DZ (2006) Feasibility of Satellite Remote Sensing Image About Spatial Resolution. Journal of Earth Sciences and Environment 28: 79–83.

25. Chu YF, Li ES, Lu J, Zhang KK (2007) The Adaptability Analysis to the Satellite Image Spatial Resolution and Mapping Scale. Hydrographic Surveying and Charting 27: 47–50.

26. Li Q, Liu C, Xi CY, Liu ML (2002) Cartographic Generalization of Digital Land Use Current Situation Map. Bulletin of Surveying and Mapping 9: 59–63.

27. Congalton RG (1991) A review of assessing the accuracy of classifications of remotely sensed data. Remote Sensing of Environment 37: 35–46.

28. Wang Y, Gao JX, Wang JS, Qiu J (2014) Value Assessment of Ecosystem Services in Nature Reserves in Ningxia, China: A Response to Ecological Restoration. PloS One 9: e89174. doi:10.1371/journal.Pone.0089174.

29. Ribeiro SC, Lovett A (2009) Associations between forest characteristics and socio-economic development: a case study from Portugal. Journal of Environmental Management 90: 2873–2881.

30. Su S, Jiang Z, Zhang Q, Zhang Y (2011) Transformation of agricultural landscapes under rapid urbanization: a threat to sustainability in Hang-Jia-Hu region, China. Applied Geography 31: 439–449.

31. Su SL, Xiao R, Jiang ZL, Zhang Y (2012) Characterizing landscape pattern and ESV changes for urbanization impacts at an eco-regional scale. Applied Geography, 34: 295–305.

32. Li H, Wu J (2004) Use and misuse of landscape indices. Landscape Ecology 19: 389–399.

33. Xie GD, Lu CX, Xiao Y, Zheng D (2003) The Economic Evaluation of Grassland Ecosystem Services in Qinghai Tibet Plateau, Journal of Mountain Science 21: 50–55. (in Chinese).

34. Xie GD, LU CX, Leng YF, Zheng D, Li SC (2008) Ecological assets valuation of Tibetan Plateau. Journal of Natural Resources 18: 190–196. (in Chinese).

35. Liu DL, Li BC, Liu Xianzhao Z, Warrington DN (2011) Monitoring land use change at a small watershed scale on the Loess Plateau, China: applications of landscape metrics, remote sensing and GIS. Environmental Earth Sciences 64: 2229–2239.

36. Pan WKY, Walsh SJ, Bilsborrow RE, Frizzelle BG, Erlien CM, et al. (2004) Farm-level models of spatial patterns of land use and land cover dynamics in the Ecuadorian Amazon. Agriculture, Ecosystems and Environment 101: 117–134.

37. de Groot RS, Wilson MA, Boumans RMJ (2002) A typology for the classification, description and valuation of ecosystem functions, goods and services. Ecological Economics 41: 393–408.

38. Kreuter UP, Harris HG, Matlock MD, Lacey RE (2001) Change in ESVs in the San Antonio area, Texas. Ecological Economics 39: 333–346.

39. Wu J, Jenerette GD, Buyantuyev A, Redman CL (2011) Quantifying spatiotemporal patterns of urbanization: the case of the two fastest growing metropolitan regions in the United States. Ecological Complexity 8: 1–8.

40. Turner RK, Paavola J, Coopera P, Farber S, Jessamya V, et al. (2003) Valuing nature: lessons learned and future research directions. Ecological Economics 46: 493–510.

41. Hein L, Koppen VK, de Groot RS, van Ierland EC (2006) Spatial scales, stakeholders and the valuation of ecosystem services. Ecological Economics 57: 209–228.

42. Konarska KM, Sutton PC, Castellon M (2002) Evaluating scale dependence of ecosystem service valuation: a comparison of NOAA-AVHRR and Landsat TM datasets. Ecological Economics 41: 491–507.

43. Limburg KE, O' Neill RV, Costanza R, Farber S (2002) Complex systems and valuation. Ecological Economics 41: 409–420.

44. Zhang MY, Wang KL, Liu HY, Zhang CH (2011) Responses of Spatial-temporal Variation of Karst Ecosystem Service Values to Landscape Pattern in Northwest of Guangxi, China. Chin. Geogra. Sci. 21: 446–453.

45. Deng L, Shangguan ZP, Li R (2012) Effects of the grain-for-green program on soil erosion in China. International Journal of Sediment Research 27: 120–127.

46. Nagendra H (2002) Opposite trends in response for the Shannon and Simpson indices of landscape diversity. Applied Geography, 22: 175–186.

47. Shrestha RP, Schmidt-Vogt D, Gnanavelrajah N (2010) Relating plant diversity to biomass and soil erosion in a cultivated landscape of the eastern seaboard region of Thailand. Applied Geography 30: 606–617.

48. Sherrouse BC, Clement JM, Semmens DJ (2011) A GIS application for assessing, mapping, and quantifying the social values of ecosystem services. Applied Geography 31: 748–760.

49. Li M, Zhu Z, Vogelmann JE, Xu D, Wen W, et al. (2011) Characterizing fragmentation of the collective forests in southern China from multitemporal Landsat imagery: a case study from Kecheng district of Zhejiang province.Applied Geography 31: 1026–1035.

50. Long JA, Nelson TA, Wulder MA (2010) Characterizing forest fragmentation: distinguishing change in composition from configuration. Applied Geography 30: 426–435.

14

Intraspecific Trait Variation Driven by Plasticity and Ontogeny in *Hypochaeris radicata*

Rachel M. Mitchell[1]*, **Jonathan D. Bakker**[1,2]

1 School of Environmental and Forest Sciences, University of Washington, Seattle, Washington, United States of America, **2** Smithsonian Environmental Research Center, Edgewater, Maryland, United States of America

Abstract

The importance of intraspecific variation in plant functional traits for structuring communities and driving ecosystem processes is increasingly recognized, but mechanisms governing this variation are less studied. Variation could be due to adaptation to local conditions, plasticity in observed traits, or ontogeny. We investigated 1) whether abiotic stress caused individuals, maternal lines, and populations to exhibit trait convergence, 2) whether trait variation was primarily due to ecotypic differences or trait plasticity, and 3) whether traits varied with ontogeny. We sampled three populations of *Hypochaeris radicata* that differed significantly in rosette diameter and specific leaf area (SLA). We grew nine maternal lines from each population (27 lines total) under three greenhouse conditions: ambient conditions (control), 50% drought, or 80% shade. Plant diameter and relative chlorophyll content were measured throughout the experiment, and leaf shape, root:shoot ratio, and SLA were measured after five weeks. We used hierarchical mixed-models and variance component analysis to quantify differences in treatment effects and the contributions of population of origin and maternal line to observed variation. Observed variation in plant traits was driven primarily by plasticity. Shade significantly influenced all measured traits. Plant diameter was the only trait that had a sizable proportion of trait variation (30%) explained by population of origin. There were significant ontogenetic differences for both plant diameter and relative chlorophyll content. When subjected to abiotic stress in the form of light or water limitation, *Hypochaeris radicata* exhibited significant trait variability. This variation was due primarily to trait plasticity, rather than to adaptation to local conditions, and also differed with ontogeny.

Editor: Eric Gordon Lamb, University of Saskatchewan, Canada

Funding: The authors have no support or funding to report.

Competing Interests: The authors have declared that no competing interests exist.

* Email: rmm57@duke.edu

Introduction

Recent and growing interest in plant functional traits has deepened our understanding of critical ecological mechanisms. For example, interspecific trait-based approaches have identified differing processes of community assembly across spatial scales [1] and in temperate and tropical ecosystems [2], shaped understanding of abiotic drivers such as altitude [3,4], climate [5,6], temperature [7,8], and resource gradients [9] on community assembly and diversity, and identified global patterns of plant-trait expression [10]. In addition, trait-based approaches have provided a mechanistic understanding of correlations between biodiversity and ecosystem function [11–13].

Although trait-based approaches have improved our understanding of ecological mechanisms, the role of intraspecific trait variation - variation from individual to individual within a species - is less well understood. Intraspecific variation in plant functional traits can be large, though it has often been ignored or its contribution underestimated [14–16]. Intraspecific variation can structure communities [17] and influence ecosystem function [18]. For example, intraspecific variation in plant functional traits has been shown to structure communities along environmental gradients [19], to drive diversity in forest ecosystems [20], and to enhance community resistance to plant invasion [21].

Changes in intraspecific variation in plant functional traits have been observed in response to a wide range of environmental conditions (e.g., flooding [19], fire [22], drought [23] and temperature [24]), but few studies have examined the mechanisms controlling these changes (but see [25–27]). Two primary processes are hypothesized to drive variation in trait expression: adaptation to local conditions, creating distinct genetic ecotypes [28], and phenotypic plasticity in response to prevailing abiotic conditions [29]. Those studies that have investigated sources of intraspecific variation have, at times, found contrasting results, depending on the species and functional trait being investigated [30–32]. For example, Firn et al. found significant plasticity in specific leaf area (SLA; leaf area per unit mass) for an exotic, invasive grass but not for co-occurring native grasses [33]. In *Quercus suber*, genetically distinct populations displayed significantly different mean values for SLA, suggesting local adaptation to climate regimes, but populations also demonstrated high phenotypic plasticity [34]. In contrast, Robson et al. found that differences in SLA in *Fagus sylvatica* were primarily driven by genetic provenance [35].

If observed intraspecific trait variation is largely driven by the genetics of a given population, abiotic or biotic filters can selectively eliminate individuals with traits that exceed the bounds of the filter. For example, species or populations with low variability in traits related to drought stress (e.g., leaf size) may be unable to cope with reduced water availability under shifting climate. However, if variation in critical plant traits is largely a plastic response, independent of genetic diversity, even rare species may respond to filtering by altering their trait expression and thus persist in the community. However, plasticity is not without cost, as very plastic species may express phenotypes unsuited to abiotic filters, or experience costs associated with the trait of plasticity [36,37].

In addition to being influenced by adaptation and plasticity, trait expression also changes due to ontogeny. Determining how traits vary ontogenetically is important because traits influence ecosystem function [11], and temporal differences in plant trait expression may facilitate species coexistence and biodiversity maintenance [38]. Smith *et al.* found that SLA decreased with age in *Populus tremuloides* [39], while Cornelissen et al. found poor correlation between greenhouse seedlings and field-measured adults for 90 woody plant species, indicating strong ontogenetic differences [40]. The effects of ontogeny can differ depending on the trait being measured. For example, wood density, but not leaf mass per unit area, was strongly affected by ontogeny in tropical *Nothofagus pumilio* [24]. Despite evidence of correlation between trait variability and ontogenetic state, this source of variability is rarely accounted for, and often explicitly avoided by use of standard collection protocols [41].

Examination of the mechanisms controlling intraspecific variation is confounded by the fact that variation is both adaptive and with cost [37,42,43]. One species, population within a species, or genetic lineage may be inherently more plastic than others for some traits. In a widely distributed species, some populations may exhibit greater plasticity in response to abiotic stress due to high environmental heterogeneity selecting for phenotypically plastic genotypes, or through genetic correlation to other traits that are under selection [44]. For example, Balaguer *et al.* [45] found that populations of *Quercus coccifera* growing in homogenous conditions displayed significantly less phenotypic plasticity in light-responsive traits than those growing in more heterogeneous conditions. Pratt & Mooney found a similar pattern for growth and flower production in *Artemisia californica* in response to precipitation variability [46].

We used *Hypochaeris radicata*, a globally widespread weed, as a model species for examining trait variation in terms of both trait means (average trait values) and trait dispersion (variability of trait values). *H. radicata* occurs in a number of ecosystems, from undisturbed grasslands to highly disturbed urban environments. Because it persists across such a wide variety of conditions, it likely exhibits considerable trait plasticity [47]. However, persistence under differing abiotic conditions may lead to distinct population-level trait differences through ecotypic adaptation or transgenerational plastic responses [14,28,38,48].

We collected and germinated seeds from three populations, exposed the seedlings to abiotic stress (shade, drought) in the greenhouse, and measured a range of traits on each individual. We used these data to experimentally test three hypotheses. First, we hypothesized that abiotic stress would significantly impact trait expression for all populations, leading to a convergence in trait values for plants experiencing abiotic stress. Second, we hypothesized that population of origin and maternal line of origin, and not random variation between individuals, would explain the majority of observed trait variation that was not explained by abiotic conditions. Third, we hypothesized that over the lifetime of a plant, individuals, maternal lines, and populations within a treatment would become more similar in terms of both trait means

and variances in order to accommodate stressful abiotic conditions.

Materials and Methods

Ethics Statement

No special permits were required for this research. Permission to conduct research at Glacial Heritage Preserve was granted by Thurston County. Permission to conduct research at Smith Prairie was granted by the non-profit owners, Pacific Rim Institute for Environmental Stewardship. Permission to collect samples at Union Bay Natural Area was granted by the University of Washington. This research did not include any populations of endangered or threatened species. Data used in this manuscript are available through the TRY database and can be accessed by submitting a data request at http://www.try-db.org.

Study Species

Hypochaeris radicata is a short-lived perennial herb species common in intact and degraded grassland ecosystems around the globe, including Europe, North America, Australia, New Zealand and Japan. It forms basal rosettes with oblong-lanceolate leaves and produces several upright, branching flower stalks during a long flowering period – June through October in the Pacific Northwest of the USA. *H. radicata* is self-incompatible [49,50] and its seeds are wind dispersed up to a few hundred meters [51].

Experimental Design

Three populations (Glacial Heritage (GL), Smith Prairie (SP), and Union Bay Natural Area (UB)) spanning a 200 km range in Washington state, USA, were sampled for this experiment. Populations occurred in exotic-dominated grasslands, but experienced different abiotic conditions in terms of annual precipitation, soil type, and land-use history (Table S1). Specific leaf area and rosette diameter were measured on 20–25 plants (maternal lines) per population, and one mature seed-head was collected from each plant. Measured plants were at least 3 m apart to prevent sampling from closely related individuals. Previous analyses detected significant differences in trait variability for SLA and rosette diameter [16]; follow-up analyses using PermDISP and PERMANOVA (see "Statistical Analysis", below) indicated significant differences in trait dispersion and mean values for diameter (Fig. 1A, $P < 0.0001$) and significantly different dispersion for SLA (Fig. 1B, $P = 0.05$) but no significant differences in seed mass (Fig. 1C).

The single seed head collected per maternal line did not contain enough seeds to allow individual maternal lines to be replicated in each abiotic treatment. Instead, maternal lines were subjected to Principal Components Analysis (PCA), and the first principal component (PC1) was used to select maternal lines spanning the range of traits expressed in each population. Thus, for each population, the three highest, three most central, and three lowest maternal lines (hereafter referred to as "high", "central", and "low") along PC1, which explained 49% of the variation, were identified (Fig. 1D). One maternal line from each group was then randomly assigned to an abiotic treatment (Figure S1).

Seeds from the selected maternal lines (3 populations ×9 maternal lines/population) were cold-stratified for 10 days and germinated at 25°C. Twenty germinants per maternal line (540 total) were planted individually into 164 ml "conetainers" (Stuewe & Sons, Oregon, USA) filled with Sunshine Mix #3 potting soil (Sun Gro Horticulture, Massachusetts, USA). Germinants were allowed to acclimate for seven days before treatments were applied.

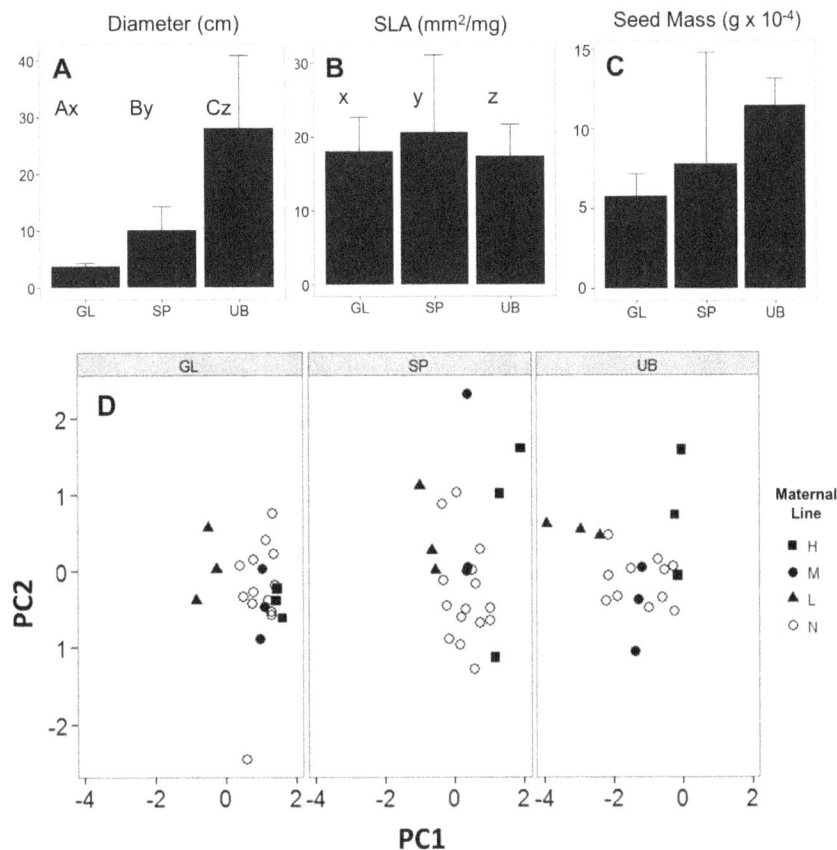

Figure 1. Mean (+ SD) of field-based measurements of three traits for the populations of origin (GL: Glacial Heritage; SP: Smith Prairie; UB: Union Bay). Significant PERMANOVA results ($\alpha = 0.05$) are indicated by different uppercase letters and significant PERMDISP results are indicated by different lowercase letters. D) Results of a Principal Components Analysis of the three traits measured on individuals from each population. Populations were analyzed together, but are shown in separate graphs for clarity. PC1 explained 49% of observed variation. Filled symbols indicate maternal "high" (H), "medium" (M) and "low" (L) maternal lines used in the greenhouse experiment, while open symbols (N) indicate maternal lines measured in the field but not included in the experiment.

Plants were grown in the greenhouse under one of three treatments: ambient conditions, 80% shade (80% of light is blocked), and drought conditions (50% of winter precipitation). The shade and drought treatment levels were selected to mimic marginal conditions where *H. radicata* is known to survive. Beginning on day seven, the high, low and central maternal lines were randomly assigned to the three treatments, and individuals were randomly arranged on the bench to account for edge effects. The ambient and shade treatments received weekly top-watering equivalent to average October-November rainfall across the three populations (53 ml per week per conetainer). The drought treatment received half as much (26 ml per week per conetainer) and visibly induced plant stress; individuals wilted between waterings. Each treatment received weekly 10-10-10 NPK fertilizer plus micronutrients per manufacturer specifications.

Mortality and rosette diameter (a measure of competitive ability similar to height [16]) were measured weekly, beginning on day seven. Relative chlorophyll content (correlated with photosynthesis and leaf life-span [52]) was measured weekly beginning on day 14 (plant leaves were too small on day seven for data collection) using a SPAD 502 plus chlorophyll meter (Spectrum Technologies, Illinois, USA). The length and width of the largest leaf on each individual were measured after five weeks to quantify whether leaves were more lanceolate or ovate, which can have consequences for a number of functions, including light interception

[53]. Plants were destructively sampled after five weeks. One leaf per individual was harvested, scanned, dried at 60°C, and weighed using a Mettler AE 163 analytical scale (Mettler Toledo, Ohio, USA). One-sided leaf area was calculated from the scanned images using ImageJ version 1.45 (http://rsb.nih.gov/ij). SLA, which is correlated with resource acquisition and leaf life-span [54,55], was calculated as leaf area (mm^2) divided by leaf dry mass (mg). Above and belowground biomass were harvested separately, dried at 60°C, weighed, and used to calculate root:shoot ratio, a measure of resource allocation.

Statistical Analysis

Differences between trait means for the populations of origin measured in the field were quantified using PERMANOVA, a non-parametric permutational form of ANOVA [56]. Differences in trait dispersion were quantified using PermDISP, a permutational analog to the Levene's test [57]. These analyses were conducted using the PERMANOVA+ add-on to PRIMER-E [58], with Euclidean distances, type III sums of squares, and 9,999 permutations. Populations were considered to differ if $P<0.05$.

For plants grown under greenhouse conditions, measured traits were checked for normality, and corrective measures were taken when traits were non-normal. Correlation between dependent variables was checked and, when response variables were strongly correlated, only one representative variable was tested (the weakest

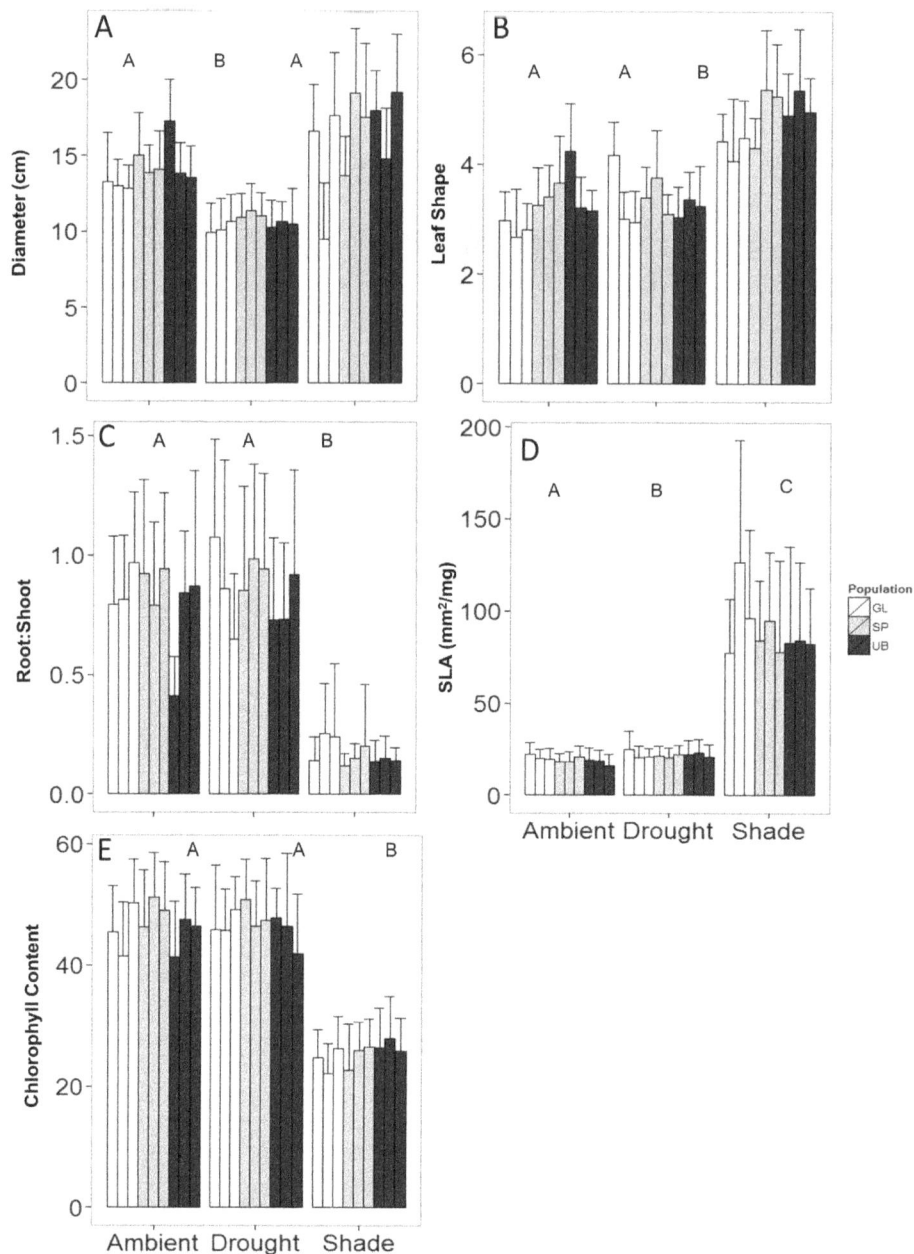

Figure 2. Mean (+ SD) values of A) final diameter, B) leaf shape (length:width ratio), C) root:shoot ratio, D) SLA, and E) final relative chlorophyll content for each maternal line (individual bar, arranged from low to high based on scores from PC1) in each population (unique color; defined in Table S1) for each treatment. Significant treatment differences ($\alpha = 0.05$) in trait means are indicated by different letters.

correlation among tested variables was $r^2 > 0.65$, see Fig S2). Five traits were selected for analysis: diameter, leaf shape (leaf length:width ratio; larger numbers indicate more lanceolate leaves and lower numbers indicate more ovate leaves), root:shoot ratio, SLA, and final relative chlorophyll content. SLA and diameter were log transformed to correct for non-normality and heteroscedacity.

We used hierarchical linear mixed effects models and variance component analysis to quantify the influence of abiotic treatment and ecotypic factors on observed traits. Population and maternal line were designated as random effects, with maternal line nested within populations. The variance attributable to population of

origin, maternal line, and individual were expressed as proportions of the total random-effects variance. Abiotic treatment (ambient, shade, or drought) was treated as a fixed effect. Models were fit with the "lme4" package in R (version 2.14.2) and P-values were generated using the "languageR" package. Significant treatment effect was followed by pairwise comparisons using TukeyHSD through the "multcomp" package (version 1.2–19), with attention focused on the differences between the ambient treatment and each of the shade and drought treatments.

Ontogenetic changes were examined by analyzing changes over time in diameter and relative chlorophyll content. These traits were selected because they did not require repeated, destructive

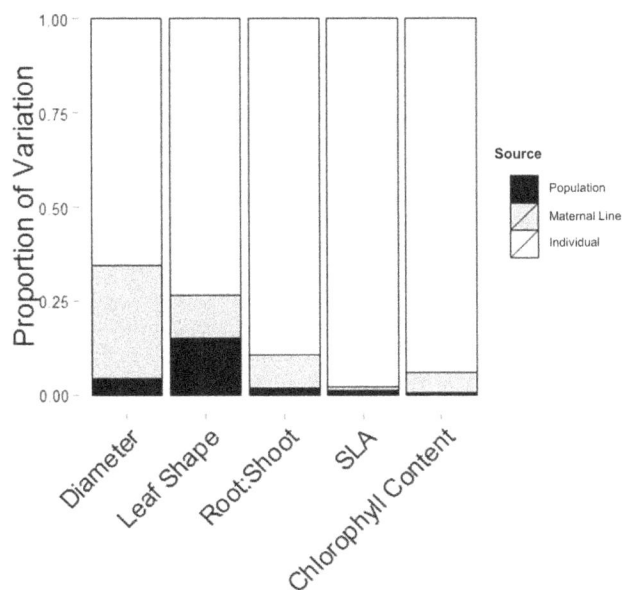

Figure 3. Proportion of total random effects variance explained by population of origin, maternal line, and individual plant (residual) for five measured traits.

measurements. Individual plants were nested within maternal lines within populations in these models, and time was included as a fixed effect. The fixed effects (abiotic treatment and time) and their interaction were iteratively added, beginning with the null model (including only random effects) and selecting the best model using Akaike's Information Criterion (AIC). Significant treatment by time effects were followed by separate analyses of each date to quantify the variance attributable to population of origin, maternal line, and individuals.

Results

Response to Abiotic Stress

Mortality was low, even for plants grown under abiotic stress: only 8 of 540 individuals died. However, the abiotic treatments affected plant trait expression (Fig. 2, See Table S2 for GLMM results). Mean final diameter was larger for plants in the shade treatment (16.4 cm) than the ambient treatment (14.1 cm), but smaller in the drought treatment (10.6 cm) (Fig. 2A). Plants in the shade treatment had significantly more lanceolate leaves (4.8) than those in the ambient (3.3) and drought treatments (3.3) (Fig. 2B), and also had significantly lower root:shoot ratios (0.17 vs 0.82 and 0.86, respectively) (Fig. 2C). Both the shade and drought treatments increased mean SLA (88.2 and 22.1 mm^2/mg, respectively) compared to those in the ambient treatment (19.4 mm^2/mg) (Fig. 3D). Relative chlorophyll content was significantly lower for plants in the shade treatment (25.6 SPAD units) compared with those in the drought (46.8 units) or ambient treatments (46.7 SPAD units) (Fig. 2E).

Contribution of Population and Maternal Line to Trait Variation

A surprisingly small amount of variation was explained by population of origin or maternal line for most measured traits (Fig. 3). Diameter was the only trait that showed an appreciable maternal contribution to observed variation; 30% of the total random effects variation was attributable to maternal lines, while

5% was attributable to populations (Fig. 3). Variation in leaf shape showed nearly equal contributions from population (15%) and maternal sources (12%). Visual inspection of the mean values for each maternal line showed no clear relationship with the maternal plant's score on PC1 (Fig. 1).

Ontogenetic Effects

There were significant interactions between treatment and time for both plant diameter and relative chlorophyll content (Fig. 4A and C). Treatment groups were indistinguishable on days 7 and 14. On day 21, plants in the shade treatment had significantly smaller diameter than those in the drought and ambient treatments ($P = 0.03$). On days 28 and 35, diameter in the drought treatment was significantly lower than the shade or ambient treatments ($P \leq 0.0001$; Fig 4A). Individuals accounted for an increasing amount of the variation in plant diameter, while variation between populations decreased over time (Fig. 4B). Variation between maternal lines remained fairly consistent.

Shading rapidly and strongly affected relative chlorophyll content, as indicated by significant treatment effects already on day 14, while drought did not ($P < 0.0001$; Fig 4C). Variation between individuals increased between days 14 and 21, and remained high throughout the experiment (Fig. 4D). Variation between maternal lines and populations decreased over the course of the experiment.

Discussion

The roles of plasticity and adaptation in shaping trait expression

We set out to determine whether observed intraspecific variation was driven by plasticity or by adaptation to local abiotic conditions. Our results indicate that, for *H. radicata*, the observed trait variation is largely attributable to plasticity. Neither shading nor drought stress were strong enough to cause appreciable mortality, confirming that *H. radicata* can tolerate a very wide range of abiotic conditions. However, plants grown under these conditions exhibited markedly different functional traits. For example, SLA values previously reported for *H. radicata* individuals have ranged from 1.7–46.1 mm^2/mg [16,59,60]; in this study, SLA ranged from 6.6–239.1 mm^2/mg. Much of the increased range of SLA in this study was due to increases in response to shading (Fig. 2D), a response which has been demonstrated in the past [61]. The inclusion of shade-grown individuals, which have significantly higher SLA than sun-grown individuals, would dramatically shift the trait distribution for this species. Inclusion of such plants would also likely influence the overall trait distribution of the plant community, and may in turn impact estimations of critical ecosystem processes like litter production and nutrient cycling [13] through differing nutrient contributions and breakdown times in the decomposition pool [62]. Standard trait collection procedures [41,63] focus trait measurements exclusively on sun-exposed plants and therefore do not account for the full trait variation of species that can tolerate varying levels of shade. The omission of individuals in partial or full-shade from trait measurements may hinder our understanding of trait mediated ecosystem processes and responses. We suggest it would be more appropriate to sample the full range of conditions experienced by a species.

The variance explained by population of origin and maternal line was surprisingly small for every measured trait. Only leaf shape and diameter had significant variation attributable to differences across populations and maternal lines respectively, and even this amount of variation (30%) was dwarfed by treatment

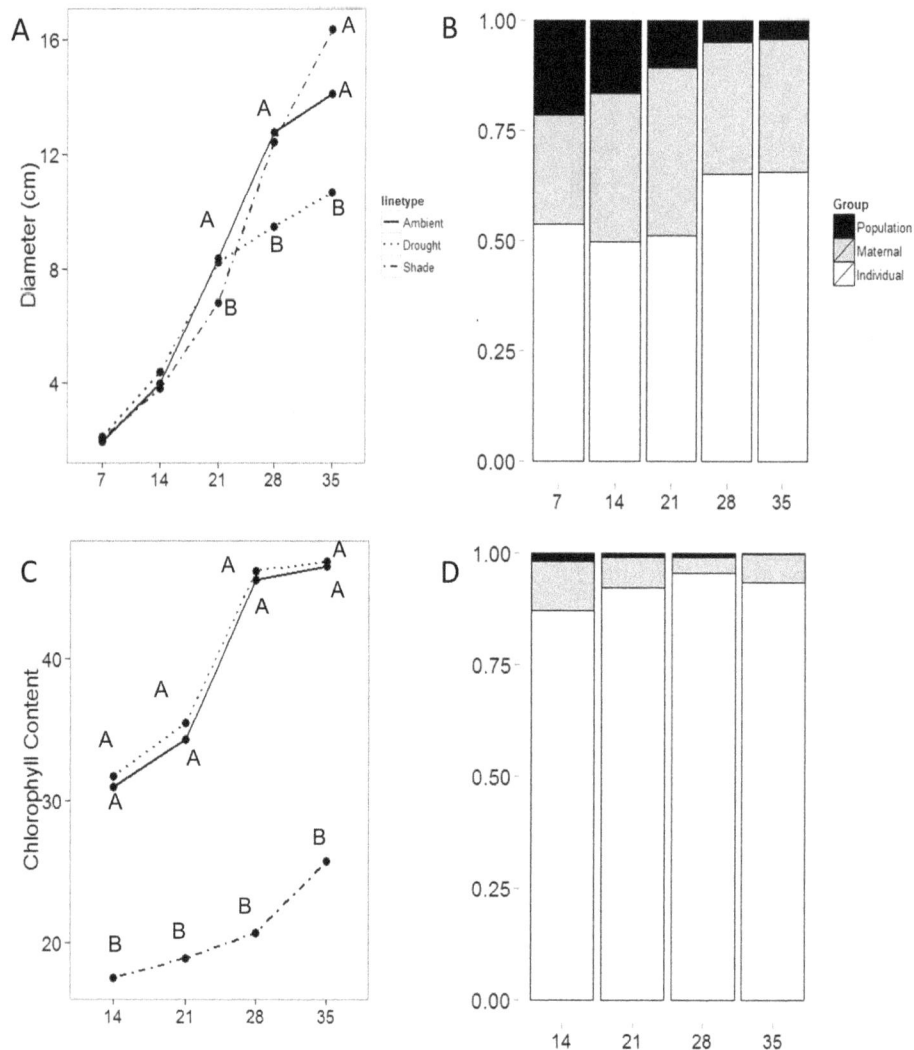

Figure 4. Change in average trait values in response to treatments (A and C; in each time period, letters indicate significantly different groups) and (B and D) the amount of variation attributed to population of origin, maternal line, and individual plant for diameter (top) and relative chlorophyll content (bottom).

effects. Virtually none of the observed variation could be attributed to population or maternal line for SLA (Fig. 3), confirming that this trait is highly plastic, a finding which supports previous analyses of this trait [61]. Furthermore, the position of each maternal line on PC1 had little correlation with the ranking of each maternal line in the greenhouse, suggesting that individuals within maternal lines exhibited significant variability and plasticity, and are not necessarily constrained by maternal genetics. The very small amount of variation attributable to population or maternal sources indicates that, for the traits measured in this experiment, plastic responses to prevailing conditions are the primary driver of observed variation.

The fact that all of the measured traits displayed significant plasticity has important implications for how traits are measured, and how they are incorporated into larger, trait-based models. Although some species are shade-intolerant, many species experience variable light conditions through interactions with neighboring plants in the community. Trait values based on fully sun-exposed leaves may inaccurately represent the trait contributions of such species, and thus may misrepresent community-level

trait values, and the ecosystem functions dependent on those trait values. In addition, attempts to model species presence or response to abiotic change may be hindered by a reliance on trait means derived from sun-exposed individuals. If species, or populations within species, are variable and highly plastic for traits critical to survival, such taxa may be present in a wider range of habitat types, and be better able to cope with a wider range of conditions [64,65] than expected from trait values measured on individuals growing in a subset of the taxa's possible growing conditions.

The role of ontogeny in shaping trait expression

Variation in trait values over the lifespan of a plant is an additional, and often unaccounted for, source of intraspecific trait variability. This source of variation may be underappreciated as collection techniques typically focus on "mature" plants and tissues [41] yet maturity can be difficult to distinguish in the field. Because plants of differing ontogenies may appear "mature" in the field, intraspecific trait variation measured in field populations could be partially attributed to differing ontogeny between individuals, rather than to genetic or environmental conditions.

Furthermore, ontogenetic differences in trait variability may be important in the structure and function of communities. For example, ontogenetic trait differences may lead to temporal niche partitioning, facilitating biodiversity maintenance through limiting similarity [66]. Ontogenetic differences leading to higher palatability can also alter plant fitness by influencing interactions with herbivores and predators [67]. Furthermore, temporal variation in plant traits may exacerbate or mediate the impacts of abiotic filters. Young plants face similar pressures during establishment, and tend to exhibit similar trait values, in both this study, and others (e.g., [68]). This lack of trait variability in young plants may increase susceptibility to stress, while increased variability in older plants may buffer the population against abiotic change. Incorporating trait values of plants at different ontogenetic stages could provide a clearer understanding of community and population dynamics, and more accurate extrapolation to ecosystem processes and functions.

Conclusion

Taken together, these results indicate that plasticity in response to abiotic conditions and ontogeny, and not adaptation and genetically distinct ecotypes, drive trait expression in *H. radicata*. However, the generality of this conclusion needs to be explored. *Hypochaeris radicata* is globally widespread, and thus may be inherently more plastic than species with restricted distributions [69,70]. To determine whether plasticity is a characteristic trait of widespread or dominant species, this research should be replicated with other widespread species and with species from restricted ranges. In addition, further research should examine how sensitive these conclusions are to the identity of the trait being examined.

The treatments applied during this experiment represented strong abiotic filters for this species, but are within the range of conditions experienced by *H. radicata*. Studies involving light or moisture gradients could assess whether the importance of population of origin or plasticity changes depending on the strength of the abiotic filter, a question which was not addressed in this study. Studies could also examine other abiotic filters, such as

soil fertility. Finally, the additive and interactive effects of multiple abiotic filters on trait expression could be tested in both field and greenhouse settings.

Despite growing interest in understanding and incorporating intraspecific variation in plant functional traits into larger community- and ecosystem-level studies, the factors driving observed variation remain poorly understood. Our results suggest that observed intraspecific variation is largely driven by trait plasticity rather than genetic factors. We also found strong evidence of variation in plant traits due to ontogeny, a source of trait variation that is often ignored.

Acknowledgments

We thank Ty Robinson for assistance in the field, and Loretta Fisher for assistance in the lab. We also thank Janneke Hille Ris Lambers, Soo-Hyung Kim, Sarah Reichard, Regina Rochefort, and Martha Groom for their advice and guidance.

Author Contributions

Conceived and designed the experiments: RMM JDB. Performed the experiments: RMM. Analyzed the data: RMM. Contributed reagents/materials/analysis tools: RMM JDB. Wrote the paper: RMM JDB.

References

1. Messier J, McGill BJ, Lechowicz MJ (2010) How do traits vary across ecological scales? A case for trait-based ecology. Ecol Lett 13: 838–848. doi:10.1111/j.1461-0248.2010.01476.x.

2. Myers JA, Chase JM, Jiménez I, Jørgensen PM, Araujo-Murakami A, et al. (2013) Beta-diversity in temperate and tropical forests reflects dissimilar mechanisms of community assembly. Ecol Lett 16: 151–157. doi:10.1111/ele.12021.

3. Long W, Zang R, Schamp BS, Ding Y (2011) Within- and among-species variation in specific leaf area drive community assembly in a tropical cloud forest. Oecologia 167: 1103–1113. doi:10.1007/s00442-011-2050-9.

4. Swenson NG, Anglada-Cordero P, Barone Ja (2011) Deterministic tropical tree community turnover: evidence from patterns of functional beta diversity along an elevational gradient. Proc R Soc B 278: 877–884. doi:10.1098/rspb.2010.1369.

5. Swenson NG, Weiser MD (2010) Plant geography upon the basis of functional traits: an example from eastern North American trees. Ecology 91: 2234–2241.

6. Thuiller W, Lavorel S, Midgley G, Lavergne S (2004) Relating plant traits and species distributions along bioclimatic gradients for 88 Leucadendron taxa. Ecology 85: 1688–1699.

7. Laughlin DC, Fulé PZ, Huffman DW, Crouse J, Laliberté E (2011) Climatic constraints on trait-based forest assembly. J Ecol 99: 1489–1499. doi:10.1111/j.1365-2745.2011.01885.x.

8. Hudson JMG, Henry GHR, Cornwell WK (2011) Taller and larger: shifts in Arctic tundra leaf traits after 16 years of experimental warming. Glob Chang Biol 17: 1013–1021. doi:10.1111/j.1365-2486.2010.02294.x.

9. Fonseca C, McC Overton J, Collins B, Westoby M (2000) Shifts in trait-combinations along rainfall and phosphorus gradients. J Ecol 88: 964–977.

10. Díaz S, Cabido M, Casanoves F (1998) Plant functional traits and environmental filters at a regional scale. J Veg Sci 9: 113–122.

11. Diaz S, Hodgson JG, Thompson K, Cabido M, Cornelissen JHC, et al. (2004) The plant traits that drive ecosystems: Evidence from three continents. J Veg Sci 15: 295–304. doi:10.1111/j.1654-1103.2004.tb02266.x.

12. Diaz S, Cabido M (2001) Vive la différence: plant functional diversity matters to ecosystem processes. 16: 646–655.

13. William KC, Johannes HCC, Kathryn A, Ellen D, Valerie TE, et al. (2008) Plant species traits are the predominant control on litter decomposition rates within biomes worldwide. Ecol Lett 11: 1065–1071.

14. Albert CH, Thuiller W, Yoccoz NG, Douzet R, Aubert S, et al. (2010) A multi-trait approach reveals the structure and the relative importance of intra- vs. interspecific variability in plant traits. Funct Ecol 24: 1192–1201. doi:10.1111/j.1365-2435.2010.01727.x.

15. Albert CH, Thuiller W, Yoccoz NG, Soudant A, Boucher F, et al. (2010) Intraspecific functional variability: extent, structure and sources of variation. J Ecol 98: 604–613. doi:10.1111/j.1365-2745.2010.01651.x.

16. Mitchell RM, Bakker JD (2014) Quantifying and comparing intraspecific functional trait variability: a case study with Hypochaeris radicata. Funct Ecol 28: 258–269. doi:10.1111/1365-2435.12167.

17. Siefert A (2012) Incorporating intraspecific variation in tests of trait-based community assembly. Oecologia 170: 767–775. doi:10.1007/s00442-012-2351-7.

18. Lecerf A, Chauvet E (2008) Intraspecific variability in leaf traits strongly affects alder leaf decomposition in a stream. Basic Appl Ecol 9: 598–605. doi:10.1016/j.baae.2007.11.003.

19. Jung V, Violle C, Mondy C, Hoffmann L, Muller S (2010) Intraspecific variability and trait-based community assembly. J Ecol 98: 1134–1140. doi:10.1111/j.1365-2745.2010.01687.x.

20. Clark JS (2010) Individuals and the variation needed for high species diversity in forest trees. Science (80-) 327: 1129–1132. doi:10.1126/science.1183506.

21. Crutsinger GM, Souza L, Sanders NJ (2008) Intraspecific diversity and dominant genotypes resist plant invasions. Ecol Lett 11: 16–23. doi:10.1111/j.1461-0248.2007.01118.x.

22. Moreira B, Tavsanoglu C, Pausas JG (2012) Local versus regional intraspecific variability in regeneration traits. Oecologia 168: 671–677. doi:10.1007/s00442-011-2127-5.

23. Jung V, Albert CH, Violle C, Kunstler G, Loucougaray G, et al. (2014) Intraspecific trait variability mediates the response of subalpine grassland communities to extreme drought events. J Ecol 102: 45–53. doi:10.1111/1365-2745.12177.

24. Fajardo A, Piper FI (2011) Intraspecific trait variation and covariation in a widespread tree species (Nothofagus pumilio) in southern Chile. New Phytol 189: 259–271. doi:10.1111/j.1469-8137.2010.03468.x.

25. Grassein F, Till-Bottraud I, Lavorel S (2010) Plant resource-use strategies: the importance of phenotypic plasticity in response to a productivity gradient for two subalpine species. Ann Bot 106: 637–645. doi:10.1093/aob/mcq154.

26. Wellstein C, Chelli S, Campetella G, Bartha S, Galiè M, et al. (2013) Intraspecific phenotypic variability of plant functional traits in contrasting mountain grasslands habitats. Biodivers Conserv 22: 2353–2374. doi:10.1007/s10531-013-0484-6.

27. Freschet GT, Bellingham PJ, Lyver PO, Bonner KI, Wardle Da. (2013) Plasticity in above- and belowground resource acquisition traits in response to single and multiple environmental factors in three tree species. Ecol Evol: n/a–n/a. doi:10.1002/ece3.520.

28. Joshi J, Schmid B, Caldeira MC, Dimitrakopoulos PG, Good J, et al. (2001) Local adaptation enhances performance of common plant species. Ecol Lett 4: 536–544. doi:10.1046/j.1461-0248.2001.00262.x.

29. Schlichting C (1986) The evolution of phenotypic plasticity in plants. Annu Rev Ecol Syst 17: 667–693.

30. Hansen C, Garcia MB, Ehlers BK (2013) Water availability and population origin affect the expression of the tradeoff between reproduction and growth in Plantago coronopus. J Evol Biol 26: 1–10. doi:10.1111/jeb.12114.

31. Grassein F, Till-Bottraud I, Lavorel S (2010) Plant resource-use strategies: the importance of phenotypic plasticity in response to a productivity gradient for two subalpine species. Ann Bot 106: 637–645. doi:10.1093/aob/mcq154.

32. Bonser SP, Ladd B, Monro K, Hall MD, Forster M (2010) The adaptive value of functional and life-history traits across fertility treatments in an annual plant. Ann Bot 106: 979–988. doi:10.1093/aob/mcq195.

33. Firn J, Prober SM, Buckley YM (2012) Plastic traits of an exotic grass contribute to its abundance but are not always favourable. PLoS One 7: e35870. doi:10.1371/journal.pone.0035870.

34. Ramírez-Valiente JA, Sánchez-Gómez D, Aranda I, Valladares F (2010) Phenotypic plasticity and local adaptation in leaf ecophysiological traits of 13 contrasting cork oak populations under different water availabilities. Tree Physiol 30: 618–627. doi:10.1093/treephys/tpq013.

35. Robson TM, Sánchez-Gómez D, Cano FJ, Aranda I (2012) Variation in functional leaf traits among beech provenances during a Spanish summer reflects the differences in their origin. Tree Genet Genomes 8: 1111–1121. doi:10.1007/s11295-012-0496-5.

36. Dewitt TJ, Sih A, Wilson DS (1998) Costs and limits of phenotypic plasticity. Trends Ecol Evol 13: 77–81.

37. Auld JR, Agrawal A, Relyea R (2010) Re-evaluating the costs and limits of adaptive phenotypic plasticity. Proc R Soc B 277: 503–511. doi:10.1098/rspb.2009.1355.

38. Violle C, Enquist BJ, McGill BJ, Jiang L, Albert CH, et al. (2012) The return of the variance: intraspecific variability in community ecology. Trends Ecol Evol 27: 244–252. doi:10.1016/j.tree.2011.11.014.

39. Smith EA, Collette SB, Boynton TA, Lillrose T, Stevens MR, et al. (2011) Developmental contributions to phenotypic variation in functional leaf traits within quaking aspen clones. Tree Physiol 31: 68–77. doi:10.1093/treephys/tpq100.

40. Cornelissen JHC, Cerabolini B, Puyravaud JP, Maestro M, Werger MJA, et al. (2003) Functional traits of woody plants: correspondence of species rankings between field adults and laboratory-grown seedlings? J Veg Sci 14: 311–322.

41. Cornelissen J, Lavorel S, Garnier E, Diaz S, Buchmann N, et al. (2003) A handbook of protocols for standardised and easy measurement of plant functional traits worldwide. Aust J Bot 51: 335–380.

42. Bradshaw AD (2006) Unravelling phenotypic plasticity-Why should we bother? New Phytol 170: 639–641. doi:10.1111/j.1469-8137.2006.01758.x.

43. Nicotra AB, Atkin OK, Bonser SP, Davidson AM, Finnegan EJ, et al. (2010) Plant phenotypic plasticity in a changing climate. Trends Plant Sci 15: 684–692. doi:10.1016/j.tplants.2010.09.008.

44. Valladares F, Gianoli E, Gómez JM (2007) Ecological limits to plant phenotypic plasticity. New Phytol 176: 749–763. doi:10.1111/j.1469-8137.2007.02275.x.

45. Balaguer L, Martinez-Ferri E, Valladares F, Perez-Corona ME, Baquedano FJ, et al. (2001) Population divergence in the plasticity of the response of Quercus coccifera to the light environment. Funct Ecol 15: 124–135. doi:10.1046/j.1365-2435.2001.00505.x.

46. Pratt JD, Mooney KA (2013) Clinal adaptation and adaptive plasticity in Artemisia californica: implications for the response of a foundation species to predicted climate change. Glob Chang Biol 19: 2454–2466. doi:10.1111/gcb.12199.

47. Davidson AM, Jennions M, Nicotra AB (2011) Do invasive species show higher phenotypic plasticity than native species and, if so, is it adaptive? A meta-analysis. Ecol Lett 14: 419–431. doi:10.1111/j.1461-0248.2011.01596.x.

48. Herman JJ, Sultan SE (2011) Adaptive transgenerational plasticity in plants: case studies, mechanisms, and implications for natural populations. Front Plant Sci 2: 102. doi:10.3389/fpls.2011.00102.

49. Pico F, Ouborg NJ, Van Groenendael JM (2004) Influence of selfing and maternal effects on life-cycle traits and dispersal ability in the herb Hypochaeris radicata (Asteraceae). Bot J Linn Soc 146: 163–170.

50. Ortiz MA, Talavera S, Garcia-Castano JL, Tremetsberger K, Stuessy T, et al. (2006) Self-Incompatibility and Floral Parameters in Hypochaers sec. Hypocaheris (Asteraceae). Am J Bot 93: 234–244.

51. Mix C, Arens PFP, Rengelink R, Smulders MJM, Van Groenendael JM, et al. (2006) Regional gene flow and population structure of the wind-dispersed plant species Hypochaeris radicata (Asteraceae) in an agricultural landscape. Mol Ecol 15: 1749–1758. doi:10.1111/j.1365-294X.2006.02887.x.

52. Poorter L, Bongers F (2006) Leaf traits are good predictors of plant performance across 53 rain forest species. Ecology 87: 1733–1743.

53. Nicotra A, Leigh A, Boyce C (2011) The evolution and functional significance of leaf shape in the angiosperms. Funct Plant … 38: 535–552.

54. Reich PB, Walters MB, Ellsworth D (1992) Leaf life-span in relation to leaf, plant and stand characteristics among diverse ecosystems. Ecol Monogr 62: 365–392.

55. Westoby M (1998) A leaf-height-seed (LHS) plant ecology strategy scheme: 213–227.

56. Anderson MJ (2001) A new method for non-parametric multivariate analysis of variance. Austral Ecol 26: 32–46. doi:10.1046/j.1442-9993.2001.01070.x.

57. Anderson MJ (2004) PERMDISP: a FORTRAN computer program for permutational analysis of multivariate dispersions (for any two-factor ANOVA design) using permutation. Department of Statistics, University of Auckland, Auckland, New Zealand.

58. Clarke K, Gorley R (2006) Primer v6: User Manual/Tutorial. Plymouth: PRIMER-E.

59. Mokany K, Ash J (2008) Are traits measured on pot grown plants representative of those in natural communities? J Veg Sci 19: 119–126. doi:10.3170/2007-8-18340.

60. Kleyer M, Bekker RM, Knevel IC, Bakker JP, Thompson K, et al. (2008) The LEDA Traitbase: a database of life-history traits of the Northwest European flora. J Ecol 96: 1266–1274. doi:10.1111/j.1365-2745.2008.01430.x.

61. Poorter H, Niinemets Ü, Poorter L, Wright IJ, Villar R (2009) Causes and consequences of variation in leaf mass per area (LMA): a meta-analysis. New Phytol 182: 565–588. doi:10.1111/j.1469-8137.2009.02830.x.

62. Cornwell WK, Cornelissen JHC, Amatangelo K, Dorrepaal E, Eviner VT, et al. (2008) Plant species traits are the predominant control on litter decomposition rates within biomes worldwide. Ecol Lett 11: 1065–1071.

63. Pérez-Harguindeguy N, Díaz S, Garnier E, Lavorel S, Poorter H, et al. (2013) New handbook for standardised measurement of plant functional traits worldwide. Aust J Bot: 167–234.

64. Sultan SE (2000) Phenotypic plasticity for plant development, function and life history. Trends Plant Sci 5: 537–542.

65. Richards CL, Bossdorf O, Muth NZ, Gurevitch J, Pigliucci M (2006) Jack of all trades, master of some? On the role of phenotypic plasticity in plant invasions. Ecol Lett 9: 981–993. doi:10.1111/j.1461-0248.2006.00950.x.

66. Diaz S, Cabido M (2001) Vive la difference: plant functional diversity matters to ecosystem processes. Trends Ecol Evol 16: 646–655.

67. Boege K (2010) Induced responses to competition and herbivory: natural selection on multi-trait phenotypic plasticity. Ecology 91: 2628–2637.

68. Mediavilla S, Escudero A (2004) Stomatal responses to drought of mature trees and seedlings of two co-occurring Mediterranean oaks. For Ecol Manage 187: 281–294. doi:10.1016/j.foreco.2003.07.006.

69. Sultan SE, Barton K, Wilczek AM (2009) Contrasting patterns of transgenerational plasticity in ecologically distinct congeners. Ecology 90: 1831–1839. doi:10.1890/08-1064.1.

70. Gitzendanner M, Soltis P (2000) Patterns of genetic variation in rare and widespread plant congeners. Am J Bot 87: 783–792.

15

Taxonomic and Functional Diversity Provides Insight into Microbial Pathways and Stress Responses in the Saline Qinghai Lake, China

Qiuyuan Huang[1,4☯], **Brandon R. Briggs**[1☯], **Hailiang Dong**[1,2,3]*, **Hongchen Jiang**[3], **Geng Wu**[3], **Christian Edwardson**[4], **Iwijn De Vlaminck**[5], **Stephen Quake**[5]

1 Department of Geology and Environmental Earth Science, Miami University, Oxford, Ohio, United States of America, **2** State Key Laboratory of Biogeology and Environmental Geology, China University of Geosciences, Beijing, China, **3** State Key Laboratory of Biogeology and Environmental Geology, China University of Geosciences, Wuhan, China, **4** Department of Microbiology, University of Georgia, Athens, Georgia, United States of America, **5** Departments of Bioengineering and Applied Physics, Stanford University and the Howard Hughes Medical Institute, Stanford, California, United States of America

Abstract

Microbe-mediated biogeochemical cycles contribute to the global climate system and have sensitive responses and feedbacks to environmental stress caused by climate change. Yet, little is known about the effects of microbial biodiversity (i.e., taxonmic and functional diversity) on biogeochemical cycles in ecosytems that are highly sensitive to climate change. One such sensitive ecosystem is Qinghai Lake, a high-elevation (3196 m) saline (1.4%) lake located on the Tibetan Plateau, China. This study provides baseline information on the microbial taxonomic and functional diversity as well as the associated stress response genes. Illumina metagenomic and metatranscriptomic datasets were generated from lake water samples collected at two sites (B and E). Autotrophic *Cyanobacteria* dominated the DNA samples, while heterotrophic *Proteobacteria* dominated the RNA samples at both sites. Photoheterotrophic *Loktanella* was also present at both sites. Photosystem II was the most active pathway at site B; while, oxidative phosphorylation was most active at site E. Organisms that expressed photosystem II or oxidative phosphorylation also expressed genes involved in photoprotection and oxidative stress, respectively. Assimilatory pathways associated with the nitrogen cycle were dominant at both sites. Results also indicate a positive relationship between functional diversity and the number of stress response genes. This study provides insight into the stress resilience of microbial metabolic pathways supported by greater taxonomic diversity, which may affect the microbial community response to climate change.

Editor: Yiguo Hong, CAS, China

Funding: This research was supported by the National Natural Science Foundation of China (Grant Nos. 41030211 and 41002123), the Scientific Research Funds for the 1000 "Talents" Program Plan from China University of Geosciences (Beijing), State Key Laboratory of Biogeology and Environmental Geology, China University of Geosciences (No. GBL11201), and the Fundamental Research Funds for National University, China University of Geosciences (Wuhan). This work was fostered by NSF IOS grant 1238801. The funders had no role in study design, data collection and analysis, decision to publish, or preparation of the manuscript.

Competing Interests: The authors have declared that no competing interests exist.

* Email: dongh@miamioh.edu

☯ These authors contributed equally to this work.

Introduction

Microorganisms have been key respondents to and drivers of global climate change by affecting the atmospheric concentrations of greenhouse gases [1]. The microbial response to future global climate change is likely controlled by the biodiversity (i.e. taxonomic and functional diversity) of the ecosystem [2,3,4]. For example, ecosystems with higher biodiversity are more likely to be stable against environmental change because of a greater likelihood of having key functioning species [5]. However, ecosystems with low diversity can be stable if the organisms have mechanisms to respond to stress [6]. For example, *Synechococcus* has multiple protective mechanisms to cope with UV stress and can maintain photosynthesis, while *Procholorcoccus* lacks these protective mechanisms and shuts down several key metabolic processes under similar UV stress [7]. Thus, studying the

biodiversity and potential stress response mechanisms can aid in understanding microbial community response to stress.

Baseline information on the microbial biodiversity and stress response mechanisms is needed in ecosystems that are sensitive to climate change. For example, the Tibetan Plateau has experienced significant warming in recent decades and is considered to be a sensitive indicator of regional and global climate change [8]. Temperature has increased 0.28°C per decade since the early 1960 s [9] causing 82% of the 46,000 glaciers to retreat [10]. The melting glaciers have caused numerous floods and altered salinity and water levels in most of the Tibetan lakes [11]. The fragility and sensitivity of the Tibetan Plateau's ecosystem to these environmental changes have resulted in loss of habitats and extinctions of endemic macrobiota [12].

Qinghai Lake, located on the Tibetan Plateau, is characterzied by oligotrophy, low temperature, moderate salinity, and high UV radiation, making it a unique ecosystem for studying microbial response to global climate change [13–15]. Previous studies on Qinghai Lake have detected novel archaea commonly found in marine environments [16], and the microbial diversity, composition, and lipid profiles all showed a response to salinity change [14,17,18]. However, the following key questions remain unanswered: (1) what is the microbial taxonomic and functional diversity for the Qinghai Lake water column, (2) what is the metabolic potential and active metabolisms related to the carbon and nitrogen cycles, and (3) what stress response genes are present in organisms involved in the carbon and nitrogen cycles? Answering these questions can provide baseline knowledge that can be used to understand the effect of biodiversity on the microbial community response to environmental stress.

An integrated approach including geochemical, metagenomic and metatranscriptomic analyses were used to answer these questions. The metagenomic and metatranscriptomic reads were annotated with both taxonomic and functional information. The synthesized cDNA was compared to the DNA retrieved from the same sample to assess the relative activity of different populations and functional gene transcription in the microbial community. In addition, the relationship between species richness and the number of stress response genes was assessed. Microbial processes in Qinghai Lake are involved in both the carbon and nitrogen biogeochemical cycles. However, certain processes (e.g., photosynthesis, denitrification) had lower diversity and fewer stress response genes, which may make them more susceptible to environmental stresses.

Materials and Methods

Site description

Qinghai Lake is a perennial lake located on the Tibetan Plateau at an elevation of 3196 m above sea level. The lake is located in a structural intermontane depression at the north-eastern corner of the Qinghai-Tibetan Plateau (Figure 1) [19]. The lake has a surface area of 4300 km^2 and lies within a catchment of limestone, sandstone, and shale. The average water depth is 19.2 m with the maximum of 28.7 m. The evaporation of the lake ($\sim$1400 mm/year) is in excess of mean annual precipitation ($\sim$400 mm/year), resulting in a mesohaline lake. Qinghai Lake is separated into two subbasins by a normal faulting horst in the middle of the lake. The northern subbasin is more dynamic than the southern subbasin because of riverine input. The southern subbasin water is stratified in winter due to ice cover, but in summer, the water column geochemistry is fairly uniform [14,16]. No specific permits were required for the described field studies because no animal or human subjects were involved in this research. The sampling locations are not privately owned or protected in any way. The field studies did not involve endangered or protected species.

Sampling and geochemical analysis

Water samples were collected from the southern subbasin mid-day in August 2011 at sites B and E (Figure 1). The water samples were pumped from water depths of 12.5 m and 13.6 m below the lake surface from sites B and E, respectively. These two water depths were selected as representatives of the microbial community in the water column. Ten to 12 L of lake water was filtered through a polyethersulfone filter with a pore size of 0.2 μm (Supor; Pall Life Sciences, Ann Arbor, MI, USA). A Horiba multi-parameter meter (W-20XD Series, HORIBA, Kyoto, Japan) was used to measure *in-situ* environmental parameters of the water

including; temperature, pH, conductivity, dissolved oxygen (DO), depth, chloride (Cl^-), and salinity at both sites. Concentrations of major ions (i.e., sulfide, sulfate, silica, nitrite, nitrate, ferrous iron, and ammonia) in the filtered water of both sites were measured using a HACH colorimeter (model CEL 850, HACH Chemical Co., Iowa, USA) as previously described [20]. Filters collected for microbial analysis were placed on dry ice immediately after filtering of lake water from each site, and stored at −80°C in the laboratory until DNA and RNA extractions were conducted. A portable global positioning system (GPS) unit (eTrex H, Garmin) was used to indicate the location of each sampling site (Figure 1).

RNA and DNA extractions

Total RNA was extracted from one half of the filter from each site following a modified version of the RNeasy kit (Qiagen, Valencia, CA, USA) as previously described [21]. Briefly, half of each frozen filter was thawed and vortexed for 10 min with 8 ml of RLT lysis buffer and 3 g of low protein binding zirconium beads (200 μm, OPS Diagnostics, Lebanon, NJ, USA). RNA was then extracted using the RNeasy kit according to manufacturer's instructions. A Turbo DNA-free kit (Ambion, Austin, TX, USA) was used to remove any residual DNA. The resulting RNA was purified and concentrated using the RNeasyMinElute Cleanup kit (Qiagen, Valencia, CA, USA) [22]. Genomic DNA was extracted from the other half of the filter using the FastDNA Spin Kit (MP Biomedical, OH, USA) as previously described [20]. The amount of DNA that was extracted from sites B and E was 14.6 and 13.8 ng μl^{-1}, indicating that DNA was extracted from roughly the same amount of biomass.

Metatranscriptomic sample preparation

Ribosomal RNA (rRNA) was removed from total RNA via a subtractive hybridization process using sample-specific biotinylated rRNA probes [23]. rRNA-subtracted RNA was amplified linearly using the MessageAmp II-Bacteria kit (Ambion, Austin, TX, USA). Amplified RNA was converted to double-stranded cDNA using the Universal RiboClone cDNA synthesis system (Promega, Madison, WI, USA) using random hexamer primers [23]. The synthesized cDNA was purified with the QIAquick PCR purification kit (Qiagen, Valencia, CA, USA).

Library preparation and sequencing

The extracted DNA and synthesized cDNA was sheared to 500 base pairs (bp) using a Covaris ultrasonicator (Covaris Inc., Woburn, MA, USA) according to manufacturer's recommendations. The sheared DNA was end-repaired, adaptor-ligated with multiplexing, and purified using the Ovation SP ultralow DR multiplex system (NuGEN Technologies Inc., CA, USA). The prepared library was then sequenced using an Illumina MiSeq (250 bp paired-end reads).

Data analysis

Sequences were paired, trimmed, and filtered using the CLC Genomics Workbench version 6 (CLC Bio, Aarhus, Denmark). Paired reads were assembled together if 25 bp overlapped. Reads were trimmed based on the length (minimum length 50 bp) and quality (quality score ≥20) [24]. Sequences were uploaded to the metagenomics RAST (MG-RAST) server [25] for annotation and are available under MG-RAST ID 4532866.3, 4532865.3, 4522126.3, and 4522125.3 for the samples B_DNA, B_RNA, E_DNA, and E_RNA, respectively.

Sequences for DNA and RNA libraries were assigned a taxonomy and function using BLASTX [26] on MG-RAST

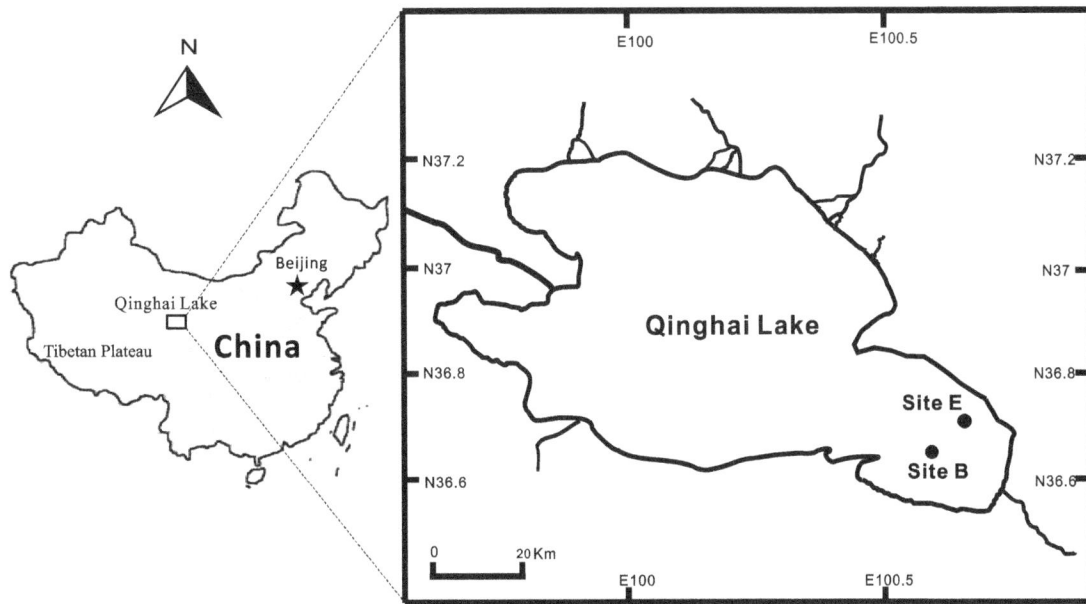

Figure 1. A geographical map showing the sampling sites in Qinghai Lake, China.

v3.3.6 against M5NR database, which integrates multiple databases (e.g., NCBI-nr, KEGG, SEED, and *etc.*). Sequences were given an annotation if it had at least a bit score cut-off of 50, E-values of 1×10^{-5}, and a minimum alignment of 15 amino acids [24]. DNA and RNA reads that were annotated with taxonomy and function were downloaded from MG-RAST and a custom R script was used to search for functional annotations involved in the carbon and nitrogen cycles, and stress responses [27]. KEGG categories were used to determine annotations that were involved in stress response. The richness of each functional gene was determined by counting the number of species or phyla that contained that particular functional gene.

Rarefaction curves were created for each of the samples to determine whether the sequencing depth was sufficient to detect the majority of species containing a functional gene. This was performed using the "rarecurve" function in the R package "Vegan" [28]. A network of correlated genes (i.e. functional genes found [DNA] or expressed [RNA] in the same organisms as stress response genes) was created using the WGCNA package [29] and the igraph package [30] in R [27], and viewed in Cytoscape [31]. Species with only one functional gene and functional genes in only a single organism were removed to reduce the complexity of the network [32]. A Spearman correlation between each functional gene and a stress response gene was calculated with a p-value adjusted for multiple comparisons. Correlations with a p-value greater than 0.05 were removed from the network.

amoA amplification

Jiang et al. [19] detected abundant ammonia oxidizing bacteria (AOB) and ammonia oxidizing archaea (AOA) in Qinghai Lake; however, the ammonia monooxygenase (*amoA*) gene was not detected in our metagenomic or metatranscriptomic datasets (see results below). Therefore, the *amoA* gene was amplified from the same DNA pool used for metagenomics. PCR amplifications were performed using FailSafe PCR System (Epicentre Biotechnologies, Madison, WI), AOB specific primer set: amoA-1F (5'-GGGGTTTCTACTGGTGGT-3') and amoA-2R (5'-CCCCTCKGSAAAGCCTTCTTC −3'), and with the same

conditions as described previously [33]. The amplicons were stained with EtBr and visualized on a 1% agarose gel.

Results and Discussion

Water geochemistry

Sites B and E shared similar geochemical profiles for most parameters (Table 1). For example, both sites had 1.4% salinity and a pH of ~9.1. The DO content was 8.9 and 8.7 ppm for sites B and E, respectively. Site E was about 1 m deeper and 2°C warmer and had a higher ammonia concentration (1 mg/L) than site B (below detection).

Descriptive statistics of DNA and RNA

After quality control of sequence reads, a total of 10,514,407 and 4,035,731 reads were retrieved in the DNA libraries for sites B and E, respectively (Table 2). The RNA libraries had a total of 1,728,111 and 2,893,577 reads for sites B and E, respectively (Table 2). Twenty-eight percent to 81% of the DNA and RNA reads that were identified to have an open reading frame were annotated with both function and organism (Table 2). Protein and rRNA features were predicted and identified for RNA and DNA libraries from both sample sites (Table 2).

Taxonomic diversity

Protein-coding genes, while not as phylogenetically robust as 16S rRNA genes, can be used to identify approximate taxonomic affiliations [34,35]. A total of 832 genera within 31 phyla were detected. The DNA samples from both sites contained more genera than were detected in the RNA samples, indicating some genera had no or low activities (Table 2). Bacterial sequences dominated both locations, with each sample containing fewer than 1% archaeal sequences. Eukaryotic sequences were removed for further analysis because of the unreliability of FragGeneScan (prokaryotic gene calling algorithm used by MG-RAST) to identify eukaryotic open reading frames [36] and our hypotheses were specific to prokaryotes.

Table 1. Sample locations and geochemistry.

Parameters	B	E
GPS location (N, E)	36.66, 100.60	36.74, 100.69
Temperature (°C)	13.7	15.6
pH	9.1	9.2
Depth (m)	12.5	13.6
Conductivity (s/m)	2.28	2.26
DO (ppm)	8.9	8.7
Salinity (%)	1.4	1.4
Cl^- (mg/L)	2930	3420
NH_4^+ (mg/L)	ND	1
NO_3^- (mg/L)	0.39	0.4
NO_2^- (mg/L)	0.13	NA
Si (mg/L)	1.43	NA
S^{2-} (mg/L)	0.05	ND
SO_4^- (mg/L)	>80	>80
Fe^{2+} (mg/L)	0.065	NA

ND: not detected. NA: not available.

At the phylum level, *Cyanobacteria* dominated the B_DNA and E_DNA samples with 52% and 63% of the total abundance, respectively (Figure 2). *Synechococcus* was the dominant genus, which has been previously detected in clone libraries from Qinghai Lake [14,15,37] and is known to be important contributors to carbon fixation in freshwater and marine ecosystems [38,39]. The second most abundant phylum in the DNA samples was *Proteobacteria,* with 28 and 24% of total abundance in B_DNA and E_DNA, respectively (Figure 2). Within the *Proteobacteria*, *Loktanella* was the dominant genus. It is an aerobic heterotroph that can supplement its energy requirements through aerobic anoxygenic photosynthesis [40,41]. Other major phyla that were detected in the DNA samples included *Actinobacteria*, *Bacteroidetes, Planctomycetes*, and *Verrucomicrobia* (Figure 2).

In the RNA samples, *Proteobacteria* were the most abundant (B_RNA: 47%, E_RNA: 70%) and *Cyanobacteria* were the second most abundant (B_RNA: 35%, E_RNA: 8%) (Figure 2). Other major phyla detected in the RNA samples were *Firmicutes, Actinobacteria, Bacteroidetes*, and *Verrucomicrobia* (Figure 2). Differences in RNA degradation rates and possible mis-annotations make it difficult to access true activity levels, but the relative activity levels can be defined as the ratio of RNA to DNA [42]. With this assumption, *Proteobacteria* were 1.6 and 2.8 fold more active than *Cyanobacteria* in samples B_RNA and E_RNA, respectively.

Metabolic Pathways in the Carbon Cycle

Functional identification of protein-coding genes identified autotrophic and heterotrophic carbon metabolisms (Figure 3). At both sites carbon fixation was performed by ribulose bis-phosphate carboxylase (RuBisCo), which is indicative of the Calvin-Benson-Bassham (CBB) cycle [43]. However, activity levels of RuBisCo (RNA:DNA 0.46 [B] and 0.2 [E]) were low compared to heterotrophic processes (see below).

Table 2. Statistic results of metagenome and metatranscriptome sequences from sites B and E in Qinghai Lake.

	B		E	
	RNA	DNA	RNA	DNA
Number of reads	1,728,111	10,514,407	2,893,577	4,035,731
Mean sequence length (bp)	211±60	194±55	200±59	185±59
Total Mbp	365	2,049	581	749
Mean GC content (%)	49±8	60±8	44±9	63±5
Reads with ORF[a]	1,463,714	10,161,795	2,098,251	3,901,755
Identified protein features	288,106	3,233,826	304,159	945,021
Identified rRNA features	413,195	10,573	113,464	2,307
% annotated reads[b]	65.0	81.2	28.3	77.3

[a]Open reading frame.
[b]% of reads with an open reading frame that were annotated by function and taxonomy.
base pairs (bp).

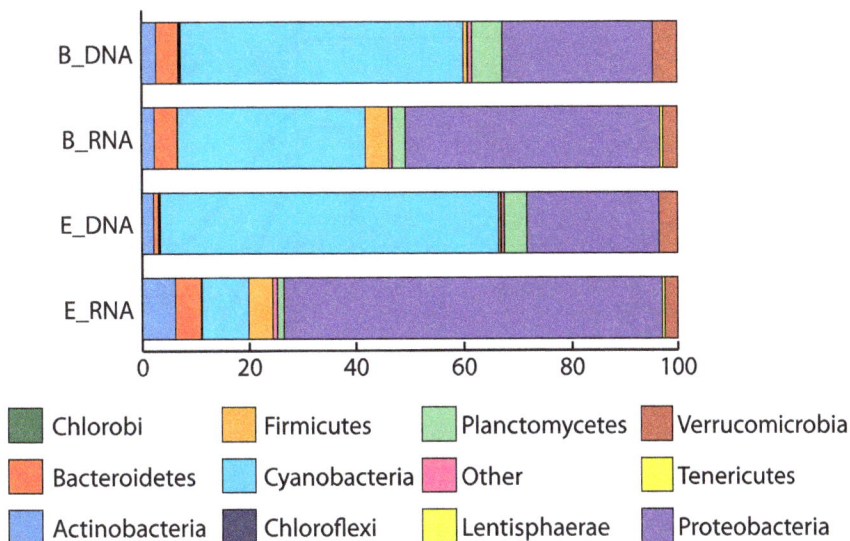

Figure 2. Distribution of phyla detected in the DNA or RNA samples from sites B and E determined by taxonomic assignment of metagenomic or metatranscriptomic reads. Phyla with <1% abundance were grouped into "other".

Photosystems I and II were detected, indicating the reliance of carbon fixation on light energy (Figure 3). Photosystem II (RNA:DNA 1.86 [B] and 0.65 [E]) was more highly expressed than photosystem I (RNA:DNA 0.52 [B] and 0.07 [E]) (Figure 3). The higher expression of photosystem II than photosystem I indicate that the *Cyanobacteria* are under UV stress [44], which is expected in Qinghai Lake because of its high UV irradiance due to the high elevation. In addition, network analysis revealed that organisms expressing photosystems I and II were also expressing orange carotenoid protein (Figure 4), which plays an important role in protecting photosynthetic organisms (such as *Cyanobacteria*) from solar radiation [45,46]. These data suggested that the

photoautotrophs in Qinghai Lake had low activities and were under UV stress.

Heterotrophic remineralization of organic carbon proceeded through the glycolysis and tricarboxylic acid cycle (TCA). The key enzymes for glycolysis (phosphofructokinase- RNA:DNA 0.67 [B] and 0.76 [E]) and the TCA cycle (pyruvate dehydrogenase-RNA:DNA 0.62 [B] and 0.66 [E]) had slightly higher RNA:DNA ratios than RuBisCo. Oxidative phosphorylation, identified by cytochrome C oxidase, was the most abundant and active (RNA:DNA 1.53 [B] and 1.99 [E]) energetic pathway. Aerobic anoxygenic phototrophic *puf* operon was also detected indicating that some energy generation came from photoheterotrophy

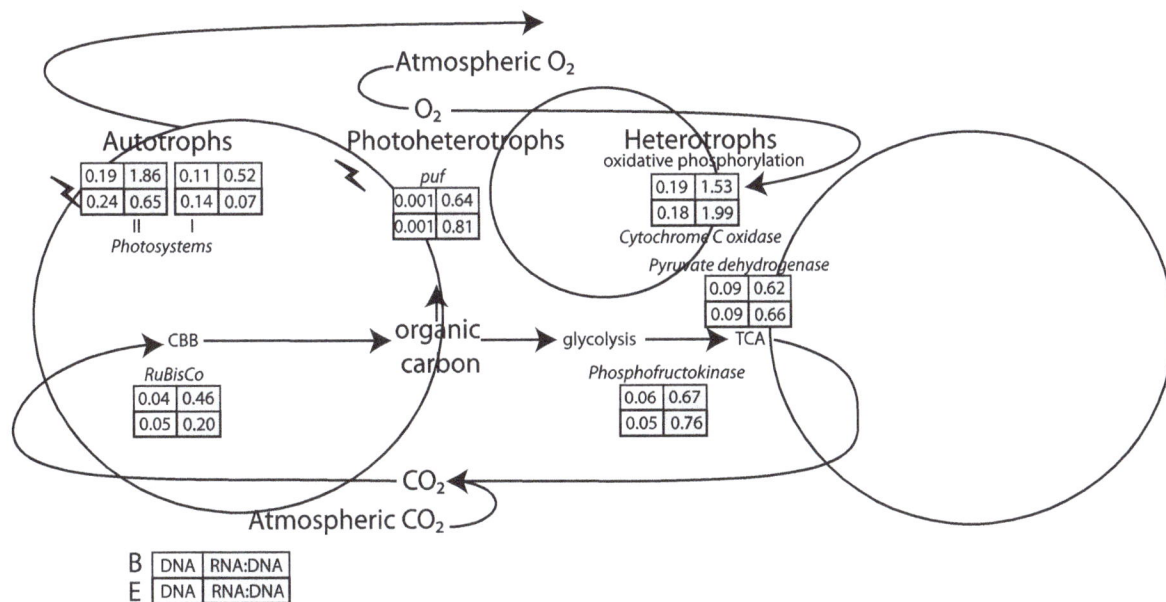

Figure 3. The carbon cycle depicted by a generalized autotroph, photohetertroph, and heterotroph in Qinghai Lake. The numbers in boxes represent the percentage and the RNA:DNA ratio of reads that were annotated within each metabolic pathway for sites B and E. The key genes used to identify a pathway was Ribulose-bisphosphate carboxylase (RuBisCo): Calvin-Benson-Bassham cycle (CBB), D-glucose 6-phosphotransferase: glycolysis, pyruvate dehydrogenase: tricarboxylic acid cycle (TCA), and cytochrome C oxidase: oxidative phosphorylation.

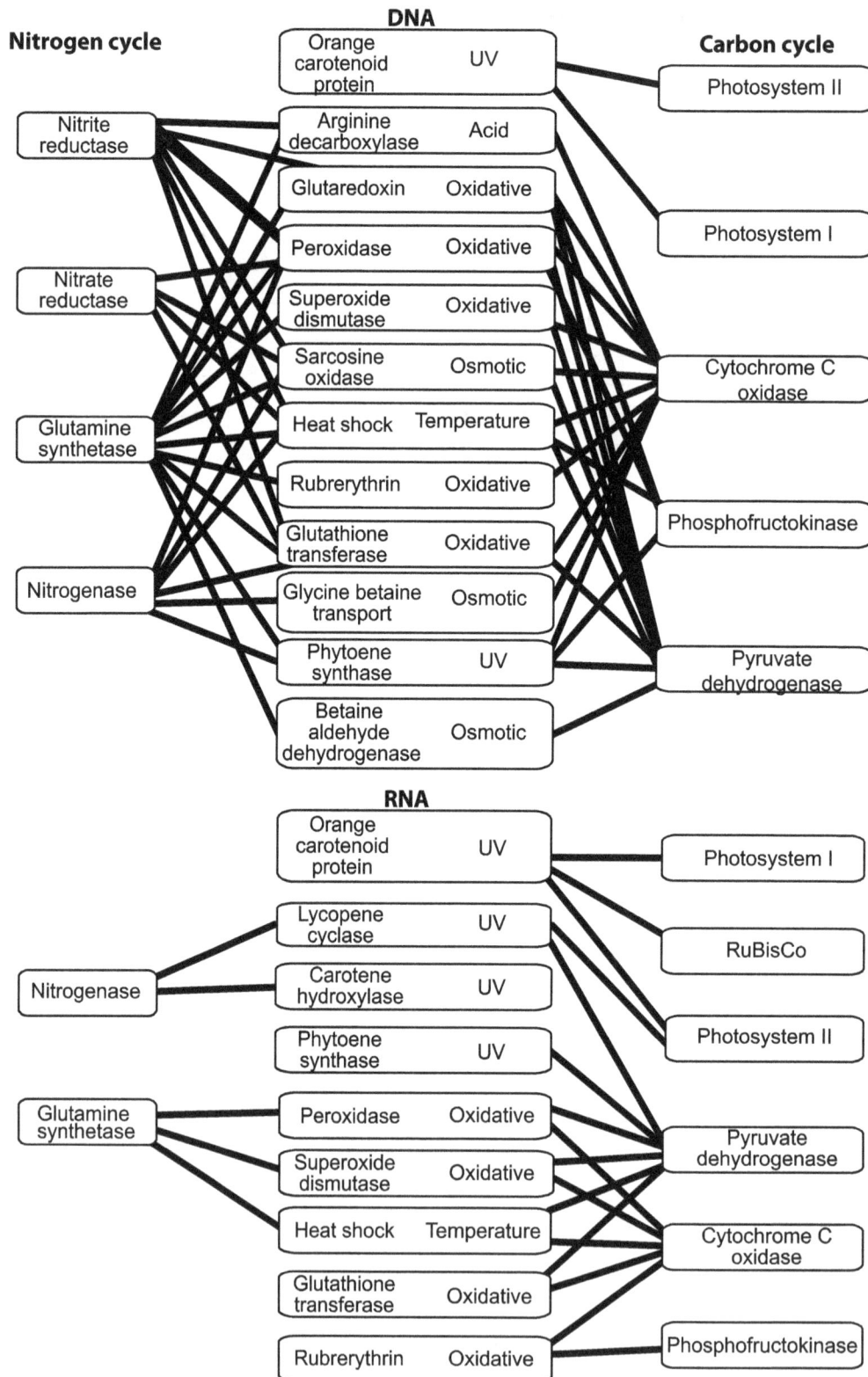

Figure 4. Network showing correlations of carbon and nitrogen cycle-related genes with stress response genes for DNA and RNA samples. Lines represent genes that are directly correlated (i.e. detected (DNA) or expressed (RNA) in the same species). Only significant correlations (p<0.05) are shown. The type of stress that each gene is involved in is also listed.

(RNA:DNA 0.64 [B] and 0.81 [E]). The high concentration of DO (8.7–8.9 ppm) suggests aerobic respiration was the most promi-

nent. Network analysis showed that these heterotrophs were also expressing genes involved in osmotic and oxidative stress

(Figure 4), again consistent with the saline and oxic nature of the water in the lake.

Metabolic Pathways in the Nitrogen Cycle

Nitrogen is biologically available in the form of ammonia. Ammonia assimilation into organic molecules occurred primarily via the glutamine synthetase pathway (Figure 5). The high functional activity of the ammonia assimilatory pathway observed probably led to the low concentration of ammonia measured at both sites ($\leq$1 mg/L). Active nitrogen fixation genes were also detected, which is possibly driven by low levels of ammonia in the lake (Figure 5). Denitrification pathways (i.e., nitric oxide reductase and nitrous oxide reductase) had very low abundance but were highly transcriptionally active (Figure 5). Denitrification requires anoxic conditions, so it was unexpected to detect denitrification transcripts in the oxic water column. It is possible that localized anoxic conditions can exist because of oxygen microgradients, shown in marine snow [48] and microbial mats [49]. These small patches of anoxic water could harbor a relatively low abundance of active anaerobic denitrifiers.

Biological ammonia oxidation is the first step in the nitrification process ($NH_3 \rightarrow NO_2^-$) and is mainly carried out by AOB and AOA that contain the *amoA* gene [50]. Putative AOB and AOA were detected at both sites. AOB genera that were detected included *Nitrosospira*, *Nitrosomonas* and *Nitrosococcus* and AOA genera that were detected included *Nitrosocaldus* and *Nitrosopumilus*. The AOB were more abundant than AOA in all samples; however, AOB and AOA comprised less than 0.1% and 0.01% of total reads, respectively. Although putative AOB- and AOA-like sequences were identified, the *amoA* gene was not detected in our metagenomic or metatranscriptomic samples. However, the *amoA* gene was detected using primers specific for *amoA*. These results indicate that, while present, the abundance of the *amoA* gene is low at the time of sampling.

In contrast, Jiang et al. [19] found that the microbial community in Qinghai lake was dominated by AOA and AOB. This discrepancy is likely because the Jiang et al., samples were collected in early summer (July 2005 and 2007). The lake is frozen during the winter and the ice cover causes stratification of the water column, resulting in gradients of salinity, light availability, and oxygen. In spring, inflow glacial meltwater and thawing permafrost reduces the lake salinity and increases the flow of organic matter and nutrients into the lake [10,51,52]. The early summer samples had a DO concentration of 5.4 ppm at the surface and 7 ppm at the bottom, salinity was 1.25%, and the ammonia concentration was 5.6 mM. High ammonia concentrations in early spring water are conducive for AOA and AOB. Late summer samples described here (late August 2011) had higher DO (8.9 ppm), higher salinity (1.4%), and little measureable ammonia. This suggests that, similar to many inland lakes and oceans [53–56], seasonal changes in the physicochemical conditions and microbiology occurs in Qinghai Lake.

Comparison of sites B and E

An additional 100 genera were detected in the B_DNA sample compared to the E_DNA sample. Both sites had similar rarefaction curve asymptotes (Figure S1), suggesting that the difference in genera is likely because more reads were obtained for the B_DNA sample (~6 million more reads) (Table 2). This also influenced the number of ORFs and protein features detected between the two sites (Table 2). Sites B and E were very similar in the taxonomy and functional genes that were detected; however, slight differences were observed in some RNA:DNA ratios. For example, differences were observed in RNA:DNA ratios of nitric oxide reductase and nitrous oxide reductase. It is difficult to conclude if this is biological or a sampling artifact. Small changes in the RNA reads annotated to these genes could translate to large RNA:DNA ratio changes. For example, nitrous oxide reductase at site E had nine DNA reads annotated and would only need four

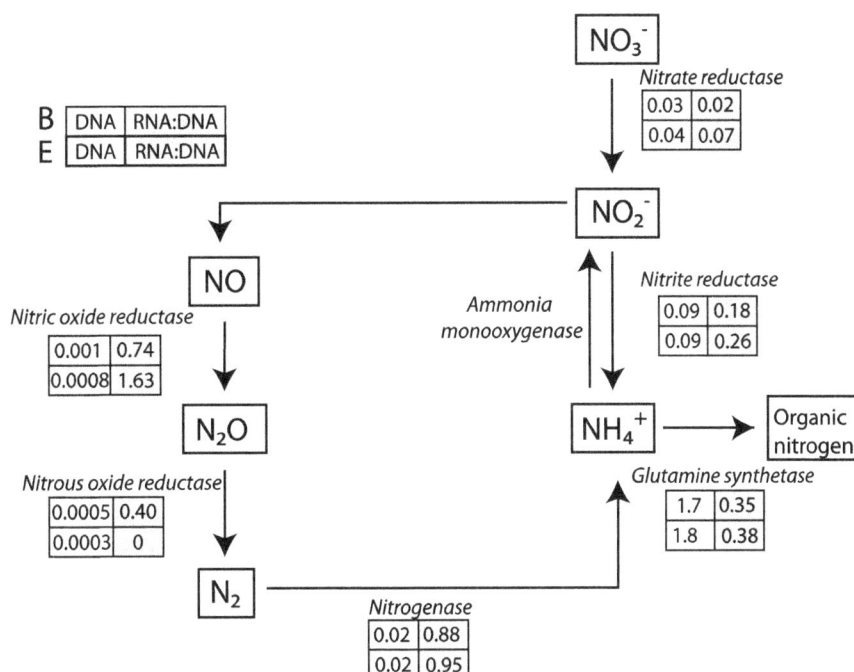

Figure 5. DNA and RNA reads detected and annotated within the nitrogen cycle. The numbers in boxes represent either the percent or the RNA:DNA ratio of reads that were annotated as genes within each pathway for sites B and E.

RNA reads to obtain a RNA:DNA ratio similar to site B. Furthermore, site E had less photosynthetic activity than site B. This is more likely a biological phenomenon because of the high abundance of reads annotated to photosystem II and I. While both sites had similar physicochemical conditions (Table 1), a possible reason for this difference is that site E was 1 m deeper than site B and as a consequence receives less photosynthetically active radiation (PAR) [7] because of the attenuation of PAR in water [47].

Biodiversity

The ability to assign both taxonomy and function to a sequence allowed for identification of the number of organisms that contained a particular functional gene. The richness of each functional gene was calculated as the number of taxa in which the gene was identified. The diversity was sufficiently sampled to detect the majority of organisms that contained a functional gene involved in the carbon and nitrogen cycle (Figure 6). The most diverse functional genes were genes that were part of the heterotrophic pathway: cytochrome C oxidase, glutamine synthetase, phosphofructokinase, and pyruvate dehydrogenase; while photosystem I, photosystem II, and genes involved in the denitrification process had low diversity (Figure 6).

Generally, there is a positive relationship between ecosystem function and taxonomic diversity, because a more diverse community increases the likelihood of finding high-performance key species or the probability of functional redundancy [57,58]. For example, in soils with a high taxonomic diversity, denitrification rates were higher and decreased more slowly with increased salinity [6]. Similar results were found for respiration [59–61], photosynthesis [62–64], and nitrogen fixation [65]. This suggests that, as a consequence of the greater diversity, the heterotrophic pathway in Qinghai Lake would be more resilient to stress.

While the greater diversity increases the probability of stress resilience, a mechanism to respond to stress is needed to continue functioning. Stress response genes are one mechanism that can be detected with metagenomics and metatranscriptomics. Genes were found that respond to UV, osmotic, and oxidative stress (Figure 4). The detection of these stress genes in Qinghai Lake was expected because of the intense UV radiation and oxidative stress (related to the high elevation of the lake) and its salinity. In addition, samples were retrieved mid-day when UV radiation was at its highest. Furthermore, the number of stress gene types increased logarithmically with the species richness for each functional gene involved in the carbon and nitrogen cycles (Figure 7). For example, in the 365 species that had a cytochrome c oxidase, there were a total of 20 different stress response genes. This observation provides a possible mechanism as to why higher biodiversity is more resilient to stress because one of the species will be able to respond to a stress.

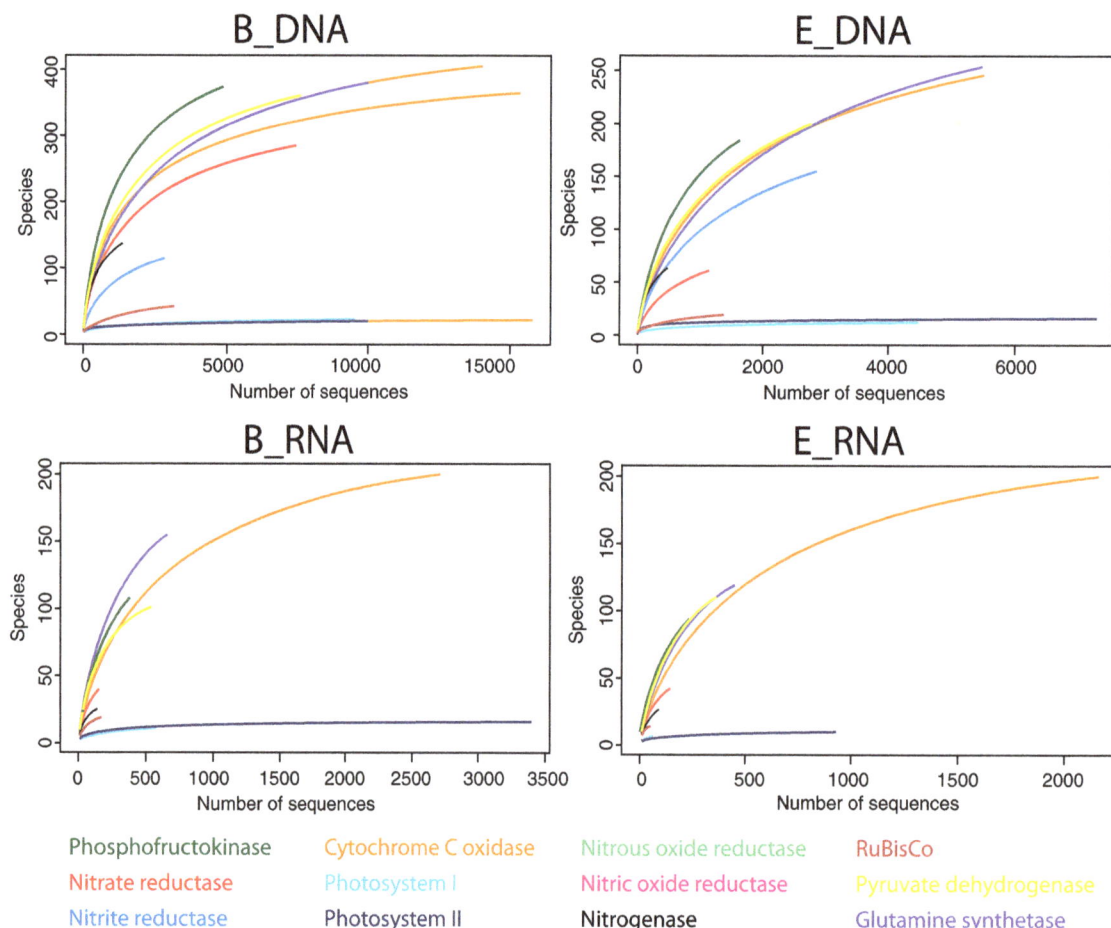

Figure 6. Rarefaction curves for the detected genes related to carbon and nitrogen cycle. Highly saturated curves (e.g. photosystem I and II) indicate that sequencing depth was sufficient to capture most species containing the corresponding gene.

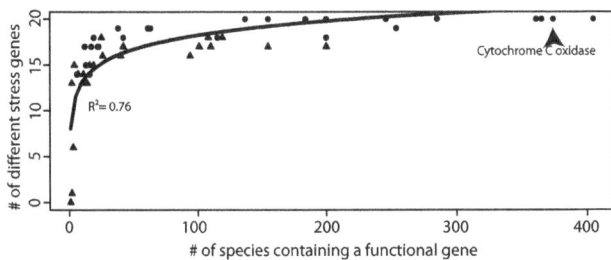

Figure 7. Plot showing the number of species that contained a particular functional gene in the carbon and nitrogen cycle and the number of stress genes that were found in all species with that functional gene (Cytochrome C oxidase is labeled as an example). Triangle symbols represent RNA samples and circle symbols represent DNA samples. Best-fit curve followed a logarithmic function.

Conclusions

A metagenomic and metatranscriptomic survey of two locations in Qinghai Lake identified microbial processes involved in the carbon and nitrogen cycle. While photoautotrophic organisms were the most abundant in the DNA samples, heterotrophic organisms were the most active at both sites. Energy generation was also supplemented by photoheterotrophy. Coupled with previous reports from early summer sampling, our data suggest that successional changes occur in the microbiota throughout the summer. Our data also shows a positive relationship between the

number of stress gene types and the species richness for each functional gene in the carbon and nitrogen cycles, suggesting that microbial processes with higher taxonomic diversity would have an increased ability to respond to a variety of stresses. Heterotrophic respiration and ammonia assimilation had the highest richness and would be expected to be more resistant to environmental change.

Acknowledgments

We thank Dr. James Hollibaugh for providing technical support and reviewing of the manuscript. We also thank Dr. Tammie Gerke and Dr. Jennifer Biddle for reviewing the manuscript. This work was fostered by NSF IOS grant 1238801. We are grateful to two anonymous reviewers whose comments improved the quality of the manuscript.

Author Contributions

Conceived and designed the experiments: BRB HD HJ. Performed the experiments: QH BRB CE IV GW. Analyzed the data: QH BRB. Contributed reagents/materials/analysis tools: CE IV SQ. Contributed to the writing of the manuscript: QH BRB HD HJ GW CE IV SQ.

References

1. Canfield DE, Glazer AN, Falkowski PG (2010) The evolution and future of Earth's nitrogen cycle. Science 330: 192–196.
2. Bardgett RD, Freeman C, Ostle NJ (2008) Microbial contributions to climate change through carbon cycle feedbacks. ISME J 2: 805–814.
3. Singh BK, Bardgett RD, Smith P, Reay DS (2010) Microorganisms and climate change: terrestrial feedbacks and mitigation options. Nat Rev Microbiol 8: 779–790.
4. Heimann M, Reichstein M (2008) Terrestrial ecosystem carbon dynamics and climate feedbacks. Nature 451: 289–292.
5. Griffiths BS, Philippot L (2013) Insights into the resistance and resilience of the soil microbial community. FEMS Microbiol Rev 37: 112–129.
6. Hallin S, Welsh A, Stenstrom J, Hallet S, Enwall K, et al. (2012) Soil functional operating range linked to microbial biodiversity and community composition using denitrifiers as model guild. PLoS One 7: e51962.
7. Mella-Flores D, Six C, Ratin M, Partensky F, Boutte C, et al. (2012) *Prochlorococcus* and *Synechococcus* have evolved different adaptive mechanisms to cope with light and UV stress. Front Microbiol 3: 285.
8. Liu X, Chen B (2000) Climatic warming in the Tibetan Plateau during recent decades. Int J Climatol 20: 1729–1742.
9. Guo D, Wang H, Li D (2012) A projection of permafrost degradation on the Tibetan Plateau during the 21st century. J Geophys Res Atmos 117: D05106.
10. Qiu J (2008) China: the third pole. Nature 468: 141–142.
11. Lei Y, Yao T, Yi C, Wang W, Sheng Y, et al. (2012) Glacier mass loss induced the rapid growth of Linggo Co on the central Tibetan Plateau. J Glaciology 58: 177–184.
12. Yang SJ, Dong H, Lei FM (2009) Phylogeography of regional fauna on the Tibetan Plateau: A review. Progr Natural Sci 19: 789–799.
13. Dong H, Jiang H, Yu B, Liu X (2010) Impacts of environmental change and human activity on microbial ecosystems on the Tibetan Plateau, NW China. GSA Today 20: 4–10.
14. Dong H, Zhang G, Jiang H, Yu B, Chapman LR, et al. (2006) Microbial diversity in sediments of saline Qinghai Lake, China: linking geochemical controls to microbial ecology. Microb Ecol 51: 65–82.
15. Xing P, Hahn MW, Wu QL (2009) Low taxon richness of bacterioplankton in high-altitude lakes of the eastern tibetan plateau, with a predominance of *Bacteroidetes* and *Synechococcus spp*. Appl Environ Microbiol 75: 7017–7025.
16. Jiang H, Dong H, Yu B, Ye Q, Shen J, et al. (2008) Dominance of putative marine benthic Archaea in Qinghai Lake, north-western China. Environ Microbiol 10: 2355–2367.
17. Jiang H, Dong H, Deng S, Yu B, Huang Q, et al. (2009) Response of archaeal community structure to environmental changes in lakes on the Tibetan Plateau, Northwestern China. Geomicrobiol J 26: 289–297.

18. Wang HY, Liu WG, Zhang CLL, Jiang HC, Dong HL, et al. (2013) Assessing the ratio of archaeol to caldarchaeol as a salinity proxy in highland lakes on the northeastern Qinghai-Tibetan Plateau. Org Geochem 54: 69–77.
19. Jiang H, Dong H, Yu B, Lv G, Deng S, et al. (2009) Diversity and abundance of ammonia-oxidizing archaea and bacteria in Qinghai Lake, Northwestern China. Geomicrobiol J 26: 199–211.
20. Huang Q, Jiang H, Briggs BR, Wang S, Hou W, et al. (2013) Archaeal and bacterial diversity in acidic to circumneutral hot springs in the Philippines. FEMS Microbiol Ecol 85: 452–464.
21. Poretsky RS, Hewson I, Sun S, Allen AE, Zehr JP, et al. (2009) Comparative day/night metatranscriptomic analysis of microbial communities in the North Pacific subtropical gyre. Environ Microbiol 11: 1358–1375.
22. Hollibaugh JT, Gifford S, Sharma S, Bano N, Moran MA (2011) Metatranscriptomic analysis of ammonia-oxidizing organisms in an estuarine bacterioplankton assemblage. ISME J 5: 866–878.
23. Stewart FJ, Ottesen EA, DeLong EF (2010) Development and quantitative analyses of a universal rRNA-subtraction protocol for microbial metatranscriptomics. ISME J 4: 896–907.
24. Jiménez DJ, Andreote FD, Chaves D, Montaña JS, Osorio-Forero C, et al. (2012) Structural and functional insights from the metagenome of an acidic hot spring microbial planktonic community in the Colombian Andes. PLoS One 7: eS2069.
25. Meyer F, Paarmann D, D'Souza M, Olson R, Glass E, et al. (2008) The metagenomics RAST server - a public resource for the automatic phylogenetic and functional analysis of metagenomes. BMC Bioinformatics 9: 386.
26. Altschul SF, Madden TL, Schäffer AA, Zhang J, Zhang Z, et al. (1997) Gapped BLAST and PSI-BLAST: a new generation of protein database search programs. Nucleic Acids Res 25: 3389–3402.
27. R Development Core Team (2012) R: A language and environment for statistical computing. Vienna: R Found. for Stat. Comput.
28. Oksanen J, Blanchet FG, Kindt R, Legendre P, Minchin PR, et al. (2011) vegan: Community ecology package. http://cran.r-project.org/, http://vegan.r-forge. r-project.org/.
29. Langfelder P, Horvath S (2008) WGCNA: an R package for weighted correlation network analysis. BMC Bioinformatics 9: 559.
30. Csardi G, Nepusz T (2006) The igraph software package for complex network research. InterJournal, Complex Systems 1695.
31. Shannon P, Markiel A, Ozier O, Baliga NS, Wang JT, et al. (2003) Cytoscape: a software environment for integrated models of biomolecular interaction networks. Genome Res 13: 2498–2504.
32. Kara EL, Hanson PC, Hu YH, Winslow L, McMahon KD (2013) A decade of seasonal dynamics and co-occurrences within freshwater bacterioplankton communities from eutrophic Lake Mendota, WI, USA. ISME J 7: 680–684.

33. Rotthauwe J-H, Witzel K-P, Liesack W (1997) The ammonia monooxygenase structural gene *amoA* as a functional marker: molecular fine-scale analysis of natural ammonia-oxidizing populations. Appl Environ Microbiol 63: 4704–4712.

34. Stewart FJ, Ulloa O, DeLong EF (2012) Microbial metatranscriptomics in a permanent marine oxygen minimum zone. Environ Microbiol 14: 23–40.

35. Urich T, Lanzen A, Qi J, Huson DH, Schleper C, et al. (2008) Simultaneous assessment of soil microbial community structure and function through analysis of the meta-transcriptome. PLoS One 3: e2527.

36. Rho M, Tang H, Ye Y (2010) FragGeneScan: predicting genes in short and error-prone reads. Nucleic Acids Res 38: e191.

37. Jiang H, Dong H, Zhang G, Yu B, Chapman LR, et al. (2006) Microbial diversity in water and sediment of Lake Chaka, an athalassohaline lake in northwest China. Appl Environ Microbiol 72: 3832–3845.

38. Joint IR (1986) Physiological ecology of picoplankton in various oceanographic provinces. Photosynthetic picoplankton, Can Bull Fish Aquat Sci 214: 287–309.

39. Fahnenstiel GL, Carrick HJ, Iturriaga R (1991) Physiological characteristics and food-web dynamics of *Synechococcus* in Lakes Huron and Michigan. Limnol 36: 219–234.

40. Newton RJ, Griffin LE, Bowles KM, Meile C, Gifford S, et al. (2010) Genome characteristics of a generalist marine bacterial lineage. ISME J 4: 784–798.

41. Jiang H, Dong H, Yu B, Lv G, Deng S, et al. (2009) Abundance and diversity of aerobic anoxygenic phototrophic bacteria in saline lakes on the Tibetan plateau. FEMS Microbiol Ecol 67: 268–278.

42. Yu K, Zhang T (2012) Metagenomic and metatranscriptomic analysis of microbial community structure and gene expression of activated sludge. PLoS One 7: e38183.

43. Berg IA (2011) Ecological aspects of the distribution of different autotrophic CO_2 fixation pathways. Appl Environ Microbiol 77: 1925–1936.

44. Campbell D, Eriksson M-J, Öquist G, Gustafsson P, Clarke AK (1998) The cyanobacterium *Synechococcus* resists UV-B by exchanging photosystem II reaction-center D1 proteins. Proc Natl Acad Sci USA 95: 364–369.

45. Armstrong GA, Alberti M, Leach F, Hearst JE (1989) Nucleotide sequence, organization, and nature of the protein products of the carotenoid biosynthesis gene cluster of *Rhodobacter capsulatus*. Mol Gen Genet 216: 254–268.

46. Moeller RE, Gilroy S, Williamson CE, Grad G, Sommaruga R (2005) Dietary acquisition of photoprotective compounds (mycosporine-like amino acids, carotenoids) and acclimation to ultraviolet radiation in a freshwater copepod. Limnol Oceanogr 50: 427–439.

47. Miller CB (2004) Biological oceanography. Malden, MA.: Blackwell Publishing. 416 p.

48. Ploug H (2001) Small-scale oxygen fluxes and remineralization in sinking aggregates. Limnol Ocean 46: 1624–1631.

49. Cravo-Laureau C, Duran R (2014) Marine coastal sediments microbial hydrocarbon degradation processes: contribution of experimental ecology in the omics'era. Front Microbiol 5: 39.

50. Thamdrup B (2012) New pathways and processes in the global nitrogen cycle. Annu Rev Ecol, Evol, and Syst 43: 407–428.

51. Li X, Cheng GD, Jin HJ, Kang E, Che T, et al. (2008) Cryospheric change in China. Global Planet Change 62: 210–218.

52. Liu J, Wang S, Yu S, Yang D, Zhang L (2009) Climate warming and growth of high-elevation inland lakes on the Tibetan Plateau. Global Planet Change 67: 209–217.

53. Pernthaler J, Glöckner F-O, Unterholzner S, Alfreider A, Psenner R, et al. (1998) Seasonal community and population dynamics of pelagic bacteria and archaea in a high mountain lake. Appl Environ Microbiol 64: 4299–4306.

54. Yannarell AC, Kent AD, Lauster GH, Kratz TK, Triplett EW (2003) Temporal patterns in bacterial communities in three temperate lakes of different trophic status. Microb Ecol 46: 391–405.

55. Gilbert JA, Steele JA, Caporaso JG, Steinbrück L, Reeder J, et al. (2012) Defining seasonal marine microbial community dynamics. ISME J 6: 298–308.

56. Treusch AH, Vergin KL, Finlay LA, Donatz MG, Burton RM, et al. (2009) Seasonality and vertical structure of microbial communities in an ocean gyre. ISME J 3: 1148–1163.

57. Yachi S, Loreau M (1999) Biodiversity and ecosystem productivity in a fluctuating environment: the insurance hypothesis. Proc Natl Acad Sci U S A 96: 1463–1468.

58. Philippot L, Spor A, Henault C, Bru D, Bizouard F, et al. (2013) Loss in microbial diversity affects nitrogen cycling in soil. ISME J 7: 1609–1619.

59. Chowdhury N, Marschner P, Burns R (2011) Response of microbial activity and community structure to decreasing soil osmotic and matric potential. Plant Soil 344: 241–254.

60. Setia R, Marschner P, Baldock J, Chittleborough D, Smith P, et al. (2011) Salinity effects on carbon mineralization in soils of varying texture. Soil Biol Biochem 43: 1908–1916.

61. Langenheder S, Lindstrom ES, Tranvik LJ (2005) Weak coupling between community composition and functioning of aquatic bacteria. Limnol Ocean 50: 957–967.

62. Allakhverdiev SI, Murata N (2008) Salt stress inhibits photosystems II and I in cyanobacteria. Photosynth Res 98: 529–539.

63. Murata N, Takahashi S, Nishiyama Y, Allakhverdiev SI (2007) Photoinhibition of photosystem II under environmental stress. Biochim Biophys Acta 1767: 414–421.

64. Takahashi S, Murata N (2008) How do environmental stresses accelerate photoinhibition? Trends Plant Sci 13: 178–182.

65. Severin I, Confurius-Guns V, Stal LJ (2012) Effect of salinity on nitrogenase activity and composition of the active diazotrophic community in intertidal microbial mats. Arch Microbiol 194: 483–491.

16

Forest Structure in Low-Diversity Tropical Forests: A Study of Hawaiian Wet and Dry Forests

Rebecca Ostertag[1]*, Faith Inman-Narahari[2], Susan Cordell[3], Christian P. Giardina[3], Lawren Sack[4]

1 Department of Biology, University of Hawai'i at Hilo, Hilo, Hawai'i, United States of America, 2 Department of Natural Resources and Environmental Management, University of Hawai'i at Mānoa, Honolulu, Hawai'i, United States of America, 3 Institute of Pacific Islands Forestry, Pacific Southwest Research Station, USDA Forest Service, Hilo, Hawai'i, United States of America, 4 Department of Ecology and Evolutionary Biology, University of California Los Angeles, Los Angeles, California, United States of America

Abstract

The potential influence of diversity on ecosystem structure and function remains a topic of significant debate, especially for tropical forests where diversity can range widely. We used Center for Tropical Forest Science (CTFS) methodology to establish forest dynamics plots in montane wet forest and lowland dry forest on Hawai'i Island. We compared the species diversity, tree density, basal area, biomass, and size class distributions between the two forest types. We then examined these variables across tropical forests within the CTFS network. Consistent with other island forests, the Hawai'i forests were characterized by low species richness and very high relative dominance. The two Hawai'i forests were floristically distinct, yet similar in species richness (15 vs. 21 species) and stem density (3078 vs. 3486/ha). While these forests were selected for their low invasive species cover relative to surrounding forests, both forests averaged 5->50% invasive species cover; ongoing removal will be necessary to reduce or prevent competitive impacts, especially from woody species. The montane wet forest had much larger trees, resulting in eightfold higher basal area and above-ground biomass. Across the CTFS network, the Hawaiian montane wet forest was similar to other tropical forests with respect to diameter distributions, density, and aboveground biomass, while the Hawai'i lowland dry forest was similar in density to tropical forests with much higher diversity. These findings suggest that forest structural variables can be similar across tropical forests independently of species richness. The inclusion of low-diversity Pacific Island forests in the CTFS network provides an ~80-fold range in species richness (15–1182 species), six-fold variation in mean annual rainfall (835–5272 mm yr^{-1}) and 1.8-fold variation in mean annual temperature (16.0–28.4°C). Thus, the Hawaiian forest plots expand the global forest plot network to enable testing of ecological theory for links among species diversity, environmental variation and ecosystem function.

Editor: Bruno Hérault, Cirad, France

Funding: The major funding for this research came from National Science Foundation's EPSCoR Grants No. 0554657 and No. 0903833 to the University of Hawai 'i. Major in-kind support was provided by the Pacific Southwest Research Station of the United States Forest Service. Logistical or financial support that was supplemental was provided for by the Smithsonian Tropical Research Institute Center for Tropical Forest Science, and the University of California, Los Angeles. The funders had no role in study design, data collection and analysis, decision to publish, or preparation of the manuscript.

Competing Interests: The authors have declared that no competing interests exist.

* Email: ostertag@hawaii.edu

Introduction

High species richness is a hallmark of many tropical forests [1,2]. Indeed, the latitudinal gradient and equatorial peak in plant diversity has attracted attention for centuries e.g., [1,3,4,5]. Numerous studies have focused on the causes of high diversity in tropical forests [1,6,7,8,9,10], and theories have been formulated to explain how species or functional diversity in turn affects ecosystem function [11,12]. However, these linkages have rarely been tested, and not all tropical forests are diverse. For example, legume-dominated swamp forests, peat forests, pine savannas, and oceanic islands that are geographically isolated can have low to very low diversity [13,14,15,16]. Such low-diversity forests are understudied, and there is no clear answer to the simple question of whether the structure of a low-diversity tropical forest would be expected to be similar to or different from that of a high-diversity tropical forest with comparable climate.

Indeed, the question of how forest structure—i.e., physiognomy, basal area, density, diameter size class distributions, biomass, and evenness—varies with species diversity is itself understudied, likely an effect of the paucity of studies of the structure of low-diversity tropical forests. Some have hypothesized that forest structure and species-richness might be related, if structure acts as a habitat scaffold or template that precedes and enables species assembly and diversity by providing an increased variety of habitat niches (e.g., nurse logs for seedlings, perches for birds that disperse seeds, climbing structures for vines [17,18]. Alternatively, higher diversity may enhance forest structure, if more species correspond to a wider variety of size classes, strata, and crown architectures [17]. Both processes are not mutually exclusive and may operate simultaneously, creating a positive feedback cycle that would enhance diversity and influence various forest structural attributes. Recent efforts have examined some structural variables, such as latitudinal trends in height across forests e.g., [19,20] and the effects of diversity and spatial scale on standing forest biomass [12], but very low-diversity tropical forests were not considered in these analyses. The tropical forests in the Hawaiian Islands represent a low-diversity extreme, as a result of its young geological origins [21] and extreme isolation from continental land masses: at approximately 4000 km from the nearest continent, Hawai'i is the world's most isolated archipelago. The resulting native flora in Hawai'i is disharmonic (i.e., missing many functional groups) and is about 90% endemic [22]. While long-term plot-based ecological measurements across the tropics have focused on high-diversity forests, there have been surprisingly few data from low-diversity tropical forests [23,24,25,26]. Such low-diversity forests present many interesting contrasts to other tropical forests, and within Hawai'i they also fall across striking environmental gradients (Table 1).

The aim of this study was to: 1) characterize and compare two extremely low-diversity Hawaiian forests, montane wet and lowland dry forest, and 2) compare the structural attributes of these two forests to more diverse tropical forests within the Center for Tropical Forest Science (CTFS) permanent plot network. Including the Hawaiian plots as part of a cross-plot analysis allows, for the first time, examination of forest structure along a diversity gradient that varies almost 80-fold across large-scale plots with consistent measurement protocols.

We used the initial census of large-scale permanent plots in Hawai'i to examine structural and floristic characteristics of two forests that are geographically close but located in widely contrasting environments. The two Hawaiian forest types examined in this first census were montane wet forest (MWF) and lowland dry forest (LDF). Many studies have shown that forests established in areas with higher rainfall or temperature have higher diversity [1,27,28], and also greater basal area, tree height, and above-ground biomass [28,29,30,31]. Further, forests in higher rainfall areas tend to have a greater representation of larger trees, but lower tree densities [32]. We therefore ask: 1) How do the two Hawaiian forests compare in terms floristic and life form composition, stand structure, species diversity, and non-native species cover? Our study was not designed to specifically examine the effects of climate on forest structure and composition, but we used this study design, to test a prediction based on the previous literature that Hawaiian dry forest would have greater stem density, lower diversity, and smaller diameter trees than wet forest [32]. To place our findings in a broader context, we also asked: 2) Can the extremely low forests of Hawai'i have similar structural attributes to more diverse tropical forests? To examine this question, we compared Hawaiian forests with others in the CTFS network enabling the comparison of forest structural variables across a range of environments and diversity levels [1,28,31,33,34,35]. If Hawaiian forests converge with other tropical forests, the importance of climate in determining forest structure is highlighted.

Materials and Methods

Study Sites

In 2008 and 2009, we established two forest dynamics plots (FDPs) on Hawai'i Island – one within montane wet forest (MWF) and one within lowland dry forest (LDF), to initiate the Hawai'i Permanent Plot Network (HIPPNET; Fig. 1). We focused our study on Hawai'i Island, because it has the greatest area of intact forests, a complete map of lava flow ages, and excellent infrastructure for ecological studies. As the youngest island in the archipelago (<700,000 years), it has had the least time for plant colonization and subsequent speciation, and thus has lower species richness relative to its size than the older islands [36]. We selected areas in excellent ecological condition that are representative of a given forest type, with high native species cover, and a commitment by ownership to long-term conservation objectives. Notably, all forests in Hawai'i are affected to some degree by altered trophic interactions due to invasion of non-native species or extinction of the native species [37], but this is not unique to Hawai'i [38]. Non-native stems that were encountered were measured for percent cover, and then controlled mechanically or

Table 1. Distinctive structural and demographic features of Hawaiian forests.

Environmental Conditions
Large variation in elevation, rainfall, temperature and soils among forests that are geographically close [1–2]
High light levels in intact wet, mesic, and dry forest (1.9–40% diffuse light transmission) [3–9]
Species Composition and Diversity Patterns
A global biodiversity hotspot due to high endemism and number of endangered species [10–11]
Same species distributed in many habitats differing in environmental conditions, demonstrating exceptional phenotypic plasticity [10–14]
Tree ferns common and often the understory dominant in wet forests at all elevations, whereas outside of Hawai'i they tend to be more restricted [10]
Monodominance by a few canopy species [15]
Autecology of Plant Species
Metrosideros polymorpha dominant in wet forests throughout succession (as pioneer and late successional species) [15–17]
Extremely slow growth of primary pioneer species, *M. polymorpha* (1–2 mm/year diameter) [18–20]
Nurse logs serve as a substrate for seedling regeneration [21]
Dieback and regeneration of canopy dominant *M. polymorpha* in cohorts contribute strongly to gap dynamics [16, 22]
Trophic Interactions
Evolution without land mammals [23, 24]
Documented extinctions of plants, pollinators and dispersers may influence present day evenness and rarity measures [23]
Animal dispersal of seeds conducted entirely by birds before human contact [24, 25]
Apparently low rates of insect herbivory [26] and seed predation [27]
Presence of invasive weeds, ungulates, and birds may alter present-day plant-animal interactions [25–26, 28]

Superscripts refer to references listed in Table S5 in File S2.

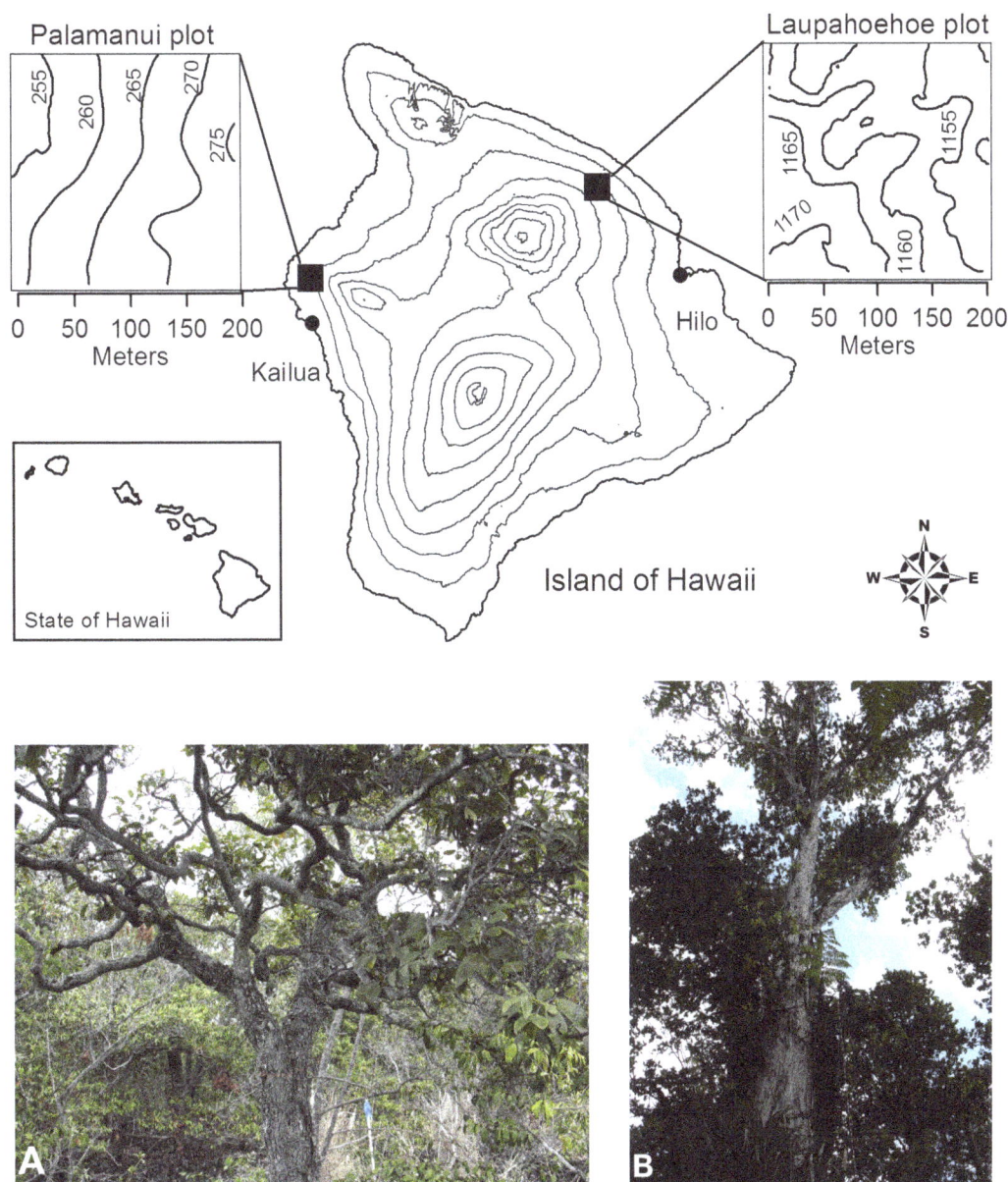

Figure 1. Contour map of the two 4-ha forest plots on Hawai'i Island. Pālamanui site in west Hawai'i is lowland dry forest (LDF; left panel showing the dominant canopy tree *Diospyros sandwicensis* and the open canopy and understory structure of small trees and shrubs); Laupāhoehoe plot in east Hawai'i is montane wet forest (MWF; right panel showing *Metrosideros polymorpha* tree and *Cibotium* spp. tree fern understory).

chemically (see "*Plot Establishment and Vegetation Measurements*" below) and were not considered in the census of stems.

Montane wet forest (MWF). The 4-ha Laupāhoehoe FDP (19°55' N, 155°17' W) is located within the state-owned Laupāhoehoe Natural Area Reserve section of the Hawai'i Experimental Tropical Forest (HETF) on the northeast slope of Mauna Kea volcano. Permits were obtained for work in the HETF through the Institute of Pacific Islands Forestry and the Hawai'i Division of Forestry and Wildlife/Department of Land and Natural Resources. The mean elevation of the plot is 1120 m.a.s.l. with slopes of 0–20%, and the overall direction of downslope is northwards towards the Pacific Ocean. The substrate within the plot is 4000-14,000 years old [39]. Soils were formed from weathered volcanic material, and are deep, rocky, and moderately well-drained silty clay loam in the Akaka series, and

classified as hydrous, ferrihydritic, isothermic Acrudoxic Hydrudands (websoilsurvey.nrcs.usda.gov). Rainfall at the MWF is dominated by tradewind-driven precipitation [40]. Interpolated mean annual precipitation, based on analysis of climate station data over 30 years, is 3440 mm with no distinct dry season [41] and mean annual air temperature is 16°C [42]. The forest consists of evergreen broad-leaved trees, and the ~25–28 m canopy is dominated by *Metrosideros polymorpha* (Myrtaceae; Fig. 1) and to a lesser extent, *Acacia koa* (Fabaceae). Vegetation at the MWF is highly representative of this forest type in Hawai'i [43] (see references in Table 1).

The dominant pre-human contact disturbance regime in this forest type was single-to multiple-tree falls, with the maximum gap size averaging 21.5 m^2 [44]. Larger openings coincide with dieback due to cohort senescence of older *M. polymorpha* stands

[45]. Following contact, large *A. koa* trees were occasionally harvested for traditional canoe building. In modern times, limited *A. koa* logging occurred in the HETF but was restricted to < 100 m of an unimproved road that traverses areas. There is no evidence of logging within the MWF [46], which >500 m from the road. Non-native wild pigs disturb soils while rooting, as well as tree ferns [47], with damage over a large area.

Lowland dry forest (LDF). The 4-ha Pālamanui FDP is an example of one of the world's most endangered forest types, and is located on a privately-owned tract of dry forest on the northwest slope of Hualālai Volcano in the district of North Kona (240 m elevation, 19°44' N, 155°59' W). A memorandum of understanding was established with the land owners and managers, the Palāmani Group, for permission to conduct research in the lowland dry forest site. The mean elevation of the plot is 240 m.a.s.l. Geological substrate in the Pālamanui area consists of 'a'ā lava with scattered pāhoehoe flows dating to 1,500–3,000 years old [48]. Soils developing at this site are shallow, rocky, highly organic, and classified as euic, isothermic, shallow Lithic Ustifolist (websoilsurvey.nrcs.usda.gov). Interpolated mean annual precipitation at the LDF site is 835 mm [41,49], with large within- and between-year variability [50]. For the LDF, major rainfall events typically occur in the winter as low pressure storms ("Kona lows") while summers tend to be dry and characterized by small convective storms. Mean daily air temperature is approximately 20°C (wrcc.dri.edu). Native vegetation consists of evergreen broad-leaved trees and shrubs that form an open-canopy forest that reaches heights of ~7–8 m dominated by *Diospyros sandwicensis* (Ebenaceae) and *Psydrax odorata* (Rubiaceae; Fig. 1). One species (*Erythrina sandwicensis*) is drought deciduous and is only represented by a few individuals.

Pre-contact disturbance regimes likely included tree falls. Following contact, selective harvesting of valuable woods (e.g., sandalwood) occurred throughout the area but we do not know of any logging that occurred within the plot. In the last 200 years, much of the lowland dry forest in Hawai'i has been subjected to grazing and browsing by exotic ungulates, with remnants impacted by wildfire carried by non-native grasses [51]. These factors have reduced the native forest to a fraction of its original extent [52]. While the area containing the FDP has not been burned or significantly browsed by ungulates, the surrounding area is a matrix of degraded LDF and open grassland, and in 2009, a fence and firebreak were installed around the area to protect it from ungulates and fire.

Plot Establishment and Vegetation Measurements

We applied field methodology developed by the Center for Tropical Forest Science global FDP network [53]. Both of our 4-ha FDPs (200×200 m) were oriented north-south and located at the center of a 16 ha buffer area, with all edges at least 100 m from any road or major trail where possible. From 2008 to 2009, we tagged all live, native woody plants ≥1 cm diameter at breast height (DBH, at 130 cm), and mapped tagged plants relative to 5 m×5 m grids installed throughout the plots. Each tagged plant was identified to species and measured for DBH. More detailed methods are in Methods S1 in File S1.

Finally, we estimated and mapped cover of abundant non-native herbaceous, shrub and tree species, which will be important for understanding long-term vegetation change. At each site, we chose six abundant focal species or life forms that were considered "invasive pests" according to their Hawai'i Weed Risk Assessment scores (Daehler 2004; www.botany.hawaii.edu/faculty/daehler/wra/full_table.asp). Percent cover within each 5×5 m subquadrat was estimated in the following categories: 0: absent, 1: <5%, 2: 5–

25%, 3: 25–50%, 4: 50–100%. Non-native trees with stems ≥ 1 cm at 130 cm were individually mapped. The DBH of the largest stem of non-native trees <5 cm was estimated to the nearest centimeter and measured to the nearest centimeter if > 5 cm. For trees with multiple stems, we counted the total number of stems ≥1 cm at 130 cm. After the non-native trees were mapped, they were girdled and sprayed with herbicide. We did not spray herbicide on the grasses in the LDF, nor the vine *Passiflora tarminiana* in the MWF.

Data Analyses

Stand structure. We determined stand structural characteristics based on DBH measurements. We considered multiple-stemmed plants as single individuals for the calculation of stem density, and summed the basal area of all stems for the calculation of basal area (m^2/ha). For each species, we calculated relative abundance (RA, %) as the number of individuals of that species/total number of individuals, relative dominance (RD, %) as the basal area of that species/total basal area, and relative frequency (RF, %) as the number of quadrats with that species/total number of quadrats.

Above-ground biomass. To estimate above-ground biomass (AGB) for the two plots, we used site-specific and species-specific information whenever possible for wood specific gravity, tree height, and DBH (equations derived from 52,54,55,56; see Table S1 in File S2). When these were not available, we compiled data from global databases, utilizing equations based on other sites, and in some cases for other species from within the same genera [56,57]. Previous studies have reported that genus means are reasonable proxies for species values for specific gravity (r^2>0.70; [58,59]).

To determine tree height, we applied species-specific equations of [54] giving the relationship of tree height vs. DBH, to each individual tree for 12 of the MWF species and 4 of the LDF species (Table S1 in File S2). For the other species, we used the general wet and dry forest equations [55] to determine tree height. We used these tree height estimates to calculate AGB for each tree using published equations that also included DBH and wood specific gravity. Hawai'i-specific equations for AGB were available for 5 MWF species and 4 LDF species. For another two species of the LDF, *D. sandwicensis* and *P. odorata*, equations were available that were developed specifically from our study site [52] (Table S1 in File S2).

Species richness and diversity. Species area curves were generated by plotting cumulative number of species against area for the 20 m×20 m quadrats. Rarefaction analyses were based on 999 permutations (PRIMER-E v. 6, PRIMER-E Ltd, Plymouth, UK), which randomized the sampling order and resulted in a robust average curve. We present several indices: Sobs (the observed number of species), Chao 1 based on rare species (nonparametric), and Michaelis-Menten (parametric), given uncertainty in the ideal estimator [60,61,62]. We used the program EstimateS 9.1.0 to calculate species diversity indices and an estimate of error. We report Fisher's alpha, Shannon diversity index, and Simpson's index (inverse form) following standard formulas [63]. Overlap in species composition between the two sites was determined using the Sørenson similarity index (SI):

$$SI = \frac{number\ of\ species\ shared\ in\ both\ sites}{(no.\ of\ species\ in\ MWF + no.\ of\ species\ in\ LDF)}$$

Table 2. Diversity and forest structure characteristics of plots in the Center of Tropical Forest Science global plot network, including the Hawaiian plots, arranged in order of descending species richness.

CTFS plot location	Plot code	Latitude	Mean annual rain (mm)	Dry season months	Mean elevation (m)	Mean annual temp (°C)	Land type	Plot size (ha)	No. of species	No. of families	Mean species/family	Fisher's α/ha	H'/ha	Trees/ha	Basal area (m²/ha)	Dom. of most common family	Above-ground biomass (Mg/ha)[7]
Lambir, Malaysia	LAM	4.19	2664	0	170	26.6	I	52	1182	83	14.2	165	2.40	6915	43.5	41.0	497.2
Yasuní, Ecuador	YAS	−0.69	3081	0	230	28.4	M	50	1114	81	13.8	187	2.44	3026	33.0	14.9	282.4
Pasoh, Malaysia	PAS	2.98	1788	1	80	28.0	M	50	814	82	9.93	124	2.31	6708	31.0	28.2	339.8
Khao Chong, Thailand	KHA	7.54	19851	32	140	27.3	M	24	593	na	na	na	Na	5063	na	na	na
Korup, Cameroon	KOR	5.07	5272	3	200	26.7	M	50	494	62	7.97	48.0	1.75	6580	32.0	16.1	na
Ituri, Dem. Rep. of Congo3	ITU	1.44	1700	3–4	780	23.1	M	404	445	na	na	na	Na	7200	na	na	na
Palanan, Philippines	PAL	17.04	3379	4–5	110	23.5	I	16	335	60	5.58	43.4	1.92	4125	39.8	52.8	290.1
Bukit Timan, Singapore	BUK	1.25	2473	0	150	26.9	I	2	329	62	5.31	60.0	1.90	5950	34.5	38.4	na
BCI, Panama	BCI	9.15	2551	3	140	26.9	M	50	299	58	5.16	34.6	1.62	4168	32.1	11.4	306.5
Mo Singto, Thailand	MOS	14.43	2200	5	770	23.0	M	30.5	262	na	na	na	Na	Na	na	na	na
Huai Kha Khaeng, Thailand	HKK	15.63	1474	6	590	24.1	M	50	251	58	4.33	23.3	1.50	1450	31.2	21.2	211.2
La Planada, Colombia	LPL	1.16	4084	0	1840	19.0	M	25	240	54	4.44	30.6	1.72	4216	29.8	14.9	177.6
Dinghushan, China	DIN	23.16	1985	0	350	20.9	M	20	210	na	na	na	Na	3581	na	na	na
Sinharaja, Sri Lanka	SIN	6.4	5016	0	500	22.6	I	25	204	46	4.43	24.4	1.53	7736	45.6	26.7	357.9
Doi Inthanon, Thailand	DOI	18.52	1908	6	1700	21.4	M	15	192	na	na	na	Na	4913	na	na	na
Luquillo, Puerto Rico	LUQ	18.33	3548	0	380	22.8	I	16	138	47	2.94	13.5	1.45	4194	38.3	17.3	276.1
Nanjenshan, Taiwan	NAN	22.06	3582	0	320	23.5	I	3	125	41	3.05	15.6	1.64	12133	36.3	32.3	na
Ilha do Cardoso, Brazil5	ILH	−25.1	2100	0	5	22.4	M	10	106	na	na	na	Na	Na	na	na	na
Mudumalai, India	MUD	11.6	1250	6	1050	22.8	M	50	72	29	2.48	6.20	0.944	510	25.5	28.2	174.2
Laupāhoehoe, USA	LAU	19.93	3440	1	1150	16.0	I	4	21	15	1.40	2.58	1.82	3078	67.36	37.4	247.9
Pālamanui, USA	PLN	19.74	835	12	240	20.0	I	4	15	15	1.00	1.18	1.00	3487	8.6	74.2	29.4

Southern hemisphere latitudes are negative; land types are island (I) and mainland (M); dry season months are as those with <100 mm precipitation (Richards 1996). Data are from [104] and ctfs.si.edu unless indicated by footnotes.

[1]Mean annual rainfall data for the nearby city of Songkhla, Thailand (www.world-climates.com)

[2]Kira T (1998) NPP Tropical Forest: Khao Chong, Thailand, 1962–1965. Data set. Available on-line [http://www.daac.ornl.gov] from Oak Ridge National Laboratory Distributed Active Archive Center, Oak Ridge, Tennessee, U.S.A.

[3]Average of 4 plots (2 monodominant forest, 2 mixed forest)

[4]Divided into four 10-ha plots

[5]Data from Ferreira de Lima RA, Oliveira AAD, Martini AMZ, Sampaio D, Souza VC, Rodrigues RR (2011) Structure, diversity, and spatial patterns in a permanent plot of a high restinga forest in Southeastern Brazil. Acta Botanica Brasilica 25: 633–645.

[6]Basal area including tree ferns; 36.1m²/ha without tree ferns

[7]Chave J and 37 others (2008) Assessing evidence for a pervasive alteration in tropical tree communities.

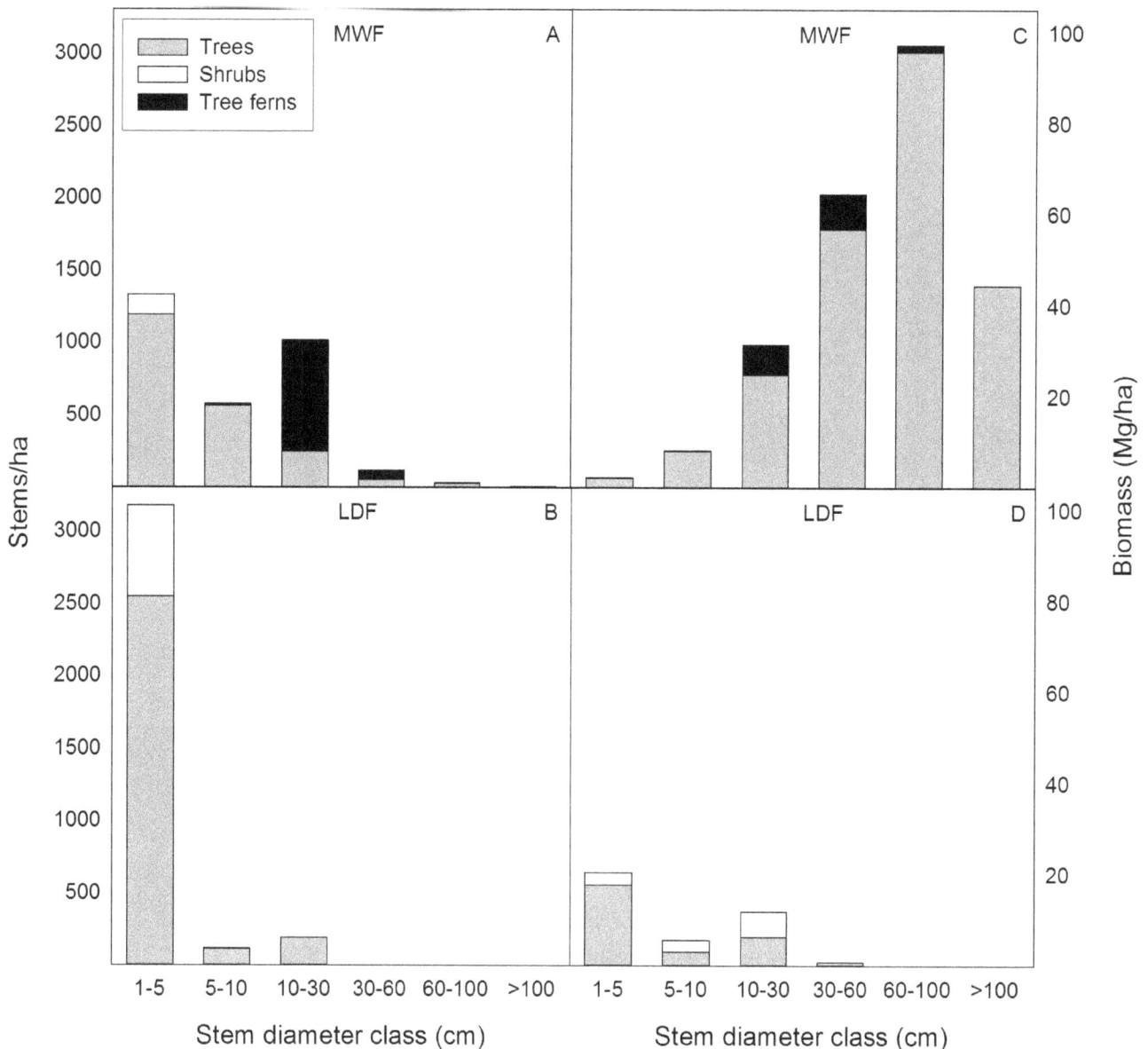

Figure 2. Life form distribution of stems and biomass by diameter size intervals. In (A) Hawaiian montane wet forest (MWF) and (B) lowland dry forest (LDF), stems represent the number of main stems (i.e., one per individual, not including other multiple stems). In (C) MWF and (D) (LDF), biomass calculations were made for all stems (including multiple stemmed individuals). Diameter classes are 1–4.99 cm, 5 - <9.99 cm, 10–29.99 cm, 30–59.99 cm, 60–99.99 cm, and ≥100 cm.

Forest type comparisons. We compiled data for 19 additional mainland and island CTFS tropical plots for which climate and structure data were available (Table 2). Differences between the Hawai'i plots and other CTFS plots were assessed using one sample *t*-tests. Differences in the characteristics of island and mainland plots were assessed using Wilcoxon signed-rank tests [64]. These statistics were analyzed with JMP v. 6 [29,65].

Results

Comparison of Floristics and Life Forms in Hawaiian Forests

The two Hawai'i forests were distinct in floristic composition (Table S2 in File S2). The plots had a very low Sørenson similarity index of 0.06 (a value of 1 would indicate complete overlap). Only

M. polymorpha occurred in both forests; it was the second most common species in MWF but was represented by only 5 individuals in the 4-ha LDF plot. Species richness was 21 in the MWF and 15 in the LDF. Fifteen families were represented at each site, and the canopy trees at the two sites were from different families, though four families were represented in the understory or the midstory at both sites (Euphorbiaceae, Fabaceae, Myrtaceae, and Rubiaceae).

The plots differed in their distribution of plant life forms. In the MWF, 68% of stems were trees, 4.5% were shrubs, and 28% were tree ferns, accounting for 45%, 8.3% and 46% of the basal area respectively. In the LDF, 82% of stems were trees and 18% were shrubs, accounting for 95% and 5% of basal area respectively (Fig. 2). In the MWF, a large proportion of stems (31%) were growing on non-soil substrates, primarily tree ferns, logs or rocks,

Table 3. Statistics on abundance, basal area, and frequency of the species in the Laupāhoehoe (montane wet forest) plot, with data displayed on an absolute and a relative basis.

Laupāhoehoe montane wet forest

Species	No. individuals	Basal area (m²/ha)	Presence (no. of quadrats)	Relative abundance (%)	Relative dominance (%)	Relative frequency (%)	IV (%)
METPOL	2631	25.2	100	21.4	37.5	10.2	69.1
CIBGLA	2274	17.7	100	18.5	26.4	10.2	55.1
CHETRI	3320	4.17	100	27.0	6.20	10.2	43.4
CIBMEN	1076	13.2	100	8.74	19.6	10.2	38.6
COPRHY	972	0.585	99	7.90	0.870	10.1	18.9
ILEANO	965	0.466	99	7.84	0.692	10.1	18.7
ACAKOA	141	5.49	57	1.15	8.16	5.84	15.1
BROARG	271	0.0454	74	2.20	0.067	7.58	9.85
MYRLES	237	0.0571	70	1.93	0.085	7.17	9.18
VACCAL	255	0.0328	51	2.07	0.0488	5.23	7.35
HEDHIL	43	0.0200	29	0.349	0.0297	2.97	3.35
PERSAN	35	0.0084	28	0.284	0.0125	2.87	3.17
CIBCHA	34	0.232	17	0.276	0.345	1.74	2.36
CLEPAR	19	0.00230	17	0.154	0.00342	1.74	1.90
MELCLU	13	0.00235	11	0.106	0.00349	1.13	1.24
PSYHAW	10	0.00219	9	0.0812	0.00326	0.922	1.01
MYRSAN	6	0.00166	6	0.0487	0.00246	0.615	0.666
PIPALB	4	0.00226	4	0.0325	0.00335	0.410	0.446
TREGRA	2	0.000107	2	0.0162	0.000159	0.205	0.221
LEPTAM	2	0.0000682	2	0.0162	0.000101	0.205	0.221
ANTPLA	1	0.000154	1	0.0081	0.000229	0.102	0.111
Total	*12311*	*67.3*					

Table 4. Statistics on abundance, basal area, and frequency of the species in the Pālamanui (lowland dry forest) plot, with data displayed on an absolute and a relative basis.

Pālamanui lowland dry forest

Species	No. individuals	Basal area (m²/ha)	Presence (no. of quadrats)	Relative abundance (%)	Relative dominance (%)	Relative frequency (%)	IV (%)
DIOSAN	2208	6.41	99	15.8	74.2	18.3	108.3
PSYODO	8640	1.27	100	62.0	14.7	18.5	95.2
DODVIS	2301	0.359	94	16.5	4.15	17.4	38.1
SOPCHR	5	0.21	4	0.0359	2.38	0.741	3.20
SANPAN	275	0.156	32	1.97	1.81	5.93	9.70
OSTANT	147	0.0900	40	1.05	1.04	7.41	9.50
WIKSAN	88	0.0890	44	0.631	1.03	8.15	9.81
EUPMUL	134	0.023800	54	0.961	0.275	10.0	11.2
SENGAU	70	0.013300	27	0.502	0.154	5.00	5.66
METPOL	12	0.013	9	0.0860	0.152	1.67	1.90
MYOSAN	54	0.007480	26	0.387	0.087	4.81	5.29
SIDFAL	1	0.0008	1	0.0072	0.00947	0.185	0.202
PLEHAW	1	0.0007	1	0.0072	0.00843	0.185	0.201
ERYSAN	2	0.000513	2	0.0143	0.00594	0.370	0.391
PITTER	8	0.00048	7	0.0574	0.00551	1.30	1.36
Total	*13946*	*8.64*					

Presence based on 100 20×20 m quadrats per plot; species sorted by importance value (IV), which is the sum of the three relative measures (max 300%); species abbreviations as in Table S2 in File S2.

Table 5. Aboveground biomass listed by species for the two Hawai'i forest plots; species abbreviations as in Table S2 in File S2.

Laupāhoehoe montane wet forest			Pālamanui lowland dry forest		
Species	Biomass (Mg/ha)	Relative biomass (%)	Species	Biomass (Mg/ha)	Relative biomass (%)
METPOL	186	74.9	PSYODO	15.3	51.9
ACAKOA	31.1	12.5	DIOSAN	10.5	35.8
CHETRI	12.4	4.99	METPOL	1.40	4.78
CIBMEN	10.9	4.39	DODVIS	0.921	3.14
CIBGLA	4.55	1.83	OSTANT	0.525	1.79
COPRHY	1.59	0.64	SANPAN	0.359	1.22
ILEANO	1.27	0.51	WIKSAN	0.181	0.615
CIBCHA	0.184	0.07	MYOSAN	0.109	0.372
MYRLES	0.109	0.04	SOPCHR	0.0446	0.152
VACCAL	0.0947	0.04	SENGAU	0.0398	0.135
HEDHIL	0.0456	0.02	EUPMUL	0.0199	0.068
BROARG	0.0394	0.02	PITTER	0.00435	0.0148
PERSAN	0.0124	0.00499	PLEHAW	0.00217	0.00740
PSYHAW	0.00384	0.00155	SIDFAL	0.00187	0.00637
MELCLU	0.00343	0.00138	ERYSAN	0.000871	0.00297
CLEPAR	0.00327	0.00132			
MYRSAN	0.00319	0.00128			
PIPALB	0.00274	0.00110	*Total*	*29.4*	
ANTPLA	0.000287	0.000116			
LEPTAM	0.00017	0.0000685			
TREGRA	0.0000703	0.0000283			
Total	*247.9*				

with 17% of all individuals growing on dead tree ferns (Table S3 in File S2). In contrast, in the LDF, all trees were growing on soil or broken lava, and tree ferns were absent.

Comparison of Stand Structure in Hawaiian Forests

The MWF had larger trees and lower stem density than the LDF (3078 ± 1.21 and 3487 ± 1.40 stems/ha respectively; Table 3). The tree size class distributions differed between the two forests as expected based on their contrasting climates: the LDF had mainly small stems and the MWF had a much more even spread of size classes (Fig. 2). Because the stems in the LDF were small, total basal area and biomass values were low. Thus, the MWF had a nearly eight-fold higher basal area than the LDF (67.3 vs. 8.6 $m^2/$ha respectively; Tables 3–4), and tree ferns accounted for 31.2 $m^2/$ha basal area. Above-ground biomass in the MWF was also more than eight times higher than the LDF (248 Mg/ha vs. 29.4 Mg/ha respectively; with 15.6 Mg/ha in the MWF accounted for by tree ferns; Table 5). The above-ground biomass value for the MWF was consistent with that previously estimated for surrounding forest in the same reserve [66]. In both forest types, the two most common canopy species represented 87–88% of biomass (Table 5). In the MWF, the very large trees (≥60 cm) made up the greatest proportion of the biomass, but in the LDF the majority of the biomass was in the 1–5 cm size class (Fig. 2). More multi-stemmed individuals make up the LDF, a mean of 3.2 stems/individual, compared to 1.4 stems/individual in the MWF plot) (Table S4 in File S2).

Comparison of Community Structure in Hawaiian Forests

For both the Hawaiian MWF and LDF, rarefaction curves indicated that a 1 ha sample was sufficient to capture 90% of the species present in the larger 4-ha area (Fig. 3). In the MWF, diversity values for the plot were 2.46 (Fisher's alpha), 1.98 (Shannon), and 5.74 (Simpson); in the LDF values were 1.66 (Fisher's alpha), 1.15 (Shannon), and 2.29 (Simpson). When viewed graphically, there was no overlap in any index value between the two forests (Fig. 4): the Hawaiian MWF was more diverse than the LDF. In the MWF, species evenness was higher than in the LDF, primarily because *P. odorata* in the LDF had a relative abundance of over 60%. In contrast the forests were similar in the abundance of uncommon species, and ~20% of species were rare, i.e., having ≤1 stem/ha (Tables 3–4).

Non-native Species in the Plots

Invasive species made up a larger presence in LDF than MWF (Figs.5–6). The grass *Pennisetum setaceum* was most widespread in the LDF and the herbaceous weed *Persicaria punctata* was most common in the MWF, where it tended to dominate low-lying boggy areas. In the LDF there was a greater overall weed cover, particularly of woody weeds (Fig. 7; see also Methods S1 in File S1). In the MWF there were only a few stems that qualified for DBH measurements (>5 cm). *F. uhdei* averaged 32.3 cm (n = 1) and *P. cattleianum* averaged 6.2 cm (n = 1). In the LDF, average DBH for *Grevillia robusta* was 20.0 ± 8.2 cm SE (n = 27), for *Leucaena leucocephalum* it was 5.9 ± 0.8 cm SE (n = 7), and for *Schinus terebinthius* it was 11.0 ± 1.5 cm SE (n = 20).

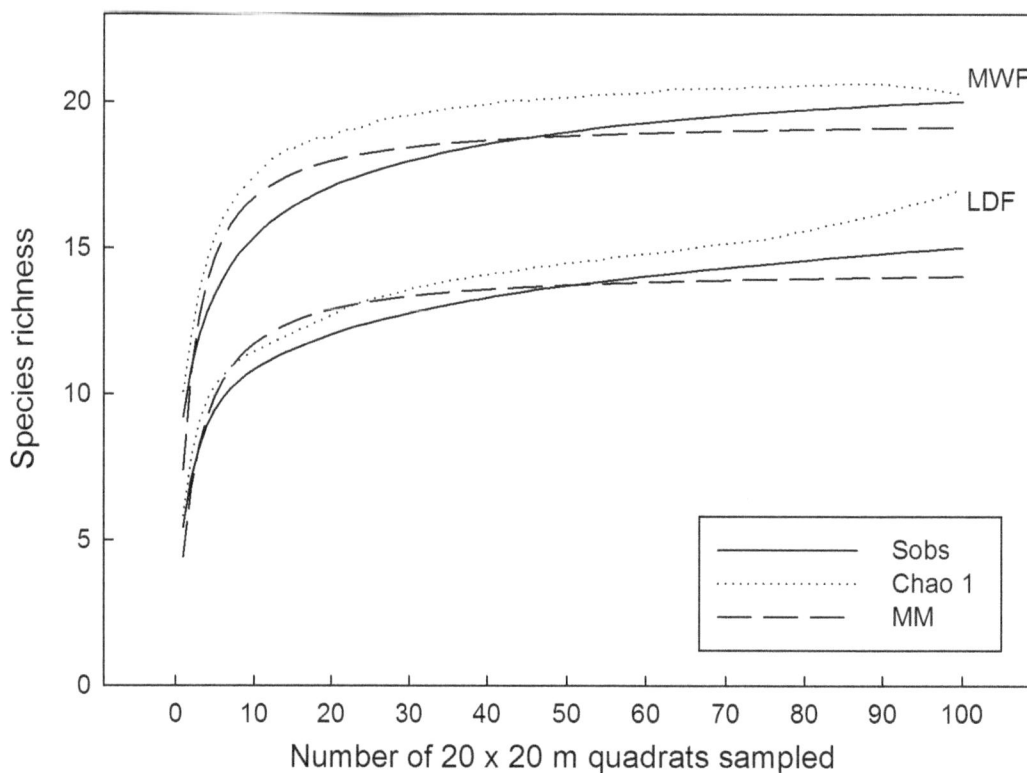

Figure 3. Species accumulation curves. Species number is shown cumulatively, as additional 20 m×20 m quadrats are sampled, until the entire 4-ha plot is represented (100 quadrats), for Hawaiian montane wet forest (MWF) and lowland dry forest (LDF). Three rarefaction techniques are used: Sobs (observed species number), Chao 1, and MM (Michaelis-Menten).

Comparison of Hawai'i to Other CTFS Plots

Diversity of both Hawai'i forests was very low relative to other CTFS forests, including those on islands and with dry climates (Fig. 8). Across the CTFS network, the mean Fisher's alpha per ha ± SE was 59.7±16.6 (n = 13), and the two Hawaiian forests were statistical outliers, with diversity values more than 2 SD lower (t = 3.45 and 3.53, P<0.005). The MWF had approximately 15% as many species as the most comparable island site with tropical wet forest (Luquillo, Puerto Rico). Compared with the next two driest CTFS sites, the Hawai'i LDF had 21% of the number of species found in the Mudumalai, India plot and just 6% of the number of species found at the Huai Kha Khaeng, Thailand plot (Table 2).

In contrast to biodiversity, the structural comparisons across the CTFS network revealed complicated patterns. The MWF was similar to other CTFS sites with respect to tree size class distribution (Fig. 9), and was not significantly different from other CTFS plots with respect to standing above-ground biomass/ha. However, the MWF had 35% lower stem density than the all-forest mean of 4733±722 SE (t = 2.29, P = 0.039, n = 14). Further, the MWF had a 92% higher basal area than the mean of other tropical FDPs due to its high tree fern abundance (t = −21.6, P< 0.0001). When tree ferns were excluded, the basal area of the Hawaiian MWF was within the range of that for other FDPs. For the LDF, stem density was not significantly different than the all-forest mean (t = 1.79, P = 0.097, n = 16, Table 2), but the LDF was an outlier with its very low basal area (t = 10.40, P<0.001, n = 19) and above-ground biomass (t = 7.12, P<0.001, n = 12). The LDF was especially distinctive in having virtually all small stems (Fig. 9).

Discussion

Comparing Hawaiian Wet and Dry Forests

Examination of how the two Hawaiian forests compared in terms of composition and structure matched and extended the paradigm for differences between mature wet and dry forests described on other tropical islands [32]. As predicted, our results for Hawai'i are in agreement with comparisons of mature wet and dry forests in Puerto Rico [32]: wet forest had larger diameter trees, greater basal area, and higher biomass than dry forest, and differences between wet and dry forest in tree density, dominance, and species richness were minor (Tables 3–4). Indeed, across many forests, biomass and basal area are typically correlated with climatic variables such as MAT, MAP, and water deficit within and across sites [31,70,71,72,73,74,75]. However, other variables such as substrate age and type [76] are likely to have contributed to structural and floristic differences between LDF and MWF and we cannot ascribe results solely to climate. While both forests occur on young lava, the much higher rainfall of the MWF and variation in substrate type and texture contributed to greater soil development. Lava flow age and substrate type are important determinants of successional stage in Hawai'i [39]. In addition, differences in disturbance regimes between the two sites may have influenced their forest structure, and their invasive species cover. In the MWF, almost a third of stems were found growing on a substrate other than soil, such as nurse logs and living tree ferns, which likely reflects preferential survivorship on those substrates. Further, canopy dieback of *Metrosideros polymorpha* [45], wind storms [44], and invasive animals [77] may be important factors influencing forest structure. Canopy gaps are larger in the MWF due to the much larger and taller trees that make up the canopy.

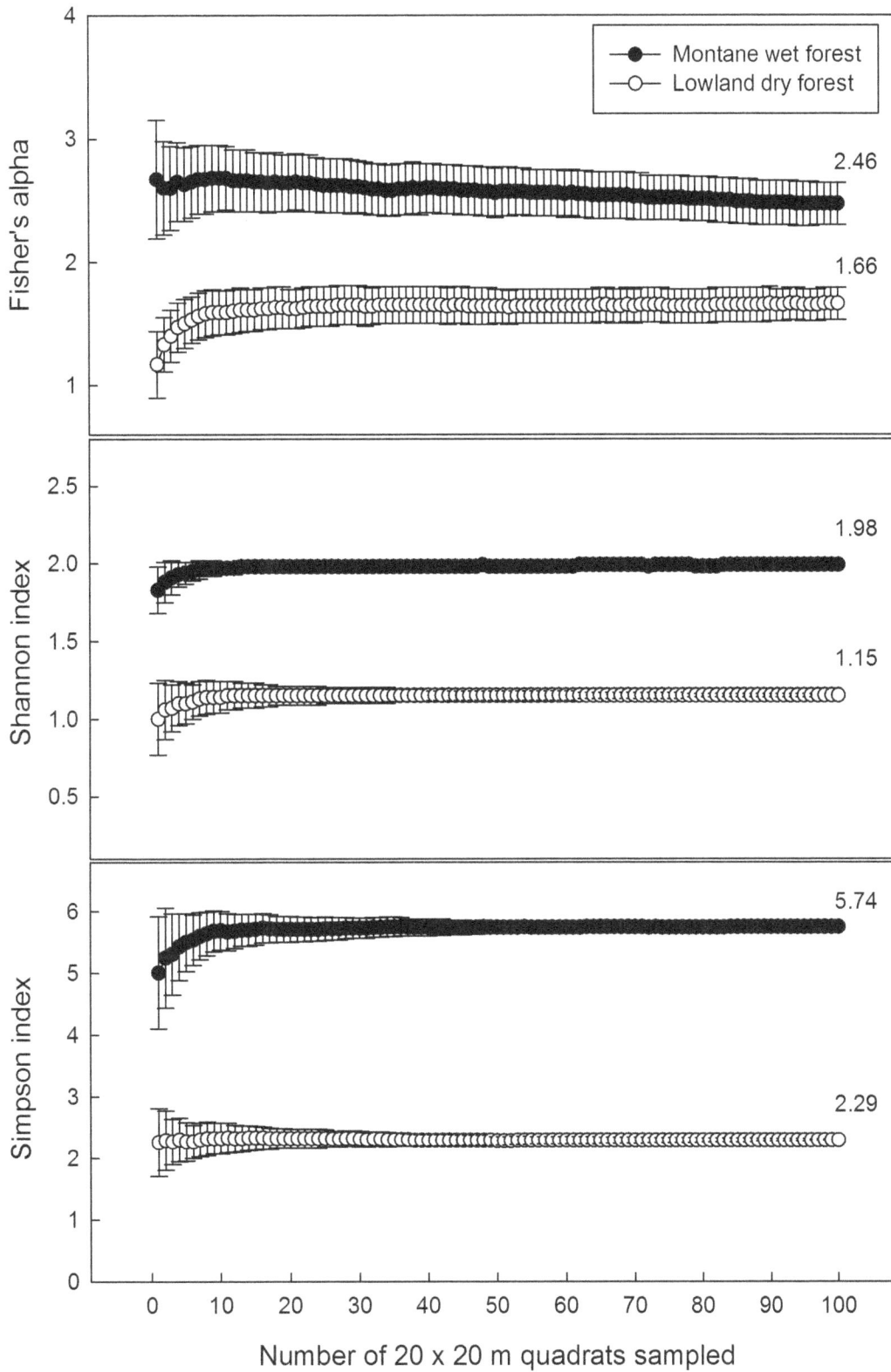

Figure 4. Species diversity indices. Fisher's alpha, Shannon index, and Simpson index for the Hawaiian montane wet forest (MWF) and lowland dry forest (LDF). Each 20×20 m subplot is shown, with the values being cumulative and number above each line representing the entire plot area (4-ha). Values are the diversity index and standard deviation, as estimated by the program EstimateS.

Fraxinus uhdei   Grass   Passiflora tarminiana

Persicaria punctata   Psidium cattleianum   Rubus ellipticus

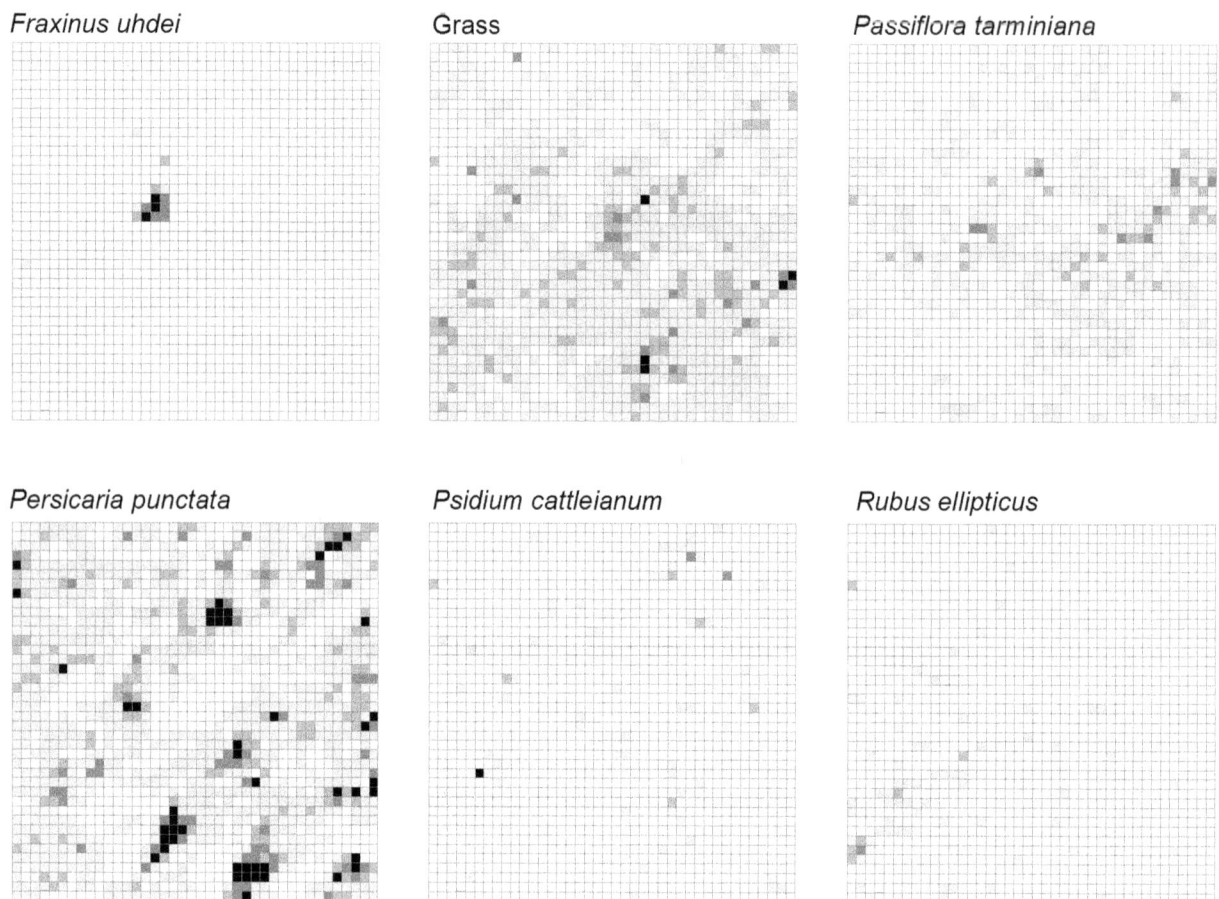

Figure 5. Invasive species cover distribution. Map showing percent cover and locations of invasive species in the MWF. Each grid square represents one 5×5-m subquadrat white: absent, light grey: present to <5%, medium grey: 5–25%, dark grey: 25–50%, black: >50% cover).

There is also evidence of pig rooting that may affect seedling regeneration preferences [47,78], explaining why many stems grow on substrates other than soil. The LDF site is currently fenced from ungulates but the large proportion of multi-stemmed trees and the higher prevalence of regeneration by sprouting suggest adaptation to disturbance [79,80,81].

The MWF site distinguishes itself in its abundance and dominance of tree ferns, which form a distinct mid-canopy layer approximately 5 m above the ground. Notably, tree ferns also make up a large proportion of stand basal area or stem density in some temperate rain forests [82,83,84] and tropical cloud forests [85,86,87,88] but not in other CTFS sites. In Hawai'i, tree ferns are common in wet forests at all elevations, and are particularly abundant in areas with more well-developed organic soils as opposed to young lava flows. While the dynamics of tree ferns have not been well studied in tropical environments [89], in Hawaiian forests tree ferns undoubtedly influence forest function, due to their long lifespans, high frond area, slow growth [90], and slow decomposition rates [91]. They also play a critical role as a substrate for tree seedlings [47,78].

Unlike other CTFS plots that are not heavily impacted by non-native plant species, all forests in Hawai'i have been invaded to some degree. We purposefully chose sites with low non-native species abundance, but cataloged cover before removal for future long-term studies. Because we are removing the invasive species after data collection, we are not examining the consequences of invasion, but previous work in Hawai'i has shown that invaders

can significantly alter forest functioning [50,51,52,92,94]. A debate in invasion biology is whether invaders owe their success to their introducing a new function to the community (e.g., N-fixing species) or are simply better competitors [92] and we argue that it is the latter case at our sites. At our sites, invasive grasses were widespread, but, woody invaders are a greater competitive threat (Figs. 7). In the LDF, *Pennisetum setaceum* is widespread, but the vegetation is still dominated by woody species with moderate canopy closure, and reduction of grass cover and fire prevention will reduce its competitive effect in the future. While non-native grasses and herbs are more common than non-native trees across the MWF, their abundance is strongly related to boggy areas, canopy openings, and pig disturbance, and these patches are not likely to expand, but rather to be shaded out in the long term. In MWF, woody invaders such as *Psidium cattleianum* represent much greater threats based on their extreme abundance elsewhere, and traits such as shade tolerance, vegetative reproduction, and animal-dispersed fruits [93,94]. At present, the MWF has limited cover of woody invaders (Fig. 7), and in that respect is in better condition that the LDF.

Clearly, site-specific properties influence the structure and species composition between the two sites, but our study also highlights that at the island scale (1 million ha), climate likely exerts a strong influence, both directly and indirectly [67]. These differences matched patterns found in continental forests, where diversity measures as well as structural measures correlate negatively with the length or severity of the dry season [68,69].

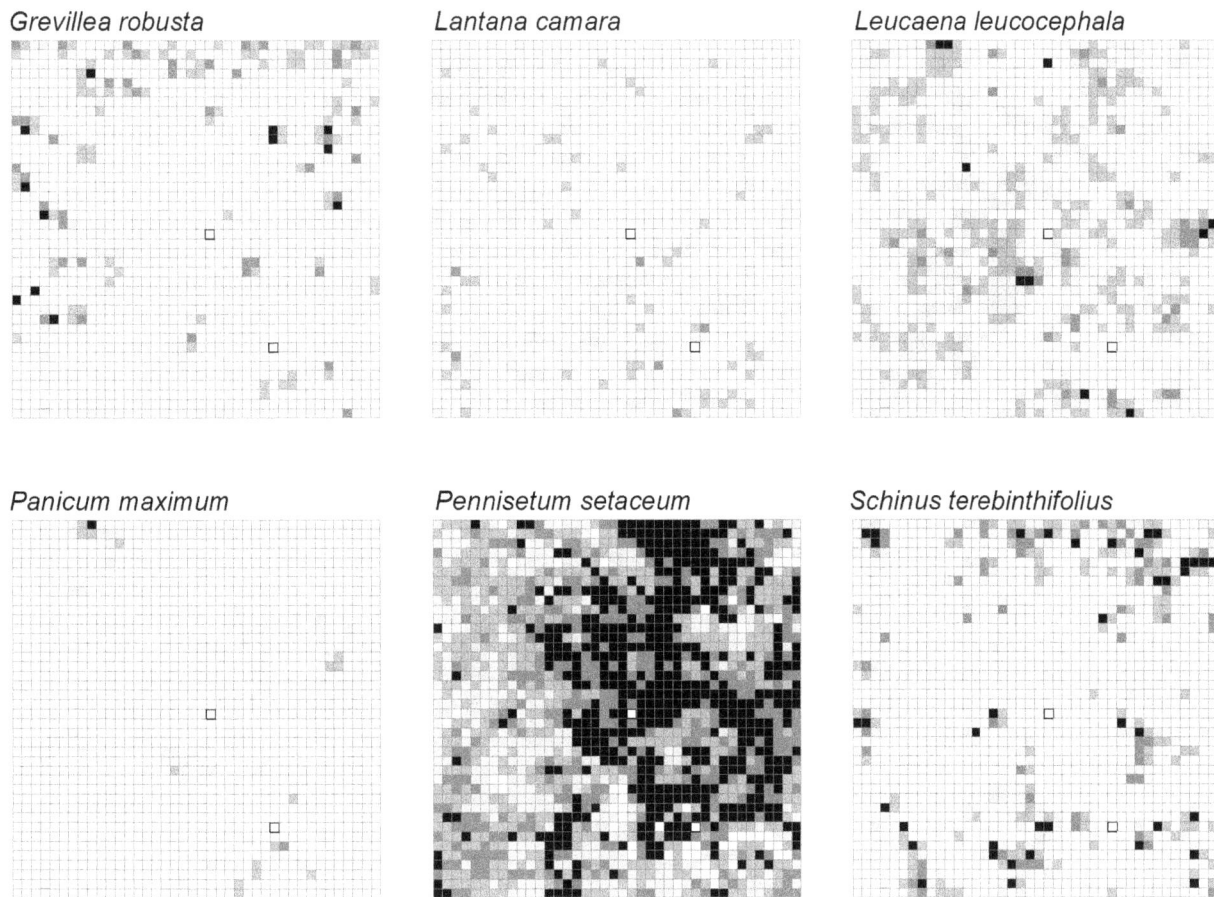

Figure 6. Map showing percent cover and locations of invasive species in the LDF. Each grid square represents one 5×5-m subquadrat (white: absent, light grey: present to <5%, medium grey: 5–25%, dark grey: 25–50%, black: >50% cover).

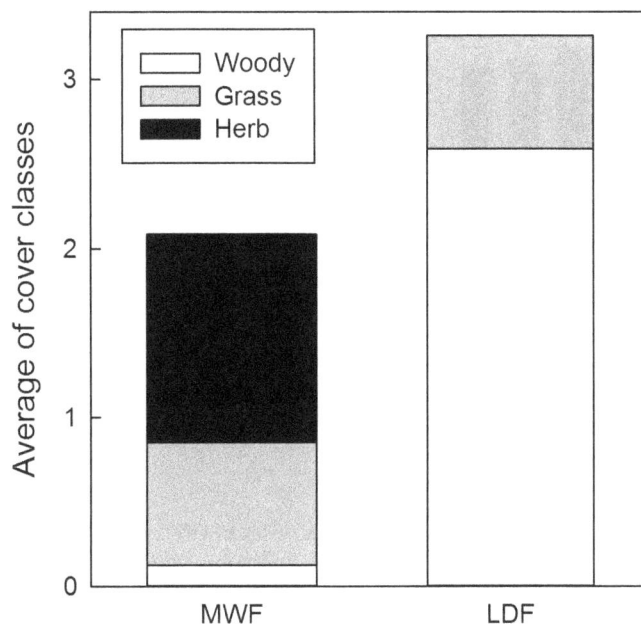

Figure 7. Combined invasive species cover. In each 5×5 m subquadrat a cover score from 0–4 was given based on cover classes (see Methods). The y axis represents the average cover class across the 400 subquadrats, separated by life form: grasses, herbaceous, or woody (shrubs and trees). The combined cover represents the species shown in Figure 5.

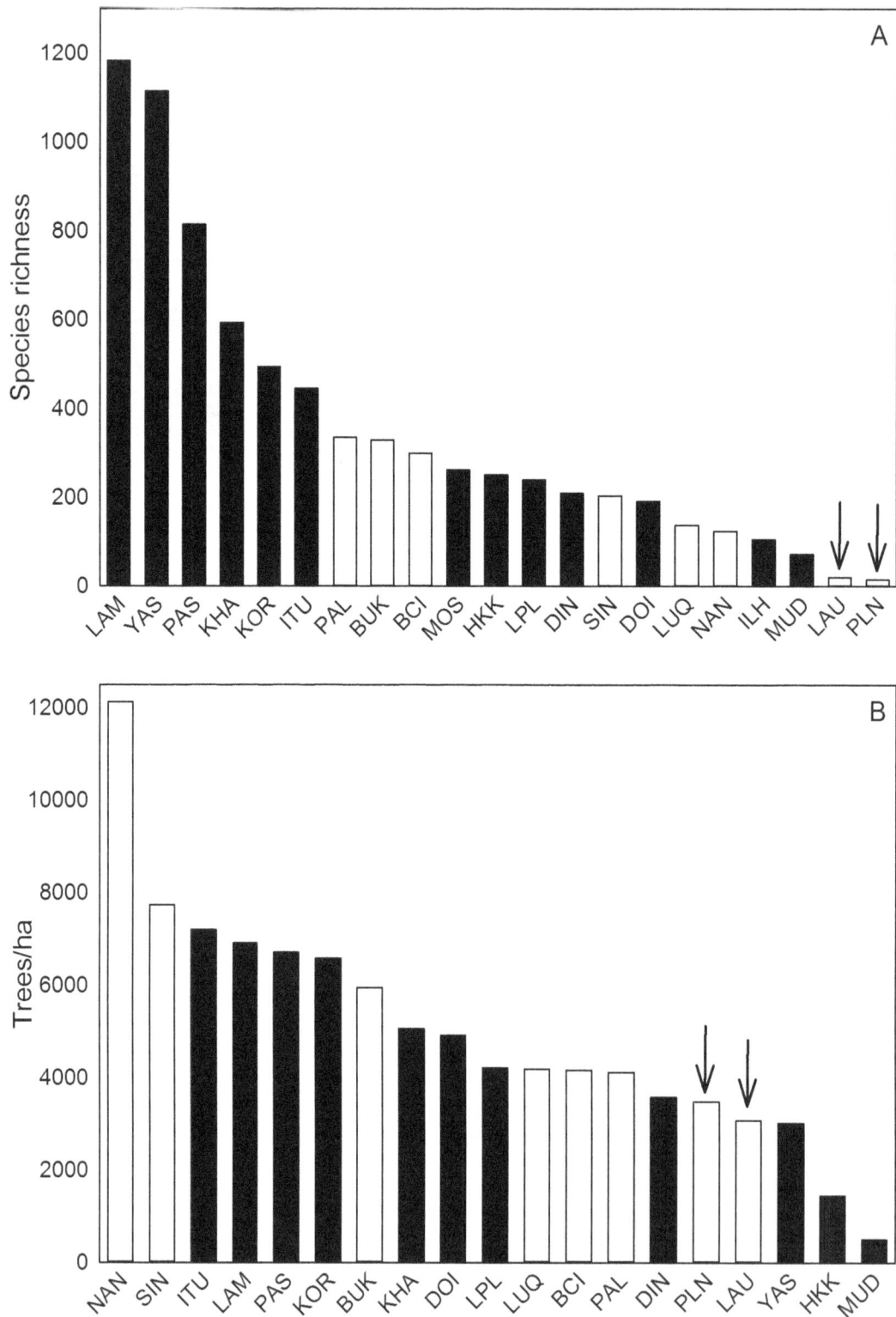

Figure 8. Comparisons of species richness and stem density across a series of CTFS plots. Black bars represent continents and open bars represent islands. Abbreviations as in Table 2. Data from Losos and Leigh, Jr. (2004) and www.ctfs.si.edu.

Structure and Diversity across Tropical Forests Globally

The Hawaiian forest data allowed for the examination of the question of how forest structure varies across species diversity gradients across a much wider range of tree species diversity than was previously available. One of the most striking conclusions of our study is that, despite the extremely low species richness of the Hawai'i FDPs, some structural variables, particularly those for the wet forest, were well within the range of values for the world's most

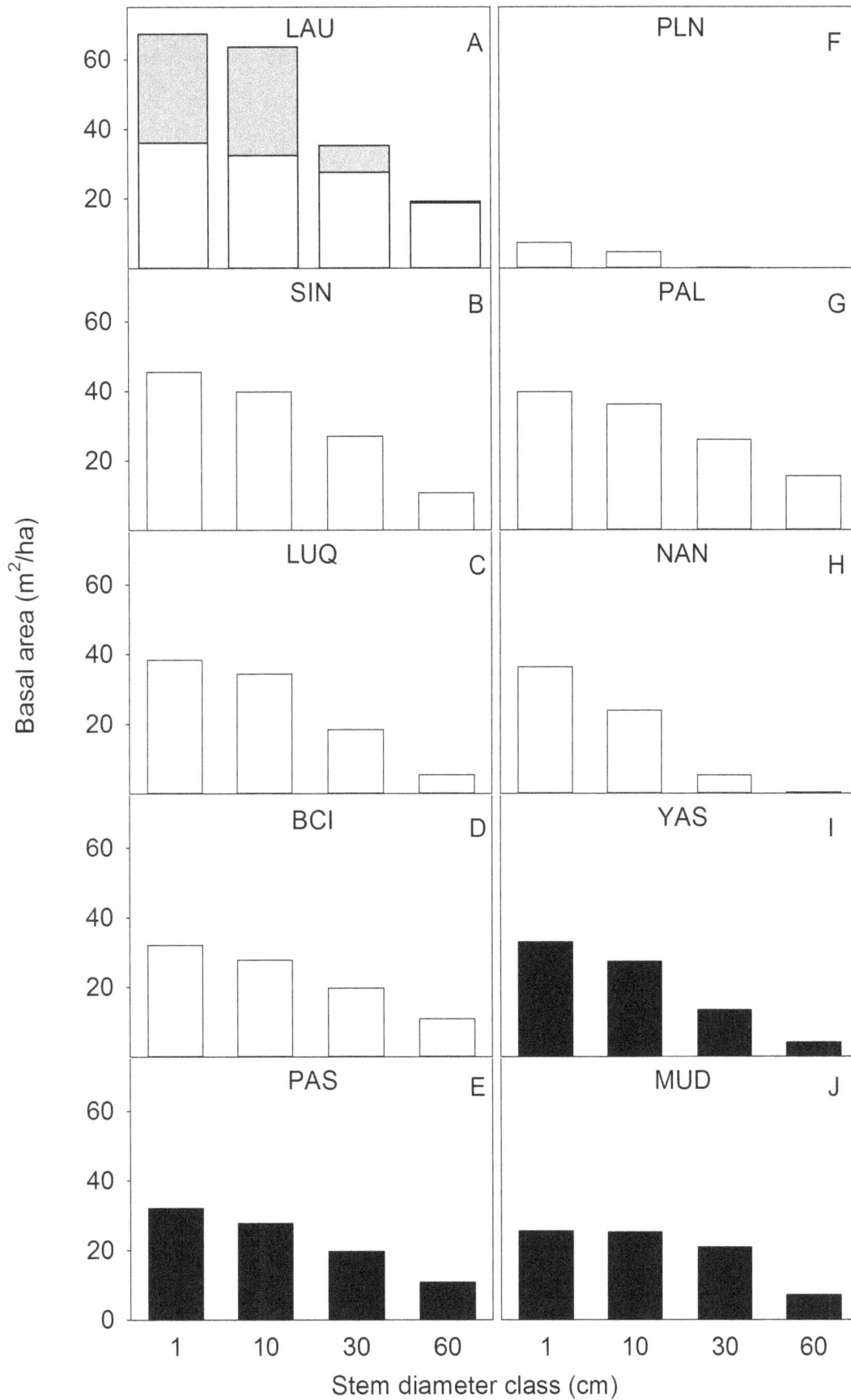

Figure 9. A reverse-cumulative distribution of basal area by size class. Size classes are: ≥ 1 cm, ≥ 10 cm, ≥ 30 cm, and ≥ 60 cm. Data shown for the Hawaiian montane wet forest (LAU) and lowland dry forest (PLN) (top row) and for selected other CTFS plots. Tree ferns (found only at LAU) are symbolized by the gray bars. Island sites are open bars and continental sites are filled bars. Abbreviations as Table 2. Data from Losos and Leigh, Jr. (2004) and www.ctfs.si.edu.

diverse tropical forests (Table 2). For example, stem densities for Hawaiian MWF and LDF were similar to those of the hyper-species-rich Yasun??? FDP (Fig. 8), while biomass and basal area of the Hawai'i MWF (excluding tree ferns) were similar to those of the higher diversity forests in the CTFS network (Fig. 9). The inclusion of tree ferns increased basal area values by 52%, but only increased biomass by 11% (Tables 3 and 4). Notably, the LDF had among the lowest basal area and biomass in the CTFS network, consistent with this site having the lowest precipitation of all FDPs (Table 2). It should also be noted that the LDF is dry year-round, while other dry sites in the CTFS network are seasonally dry (Table 2).

The low floristic richness and population structure of the Hawaiian forest plots represented strong convergence with other island forests. Hawaiian forests had fewer species per family and greater average population densities for each species, as seen in other very isolated sites [95]. High relative dominance values were consistent with island forests having greater dominance by the most common family than mainland tropical forests (Table 2). In the MWF and LDF, 37% and 74% of basal area respectively were accounted for by a single canopy dominant species. On average 20% of species were rare in Hawaiian forests (defined as ≤ 1 tree/ha), by contrast with 42% on average across other high-diversity forests [96]. It is likely that the patterns of high basal area dominance in Hawaiian forests arose due to the biogeographic consequences of isolation, but we cannot rule out species loss due to human disturbance and invasive species of multiple trophic levels [97,98,99,100,101].

In conclusion, Hawaiian forests have among the lowest species richness and highest endemism rates globally, but in a number of key structural variables both of these forests were similar to even the highest diversity tropical forests in the CTFS network. Future work could examine the evolutionary consequences of such a limited species pool. Biodiversity theory developed in high-diversity tropical forests emphasizes that competitive interactions among species are unlikely on evolutionary time scales because any given two species are rarely consistent neighbors [9]. However, in low-diversity forest any two given species have far greater potential for competitive interactions than in high-diversity tropical forests [102,103]. The addition of Hawai'i to the global plot network enables investigations of the consequences of such

differences across a very wide range in species diversity and environmental gradients.

Supporting Information

File S1 Methods S1. Detailed methods and description of situations where field site conditions dictated a different or entirely new methodology by adopted than standardized CTFS protocol in [1].

File S2 Supporting tables. Table S1. Values and equations used for estimating aboveground biomass (AGB) in the montane wet forest (MWF) and lowland dry forest sites (LDF). **Table S2.** Species ≥ 1 cm diameter at breast height recorded in Laupāhoehoe (montane wet forest) plot with canopy dominants in bold. **Table S3.** Percentage of individuals in the Laupāhoehoe (montane wet forest) plot growing on each substrate type. **Table S4.** Size and multiple stem characteristics of the species species in Laupāhoehoe (montane wet forest) and Pālamanui (lowland dry forest) plots; species abbreviations as in Table S4. **Table S5.** References from Table 1.

Acknowledgments

The Hawai'i Permanent Plot Network thanks the USFS Institute of Pacific Islands Forestry (IPIF) and the Hawai'i Division of Forestry and Wildlife/Department of Land and Natural Resources for permission to conduct research within the Hawai'i Experimental Tropical Forest; the Palāmanui Group, especially Roger Harris, for access to the lowland dry forest site. We thank N. DiManno, L. Ellsworth, B. Hwang, R. Moseley, M. Murphy, K. Nelson-Kaula, M. Nullet, C. Perry, J. Schulten, M. Snyder, and J. VanDeMark for logistical assistance, among the many others who served as project interns and volunteers (see hippnet.hawaii.edu); G. Asner, S. Davies, T. Giambelluca, J. Mascaro, D. Metcalfe, J. Michaud, and J. Thompson for technical advice and/or comments on the manuscript

Author Contributions

Conceived and designed the experiments: RO FIN SC CPG LS. Performed the experiments: RO FIN SC CPG LS. Analyzed the data: RO FIN LS. Contributed reagents/materials/analysis tools: RO FIN SC CPG LS. Wrote the paper: RO FIN SC CPG LS.

References

1. Gentry AH (1988) Changes in plant community diversity and floristic composition on environmental and geographical gradients. Annals of the Missouri Botanical Garden 75: 1–34.
2. Givnish TJ (1999) On the causes of gradients in tropical tree diversity. Journal of Ecology 87: 193–210.
3. Phillips O, Gentry AH, Hall P, Sawyer S, Vasquez R (1994) Dynamics and species richness of tropical rain forests. Proceedings of the National Academy of Sciences 91: 2805–2809.
4. Mittelbach GG, Schemske DW, Cornell HV, Allen AP, Brown JM, et al. (2007) Evolution and the latitudinal diversity gradient: speciation, extinction and biogeography. Ecology Letters 10: 315–331.
5. Hawkins BA, Rodriguez MÁ, Weller SG (2011) Global angiosperm family richness revisited: linking ecology and evolution to climate. Journal of Biogeography 38: 1253–1266.
6. Condit R (1998) Tropical forest densus plots: Methods and results from Barro Colorado Island, Panama and a comparison with other plots. Berlin, Germany, and Georgetown, Texas, USA: Springer-Verlag and R. G. Landes Company. 211 p.

7. Rosenzweig ML (1995) Species diversity in space and time. Cambridge, UK: Cambridge University Press.
8. Givnish TJ (1999) On the causes of gradients in tropical tree diversity. Journal of Ecology 87: 193–210.
9. Hubbell SP (2001) The unified neutral theory of biodiversity and biogeography. Princeton, New Jersey, USA: Princeton University Press.
10. Zimmerman JK, Thompson J, Brokaw N (2008) Large tropical forest dynamics plots: testing explanations for the maintenance of species diversity. Chichester, UK: Wiley-Blackwell.
11. Hooper DU, Chapin FS, Ewel JJ, Hector A, Inchausti P, et al. (2005) Effects of biodiversity on ecosystem functioning: A consensus of current knowledge. Ecological Monographs 75: 3–35.
12. Chisholm RA, Muller-Landau HC, Abdul Rahman K, Bebber DP, Bin Y, et al. (2013) Scale-dependent relationships between tree species richness and ecosystem function in forests. Journal of Ecology 101: 1214–1224.
13. Torti SD, Coley PD, Kursar TA (2001) Causes and consequences of monodominance in tropical lowland forests. American Naturalist 157: 141–153.

14. Ghazoul J, Sheil D (2010) Tropical rain forest ecology, diversity, and conservation. Oxford: Oxford University Press.

15. Whittaker RJ, Triantis KA, Ladle RJ (2008) A general dynamic theory of oceanic island biogeography. Journal of Biogeography 35: 977–994.

16. Kier G, Kreft H, Lee TM, Jetz W, Ibisch PL, et al. (2009) A global assessment of endemism and species richness across island and mainland regions. Proceedings of the National Academy of Sciences 106: 9322–9327.

17. Terborgh J (1985) The vertical component of plant species diversity in temperate and tropical forests. The American Naturalist 126: 760–776.

18. Spies TA (1998) Forest structure: A key to the ecosystem. Northwest Science 72: 34–39.

19. Whitmore TC (1998) An introduction to tropical rain forests. Oxford, UK: Oxford Press.

20. Moles AT, Bonser SP, Poore AGB, Wallis IR, Foley WJ (2011) Assessing the evidence for latitudinal gradients in plant defence and herbivory. Functional Ecology 25: 380–388.

21. Price JP, Clague DA (2002) How old is the Hawaiian biota? Geology and phylogeny suggest recent divergence. Proceedings of the Royal Society of London, Series B Biological Sciences 269: 2429–2435.

22. Wagner WL, Herbst DR, Sohmer S (1999) Manual of the flowering plants of Hawaii. 2nd ed. Honolulu, HI: Bishop Museum.

23. Condit R (1995) Research in large, long-term tropical forest plots. Trends in Ecology and Evolution 10: 18–21.

24. Connell JH, Green PT (2000) Seedling dynamics over thirty-two years in a tropical rain forest. Ecology 81: 568–584.

25. Hobbie JE, Carpenter SR, Grimm NB, Gosz JR, Seastedt TR (2003) The US long term ecological research program. BioScience 53: 21–32.

26. Hubbell SP (2004) Two decades of research on the BCI Forest Dynamics Plot. In: Losos EC, Leigh EG Jr, editors. Tropical forest diversity and dynamism: Findings from a large-scale plot network. Chicago: University of Chicago Press. pp. 8–30.

27. Clinebell RR, Phillips OL, Gentry AH, Stark N, Zuuring H (1995) Prediction of neotropical tree and liana species richness from soil and climatic data. Biodiversity and Conservation 4: 56–90.

28. Malhi Y, Wood D, Baker TR, Wright J, Phillips OL, et al. (2006) The regional variation of aboveground live biomass in old-growth Amazonian forests. Global Change Biology 12: 1107–1138.

29. Saatchi SS, Houghton RA, Dos Santos Alvalá RC, Soares JV, Yu Y (2007) Distribution of aboveground live biomass in the Amazon basin. Global Change Biology 13: 816–837.

30. Slik JWF, Aiba S-I, Brearley FQ, Cannon CH, Forshed O, et al. (2010) Environmental correlates of tree biomass, basal area, wood specific gravity and stem density gradients in Borneo's tropical forests. Global Ecology and Biogeography 19: 50–60.

31. Toledo M, Poorter L, Peña-Claros M, Alarcón A, Balcázar J, et al. (2011) Climate and soil drive forest structure in Bolivian lowland forests. Journal of Tropical Ecology 27: 333–345.

32. Lugo AE, Scatena FN, Silver WL, Molina Colón S, Murphy PG (2002) Resilience of tropical wet and dry forests in Puerto Rico. In: Gunderson LH, Pritchard L Jr, editors. Resilience and the behavior of large-scale systems. Washington, DC: Island Press. pp. 195–225.

33. Bellingham PJ, Stewart GH, Allen RB (1999) Tree species richness and turnover throughout New Zealand forests. Journal of Vegetation Science 10: 825–832.

34. Sagar R, Singh JS (2006) Tree density, basal area and species diversity in a disturbed dry tropical forest of northern India: implications for conservation. Environmental Conservation 33: 256–262.

35. Stegen JC, Swenson NG, Enquist BJ, White EP, Phillips OL, et al. (2011) Variation in above-ground forest biomass across broad climatic gradients. Global Ecology and Biogeography: 744–754.

36. Price JP, Wagner WL (2004) Speciation in Hawaiian angiosperm lineages: cause, consequence, and mode. Evolution 58: 2185–2200.

37. Ziegler AC (2002) Hawaiian natural history, ecology, and evolution. Honolulu, HI: University of Hawaii Press.

38. Harrison RD (2011) Emptying the forest: Hunting and the extirpation of wildlife from tropical nature reserves. BioScience 61: 919–924.

39. Vitousek PM, Farrington H (1997) Nutrient limitation and soil development: experimental test of a biogeochemical theory. Biogeochemistry 37: 63–75.

40. Juvik SP, Juvik JO, Paradise TR (1998) Atlas of Hawaii, 3rd edition. Honolulu, HI: University of Hawaii Press.

41. Giambelluca TW, Chen Q, Frazier AG, Price JP, Chen Y-L, et al. (2013) Online Rainfall Atlas of Hawai'i. Bulletin of the American Meteorological Society 94: 313–316.

42. Crews TE, Kitayama K, Fownes JH, Riley RH, Herbert DA, et al. (1995) Changes in soil phosphorus fractions and ecosystem dynamics across a long chronosequence in Hawaii. Ecology 76: 1407–1424.

43. Tosi JAJ, Watson V, Bolaños R (2002) Life zone maps Hawaii, Guam, American Samoa, Northern Mariana Islands, Palau, and the Federated States of Micronesia. San Jose, Costa Rica and Hilo, Hawaii, USA: Tropical Science Center and the Institute of Pacific Islands Forestry, USDA Forest Service.

44. Kellner JR, Asner GP (2009) Convergent structural responses of tropical forests to diverse disturbance regimes. Ecology Letters 12: 1–11.

45. Mueller-Dombois D (2000) Rain forest establishment and succession in the Hawaiian Islands. Landsc Urban Plann 51: 147–157.

46. Friday JB, Scowcroft PG, Ares A (2008) Responses of native and invasive plant species to selective logging in an *Acacia koa-Metrosideros polymorpha* forest in Hawai'i. Applied Vegetation Science 11: 471–482.

47. Murphy M, Inman-Narahari F, Ostertag R, Litton CM (2014) Invasive feral pigs impact native tree ferns and woody seedlings in Hawaiian forest. Biological Invasions 16: 63–71.

48. Moore RB, Claque DA, Rubin M, Bohrson WA (1987) Hualalai volcano: A preliminary summary of geologic, petrologic, and geophysical data. In: Decker RW, Wright TL, Stauffer PH, editors. Volcanism in Hawaii. Washington, DC: U.S. Geological Service Professional Paper 1350, U'. Government Printing Office. pp. 571–585.

49. Giambelluca T, Chen Q, Frazier A, Price J, Chen Y-L, et al. (2011) The Rainfall Atlas of Hawai'i. Available: http://rainfall.geography.hawaii.edu. Accessed 2014 Aug 10.

50. Thaxton JM, Cole TC, Cordell S, Cabin RJ, Sandquist DR, et al. (2010) Native species regeneration following ungulate exclusion and nonnative grass removal in a remnant Hawaiian dry forest. Pacific Science 64: 533–544.

51. Freifelder RR, Vitousek PM, D'Antonio CM (1998) Microclimate change and effect on fire following forest-grass conversion in seasonally dry tropical woodland. Biotropica 30: 286–297.

52. Litton CM, Sandquist DR, Cordell S (2006) Effects of non-native grass invasion on aboveground carbon pools and tree population structure in a tropical dry forest of Hawaii. Forest Ecology and Management 231: 105–113.

53. Condit R (1998) Tropical forest census plots. Berlin: Springer-Verlag, Berlin, Germany.

54. Asner GP, Hughes RF, Mascaro J, Uowolo AL, Knapp DE, et al. (2011) High-resolution carbon mapping on the million-hectare island of Hawai'i. Frontiers in Ecology and the Environment 9: 434–439.

55. Chave J, Andalo C, Brown S, Cairns MA, Chambers JQ, et al. (2005) Tree allometry and improved estimation of carbon stocks and balance in tropical forests. Oecologia 145: 87–99.

56. Zanne AE, Lopez-Gonzalez G, Coomes DA, Ilic J, Jansen S, et al. (2009) Towards a worldwide wood economics spectrum. Dryad Digital Repository doi: 10.5061/dryad.5234.

57. Chave J, Coomes D, Jansen S, Lewis SL, Swenson NG, et al. (2009) Towards a worldwide wood economics spectrum. Ecology Letters 12: 351–366.

58. Slik JWF (2006) Estimating species-specific wood density from the genus average in Indonesian trees. Journal of Tropical Ecology 22: 481–482.

59. Chave J, Alonso D, Etienne RS (2006) Theoretical biology: Comparing models of species abundance. Nature 441: E1.

60. Gotelli NJ, Colwell RK (2001) Quantifying biodiversity: procedures and pitfalls in the measurement and comparison of species richness. Ecology Letters 4: 379–391.

61. Chiarucci A, Enright NJ, Perry GLW, Miller BP, Lamont BB (2003) Performance of nonparametric species richness estimators in a high diversity plant community. Diversity and Distributions 9: 283–295.

62. Colwell RK, Mao CX, Chang J (2004) Interpolating, extrapolating, and comparing incidence-based species accumulation curves Ecology 85: 2717–2727.

63. Magurran A (2004) Measuring biological diversity. Oxford, UK: Blackwell Publishing.

64. Sokal RR, Rohlf FJ (1994) Biometry: The principles and practice of statistics in biological research. 3rd edition. New York, New York: W.H. Freeman.

65. SAS Institute (2005) JMP introductory guide, release 6. Cary, NC: SAS Institute.

66. Asner GP, Hughes RF, Varga TA, Knapp DE, Kennedy-Bowdoin T (2009) Environmental and biotic controls over aboveground biomass throughout a tropical rain forest. Ecosystems 12: 261–278.

67. Brenes-Arguedas T, Roddy A, Coley P, Kursar T (2010) Do differences in understory light contribute to species distributions along a tropical rainfall gradient? Oecologia: 1–14.

68. Ashton PS (2004) Floristics and vegetation of the forest dynamics plots. In: Losos E, Leigh EG Jr, editors. Tropical forest diversity and dynamism: Findings from a large-scale plot network. Chicago, Illinois, USA: University of Chicago Press.

69. Davidar P, Puyravaud JP, Leigh EG Jr (2005) Changes in rain forest tree diversity, dominance and rarity across a seasonality gradient in the Western Ghats, India. Journal of Biogeography 32: 493–501.

70. Clark JS, McLachlan JS (2003) Stability of forest diversity. Nature 423: 635–638.

71. Clark DA, Piper SC, Keeling CD, Clark DB (2003) Tropical rain forest tree growth and atmospheric carbon dynamics linked to interannual temperature variation during 1984–2000. Proceedings of the National Academy of Sciences 100: 5852–5857.

72. Feeley KJ, Wright SJ, Supardi MNN, Kassim AR, Davies SJ (2007) Decelerating growth in tropical forest trees. Ecology Letters 10: 1–9.

73. Clark DB, Clark DA, Oberbauer SF (2010) Annual wood production in a tropical rain forest in NE Costa Rica linked to climatic variation but not to increasing CO_2. Global Change Biology 16: 747–759.

74. Stegen JC, Swenson NG, Enquist BJ, White EP, Phillips OL, et al. (2011) Variation in above-ground forest biomass across broad climatic gradients. Global Ecology and Biogeography 20: 744–754.

75. Martinez-Yrizar A (1995) Biomass distribution and primary productivity of tropical dry forests. In: Bullock SH, Mooney HA, Medina E, editors. Seasonally

Dry Tropical Forests. Cambridge, UK: Cambridge University Press. pp. 326–345.

76. Vitousek P, Asner GP, Chadwick OA, Hotchkiss S (2009) Landscape-level variation in forest structure and biogeochemistry across a substrate age gradient in Hawaii. Ecology 90: 3074–3086.

77. Cole RJ, Litton CM, Koontz MJ, Loh RK (2012) Vegetation recovery 16 years after feral pig removal from a wet Hawaiian forest. Biotropica 44: 463–471.

78. Inman-Narahari F, Ostertag R, Cordell S, Giardina CP, Nelson-Kaula K, et al. (2013) Seedling recruitment factors in low-diversity Hawaiian wet forest: towards global comparisons among tropical forests. Ecosphere 4: 24. http://dx.doi.org/10.1890/ES1812-00164.00161.

79. Busby PE, Vitousek PM, Dirzo R (2010) Prevalence of tree regeneration by sprouting and seeding along a rainfall gradient in Hawai'i. Biotropica 42: 80–86.

80. Kammesheidt L (1999) Forest recovery by root suckers and above-ground sprouts after slash-and-burn agriculture, fire and logging in Paraguay and Venezuela. Journal of Tropical Ecology 15: 143–157.

81. Bellingham PJ, Sparrow AD (2009) Multi-stemmed trees in montane rain forests: their frequency and demography in relation to elevation, soil nutrients, and disturbance. Journal of Ecology 97: 472–483.

82. Lehmann A, Leathwick JR, Overton JM (2002) Assessing New Zealand fern diversity from spatial predictions of species assemblages. Biodiversity and Conservation 11: 2217–2238.

83. Coomes DA, Allen RB, Bentley WA, Burrows LE, Canham CD, et al. (2005) The hare, the tortoise and the crocodile: the ecology of angiosperm dominance, conifer persistence and fern filtering. Journal of Ecology 93: 918–935.

84. Bellingham PJ, Richardson SJ (2006) Tree seedling growth and survival over 6 years across different microsites in a temperate rain forest. Canadian Journal of Forest Research 36: 910–918.

85. Tanner EVJ (1983) Leaf demography and growth of the tree-fern *Cyathea pubescens* Mett ex Kuhn in Jamaica Bot J Linn Soc 87: 213–227.

86. Bernabe N, Williams-Linera G, Palacios-Ríos M (1999) Tree ferns in the interior and at the edge of a Mexican cloud forest remnant: Spore germination and sporophyte survival and establishment. Biotropica 31: 83–88.

87. Arens NC (2001) Variation in performance of the tree fern *Cyathea caracasana* (Cyatheaceae) across a successional mosaic in an Andean cloud forest. American Journal of Botany 88: 545–551.

88. Williams-Linera G, Palacios-Ríos M, Hernández-Gómez R (2005) Fern richness, tree species surrogacy and fragments complementarity in a Mexican tropical montane cloud forest. Biodiversity and Conservation 14: 119–133.

89. Jones MM, Olivas Rojas P, Tuomisto H, Clark DB (2007) Environmental and neighbourhood effects on tree fern distributions in a neotropical lowland rain forest. Journal of Vegetation Science 18: 13–24.

90. Durand LZ, Goldstein G (2001) Growth, leaf characteristics, and spore production in native and invasive tree ferns in Hawaii. American Fern Journal 91: 25–35.

91. Amatangelo KL, Vitousek PM (2009) Contrasting predictors of fern versus angiosperm decomposition in a common garden. Biotropica 41: 154–161.

92. Mack MC, D'Antonio CM, Ley RE (2001) Alteration of ecosystem nitrogen dynamics by exotic plants: A case study of C_4 grasses in Hawaii. Ecological Applications 11: 1323–1335.

93. Huenneke LF, Vitousek PM (1989) Seedling and clonal recruitment of the invasive tree, *Psidium cattleianum*: implications for management of native Hawaiian forests. Biological Conservation 53: 199–211.

94. Hughes RF, Denslow JS (2005) Invasion by an N2-fixing tree, *Falcataria moluccana*, alters function, composition, and structure of wet lowland forests of Hawai'i. Ecological Applications 15: 1615–1628.

95. Gravel D, Canham CD, Beaudet M, Messier C (2006) Reconciling niche and neutrality: The continuum hypothesis. Ecology Letters 9: 399–409.

96. Hubbell SP, Foster RB (1986) Commonness and rarity in a neotropical forest: Implications for tropical tree conservation. In: Soule ME, editor. Conservation biology: The science of scarcity and diversity: Sunderland, Mass, USA Illus Paper Maps. Sunderland, Massachussetts: Sinauer Associates, Inc. pp. 205–231.

97. Lach L (2003) Invasive ants: Unwanted partners in ant-plant interactions? Annals of the Missouri Botanical Garden 90: 91–108.

98. Nogueira SSdC, Nogueira-Filho SLG, Bassford M, Silvius K, Fragoso JMV (2007) Feral pigs in Hawai'i: Using behavior and ecology to refine control techniques. Appl Anim Behav Sci 108: 1–11.

99. Nogueira-Filho S, Nogueira S, Fragoso J (2009) Ecological impacts of feral pigs in the Hawaiian Islands. Biodiversity and Conservation 18: 3677–3683.

100. Shiels A, Drake D (2010) Are introduced rats (*Rattus rattus*) both seed predators and dispersers in Hawaii? Biological Invasions: 1–12.

101. Shiels AB (2011) Frugivory by introduced black rats (*Rattus rattus*) promotes dispersal of invasive plant seeds. Biological Invasions 13: 781–792.

102. Gilbert B, Lechowicz MJ (2004) Neutrality, niches, and dispersal in a temperate forest understory. Proceedings of the National Academy of Sciences of the United States of America 101: 7651–7656.

103. Hubbell SP (2005) Neutral theory in community ecology and the hypothesis of functional equivalence. Functional Ecology 19: 166–172.

104. Losos E, Leigh EG Jr, editors (2004) Tropical forest diversity and dynamism: Findings from a large-scale plot network. Chicago, Illinois: University of Chicago Press.

17

Unimodal Latitudinal Pattern of Land-Snail Species Richness across Northern Eurasian Lowlands

Michal Horsák*, Milan Chytrý

Department of Botany and Zoology, Masaryk University, Brno, Czech Republic

Abstract

Large-scale patterns of species richness and their causes are still poorly understood for most terrestrial invertebrates, although invertebrates can add important insights into the mechanisms that generate regional and global biodiversity patterns. Here we explore the general plausibility of the climate-based "water-energy dynamics" hypothesis using the latitudinal pattern of land-snail species richness across extensive topographically homogeneous lowlands of northern Eurasia. We established a 1480-km long latitudinal transect across the Western Siberian Plain (Russia) from the Russia-Kazakhstan border (54.5°N) to the Arctic Ocean (67.5°N), crossing eight latitudinal vegetation zones: steppe, forest-steppe, subtaiga, southern, middle and northern taiga, forest-tundra, and tundra. We sampled snails in forests and open habitats each half-degree of latitude and used generalized linear models to relate snail species richness to climatic variables and soil calcium content measured in situ. Contrary to the classical prediction of latitudinal biodiversity decrease, we found a striking unimodal pattern of snail species richness peaking in the subtaiga and southern-taiga zones between 57 and 59°N. The main south-to-north interchange of the two principal diversity constraints, i.e. drought stress vs. cold stress, explained most of the variance in the latitudinal diversity pattern. Water balance, calculated as annual precipitation minus potential evapotranspiration, was a single variable that could explain 81.7% of the variance in species richness. Our data suggest that the "water-energy dynamics" hypothesis can apply not only at the global scale but also at subcontinental scales of higher latitudes, as water availability was found to be the primary limiting factor also in this extratropical region with summer-warm and dry climate. A narrow zone with a sharp south-to-north switch in the two main diversity constraints seems to constitute the dominant and general pattern of terrestrial diversity across a large part of northern Eurasia, resulting in a subcontinental diversity hotspot of various taxa in this zone.

Editor: Maura (Gee) Geraldine Chapman, University of Sydney, Australia

Funding: This study was funded by the Czech Science Foundation (P504/11/0454). The funders oversaw the collection of data and preparation of the manuscript, but had no role in the study design, data analysis, or decision to publish.

Competing Interests: The authors have declared that no competing interests exist.

* Email: horsak@sci.muni.cz

Introduction

The overall decline in species diversity with increasing latitude is one of the most prominent features of the natural world (e.g. [1], [2], [3], [4]), but the causes of this gradient remain insufficiently explained in spite of many hypotheses proposed and extensive discussions (e.g. [5], [6], [7]). Previous research was biased to some taxa and regions, and little evidence still exists for terrestrial invertebrates (e.g. [2], [8]). While recent studies stress the greater importance of climate than geographical location (e.g. [9], [10]), the mechanisms by which latitudinal variation in climate determines species numbers are still poorly understood ([7], [11]). As latitude correlates with a number of interacting and inter-correlated environmental gradients (e.g. temperature, precipitation, seasonality, evapotranspiration), direct tests of the hypotheses are difficult and can be controversial ([2]). Although most studies have recognized the "classical" pattern of decreasing species richness towards the poles, both positive and unimodal relationships were repeatedly revealed in several taxa (see [2]). These exceptions were found to be scale dependent. Almost all positive relationships were found across small latitudinal extents (<20° latitude), being associated with regional patterns of heterogeneity in local topography, geology, hydrology, or historical factors ([2]). In contrast, unimodal relationships with a mid-latitudinal peak of diversity were all but one found across broad extents (>20° latitude). The examples came mostly from studies of terrestrial insects (e.g. [12], [13]), but there is also evidence for aquatic invertebrates, herbs, marine and terrestrial birds, and mammals (e.g. [11], [14]). Several explanations were suggested, e.g. mid-latitudinal peak in host density ([15]), mid-domain effect ([13]), habitat specificity ([12]), topographical heterogeneity ([16]) or an increase in resources ([14]). Although the mechanism underlying such unimodal patterns remain poorly known, hardly any doubt exists that climate influences large-scale patterns of species richness. The climatically based "energy hypothesis" has been postulated in several versions and received a considerable attention over the last three decades (e.g. [11], [17], [18]).

In this study we focus on land snails, an invertebrate taxon with as yet poorly explored patterns of latitudinal diversity. Land snail diversity across large scales can be climatically controlled by two principal ecological factors, winter temperature and moisture. Many land snail species apparently do not have any cryoprotective chemicals ([19]), which results in poorly evolved cold-hardiness within this taxon ([20]). The original "freezing tolerance" hypothesis of von Humboldt ([21]) predicts that the number of species is reduced at higher latitudes by the inability of many organisms to withstand low winter temperature ([11]). A commonly recognized decline in land snail species richness towards colder climate at higher elevations ([22], [23]) or in low-productive and environmentally harsh systems with a lack of shelters for overwintering ([8], [24]) suggest that exposure to winter frosts can be a dominant factor shaping latitudinal pattern of land snail diversity. However, there is also a limited number of drought-adapted snail species (e.g. [25]), therefore the number of species generally decreases as conditions become drier (e.g. [26]). Independently of winter temperature, water availability may thus represent the second critical factor, especially in drier areas. While temperature generally decreases with latitude, water availability depends on precipitation, which is independent of latitude, being determined by various systems of atmospheric circulation. Liquid-water availability is also determined by the interaction between temperature and precipitation, because in warmer areas more water is lost due to evapotranspiration. Recent global studies of woody plants (e.g. [27], [28]) indicate that most geographical variation in species richness can be attributed to liquid water-energy dynamics, caused by water doing work ([29]). This mechanism can explain co-variation between climate and richness over space and time ([30], [31]).

In this study we ask (1) what is the latitudinal pattern of land-snail diversity across northern Eurasian lowlands, (2) whether it is determined by low-temperature and drought stress, (3) if so, which of these factors is more important, or whether they interact? To answer these questions, we established a latitudinal transect across Western Siberia spanning 13° from the steppe zone through the forest (taiga) zone to tundra. Western Siberia is the most suitable area for such a study in northern Eurasia, because it is flat lowland with very limited topographic and geological heterogeneity and small human impact, factors which might confound the effects of latitude and macroclimate on species richness in mountainous or densely populated regions. Based on the expected limitation of snail species richness by low temperatures and water deficit, we expect low richness in both dry southern areas of the steppe zone and cold northern areas of the tundra zone. Therefore we hypothesize a unimodal species richness pattern with a peak in the forest zone.

Materials and Methods

Study area and sites

We studied land snail species richness across a latitudinal transect from the northern steppe zone at the Russia-Kazakhstan border SW of the city of Omsk at 54.5°N to the southern tundra zone at the Arctic Ocean coast near the town of Tazovskii at 67.5°N (Fig. 1). The straight-line distance between the southern-most and northernmost site was 1480 km. This transect crossed seven latitudinal vegetation zones ([32]): (**1**) **steppe zone**, dominated by grasses such as *Stipa* and *Festuca*, with rare occurrence of small woodland and shrubland patches in terrain depressions; (**2**) **forest-steppe zone**, consisting of a mosaic of dry to mesic grasslands at the flatland and small open woodlands dominated by birches (*Betula pendula* and *B. pubescens*) and aspen (*Populus tremula*) with many temperate light-demanding species in the herb layer, occurring in shallow terrain depressions; (**3**) **subtaiga zone**, dominated by forests with birches and aspen, with locally admixed fir (*Abies sibirica*) and lime (*Tilia cordata*), and Scots pine (*Pinus sylvestris*) on sandy soils, with a herb layer

composed mainly of temperate forest herbs; wet and mesic grasslands occur in scattered patches in a predominantly forested landscape; (**4**) **southern taiga zone**, with the same trees as in the subtaiga zone, but with an additional occurrence of spruce (*Picea obovata*) and Siberian pine (*Pinus sibirica*); grasslands disappear from the landscape, and patches of deep bogs with open low growing stands of Scots pine appear, (**5**) **middle taiga zone**, consisting of a mosaic of coniferous forests with Siberian pine and spruce (on loamy soils) or Scots pine (on sandy soils), frequent occurrence of birch (*Betula pubescens*) and extensive deep bogs; the herb layer of forests is dominated by boreal dwarf shrubs and herbs, bryophytes and lichens, (**6**) **northern taiga zone**, a mosaic of forests with Siberian pine, Scots pine, larch (*Larix sibirica*), spruce and birch (*Betula pubescens*) on permafrost, and shallow bogs including palsas, (**7**) **forest-tundra zone**, semi-open landscape with patches of larch woodlands, extensive stands of dwarf birch (*Betula nana*), grasslands, mires and willow scrub at wet sites; (**8**) **tundra zone**, a treeless landscape with herbaceous and dwarf shrub vegetation, dwarf birch stands, mires and willow scrub.

The whole transect run across the flatland of the West Siberian Plain, ranging in altitude from 148 m (in the south) to 4 m a.s.l. (in the north). Soils were loamy from the steppe zone to the south of the middle taiga zone, and also in the forest-tundra and tundra zones. In contrast, across most of the middle and northern taiga zones soils were predominantly sandy. According to the World-Clim dataset ([33]), mean January temperature ranged from −17°C (in the south) to −26°C (in the north) and mean July temperature from 20°C to 13°C. Annual precipitation ranged from 360 mm (in the south) to 570 mm (in the middle taiga zone).

Sampling sites (n = 29) were placed systematically along the transect, with one site each half degree of latitude. Only in the northernmost part, four sites were spaced by a quarter degree in order to capture the rather narrow zone of forest-tundra. Disturbed sites such as arable land, areas affected by oil or gas extraction or roadsides were avoided. Large river floodplains were avoided too in order to focus on zonal habitats. In some cases, it was necessary to shift the sampling site slightly to the north or south from the target latitude to avoid disturbed sites, floodplains or to sample both forest and open habitats at the same site. However, these shifts were never larger than 5 minutes of latitude.

Species sampling

At each site we randomly chose three squared sampling plots of 100 m² representing locally predominant habitat types at the site. Because at most sites overall habitat heterogeneity was very low, represented by few homogeneous habitats of large spatial extent, the three plots were sufficient to cover the main habitat heterogeneity. At some sites we had enough time to sample one or two additional plots, which resulted in four and five plots sampled at seven and two sites, respectively (Table 1). This gave a total of 98 plots investigated. In all plots snails were sampled by a single person (M. Horsák). To record as complete inventory of snail assemblage of each plot as possible, various sampling techniques were applied depending on the habitat type investigated. At non-wetland habitats, snails were carefully searched by eye in all microhabitats of the plot for one hour. At species-poor sites (e.g. dry steppes on loess) the sampling was stopped when all recorded species were represented by more than two individuals and at the same time no other species was found for ca. 15 minutes (but sampling was never shorter than 30 minutes). The same stopping rules were used in richer habitats, where sampling took longer time, but not longer than another 30 minutes (in cases of the richest forest plots in the southern taiga). In damper habitat

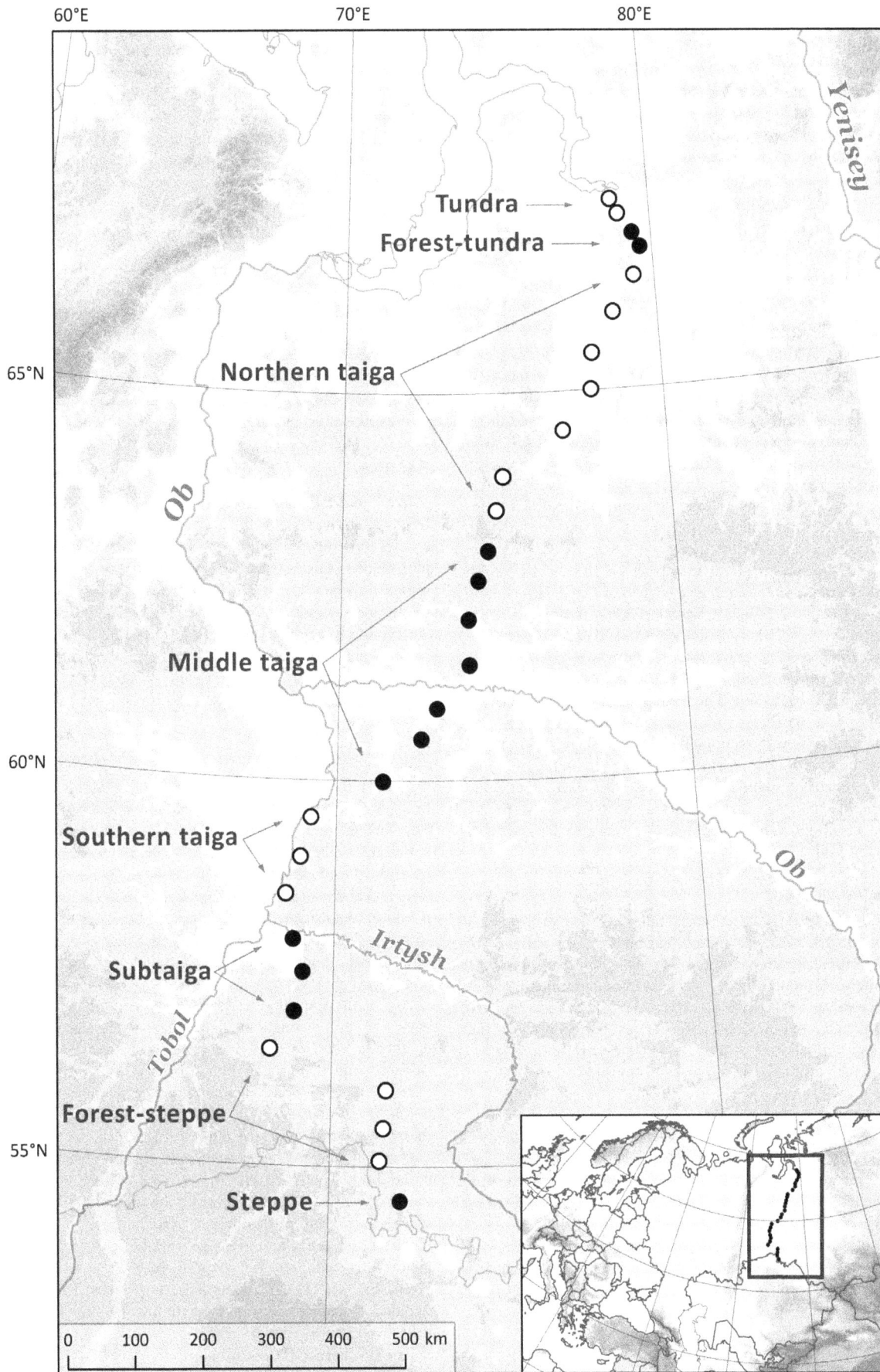

Figure 1. Study area with the position of 29 sampling sites along the transect.

types, i.e. mires, fens and wet tundra, a 12 l sample of the top layer of the ground surface including topsoil, litter, bryophytes and herbaceous vegetation was collected and processed in the field using the wet sieving method ([34]). All recorded shells and live shelled snails were collected and kept dry. Few recorded individuals of slugs were preserved in 70% ethanol. Samples were identified under a dissecting microscope in the laboratory using available identification literature ([35], [36], [37], [38]) and the first author's comparative collection of Siberian samples. Nomenclature mainly follows [38], but [35] and [37] were used for species of non-European distribution. We declare that no permission was needed as the samples were not collected on privately owned or protected lands and no legally protected species were sampled.

Explanatory variables

For each site, annual precipitation sum and mean January and July temperature were obtained from the WorldClim database ([33], www.worldclim.org) using the ArcGIS 8.3 program (www.esri.com). Annual potential evapotranspiration, calculated from the WorldClim data, was obtained from the website of CGIAR-CSI Consortium for Spatial Information (http://csi.cgiar.org/Aridity/, [39]). We also expressed water balance as the difference between annual precipitation sum and annual potential evapotranspiration ([40]). As land snail species richness and composition is tightly related to calcium availability (e.g. [41]), we also measured calcium (Ca) content in topsoil (or in peat in mires). Although the amount of calcium is strongly correlated with climate due to higher cation leaching in wetter conditions, any heterogeneity in local geology and vegetation cover ([42]) can have positive effects on snail species richness. Therefore we collected soil samples from the mineral topsoil horizon at a depth of 5–10 cm in four places within each plot. These four subsamples were mixed and sieved at a mesh size of 1 mm. Available calcium was extracted from the sieved soil using the Mehlich III method (strong acid extraction with ion complex) and determined by atomic absorption spectrophotometry (AAS 933 Plus, GBC Scientific Equipment, Melbourne, Australia). The analysis was done by AgroLab, Troubsko, Czech Republic according to the methods described by [43].

Statistical analyses

All basic data used in the analyses and the graphs are provided in Table S1. Correlations among explanatory variables were inspected graphically (Figs. S1 and S2) and quantified using the Spearman correlation coefficient. As tight correlations were found among mean annual temperature, mean July and January temperature (Fig. S1), we used only January temperature as the most ecologically relevant measure of temperature for land snails due to a poorly evolved cold-hardiness within this taxon ([20]).

For each of 29 sampling sites we calculated mean number of species recorded at three to five sampling plots. As the mean number of species at a site tightly correlated with the total number of species recorded at all plots of a site ($r_S = 0.98$, P<0.001), we used only mean numbers (for correlations see Fig. S2). As the majority of temperate and boreal land snail diversity is confined to forest environments ([36], [37]), we also counted total and mean numbers of species recorded at forest and open-habitat plots separately. Relationships between the numbers of species recorded at each of 29 sites and explanatory variables were explored only graphically as there were no systematic trends except for water balance. Therefore, we modelled only the response of mean numbers of species recorded at each site to water balance using a generalized linear model with a Poisson error structure (GLM-p). To correct for under- or over-dispersion, we used a quasi-GLM-p

with the variance given by $\phi \times \mu$, where ϕ is the dispersion parameter and μ is the mean. Both linear and quadratic terms were included into the model and their significance was tested using the F-test. The final model was inspected using the distribution of residuals and standardized residuals against predicted values, and Cook's distances ([44]). The same procedures were used to construct the most parsimonious model and to find a set of uncorrelated significant predictors. The minimal adequate model, starting with a full model that included both linear and quadratic terms of all predictors (except water balance that combines precipitation and evapotranspiration) and all two-way interactions, was established using a stepwise deletion procedure based on F-tests. To visualize changes of diversity with the most important climatic and soil variables, the changes of these variables with latitude were expressed using the locally-weighted polynomial regression. The obtained regression lines were overlaid with the diversity changes in a single plot. The regression was calculated using the function "*lowess*" ([45]). All graphics and calculations were performed using the R program ([46]).

Results

We found a total of 33 land snail species (Table 1) represented by 2,788 individuals in 98 studied plots. Cumulative numbers of species recorded per site varied from 0 to 16 and the average per plot ranged between 0 and 11 (Fig. S2). Forest plots were notably richer in species than open plots at almost all sites except for those in the forest-tundra zone (Fig. 2A). Median numbers of species were 4 and 0 (mean 4.1 and 1.6) for forest and open-habitat plots, respectively. However, the difference in the cumulative lists of species recorded in these two contrasting habitat types was not high, as 28 species were found in 54 forest plots and 22 species in 44 open plots.

We found a pronounced unimodal pattern of species richness along the latitudinal gradient, peaking between 56.5° and 59.0°N (Fig. 2). This pattern was detected for both cumulative and mean number of species as well as for both forest and open habitats (Fig. S2). Taking forest plots separately, the peak was in the subtaiga and southern taiga zones (Fig. 2A). In the southern part of the transect, the number of species increased linearly with precipitation, especially in the steppe and forest-steppe zones, reaching the maximal mean value at the transition between the subtaiga and southern taiga zones (Fig. 3B). After that species richness declined, reflecting the appearance of acidic mires with no snails recorded (Fig. 2A). For forest species, a linear decrease of species richness started at the transition between southern taiga and middle taiga and continued throughout the latter zone (Fig. 2A). The number of species was linearly decreasing with temperature and soil calcium content (Fig. 2B). Harsh conditions, strengthened by the presence of sandy soils, kept the number of species very low from the northern margin of the middle taiga zone across the whole of the northern taiga zone. At the transition to the forest-tundra zone and within this zone an increase in the number of species was found, likely associated with the increase in calcium content in the soils (Figs. 2B and S2). However, in the tundra zone the number of species linearly dropped down to zero.

Water balance was the single variable that expressed a tight relationship with species richness along the whole latitudinal extent (Fig. 3D). Using a generalized linear model we found highly significant quadratic response of the mean number of species to water balance. This single predictor was able to explain 81.7% of the total variance in mean numbers of species (quasi-GLM-p, p<<0.001). The most parsimonious model (with water balance not included, see Methods) consisted of annual precipitation (linear

Table 1. List of all land-snail species recorded in 98 plots at 29 sites on a latitudinal transect across Western Siberia.

Species/Latitude (°N)	54.5	55.0	55.5	56.0	56.5	57.0	57.5	58.0	58.5	59.0	59.5	60.0	60.5	61.0	61.5	62.0	62.5	63.0	63.5	64.0	64.5	65.0	65.5	66.0	66.5	66.75	67.0	67.25	67.5
Vallonia pulchella (Müller)	+	+	+	+	+	.	.	.	.	.	.	.	.	.	.	.	.	.	.	.	.	.	.	.	.	.	.	.	.
Vallonia costata (Müller)	+	+	+	+	+	+	+	+	.	.	.	.	.	.	.	.	.	.	.	.	.	.	.	.	.	.	.	.	.
Cochlicopa lubricella (Porro)	+	+	+	+	+	+	+	+	+	+	.	.	.	.	.	.	.	.	.	.	.	.	.	.	.	.	.	.	.
Nesovitrea hammonis (Ström)	+	.	+	.	+	+	+	+	+	+	+	+	+	+	+	.	.	.	+	+	+	.	.	.	+	+	+	.	.
Succinella oblonga (Draparnaud)	.	+	.	.	+	.	.	.	.	.	.	.	.	.	.	.	.	.	.	.	.	.	.	.	.	.	.	.	.
Deroceras altaicum (Simroth)	.	+	+	.	+	.	.	.	.	.	.	.	.	.	.	.	.	.	.	.	.	.	.	.	.	.	.	.	.
Zonitoides nitidus (Müller)	.	+	+	.	.	+	.	.	.	.	.	+	.	.	.	.	.	.	.	.	.	.	.	.	.	.	.	.	.
Punctum pygmaeum (Draparnaud)	.	+	+	+	+	+	+	+	+	+	+	.	.	.	.	.	.	.	.	.	.	.	.	.	.	.	.	.	.
Euconulus fulvus (Müller)	.	+	+	+	.	+	+	+	+	+	+	+	.	+	+	+	+	.	+	.	+	+	+	+	+	+	+	+	.
Vertigo antivertigo (Draparnaud)	.	.	+	.	.	+	.	.	.	.	.	.	.	.	.	.	.	.	.	.	.	.	.	.	.	.	.	.	.
Vertigo pygmaea (Draparnaud)	.	.	+	+	+	+	+	.	.	.	.	.	.	.	.	.	.	.	.	.	.	.	.	.	.	.	.	.	.
Euconulus praticola (Reinhardt)	.	.	+	.	.	+	.	.	.	.	.	+	.	.	.	.	.	+	.	.	.	.	.	.	.	.	.	.	.
Vitrina pellucida (Müller)	.	.	+	+	+	+	+	+	+	.	.	.	.	.	.	.	.	.	.	.	.	.	.	.	.	.	.	.	.
Fruticicola schrenckii (Middendorff)	.	.	.	+	+	+	+	.	+	.	.	.	+	.	.	.	.	.	.	.	.	.	.	.	.	.	.	.	.
Discus ruderatus (Hartmann)	.	.	.	+	.	.	+	.	+	+	+	.	+	.	.	.	.	.	+	.	.	.	.	.	.	.	.	.	.
Carychium minimum Müller	.	.	.	.	.	+	.	.	.	.	.	.	.	.	.	.	.	.	.	.	.	.	.	.	.	.	.	.	.
Oxyloma elegans (Risso)	.	.	.	.	.	+	.	.	+	+	+	.	.	.	.	.	.	.	.	.	.	.	.	.	.	.	.	.	.
Vertigo angustior Jeffreys	.	.	.	.	.	+	.	.	.	.	.	.	.	.	.	.	.	.	.	.	.	.	.	.	.	.	.	.	.
Vertigo substriata (Jeffreys)	.	.	.	.	.	+	.	.	+	+	+	.	.	.	.	.	.	.	.	.	.	.	.	.	.	.	.	.	.
Columella edentula (Draparnaud)	.	.	.	.	.	+	+	.	+	+	.	+	+	.	.	.	.	.	.	.	.	.	.	.	.	.	.	.	.
Vertigo pusilla Müller	.	.	.	.	.	.	.	.	+	.	.	.	.	.	.	.	.	.	.	.	.	.	.	.	.	.	.	.	.

Table 1. Cont.

Species/Latitude (°N)	54.5	55.0	55.5	56.0	56.5	57.0	57.5	58.0	58.5	59.0	59.5	60.0	60.5	61.0	61.5	62.0	62.5	63.0	63.5	64.0	64.5	65.0	65.5	66.0	66.5	66.75	67.0	67.25	67.5
Cochlicopa lubrica (Müller)	.	.	.	.	.	.	+	.	+	.	.	.	.	.	.	.	.	.	.	.	.	.	.	.	.	.	.	.	.
Succinea putris (Linné)	.	.	.	.	.	.	+	.	.	.	.	+	.	.	.	.	.	.	.	.	.	.	.	.	+	.	.	.	.
Nesovitrea petronella (Pfeiffer)	.	.	.	.	.	.	+	+	+	.	+	+	+	+	.	.	.	+	+	.	.	.	.	.	.	.	.	.	.
Vertigo ronnebyensis (Westerlund)	.	.	.	.	.	+	+	.	+	+	+	+	.	+	.	.	.	+	+	.	.	.	.	.	+	.	.	.	.
Arion fuscus (Müller)	.	.	.	.	.	.	.	+	.	.	.	.	.	.	.	.	.	.	.	.	.	.	.	.	.	.	.	.	.
Vertigo aff. gouldi Binney	.	.	.	.	.	.	.	+	.	.	.	.	.	.	.	.	.	.	.	.	.	.	.	.	.	.	.	.	.
Vertigo modesta aff. hoppii (Möller)	.	.	.	.	.	.	.	.	.	.	+	.	.	.	.	.	.	.	.	.	.	.	.	.	.	+	.	+	.
Deroceras laeve (Müller)	.	.	.	.	.	.	.	.	.	.	+	.	+	.	.	.	.	.	.	.	.	.	.	.	.	.	.	.	.
Zoogenetes harpa (Say)	.	.	.	.	.	.	.	.	.	.	.	.	.	+	.	.	.	.	.	.	.	.	.	+	.	.	.	.	.
Vertigo lilljeborgi (Westerlund)	.	.	.	.	.	.	.	.	.	.	.	.	.	.	.	.	.	.	.	.	.	.	.	.	+	.	.	.	.
*Vertigo extima (Westerlund)	.	.	.	.	.	.	.	.	.	.	.	.	.	.	.	.	.	.	.	.	.	.	.	.	.	(+)	.	.	.
Total no. of plots	**5**	**4**	**3**	**4**	**4**	**3**	**3**	**3**	**3**	**4**	**3**	**4**	**3**	**3**	**4**	**3**	**4**	**3**	**5**	**3**	**4**	**3**	**3**	**3**	**3**	**3**	**3**	**3**	**3**
Total no. of species	**4**	**8**	**11**	**7**	**16**	**15**	**15**	**9**	**10**	**8**	**8**	**9**	**5**	**5**	**3**	**1**	**2**	**0**	**5**	**2**	**1**	**1**	**6**	**3**	**2**	**3**	**2**	**2**	**0**

*this species was recorded at the site but not in the studied plots.

Individual zones (from left to right): steppe, forest-steppe, subtaiga, southern taiga, middle taiga, northern taiga, forest-tundra, and tundra, indicated using alternating bold style. Presence of a species is marked by crosses; species are ordered based on their first finding along the transect from south to north.

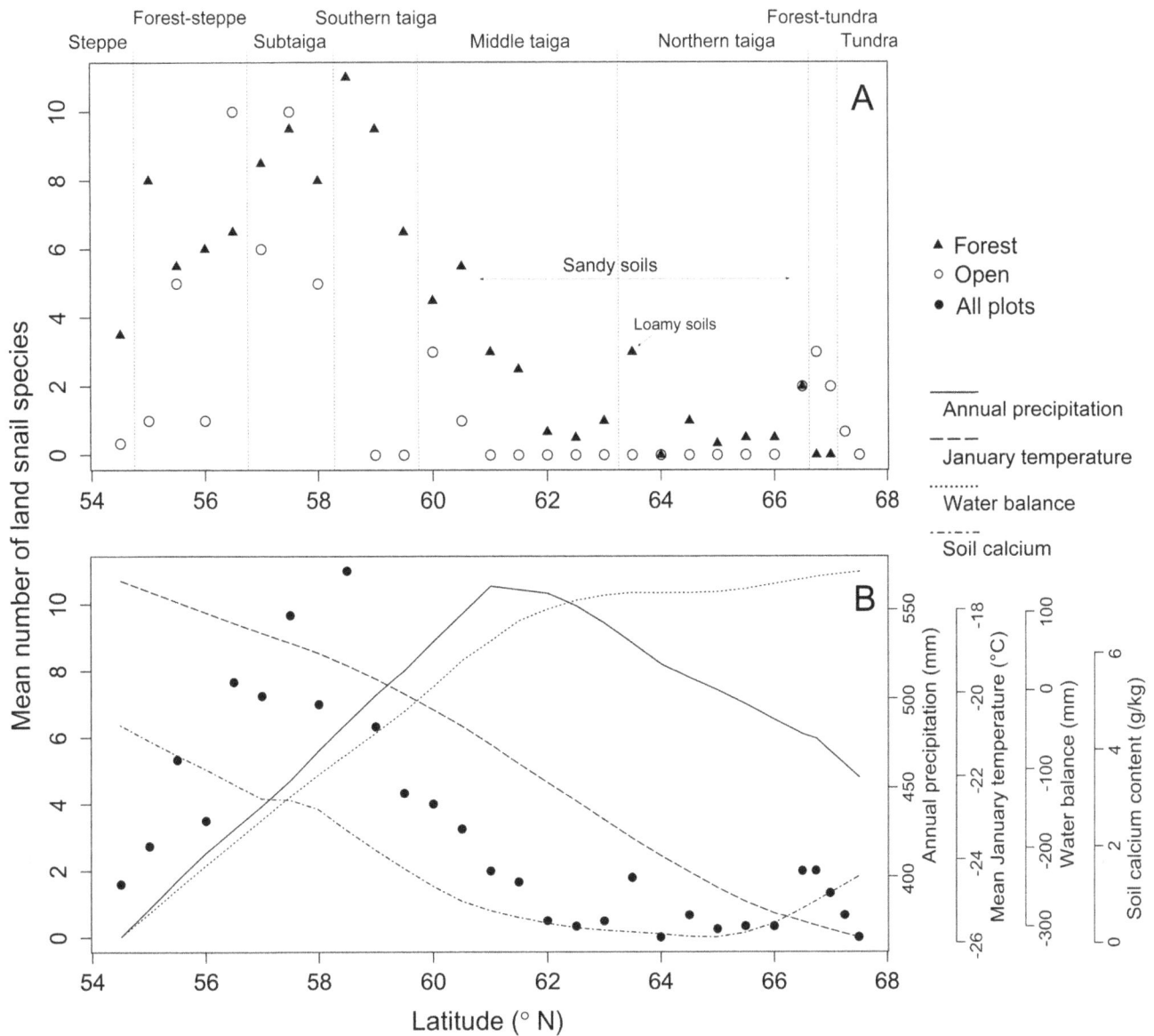

Figure 2. Mean numbers of land snail species recorded in 100 m² plots at 29 sites located along a latitudinal transect in the West Siberian Plain between 54.5° and 67.5°N. A, species from forest and open habitats counted separately; **B**, means from all plots at each site shown with patterns of four environmental variables visualized using the locally-weighted polynomial regression.

and quadratic term), mean January temperature (linear term), soil calcium content (linear term) and interaction between the quadratic term of annual precipitation and soil calcium. The model explained 87.3% of the total variance in species richness (quasi-GLM-p, p≪0.001).

Discussion

Our data suggest a climatically controlled unimodal pattern of latitudinal land snail diversity across Western Siberia. Previous reports of unimodal diversity pattern along latitude originated from larger spatial extents ([12], [13], [16]) and were explained by various mechanisms, although climatically related. The pattern observed for land snails in Western Siberia was controlled by two main climatic constraints that were, however, operating at different latitudes: drought stress in the south and cold stress in

the north. Hawkins et al. [11] analysed global diversity of terrestrial birds and found that linear association of climate with species richness at continental extents is rather exceptional. The data from the former USSR reported in their study (see Figure six in [11]) suggest the same unimodal pattern as we found for land snail diversity, with a peak in the subtaiga and southern taiga zones. A similar pattern was also shown for amphibians (Figure one in [47]), which seems to be more prominent towards the western Palaearctic. As we observed the same pattern along the studied transect also for vascular plant diversity (unpubl. data), which partly coincides with the coarse-scale map of plant species numbers per 1000 km² ([48]), the climatically driven latitudinal change in diversity seems to be universal for this part of Eurasia. However, the mechanisms shaping these unimodal diversity changes can differ among individual taxa. With a high certainty

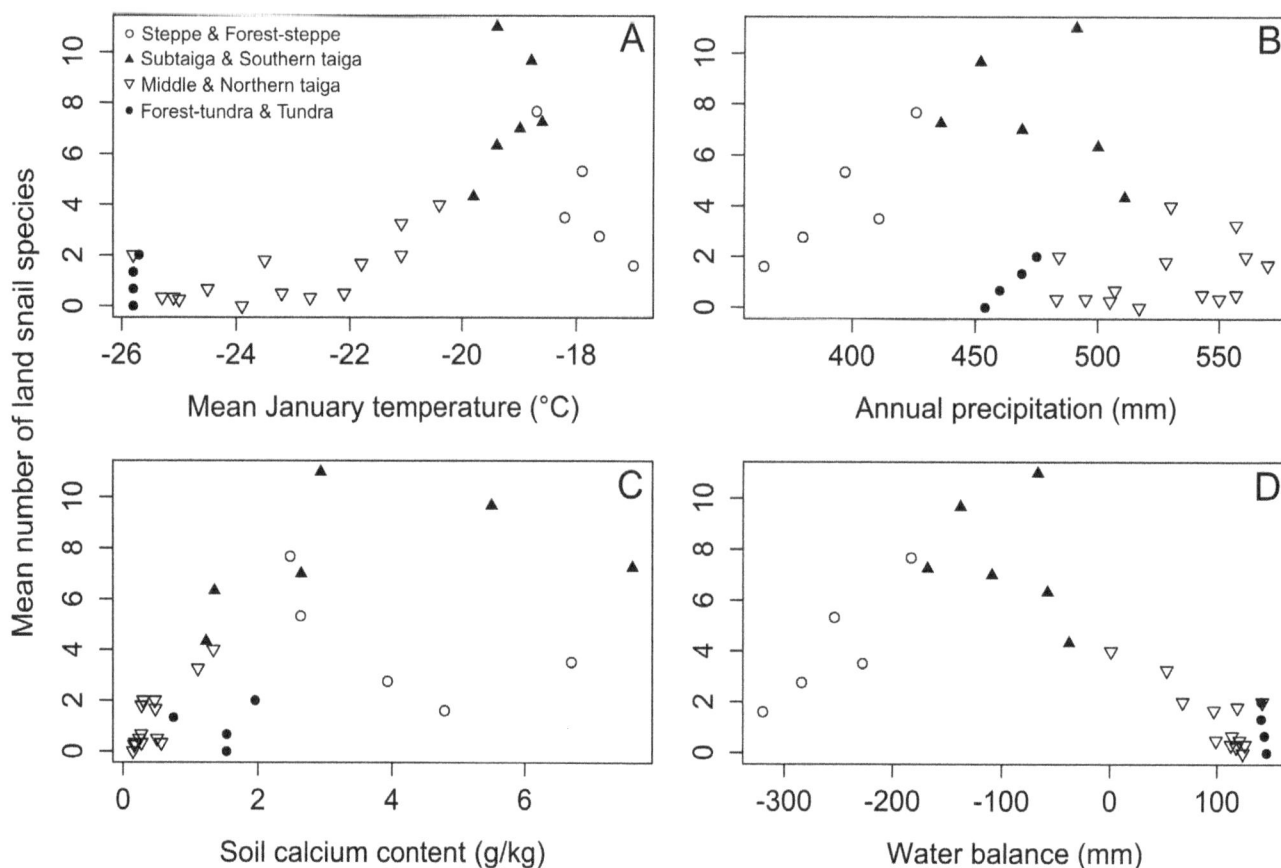

Figure 3. Changes in mean numbers of land snail species in relation to four environmental variables.

we can reject Rosenzweig's ([49]) "habitat heterogeneity" hypothesis, likely explaining a unimodal latitudinal pattern of plant diversity in continental Chile ([16]), as the observed changes in the number of land snail species in Western Siberia occurred on flatland, in absence of any significant topographical variance along the transect (elevation of studied plots along the whole transect varied between 148 and 4 m a.s.l.).

There was no clear relation between the heterogeneity of habitat types in the landscape and local species richness at particular sites, either mean or cumulative. We found southern taiga zone to be the richest in species, although the landscape was dominated by a single habitat type, mixed broadleaf-conifer hemiboreal forest. Among forest assemblages, compositional pattern was nested, with species accumulating towards the southern taiga zone from both latitudinal directions. However, the diversity pattern was notably different between forest and open habitats. Open habitat types changed more dramatically than forest types with latitude, continuously replacing one another along the transect, from dry steppe through mesic and wet grasslands and bogs to herbaceous and shrubby tundra. This constituted a considerable compositional turnover with no shared snail species for the steppe and tundra zones. This difference also explains why the cumulative numbers of species recorded in all forest and open plots were not much different, despite often high differences in the numbers for individual plots.

The observed changes in species richness can possibly result from different colonization histories along the transect. However, there are no obvious geographical barriers for land snails that

could hamper species migration as almost all recorded snails belong to minute species known as very effective passive dispersers (e.g. [50], [51]). Further, the whole area was ice-free during the last glacial maximum ([52]), so there was enough time for species to saturate all favourable habitats. Further, almost all recorded species have wide distributions (Holarctic, Palaearctic, Siberian) with ranges extending across the temperate and boreal zones of the continent. Thus, there seems to be no significant imprint of the postglacial colonization history in this part of Eurasia, as known from western and central Europe ([51], [53]).

We observed a sharp drop of species numbers with decreasing water availability in areas of high ambient energy that supports the "water-energy dynamics" hypothesis developed for plant diversity ([27], [29]). As the major predictions of this hypothesis were fulfilled also by the global diversity pattern of terrestrial birds, Hawkins et al. [11] suggested this hypothesis as a unifying hypothesis for terrestrial diversity gradients. It postulates that measures of ambient energy are the best predictors of diversity patterns in high-latitude regions, while water-related variables best explain diversity gradients in the subtropics and tropics, i.e. areas with high energy input ([30]). Our data suggest a broader plausibility of this prediction, as water availability can be the primary limiting factor also in extratropical regions with warm and dry climate. Although the ambient energy was proposed as the ultimate constraint for most terrestrial organisms above ca. 50°N (e.g. [17], [54]), we found the main interchange of the two principal diversity constraints (cold stress vs. drought stress) between 57 and 59°N. It seems to be a dominant and general pattern of diversity across a great part of northern Eurasia, from Eastern Europe to Eastern Siberia (see Figure six in [11]). Water-related decrease in bird diversity from the southern taiga towards the dry steppe zone of Kazakhstan, Uzbekistan and Turkmenistan, occurs between 38°and 58° of latitude. Our transect ended at the northern margin of this zone, but further continuous decrease in diversity towards southern part of this warm and dry region is very likely. Dry steppes or saline habitats in this region typically contain no snails in small plots. The unimodal change in snail species richness, though caused by different limiting factors, is very well described by water balance, a variable describing moisture availability resulting from the counteracting effects of precipitation and evapotranspiration. However, a more complex model incorporating both precipitation and temperature as separate variables, accompanied with soil calcium, was able to explain slightly more variation in species data. Likewise, the balance of temperature and precipitation (included in both linear and quadratic terms) was the best overall predictor of the variation in global amphibian richness ([47]).

The question remains, however, which mechanism causes the decline of species richness from the mixed forests of southern taiga towards the north. Decreasing ambient energy can control diversity by low temperature stress (freezing) or low productivity that translates into lower availability of food and shelters for overwintering at higher latitudes. There is not only a strong collinearity but also a strong positive feedback between these two factors. In ecosystems with very cool winters, distribution of terrestrial snails is constrained by the lack of shelters for overwintering ([8]), which becomes more pronounced where herbaceous biomass production is low ([24]). Considering the known physiological constraints of land snails (e.g. [19]), low temperature stress is a more parsimonious explanation than the limited food availability. Some species frequently inhabiting low-productive steppe ecosystems (e.g. *Vallonia pulchella*) are not able to cope with open tundra habitats of similar productivity levels ([55]). The freezing intolerance hypothesis is also supported by a limited distribution of large-bodied species towards higher altitudes, as smaller species have better supercooling abilities ([20]). Although the freezing hypothesis was repeatedly rejected for terrestrial birds (e.g. [11]), temperature can be more serious constraint for ectotherms, as supported by the analysis of global amphibian diversity ([47]). In land snails particularly, we need to consider collinearity between the poleward decline in temperature and soil calcium content, likely occurring in most regions due to lower evaporation of soil water and consequent higher cation leaching at higher latitudes. Content of soil calcium showed a more complex pattern along our latitudinal transect. The amount of calcium dropped steeply between the southern and middle taiga zones, being very low in most of the middle and northern taiga zones (a zone with predominant sandy soils), but increased again in the forest-tundra and tundra zones (Fig. 2A). As land snail species richness typically increases with increasing availability of calcium ([41], [42], [56], [57]), more acidic conditions at higher latitudes can contribute to the decline in species richness independently of temperature. Although the causality of these two factors cannot be disentangled using our data, we suggest that the main effect is that decreasing species richness at higher latitude is primarily determined by low winter temperature, but it is further modified by low calcium levels, which may cause an additional reduction of the number of species. Although extremely calcium-poor conditions occurred in the middle and northern taiga zone due to the predominance of acidic sandy soils, smaller areas of loamy soils can be found there. At 63.5°N we sampled four adjacent forest plots, two on sandy and two on loamy soils. Loamy-soil plots were

richer in both the calcium content (ca. 0.51 g/kg) and the number of species (five and three) in contrast to sandy-soil plots with ca. 0.15 g/kg of soil calcium, which hosted only one or no snail species. Higher content of calcium (ca. 1.5 g/kg) in the forest-tundra and tundra soils in comparison with those in the southern taiga zone (ca. 1.3 g/kg) clearly demonstrates that higher calcium content is necessary, but not sufficient condition for achieving a given level of species richness (Fig. 3C). Climatically driven decline in diversity was notable particularly in forests of the forest-tundra zone, which remained very poor in snails even under more calcium-rich conditions (Fig. 2A). The increase in species richness in open habitats of this zone was associated with the first appearance of the arctic *Vertigo* species (Table 1), recorded in wet willow scrub or fen tundra vegetation. These species belong to the smallest land snails, which are able to survive under very low temperature, contrary to the majority of forest species.

The observed responses of land snail species richness to two largely independent factors, water availability and winter temperature, are important for considerations about future changes of latitudinal diversity patterns induced by future climate change ([58]). As geographical patterns of precipitation and temperature shift rather independently, future shifts in the distribution of species and whole ecosystems can be rather complex. This can possibly result in a disintegration of the most suitable zone of the highest diversity, i.e. subtaiga and southern taiga, which are now both warm and wet. Containing nearly 80% of the total snail species number recorded from the steppe to the tundra zone, these two rather narrow zones represent the main diversity hot-spot in this part of northern Eurasia. In this study, we also provided evidence that no single factor is sufficient to explain the unimodal latitudinal gradient in diversity. This adds to a growing body of evidence that multiple factors vary in concert with latitude and interact to cause the latitudinal gradient of species richness both on global and continental scales (e.g. [2], [10], [11], [27], [47], [59]). Our data extend plausibility of this observation also to smaller spatial extents.

Supporting Information

Figure S1 Patterns of selected climatic and environmental variables along the studied latitudinal transect and their pairwise relationships. The upper right part shows values of Spearman correlation coefficient.

Figure S2 Changes in land-snail species numbers measured in different habitats and different ways along the studied latitudinal transect and their pairwise relationships. The upper right part shows values of Spearman correlation coefficient.

Table S1 Data used in the analyses with numbers of all recorded land snail species (All Sum) at each of the 29 studied sites, mean numbers of species (All Mean), and means of species recorded at open (Open Mean) and forest (Forest Mean) habitats separately at each site; numbers of open and forest plots sampled is given in the second column (Open/Forest). NA refers to the situations where no open or forest habitats were sampled. Values of all explanatory variables used in the study are also provided.

Acknowledgments

Many thanks go to Nikolai Lashchinskyi with his Russian colleagues and Jiří Danihelka, Petra Hájková, Martin Kočí, Svatava Kubešová and Pavel Lustyk for their logistic support and help in the field, Ondřej Hájek for preparing the climate data and the map, and Robert A. D. Cameron, Richard Field, Bernhard Hausdorf and Hanna Tuomisto for helpful comments on previous versions of the manuscript; Robert Cameron also improved the English.

References

1. Rosenzweig ML (1995) Species diversity in space and time. Cambridge: Cambridge University Press. 436 p.
2. Willig MR, Kaufmann DM, Stevens RD (2003) Latitudinal gradients of biodiversity: pattern, process, scale and synthesis. Annu Rev Ecol Syst 34: 273–309.
3. Pimm SL, Brown JH (2004) Domains of diversity. Science 304: 831–833.
4. Nekola JC (2005) Latitudinal richness, evenness, and shell size gradients in eastern North American land snail communities. Rec West Aust Mus Suppl. 68: 39–51.
5. Bromham L, Cardillo M (2003) Testing the link between the latitudinal gradient in species richness and rates of molecular evolution. J Evol Biol 16: 200–207.
6. Hillebrand H (2004) On the generality of the latitudinal diversity gradient. Amer Nat 163: 192–211.
7. Cardillo M, Orme CDL, Owens IPF (2005) Testing for latitudinal bias in diversification rates: an example using New World birds. Ecology 86: 2278–2287.
8. Horsák M, Chytrý M, Axmanová I (2013). Exceptionally poor land snail fauna of central Yakutia (NE Russia): climatic and habitat determinants of species richness. Polar Biol 36: 185–191.
9. Francis AP, Currie DJ (2003) A globally consistent richness–climate relationship for angiosperms. Amer Nat 161: 523–536.
10. Field R, Hawkins BA, Cornell HV, Currie DJ, Diniz-Filho AJF, et al. (2009) Spatial species-richness gradients across scales: a meta-analysis. J Biogeogr 36: 132–147.
11. Hawkins BA, Porter EE, Diniz-Filho JAF (2003) Productivity and history as predictors of the latitudinal diversity gradient of terrestrial birds. Ecology 84: 1608–1623.
12. Davidowitz G, Rosenzweig ML (1998) The latitudinal gradient in species diversity among North American grasshoppers (Acrididae) within a single habitat: a test of the spatial heterogeneity hypothesis. J Biogeogr 25: 553–560.
13. Skillen EL, Pickering J, Sharkey MJ (2000) Species richness of the Campopleginae and Ichneumoninae (Hymenoptera: Ichneumonidae) along a latitudinal gradient in eastern North America old-growth forests. Environ Entomol 29: 460–466.
14. Chown SL, Gaston KJ, Williams PH (1998) Global patterns in species richness of pelagic seabirds: the Procellariiformes. Ecography 21: 342–350.
15. Janzen DH (1981) The peak in North American ichneumonid species richness lies between 38° and 42° North. Ecology 62: 532–537.
16. Bannister JR, Vidal OJ, Teneb E, Sandoval V (2012) Latitudinal patterns and regionalization of plant diversity along a 4270-km gradient in continental Chile. Austral Ecol 37: 500–509.
17. Currie DJ (1991) Energy and large-scale patterns of animal and plant-species richness. Amer Nat 137: 27–49.
18. Mittelbach GG, Steiner CF, Scheiner SM, Gross KL, Reynolds HL, et al. (2001) What is the observed relationship between species richness and productivity? Ecology 82: 2381–2396.
19. Riddle WA (1983) Physiological ecology of snails and slugs. In: Russell-Hunter WD, editor. The Mollusca, Vol. 6: Ecology. London: Academic Press, 431–461.
20. Ansart A, Vernon P (2003) Cold hardiness in molluscs. Acta Oecol 24: 95–102.
21. von Humboldt A (1808) Ansichten der Natur, mit wissenschaftlichen Erläuterungen. Tübingen: J. G. Cotta, 474 p.
22. Cameron RAD, Greenwood JJD (1991) Some montane and forest molluscan faunas from eastern Scotland: effects of altitude, disturbance and isolation. In: Meier-Brook C, editor. Proceedings of the 10th International Malacological Congress. Tübingen: Unitas Malacologica, 437–442.
23. Horsák M, Cernohorsky N (2008) Mollusc diversity patterns in Central European fens: hotspots and conservation priorities. J Biogeogr 35: 1215–1225.
24. Schamp B, Horsák M, Hájek M (2010) Deterministic assembly of land snail communities according to species size and diet. J Anim Ecol 79: 803–810.
25. Ložek V (1964) Quartärmollusken der Tschechoslowakei. Praha: Nakladatelství Československé akademie věd. 374 p.
26. Martin K, Sommer M (2004) Relationships between land snail assemblage patterns and soil properties in temperate-humid forest ecosystems. J Biogeogr 31: 531–545.
27. O'Brien EM (1998) Water–energy dynamics, climate, and prediction of woody plant species richness: an interim general model. J Biogeogr 25: 379–398.
28. O'Brien EM, Whittaker RJ, Field R (1998) Climate and woody plant diversity in southern Africa: relationships at species, genus and family levels. Ecography 21: 495–509.
29. O'Brien EM (2006) Biological relativity to water–energy dynamics. J Biogeogr 33: 1868–1888.
30. Hawkins BA, Field R, Cornell HV, Currie DJ, Guegan JF, et al. (2003) Energy, water, and broad-scale geographic patterns of species richness. Ecology 84: 3105–3117.

31. Field R, O'Brien EM, Whittaker RJ (2005) Global models for predicting woody plant richness from climate: development and evaluation. Ecology 86: 2263–2277.
32. Walter H (1974) Die Vegetation Osteuropas, Nord- und Zentralasiens. Stuttgart: Gustav Fischer Verlag. 452 p.
33. Hijmans RJ, Cameron SE, Parra JL, Jones PG, Jarvis A (2005) Very high resolution interpolated climate surfaces for global land areas. Inter J Climatol 25: 1965–1978.
34. Horsák M (2003) How to sample mollusc communities in mires easily. Malacologica Bohemoslovaca 2: 11–14.
35. Pilsbry HA (1948) Land Mollusca of North America north of Mexico. Vol. II, part 2. Philadelphia: The Academy of Natural Sciences of Philadelphia. 1113 p.
36. Kerney MP, Cameron RAD, Jungbluth JH (1983) Die Landschnecken Nord- und Mitteleuropas. Hamburg/Berlin: Parey Verlag. 384 p.
37. Sysoev A, Schileyko A (2009) Land snails of Russia and adjacent countries. Sofia/Moscow: Pensoft. 312 p.
38. Horsák M, Juřičková L, Picka J (2013) Molluscs of the Czech and Slovak Republics. Zlín: Kabourek. 264 p.
39. Trabucco A, Zomer RJ, Bossio DA, van Straaten O, Verchot LV (2008) Climate change mitigation through afforestation/reforestation: A global analysis of hydrologic impacts with four case studies. Agricul Ecosyst Environ 126: 81–97.
40. Churkina G, Running SW, Schloss AL, the participants of the Potsdam NPP model intercomparison (1999) Comparing global models of terrestrial net primary productivity (NPP): the importance of water availability. Glob Change Biol 5 (Suppl. 1): 46–55.
41. Juřičková L, Horsák M, Cameron R, Hylander K, Míkovcová A, et al. (2008) Land snail distribution patterns within a site: the role of different calcium sources. Eur J Soil Biol 44: 172–179.
42. Wäreborn I (1969) Land molluscs and their environments in an oligotrophic area in southern Sweden. Oikos 20: 461–479.
43. Zbíral J (1995) [Analysis of plant material. Unified techniques]. Brno: Central Institute for Supervising and Testing in Agriculture. 192 p. [In Czech].
44. Cook R D (1977) Detection of influential observation in linear regression. Technometrics 19: 15–18.
45. Cleveland WS (1979) Robust locally weighted regression and smoothing scatterplots. J Amer Stat Assoc 74: 829–836.
46. R Core Team (2012) R: A language and environment for statistical computing. R Foundation for Statistical Computing, Vienna- URL http://www.R-project.org/.
47. Buckley LB, Jetz W (2007) Environmental and historical constraints on global patterns of amphibian richness. Proc R Soc Lond B 274: 1167–1173.
48. Malyshev LI (1993) Ecological background of the floristic diversity in northern Asia. Fragm Flor Geobot Suppl. 2(1): 331–342.
49. Rosenzweig ML (1992) Species diversity gradients: we know more and less than we thought. J Mammal 73: 715–730.
50. Nekola JC (2009) Big ranges from small packages: North American vertiginids more widespread than thought. The Tentacle 17: 26–27.
51. Cameron RAD, Pokryszko BM, Horsák M (2010) Land snail faunas in Polish forests: patterns of richness and composition in a post-glacial landscape. Malacologia 53: 77–134.
52. Ehlers J, Gibbard PL, editors (2004) Quaternary glaciations - extent and chronology, Part III: South America, Asia, Africa, Australasia, Antarctica. Amsterdam: Elsevier. 903 p.
53. Hausdorf B, Hennig C (2003) Nestedness of northwest European land snail ranges as a consequence of differential immigration from Pleistocene glacial refuges. Oecologia 135: 102–109.
54. Lennon JJ, Greenwood JJD, Turner JRG (2000) Bird diversity and environmental gradients in Britain: a test of the species–energy hypothesis. J Anim Ecol 69: 581–598.
55. Del Grosso S, Parton W, Stohlgren T, Zheng D, Bachelet D, et al. (2008) Global potential net primary production predicted from vegetation class, precipitation, and temperature. Ecology 89: 2117–2126.
56. Hylander K, Nilsson C, Jonsson BG, Göthner T (2005) Differences in habitat quality explain nestedness in a land snail metacommunity. Oikos 108: 351–361.
57. Horsák M (2006) Mollusc community patterns and species response curves along a mineral richness gradient: a case study in fens. J Biogeogr 33: 98–107.
58. IPCC (2007) Climate change 2007: The physical science basis. Contribution of Working Group I to the Fourth Assessment Report of the Intergovernmental Panel on Climate Change. Cambridge: Cambridge University Press. 582 p.
59. Willis KJ, Whittaker RJ (2002) Species diversity-scale matters. Science 295: 1245–1248.

Author Contributions

Conceived and designed the experiments: MC MH. Analyzed the data: MH. Contributed to the writing of the manuscript: MH MC.

18

Decreasing Abundance, Increasing Diversity and Changing Structure of the Wild Bee Community (Hymenoptera: Anthophila) along an Urbanization Gradient

Laura Fortel[1]*, Mickaël Henry[1,2], Laurent Guilbaud[1], Anne Laure Guirao[1], Michael Kuhlmann[3], Hugues Mouret[4], Orianne Rollin[2,5], Bernard E. Vaissière[1,2]

1 INRA, UR 406 Abeilles et Environnement, Avignon, France, 2 UMT Protection des Abeilles dans l'Environnement, Avignon, France, 3 Department of Life Sciences, Natural History Museum, London, United Kingdom, 4 Arthropologia, Ecocentre du Lyonnais, La Tour de Salvagny, France, 5 ACTA, Site Agroparc, Avignon, France

Abstract

Background: Wild bees are important pollinators that have declined in diversity and abundance during the last decades. Habitat destruction and fragmentation associated with urbanization are reported as part of the main causes of this decline. Urbanization involves dramatic changes of the landscape, increasing the proportion of impervious surface while decreasing that of green areas. Few studies have investigated the effects of urbanization on bee communities. We assessed changes in the abundance, species richness, and composition of wild bee community along an urbanization gradient.

Methodology/Principal Findings: Over two years and on a monthly basis, bees were sampled with colored pan traps and insect nets at 24 sites located along an urbanization gradient. Landscape structure within three different radii was measured at each study site. We captured 291 wild bee species. The abundance of wild bees was negatively correlated with the proportion of impervious surface, while species richness reached a maximum at an intermediate (50%) proportion of impervious surface. The structure of the community changed along the urbanization gradient with more parasitic species in sites with an intermediate proportion of impervious surface. There were also greater numbers of cavity-nesting species and long-tongued species in sites with intermediate or higher proportion of impervious surface. However, urbanization had no effect on the occurrence of species depending on their social behavior or body size.

Conclusions/Significance: We found nearly a third of the wild bee fauna known from France in our study sites. Indeed, urban areas supported a diverse bee community, but sites with an intermediate level of urbanization were the most speciose ones, including greater proportion of parasitic species. The presence of a diverse array of bee species even in the most urbanized area makes these pollinators worthy of being a flagship group to raise the awareness of urban citizens about biodiversity.

Editor: M. Alex Smith, University of Guelph, Canada

Funding: This research was funded with the contribution of the LIFE financial instrument of the European Community for the project "URBANBEES - URBAN BEE biodiversity action planS" (LIFE08 NAT/F/000478). Additional funding was provided by the French Ministry of Ecology (MEDDAT - Bureau de la Faune et de la Flore Sauvages), the region Rhône-Alpes, the Communauté Urbaine du Grand Lyon, and Botanic sas. The funders had no role in study design, data collection and analysis, decision to publish, or preparation of the manuscript.

Competing Interests: Botanic sas provided funding towards this study. There are no patents, products in development or marketed products to declare.

* Email: laura.fortel@gmail.com

Introduction

Urbanization is one of the main human activities that causes drastic and irreversible habitat alterations, and it is likely to increase in the coming years [1]. Urban environments are defined as mosaics of impervious and permeable surfaces that harbor regularly disturbed habitats [2]. In urbanized landscapes, green areas decrease with a corresponding increase of impervious surface, which includes buildings, roads and industrial areas. An urban environment can thus be characterized by its proportion of impervious surface and the level of connectivity among its patches of permeable surface, both of which have an impact on the fauna [3–5].

Even if urbanization has negative impacts on the insect fauna [6–9], many bee species are common within urban areas [3,8–11]. Indeed, man-made environments like urban habitats and gardens can host a rich and abundant wild bee fauna [12–14]. For example, 262 bee species were recorded within the city of Berlin, Germany, over five years [9]. Matteson *et al.* (2008) collected 54 bee species in 19 urban gardens, and Fetridge *et al.* (2008) recorded 110 species in 21 residential gardens, both studies were conducted over two years in New York City during the summer months [13,15]. For a bee species to be present in a given habitat,

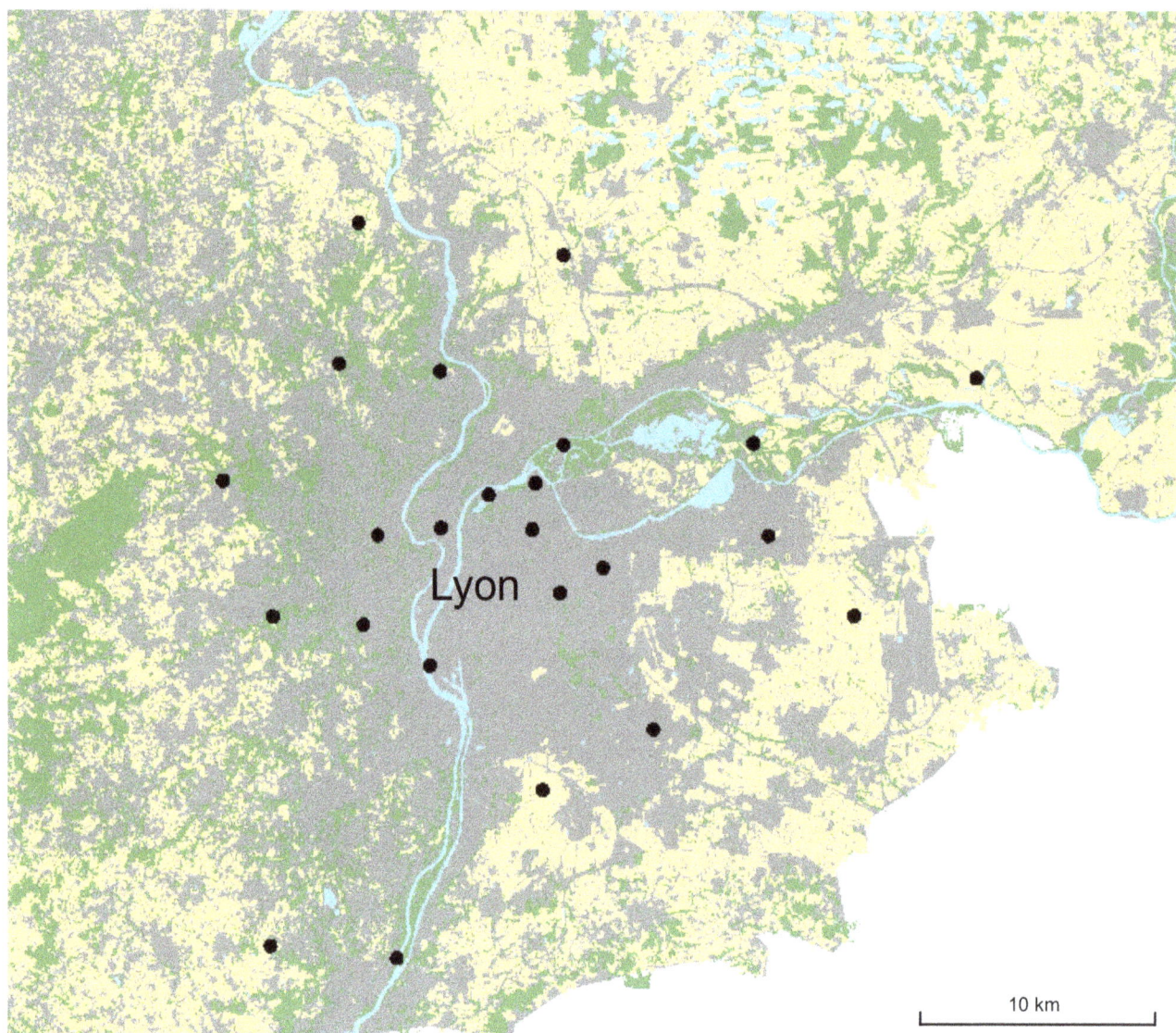

Figure 1. Distribution of the 24 sites along the urbanization gradient around Lyon, France. Base map colors represent: impervious surface (grey), agricultural landscape (yellow), semi-natural habitat (green) or water (blue).

it must be able to find food and nesting substrate within its species specific range of activity [16]. Urban and periurban sites can provide high quantities of flowers all year long [15], they show a high diversity of land-cover types, and are often warmer than surrounding landscapes [17]. Also, such habitats are seldom treated with pesticides [10] which are involved in the decline of bees elsewhere [18].

Williams *et al.* (2010) demonstrated that ecological traits can be used to predict bee responses to a variety of disturbance types [19]. Indeed, the presence of a bee species may be jeopardized by the fragmented nature of urban habitats because of its limited flight ability. Concerning the nesting behavior, some bees are soil-nesting, while others nest above ground in stems, dead wood or walls (cavity-nesting species). The regular disturbance in urban habitats (e.g. mowing, weeding or soil plowing) may prevent the long-term establishment of soil-nesting bee species [13], which represent over 80% of the bee fauna worldwide [20]. There is also some evidence that cavity-nesting species are over-represented in urban bee communities [3], defined as the assemblage of species

populations that occur together in space and time [21]. Every species has its own functional traits and will respond accordingly to habitat alteration that characterizes urban environments [22]. Therefore, the species and its functional traits are essential elements to study the impact of urbanization on wild bee community structure, defined as the species diversity found in a given area. Indeed, several studies have documented the changes in wild bee community structure in urban environments [6,23,24].

It is unknown whether, and if so how, the proportion of impervious surface and the level of connectivity among permeable surfaces combine to affect the structure of wild bee communities. Only few studies have surveyed bee communities along a gradient of urbanization [23,24]. In most cases, the effect of urbanization on bee communities was analyzed using different categories of landscapes such as urban, periurban or natural areas [6,13,15]. We did not choose this approach, but rather we followed McDonnell and Hahs (2008) and McDonnell and Pickett (1990) and used a gradient to assess the effects of urbanization [25,26]. Our objectives were to 1) assess the wild bee community structure along an urbanization gradient; 2) test the effects of the proportion of impervious surface and the level of connectivity among permeable surfaces on the wild bee abundance and species richness; and 3) investigate the changes of composition in the wild bee community along the gradient in relation to functional traits.

Materials and Methods

Study sites

The study was conducted in the urban community of Grand Lyon, France, which includes 58 towns around Lyon (45° 46′N, 4° 50′E) and covers an area of 516 km^2. With approximately 1.3 million inhabitants [27], this urban community consists of diverse ecosystems ranging from densely populated urban areas to intensive agricultural landscapes or semi-natural grasslands. The climate of Lyon is at the temperate-Mediterranean interface. Located in the Rhône valley, the wind commonly blows from the south. The 30-year annual average temperature is 12°C with a minimum of 3°C in January and a maximum of 21°C in July [28].

We selected twenty-four sites following a increasing gradient of impervious surface (from 10 to 95%) over a two kilometer radius in different directions from the downtown Lyon area (Figure 1), and secured appropriate authorizations from the different authorities for each of them (farmer, city,…; see Table S1). Thus, eight sites were covered by less than 30% of impervious surface, eight by a proportion between 30 and 70%, and the remaining eight by more than 70% of impervious surface. For part of the surveys, we captured bees on flowers, so sites were chosen in green areas, parks or gardens. All sites were distant by more than two kilometers from each other to prevent overlapping bee communities [29].

Wild bee surveys

We used both pan traps and insect nets to assess the bee community at each site in 2011 and 2012 [30,31]. Pan trapping is a standard method for catching bees [30], though it is known to perform poorly for some taxa [32]. It is a passive method based on the visual attraction to colored pan traps and it provides quantitative data on the abundance of a large part of the wild bee fauna without the bias associated with the difference in capture efficiency among observers using active collecting methods (e.g. netting) [30,33–36]. We used 500 ml plastic bowls painted with yellow, blue or white fluorescent paint (Rocol Top, France) [30,31]. Pan traps were arranged in triplets, with each triplet consisting of a pan of each of the three colors randomly distributed either at the corners of a three meters side equilateral triangle, or, when space did not permit otherwise, linearly with three meters between two adjacent bowls. The pan traps were set at a height slightly above that of the average vegetation, and they were activated by filling them with 400 ml of water with a drop of detergent, and left active for 24 hours. Pan trapping is very sensitive to the immediate environment [37]. In order to take this effect into account, we set two triplets of pan traps separated by 20 to 40 m from each other [38], one being in an open area and the other along the sunniest side of a vertical landscape element (edge, wall, or tree). From March until October, we sampled bees on the same day for all 24 sites on a monthly basis.

Net surveys were done from March until September on a monthly basis also right after pan trapping by a range of observers so that it lasted between five and eight days (weather did not permit to do these observations in October in both years). At each study site, we surveyed all flowering plant species in bloom within a radius of 100 m around the centroid of pan traps, except for grasses since we found no records of wild bees foraging on flowers in the Poaceae family in Europe. For each species, flowers were observed for up to two minutes. Observation then stopped if no foraging activity was detected. Else, the first bee observed was caught and net catching lasted for five minutes after this first capture. Sampling took place alternately in the morning and in the afternoon at each site to cover the whole foraging bee population [39].

The PLANT DIVERSITY was recorded for each site in April and July 2012, over two perpendicular transects of 50 m each centered on the centroid of the pan trap triplets. One transect was aligned along the centers of the two pan trap triplets and the other one was perpendicular. At each date, all plants (in bloom or not) on these transects were identified to species by professional botanists. In that way, we had a standardized and exhaustive estimation of the plant diversity of each site.

Pan trapping and net sampling were performed only during periods of good weather for foraging activity (maximum temperature ≥15°C, sunny sky or with scattered clouds only, and wind speed ≤15 km/h [40]). Specimens collected in pan traps were first stored in 70% ethanol (w/w) until washed and dried following Lebuhn (2013). All these specimens as well as sweep samples were frozen for later processing. Individuals were then pinned, labeled, and sent for identification to species to the respective authority for each genus (see Acknowledgements). All voucher specimens are now deposited in the bee collection of INRA Avignon. For taxonomy, we followed the nomenclature of Kuhlmann et al. [41] (see Table S2 for the entire species list). Honey bees (Apis mellifera) were caught in pan traps and observed during net sampling, but they were not considered in this study so that 'bees' will be used synonymously with 'wild bees' in the following unless stated otherwise.

Landscape structure

To characterize the landscape surrounding each study site, we used the Geographic Information System Arcgis v 9.3 and Fragstat software [42]. Landscape characteristics were analyzed at the three radii of 500 m, 1000 m, and 2000 m centered on the centroid of the two pan-trap triplets. These radii were chosen because flight distance of wild bees are estimated between a few hundred meters to several kilometers depending on the species [43–49]. The minimum size of habitat patches was defined by the spatial resolution of our raster, which was of 256 m^2 (i.e. 16 m × 16 m). We used seven mutually exclusive land-cover types: roads, buildings, industrial areas, agricultural land, wooded areas (e.g. forests, hedgerows), open areas (e.g. meadows, bare soils areas), and water. Based on principal component analyses of the proportion of land-cover types at each site, the proportion of roads, buildings, and industrial areas were strongly correlated with the first axis (see Figure S1). These three variables were therefore pooled together as the proportion of impervious surface (IMPERVIOUS SURFACE) for further analyses. There was a clear gradient in the proportion of impervious surface among the sites that ranged between 0–98%, 1–98%, and 12–93% at the radii of 500 m, 1000 m, and 2000 m, respectively. In addition to land-cover uses, we calculated the variables CONNECTIVITY OF OPEN AREA and CONNECTIVITY OF WOODED AREA. Landscape connectivity is defined as the degree to which the landscape facilitates or impedes movements among resource patches [50]. In this study, connectivity is defined as the number of functional joinings between patches of the same type, where each pair of patches is either connected or not, based on a user specified distance criterion (here 100 m, that is the radius surveyed for net captures) [42]. Connectivity is the percentage of patches of a given land-cover distant from each other by a maximum of 100 m (connectivity = 100 when all patches in the landscape are connected; [42]).

Data analyses

Bee community parameters were computed separately for each of the two consecutive years. Species diversity was characterized by species richness (using EstimateS v 9.1.0 [51]) and rank abundance distribution (using BiodiversityR package in R v 2.15.2 software [52,53]). The observed cumulative species richness curve and the total expected species richness were computed using a bootstrapping procedure with 1000 random reorganizations of sampling order. Total expected species richness was assessed using the Jack1 and the Chao2 estimators because they are the least biased estimators for species-rich assemblages [54]. The proportions of singletons (species represented by a single specimen) and of species for each modality of the functional traits were further compared for each year by means of Chi-square tests.

Pearson correlation coefficients were calculated to quantify how the landscape variables were correlated with each other (see Table S3 for further information). When variables were significantly correlated with IMPERVIOUS SURFACE, we kept only this latter variable for final analyses. Because of the high correlation between the measurements at the three radii ($p<0.001$), the analyses were performed separately for each radius. After correlation analyses, we examined the effect of landscape variables on bee richness and abundance using generalized linear models (GLM). Pan-trapping data were used to analyze abundance and data from both sampling methods were used to analyze species richness and composition [30]. Normality of the abundance and richness data was tested by Shapiro tests. As abundance data were skewed to the right, a log-transformation was performed to normalize data before analyses. At each radius, models were simplified by forward selection based on AIC (Akaike Information Criterion) values. We then considered the model with the lowest AIC value as the most parsimonious one.

To further determine which ecological processes would best explain changes in species composition along the urbanization gradient, we performed complementary analyses that incorporated species-specific information on functional traits [19,55]. We first compared the response of parasitic vs. non-parasitic species to landscape variables. Then, for non-parasitic species, we gathered information on tongue length, nesting behavior, and social behavior from published information [20,56–61]. Pollen diet specialization will be analyzed elsewhere in relation with the composition of the local flora. Species of the families Apidae and Megachilidae were considered as long-tongued and the others as short-tongued. Species were divided into the following binary ecological categories: soil-nesting or cavity-nesting for the nesting behavior, and solitary (each female constructs her own nest and provides food for her offspring) or social (from gregarious to eusocial) for social behavior [20,62]. We also used body size by measuring the inter-tegular distance (ITD) with a dissecting microscope and calibrated ocular micrometer on a sample of 3 to 10 randomly selected female specimens per species. The ITD measures the width of the thorax, which contains the flight muscles, and it is related to dry body mass and also to foraging distance [44,63]. A total of 58 species could not be included in these analyses due to partly missing information on functional traits. GLMs were performed on the occurrence frequency of bee species in all sites based on landscape variables in interaction with functional traits. In all GLMs, the effect of each landscape variable was nested in the year to account for interannual variations.

Whenever a large number of different tests are conducted, one uses a correction for multiple comparisons (often the Bonferroni adjustment [64]) because series of non-independent tests increase the probability of significant results due to chance only. Thus, we used a three-fold Bonferroni correction for abundance and richness analyses repeated throughout the three spatial scales and a five-fold correction for species occurrence analyses repeated along the five functional trait categories.

Results

Characterization of the bee fauna

Over the two years of survey, a total of 12872 bee specimens were collected, 7187 in 2011 and 5685 in 2012. They belonged to six families (Andrenidae, Apidae, Colletidae, Halictidae, Megachilidae, Melittidae), 34 genera and 291 species (256 in 2011 and 226 in 2012). Halictidae had the largest diversity with 59 different species, while there were only two species in the Melittidae. A total of 100 species were collected only in one of the two years (65 in 2011 and 35 in 2012), which represents 34% of the recorded species. Species accumulation curves did not reach saturation, which indicates that we did not capture all the species potentially present in our study area (Figure 2). Using EstimateS, the predictor of estimated species richness over both years pooled together was 366.7 for Chao2 and 367.7 for Jack1 (Table 1). Thus nearly 79% of the estimated number of bee species present in the study area were recorded for the two methods combined over the two years.

The proportion of singletons was not significantly different between the two years ($\chi^2 = 1.26$, $df = 1$, $p = 0.26$), nor were the proportions of species among each modality of the functional traits ($\chi^2 \leq 0.69$, $df = 1$, $p \geq 0.4$). Overall, 57 species (20% of the total) were recorded as singletons and 37 (13%) as doubletons. Among singletons, 11 species (19.5%) were parasitic and among all species, there were 49 parasitic ones (17%) and 242 non-parasitic ones. Non-parasitic species were dominated by solitary species (74%), short-tongued species (67%) and soil-nesting species (69%). Twenty-two species represented each from 1% to 4% of the total number of specimens (138 to 565 specimens). Twelve of those species were social and soil-nesting (*Bombus* spp. (Apidae), *Andrena* spp. (Andrenidae), *Halictus* spp. and *Lasioglossum* (*Evylaeus*) spp. (Halictidae)). Eight were solitary and soil-nesting (*Andrena bicolor* and *A. minutula* (Andrenidae), *Anthophora plumipes* and *Tetralonia malvae* (Apidae), *H. scabiosae*, *L. villosulum*, *L. nitidulum* and *L. leucozonium* (Halictidae)) and two were solitary and cavity-nesting (*Hylaeus communis* (Colletidae) and *Osmia cornuta* (Apidae)). The three most abundant species were *Lasioglossum politum* (1045 specimens; 8% of the total), *L. malachurum* (837 specimens; 6.5%), and *L. pauxillum* (566 specimens; 4.5%; Figure 2). Those three species are social, short-tongued, and soil-nesting.

Abundance and species richness

Based upon correlation analyses, among each set of significantly correlated variables, we retained only the one that gave the lowest AIC to explain abundance and species richness. In doing so, IMPERVIOUS SURFACE, CONNECTIVITY OF OPEN AREA, PLANT DIVERSITY and CONNECTIVITY OF WOODED AREA were the sole variables that were retained in models and these three were not correlated among one another. We further introduced a quadratic term in our model (IMPERVIOUS SURFACE^2) to account for a non-linear pattern of the observed relationship between species richness and IMPERVIOUS SURFACE. The forward selection based on AIC enabled us to keep the variables with the greatest explanatory power in our models (Table 2). IMPERVIOUS SURFACE had a negative linear effect on abundance and a quadratic effect on species richness within the 500 m and 1000 m radii (Figure 3). Based on the quadratic models with IMPERVIOUS SURFACE only, the maximum predicted number of bee species was 69 species at a site with 53%

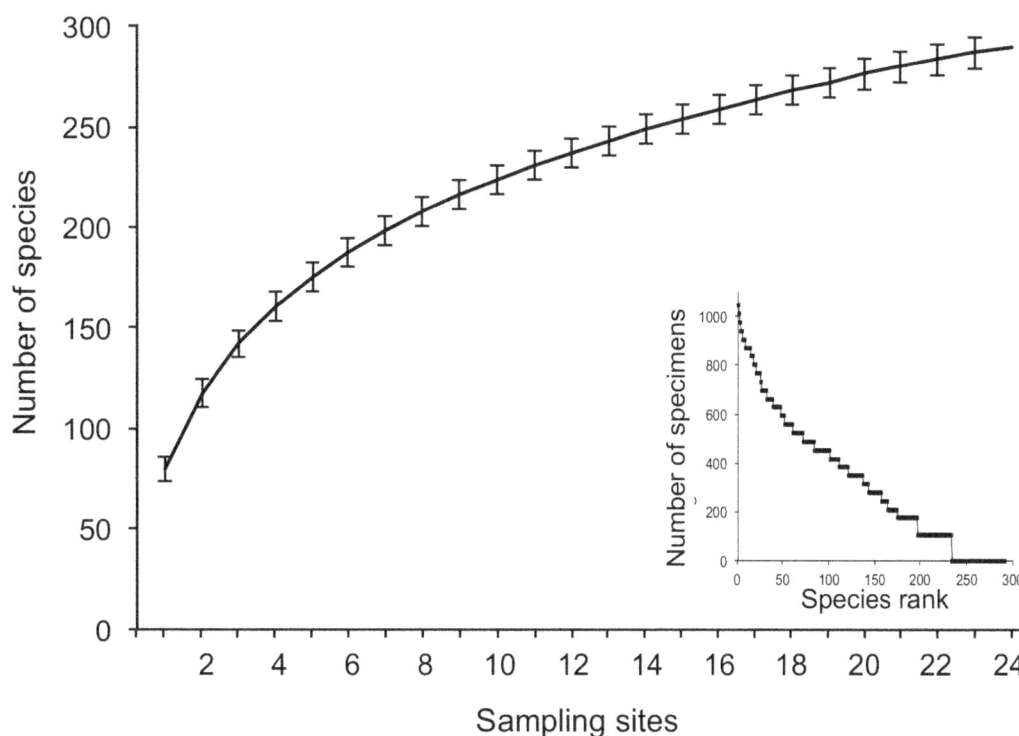

Figure 2. Mean species accumulation curve for pooled data from 2011 and 2012 (1000 randomizations).

impervious surface within 500 m in 2011 and 60 species at a site with 47% impervious surface within 500 m in 2012 (Figure 3.B). Three of the four sites with the lowest species richness over both years had low proportions of impervious surface (<12%), and high proportions of agricultural land cover (70% to 94%). CONNECTIV-ITY OF OPEN AREA had a positive effect on species richness within 1000 m (Table 2). Within 2000 m, the quadratic effect of IMPERVIOUS SURFACE on species richness was not significant, but the linear effect was, and the variable with the highest explanatory power for abundance was CONNECTIVITY OF WOODED AREA (Table 2). PLANT DIVERSITY was not significant in any model.

After the Bonferroni correction ($p \times 3$), the effect of IMPERVIOUS SURFACE on abundance was still significant within 500 m but not anymore within 1000 m. For species richness, the factors with a significant effect after the Bonferroni correction were the quadratic function of IMPERVIOUS SURFACE within 500 m and the CONNEC-TIVITY OF OPEN AREA within 1000 m. The best model fit was achieved for the 1000 m radius model (AIC = 377.23), though the low ΔAIC between the 1000 m and the 500 m models (<2, Table 2) indicates that both models are equally well supported by

the data. For subsequent analyses, we kept IMPERVIOUS SURFACE and IMPERVIOUS SURFACE^2 as explanatory variables, and 500 m as the most relevant focus scale.

Bee community composition and structure

The occurrence frequency of bee species based on their functional traits was analyzed with selected GLM at the 500 m radius also (Table 3). The occurrence frequency of bees depending on their nesting behavior and their parasitism had a quadratic relation with IMPERVIOUS SURFACE (Figure 4.A and 4.B). The effect was higher for cavity-nesting than for soil-nesting species. The occurrence frequency of bees was highest in sites with an average of 50% impervious surface for parasitic species (Figure 4.B) and of 56% impervious surface for cavity-nesting species. The occurrence frequency of bees depending on their tongue length changed with increasing IMPERVIOUS SURFACE as there were more long-tongued species ($F_{2,4463} = 4316.7$, $p < 0.001$) in urbanized sites (Figure 4.C). CONNECTIVITY OF OPEN AREA had no effect on any functional traits. There was no effect of any landscape variable on social behavior and body size (ITD).

Table 1. Observed and estimated species richness.

Year	Sobs*±SD**	Chao 2±SD (completeness)	Jack 1±SD (completeness)
2011–2012	291±7.87	366.71±22.49 (79.35)	367.67±11.56 (79.15)
2011	256±8.43	350.09±28.23 (73.12)	333.63±11.32 (76.73)
2012	226±7.96	309.95±26.51 (72.91)	295.96±11.2 (76.36)

*Sobs = observed species richness.
**SD = standard deviation.

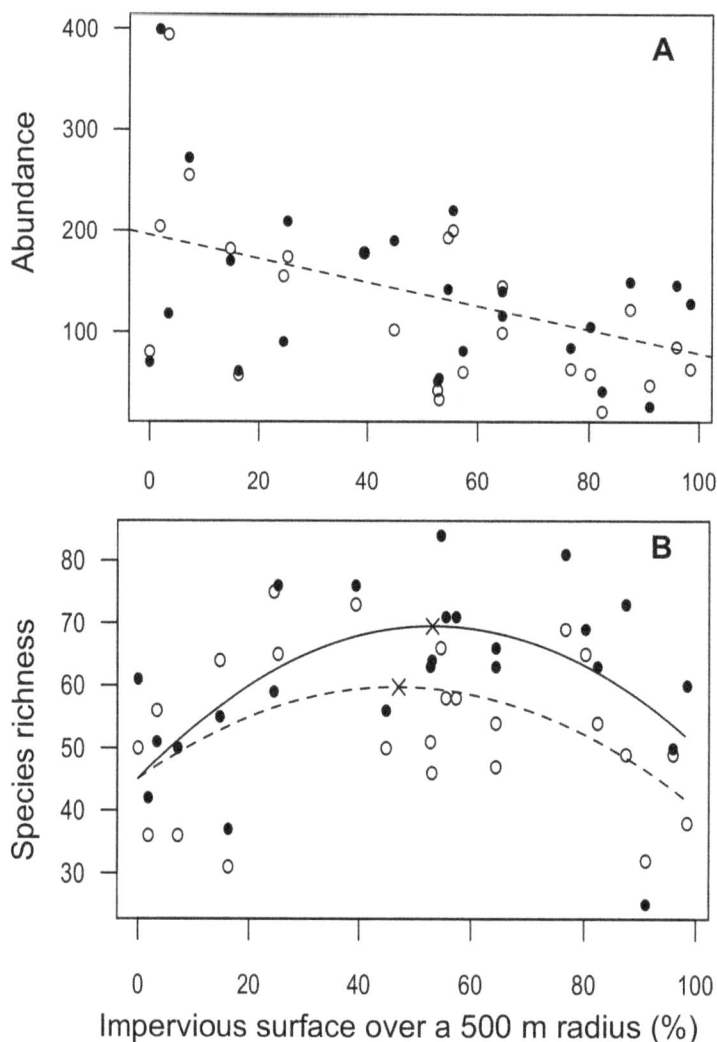

Figure 3. Effect of impervious surface percentage within 500 m on the abundance and species richness of bees. A. Abundance of bees (filled circles = 2011, open circles = 2012); B. Species richness of bees (filled circles and full line = 2011, open circles and dashed line = 2012). Model for species richness = Impervious surface (Year) + Impervious surface^2 (Year).

Discussion

Our study aimed to investigate the impact of urbanization on wild bee communities. We found there were fewer individuals in sites with higher levels of urbanization, and there were more species in sites with an intermediate proportion of impervious surface. In addition, the composition of the wild bee community changed in relation to the nesting behavior of the species along the urbanization gradient.

Over two years of survey using both sweep nets and pan traps to study the effect of urbanization on the wild bee community, 291 bee species were recorded, which represents nearly 79% of the predicted number of species in the study area. Intensive sampling of bees usually leads to low number of singletons because the numbers of bee specimens and that of singletons are negatively correlated [65]. Indeed, our number of singletons represented 20% of our total number of species, which is low compared to the average of 28% (range 9-54%) recorded in 44 studies of bee communities over a range of temporal and spatial scales [65]. This suggests that the bee fauna in Grand Lyon was thoroughly surveyed or that the requirements of rare bee species (floral or

nesting resources) may not be present in our study area, so these species were not detected even as singletons.

This figure of 291 accounts for nearly a third of the 912 wild bee species known in France [66]. In comparison, 262 bee species were recorded by net-collecting over 5 years in about 20 localities within the city of Berlin in Germany [9], that is 46% of the reportedly 574 wild bee species in this country [66]. In the city center and suburbs of Poznań, Poland, 104 bee species (or 19% of the national total of 537 [66]) were collected by sampling bees with yellow pan traps and insect nets every 7–10 days from April to September for 3 years (2006–2008) [23]. While direct comparison between these figures and ours is not possible due to the differences in the methodology used, it indicates nevertheless that the Lyon area did harbor a diverse bee fauna. This result may be linked to the geographical location of the Grand Lyon which is at the temperate-Mediterranean interface [67]. Climate has an important role in the establishment of wild bee communities and Mediterranean climate is known to be favorable for wild bees [68].

Parasitic bee community structure follows that of the remaining bee community, since their species richness and abundance depend on those of their hosts [69]. Indeed, several studies suggest

Table 2. Generalized linear models for bee abundance and species richness depending on landscape variables.

Dependent variable	Radius (m)	AIC*	Independent variable	F value	p
Abundance	500	13.14	Impervious surface	$F_{2,45} = 6.54$	**0.003 (−)**
			Impervious surface^2		
			Connectivity of open area		
			Connectivity of wooded area		
			Plant diversity		
Abundance	1000	18.31	Impervious surface	$F_{2,45} = 3.57$	0.036 (−)
			Impervious surface^2		
			Connectivity of open area		
			Connectivity of wooded area		
			Plant diversity		
Abundance	2000	19.19	Impervious surface		
			Impervious surface^2		
			Connectivity of open area		
			Connectivity of wooded area	$F_{2,45} = 3.1$	0.055 (+)
			Plant diversity		
Species richness	500	378.8	Impervious surface	$F_{2,45} = 3.4$	0.043 (+)
			Impervious surface^2	$F_{2,43} = 7.8$	**0.001 (−)**
			Connectivity of open area		
			Connectivity of wooded area		
			Plant diversity		
Species richness	1000	377.23	Impervious surface	$F_{2,45} = 3.5$	0.039 (+)
			Impervious surface^2	$F_{2,43} = 3.36$	0.045 (−)
			Connectivity of open area	$F_{2,41} = 7.66$	**0.002 (+)**
			Connectivity of wooded area		
			Plant diversity		
Species richness	2000	388.45	Impervious surface	$F_{2,45} = 3.25$	0.048 (+)
			Impervious surface^2		
			Connectivity of open area		
			Connectivity of wooded area		
			Plant diversity		

Results of generalized linear models with abundance or species richness as dependent variables and landscape variables as independent variables. The effect of independent variables was nested in the year to account for interannual.
*AIC = Akaike Information Criterion.
P-value significant after the Bonferroni correction (i.e. $p \times 3$) has been applied are written in bold.

that parasitic species are good indicators of ecosystem health and stability [70–75]. In our study, parasitic species represented 17% of all species. By comparison, Banaszak-Cibicka and Żmihorski (2012) found 12% parasitic species over a total of 104 species in the city of Poznań, Poland, which has 560 000 inhabitants over 261.8 km^2 and is distant of 1469 km from Lyon [23]. The proportion of parasitic species at a national level is similar in Poland (23%, 122 species) and in France (21%, 195) ($\chi^2 = 0.27$, $df = 1$, $p = 0.6$). However, the proportion of parasitic species captured in urban areas, with respect to the species proportions at the national scale, was significantly greater in our study in France compared to the Polish one (Mantel-Haenszel: $\chi^2 = 7.3$, $df = 1$, $p < 0.01$). The relationship between the number of parasitic bee species and the proportion of impervious surface was curvilinear with a maximum at an intermediate proportion of impervious surface (50%). Guild profiles are specific to habitats, and disturbance do not have the same effect on different guilds [55,76,77]. Parasitic bees play a stabilizing role in bee commu-

nities [69,70]. They are the first to respond to disturbances. Therefore, a high diversity of parasitic species may reflect a higher stability and a higher diversity of habitats in these landscapes.

We found that an increasing proportion of impervious surface negatively affected bee abundance. Soil-nesting bees represented 86% of the total number of specimens recorded in our study and also the largest number of species. Indeed, these species represented 63% of the total species richness along our urbanization gradient, even if the occurrence frequency of soil-nesting bees slowly decreased with increasing proportion of impervious surface. In urban sites, resources for ground-nesting bees are less abundant because of the predominance of impervious surface and this would likely jeopardize the establishment of soil-nesting bees. Furthermore, 15 of the 25 most abundant species were soil-nesting and social, so that these species may be over-represented in our pan trap captures simply owing to their social behavior. Indeed, social bee species tend to be active for a longer period than solitary species. The attractiveness pattern of pan traps may also explain

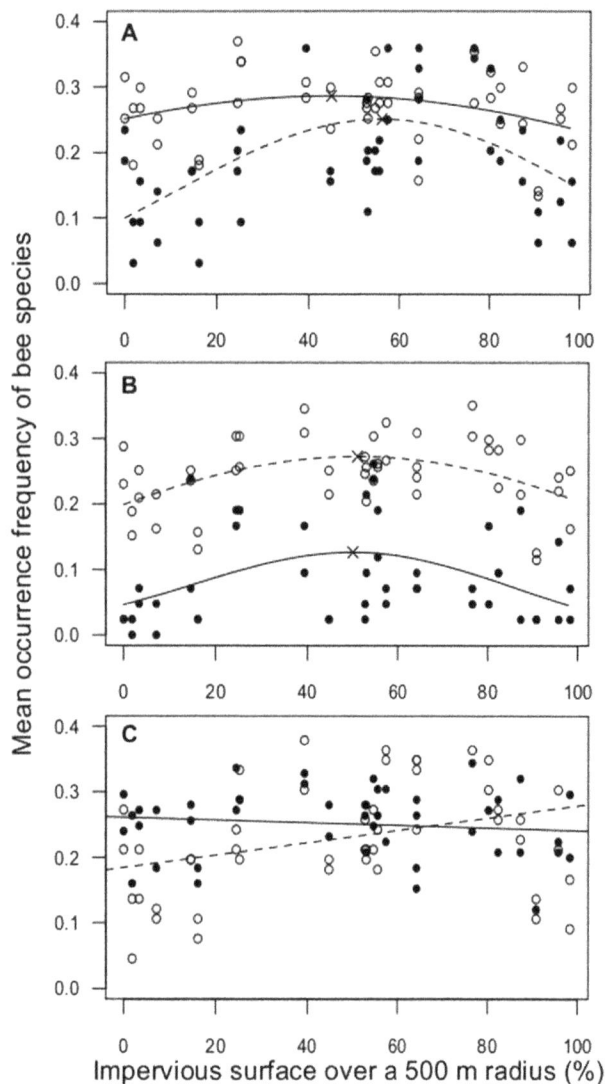

Figure 4. Effect of the proportion of impervious surface within 500 m on species occurrence based upon functional traits. A. Nesting behavior (filled circles and full line = cavity-nesting species, open circles and dashed line = soil-nesting species); B. Parasitic or host behavior (filled circles and full line = parasitic species, open circles and dashed line = host species) species; C. Tongue length (filled circles and full line = short-tongued species, open circles and dashed line = long-tongued species).

this negative relationship between bee abundance and urbanization. The effectiveness of pan traps is inversely related to the abundance of flowers in their surroundings [35,78]. In urban green areas where we exposed our pan traps, flowers were concentrated in flowerbeds that usually provide a large and year-long floral display to bees [15]. But in sites with less impervious surface, bees probably had to fly longer distances between adjacent forage resources and pan traps attractiveness may therefore have been better in these habitats.

To predict diversity and species composition changes in urban systems, urban areas can be modeled using the disturbance heterogeneity model ([79] [80]. This model specifically incorporates spatial (as opposed to temporal) disturbances to account for increased habitat diversity and suggests that when the proportion

of disturbed habitat reaches 50%, the area has maximal heterogeneity [80]. When the proportion of disturbed habitat increases or decreases beyond this value, the area becomes more homogeneous. Following this disturbance heterogeneity model, maximum heterogeneity should lead to peak species diversity at 50% impervious surface [80], since such surfaces can be considered as disturbed and mainly unusable habitats for bees, especially ground-nesting ones. Indeed, urban disturbances eliminate potential ground nesting habitats because of impervious surface [80]. In our study, the response of bee diversity to urbanization was consistent with this model with maximum species diversity at 53% impervious surface in 2011 and 47% in 2012. The city center is largely composed of abiotic elements such as paved streets, sidewalks, and buildings with planted trees and flowerbeds usually as sole green elements. In contrast, the periurban landscape, although heavily disturbed too, usually includes many gardens and green recreation areas, as well as roadsides with vegetation that provide more suitable habitats for ground-nesting bees. Fully urbanized areas may thus provide fewer resources for bees in comparison with periurban areas that have around 50% impervious surface and, thus, can harbor more diverse floral and nesting resources [15].

In our study, we took botanical information into account by recording plant species richness over two 50 m perpendicular transects at each site. This variable, which included all flowering plant species (Spermaphytes), had no effect on bee species richness, which was surprising given the importance of floral diversity on bee diversity [96]. We probably should have focused on the diversity of flowers that are actually visited by bees to better assess the importance of this factor.

In addition to richness and abundance, we studied the changes of the community structure along the urbanization gradient by the studying functional traits of bee species. Within all families, bees present a diverse assemblage of functional traits [19,20], which makes it difficult to characterize the community as a whole, especially when habitat comparisons are the topic of investigation [69]. Urbanized landscapes usually include some green areas that can provide forage resources for a diversity of wild bees [11,81]. These landscapes may also contain diverse nesting opportunities, such as bare soil, dead stems and manmade cavities [12,24]. In our study, soil-nesting and short-tongued bees were little affected by urbanization, whereas cavity-nesting species and long-tongued species were more numerous in moderately and highly urbanized areas, respectively. For nesting behavior, our result is in agreement with several studies that report a greater abundance of cavity-nesting bee species in periurban and urban areas compared to sites with less impervious surface [3,7,13]. Even if cavity-nesting species richness reached a maximum in sites with intermediate proportion of impervious surface, there were more cavity-nesting bee species in urbanized areas than in more natural ones. The hypothesis here is that cavity-nesting bees may find more nesting resources in urbanized habitats because of manmade cavities [12,82]. Concerning tongue-length, long-tongued species can visit flowers with short or long corolla [83], so they may be less affected than short-tongued species by the changes in floral resources that may occur over an urbanization gradient. Overall, these patterns were not unexpected, since nesting behavior and tongue length are not independent functional traits. Indeed, most ground nesting species were Halictidae and Andrenidae, which are also short-tongued, while cavity nesting species were Megachilidae, which are mostly long-tongued.

Flight distance is related to body size [43,44,84], and it influences the ability of bees to recolonize disturbed sites [19]. Thus, we expected larger species to be less affected by

Decreasing Abundance, Increasing Diversity and Changing Structure of the Wild Bee Community...

191

Table 3. Generalized linear models for the occurrence frequency of bee species depending on functional traits and landscape variables within 500 m.

Functionnal traits	Landscape variables	Residual deviance	p
Body size	Impervious surface	NS*	NS
	Impervious surface^2	NS	NS
Nesting	Impervious surface	$F_{2,9159} = 10079$	**0.003 (+)**
	Impervious surface^2	$F_{2,9157} = 10063$	**<0.001 (−)**
Parasitism	Impervious surface	NS	NS
	Impervious surface^2	$F_{2,11173} = 11342$	**0.0076 (−)**
Sociality	Impervious surface	NS	NS
	Impervious surface^2	NS	NS
Tongue length	Impervious surface	$F_{2,9159} = 10125$	**<0.001 (−)**
	Impervious surface^2	$F_{2,9157} = 10118$	0.032 (+)

Results of generalized linear models with the occurrence frequency of bee species as dependent variables and landscape variables in interaction with functional traits as independent variables. The effect of independent variables was nested in the year to account for interannual. The effect of quadratic term of impervious surface proportion (IMPERVIOUS SURFACE^2) was higher on cavity-nesting than on soil-nesting bee species, and on non-parasitic than on parasitic bees species. The effect of impervious surface proportion was higher for long-tongued than for short-tongued species.
*NS = non significant.
P-value significant after the Bonferroni correction has been applied (i.e. $p \times 5$) are written in bold.

urbanization or by the connectivity of open area [85]. Yet, none of the landscape variables had a significant effect on the body size of bees along our urbanization gradient. Although, this functional trait is important for determining species responses to landscape changes, there are opposing predictions for these responses [3,85–87]. Even if small species (<3 mm; [85]) have limited abilities to recolonize disturbed habitats, this may be counterbalanced by the fact that they require less food resources than large species (> 5 mm; [85]) and so may be better able to maintain their populations in disturbed habitats, such as urbanized areas [19]. It is known that social bees have a better adaptability to disturbance than solitary species [23], and that solitary species are more sensitive to disturbance in temperate grasslands [88]. However, none of the landscape variables had a significant effect on the proportion of social species. In our study, most of the social bees were soil-nesting (94%), and we found that cavity-nesting species were more numerous in urbanized sites, thus this soil-nesting preference may counterbalance the social status.

Among many human activities that promote biotic homogenization, urbanization is one of the strongest [1]. Urban biotic communities reflect adaptations to the physical environment as well as the biotic interactions (such as predation and competition) that occur in these environments [89,90]. Species along an urban gradient can be classified into three distinct categories reflecting their response to urbanization [91,92]: avoidance, adaptation, and exploitation [93]. Witte et al. (1985) even use the terms 'urbanophobes' and 'urbanophiles' to describe negative and positive responses to urbanization, respectively, and Kuhn et al. (2004) added the term 'moderately urbanophilic' species that are most abundant in sites with intermediate proportion of impervious surface [92,94]. Following this terminology, parasitic species and cavity nesting could be qualified as 'moderately urbanophilic', and long-tongued species as 'urbanophiles'.

Urbanization and agricultural intensification are two human activities that result in extensive changes of the landscape and its environment, and lead to the destruction or the fragmentation of natural habitats [24]. In our study, three of the four sites with the lowest species richness had a high proportion of agricultural land cover (range 70–94%). Our urbanized sites thus seemed more favorable to a diverse wild bee fauna than agricultural ones. High spatial and temporal instability of agricultural sites, associated with intensive agricultural practices (e.g. soil plowing, pesticide use, crop rotation, landscape simplification) are the main causes of bee diversity loss in farmland areas [95,96]. Further studies are needed to test the hypothesis that, in a given context of fragmentation, urbanized landscapes are more favorable to a species-rich wild bee community than agricultural ones.

Overall, our results suggest that urbanized sites can provide forage and nesting resources for a large community of wild bee species, even if the landscapes with an intermediate proportion of impervious surface have a more diverse and abundant bee fauna. Flagship species are defined as 'known charismatic species that serve as a symbol or focus point to raise environmental consciousness' [97]. Although their individual species may be difficult to identify [98], bees can collectively be considered as a flagship group of species and used to raise the awareness of city-dwellers to biodiversity, as we observed in this study (http://www.urbanbees.eu). Indeed, the loss of a charismatic species can affect people more than the loss of habitat, even when the loss of habitat is the very threat to the species [99]. Also, because bees are a key group of pollinators worldwide for both wild and cultivated entomophilous plants [100,101], bees can be readily used to illustrate the importance of ecosystem services, ecosystem functions and natural capital. Focusing public attention on city-dwelling species such as wild bees provides great opportunities to demonstrate the importance of conservation to society. The perception of wildlife by society is crucial for effective conservation of biodiversity [102,103], and, since today 74% of the Europe's population lives in cities [104], it is both essential and urgent to raise the awareness of urban citizens on the importance for biodiversity conservation [105].

Supporting Information

Figure S1 Results of the principal component analyses on the landscapes variables over a 500 m radius.

Table S1 Information on the 24 sites of the study.

Table S2 List of recorded bee species list and their functional traits.

Table S3 Significant correlation between landscape variables.

Acknowledgments

We are grateful to all the specialists who identified our bee specimens: Holger Dathe for *Hylaeus* spp., Eric Dufrêne for *Nomada* spp. and *Sphecodes* spp., David Genoud for *Andrena* spp., Gerard Le Goff for *Anthophora* spp., *Amegilla* spp. and Megachilidae, Denis Michez for Melittidae, Alain Pauly for *Halictus* spp. and *Lasioglossum* spp., Stephan Risch for *Tetralonia* spp. and *Eucera* spp., Erwin Scheuchl for *Andrena* spp., and Robert Fonfria for all other genera. We are grateful to Didier Betored for his help with the geographical data. We also thank Charlotte Visage for her great help and skills as coordinator of the Urbanbees project. We are grateful to Vincent Létoublon and the botanists of "Jardin Botanique de Lyon" for their plant identifications. Finally, we thank the persons who helped us in the field and contributed to the project: Jean Aptel, Frédéri Bac, Stan Chabert, Lucia Corredor, Lolita Domon, Fabrice Lafond, Nicolas Morison, Lola Motino and Frédéric Vyghen.

Author Contributions

Conceived and designed the experiments: LF MH MK HM BV. Performed the experiments: LF LG ALG MK HM BV. Analyzed the data: LF MH ALG OR BV. Contributed reagents/materials/analysis tools: LF LG ALG HM OR. Wrote the paper: LF MH MK BV.

References

1. McKinney ML (2006) Urbanization as a major cause of biotic homogenization. Biol Conserv 127: 247–260. doi:10.1016/j.biocon.2005.09.005
2. Sattler T, Borcard D, Arlettaz R, Bontadina F, Legendre P, et al. (2010) Spider, bee, and bird communities in cities are shaped by environmental control and high stochasticity. Ecology 91: 3343–3353. doi:10.1890/09-1810.1
3. Cane JH, Minckley RL, Kervin LJ, Roulston TH, Williams NM (2006) Complex responses within a desert bee guild (Hymenoptera: Apiformes) to urban habitat fragmentation. Ecol Appl 16: 632–644. doi:10.1890/1051-0761(2006)016[0632:CRWADB]2.0.CO;2
4. Clergeau P, Croci S, Jokimäki J, Kaisanlahti-Jokimäki M-L, Dinetti M (2006) Avifauna homogenisation by urbanisation: analysis at different European latitudes. Biol Conserv 127: 336–344. doi:10.1016/j.biocon.2005.06.035
5. Williams NM, Kremen C (2007) Resource distributions among habitats determine solitary bee offsrping production in a mosaic landscape. Ecol Appl 17: 910–921. doi:10.1890/06-0269
6. Bates AJ, Sadler JP, Fairbrass AJ, Falk SJ, Hale JD, et al. (2011) Changing bee and hoverfly pollinator assemblages along an urban-rural gradient. PLoS ONE 6: e23459. doi:10.1371/journal.pone.0023459
7. Zanette LRS, Martins RP, Ribeiro SP (2005) Effects of urbanization on Neotropical wasp and bee assemblages in a Brazilian metropolis. Landsc Urban Plan 71: 105–121. doi:10.1016/j.landurbplan.2004.02.003
8. Tommasi D, Miro A, Higo HA, Winston ML (2004) Bee diversity and abundance in an urban setting. Can Entomol 136: 851–869. doi:10.4039/n04-010
9. Saure C (1996) Urban habitats for bees: the example of city of Berlin. The conservation of bees. London: Aca. pp. 47–52.
10. McIntyre NE, Hostetler ME (2001) Effects of urban land use on pollinator (Hymenoptera: Apoidea) communities in a desert metropolis. Basic Appl Ecol 2: 209–218.
11. Frankie GW, Thorp RW, Schindler M, Hernandez J, Ertter B, et al. (2005) Ecological patterns of bees and their host ornamental flowers in two northern California cities. J Kans Entomol Soc 78: 227–246. doi:10.2307/25086268
12. McFrederick QS, LeBuhn G (2006) Are urban parks refuges for bumble bees *Bombus* spp. (Hymenoptera: Apidae)? Biol Conserv 129: 372–382.
13. Matteson KC, Ascher JS, Langellotto GA (2008) Bee richness and abundance in New York city urban gardens. Ann Entomol Soc Am 101: 140–150. doi:10.1603/0013-8746(2008)101[140:BRAAIN]2.0.CO;2
14. Matteson KC, Langellotto GA (2009) Bumble bee abundance in New York city community gardens: implications for urban agriculture. Cities Environ CATE 2: 5.
15. Fetridge ED, Ascher JS, Langellotto GA (2008) The bee fauna of residential gardens in a suburb of New York city (Hymenoptera: Apoidea). Ann Entomol Soc Am 101: 1067–1077.
16. Westrich P (1996) Habitat requirements of central European bees and the problems of partial habitats. The conservation of bees. London: Academic press. pp. 2–15.
17. Collins JP, Kinzig A, Grimm NB, Fagan WF, Hope D, et al. (2000) A new urban ecology modeling human communities as integral parts of ecosystems poses special problems for the development and testing of ecological theory. Am Sci 88: 416–425.
18. Potts SG, Biesmeijer JC, Kremen C, Neumann P, Schweiger O, et al. (2010) Global pollinator declines: trends, impacts and drivers. Trends Ecol Evol 25: 345–353. doi:10.1016/j.tree.2010.01.007
19. Williams NM, Crone EE, Roulston TH, Minckley RL, Packer L, et al. (2010) Ecological and life-history traits predict bee species responses to environmental disturbances. Biol Conserv 143: 2280–2291. doi:10.1016/j.biocon.2010.03.024
20. Michener CD (2007) The bees of the world. 2nd revised edition. Baltimore and London: The Johns Hopkins University Press. 913 p.
21. Begon M, Townsend CR, Harper JL (2006) Ecology: from individuals to ecosystems. Malden, MA: Blackwell Publishing. 738 p.
22. McIntyre NE, Rango J, Fagan WF, Faeth SH (2001) Ground arthropod community structure in a heterogeneous urban environment. Landsc Urban Plan 52: 257–274. doi:10.1016/S0169-2046(00)00122-5
23. Banaszak-Cibicka W, Zmihorski M (2012) Wild bees along an urban gradient: winners and losers. J Insect Conserv 16: 331–343.
24. Ahrné K, Bengtsson J, Elmqvist T (2009) Bumble bees (*Bombus* spp.) along a gradient of increasing urbanization. PLoS ONE 4: e5574. doi:10.1371/journal.pone.0005574
25. McDonnell MJ, Hahs AK (2008) The use of gradient analysis studies in advancing our understanding of the ecology of urbanizing landscapes: current status and future directions. Landsc Ecol 23: 1143–1155. doi:10.1007/s10980-008-9253-4
26. McDonnell MJ, Pickett STA (1990) Ecosystem structure and function along urban-rural gradients: an unexploited opportunity for ecology. Ecology 71: 1232–1237. doi:10.2307/1938259
27. Insee Rhône-Alpes - Agglo Grand Lyon (ZT9GL) (2013). Available: http://www.insee.fr/fr/regions/rhone-alpes/default.asp?page=themes/dossiers_electroniques/tableau_bord/cdra/cdra_grand_lyon.htm. Accessed 22 July 2013.
28. InfoClimat (2011). Available: http://www.infoclimat.fr/stations-meteo/climato-moyennes-records.php?staid=07481&from=1981&to=2010&redirect=1. Accessed 29 August 2013.
29. Zurbuchen A, Landert L, Klaiber J, Müller A, Hein S, et al. (2010) Maximum foraging ranges in solitary bees: only few individuals have the capability to cover long foraging distances. Biol Conserv 143: 669–676.
30. Westphal C, Bommarco R, Carré G, Lamborn E, Morison N, et al. (2008) Measuring bee diversity in different European habitats and biogeographical regions. Ecol Monogr 78: 653–671.
31. Nielsen A, Steffan-Dewenter I, Westphal C, Messinger O, Potts SG, et al. (2011) Assessing bee species richness in two Mediterranean communities: importance of habitat type and sampling techniques. Ecol Res 26: 969–983. doi:10.1007/s11284-011-0852-1
32. Wilson JS, Griswold T, Messinger OJ (2008) Sampling bee communities (Hymenoptera: Apiformes) in a desert landscape: are pan traps sufficient? J Kans Entomol Soc 81: 288–300. doi:10.2307/25086445
33. Cane JH, Minckley RL, Kervin LJ (2000) Sampling bees (Hymenoptera: Apiformes) for pollinator community studies: pitfalls of pan-trapping. J Kans Entomol Soc 73: 225–231.
34. Campbell JW, Hanula JL (2007) Efficiency of Malaise traps and colored pan traps for collecting flower visiting insects from three forested ecosystems. J Insect Conserv 11: 399–408. doi:10.1007/s10841-006-9055-4
35. Roulston TH, Smith SA, Brewster AL (2007) A Comparison of pan trap and intensive net sampling techniques for documenting a bee (Hymenoptera: Apiformes) fauna. J Kans Entomol Soc 80: 179–181. doi:10.2317/0022-8567(2007)80[179:ACOPTA]2.0.CO;2
36. Toler TR, Evans EW, Tepedino VJ (2005) Pan-trapping for bees (Hymenoptera: Apiformes) in Utah's West Desert: the importance of color diversity. Pan-Pac Entomol 81: 103–113.
37. Dauber J, Hirsch M, Simmering D, Waldhardt R, Otte A, et al. (2003) Landscape structure as an indicator of biodiversity: matrix effects on species richness. Agric Ecosyst Environ 98: 321–329. doi:10.1016/S0167-8809(03)00092-6
38. Morandin LA, Winston ML (2005) Wild bee abundance and seed production in conventional, organic, and genetically modified canola. Ecol Appl 15: 871–881.
39. Pouvreau A (2004) Les insectes pollinisateurs. Paris: Delachaux et Niestlé.
40. Kevan PG, Baker HG (1983) Insects as flower visitors and pollinators. Annu Rev Entomol 28: 407–453.
41. Kuhlmann M (2013) Checklist of the western palaearctic bees (Hymenoptera: Apoidea: Anthophila). Available: http://westpalbees.myspecies.info/. Accessed 10 September 2013.

42. FRAGSTATS: Spatial Pattern Analysis Program for Categorical Maps (2012). Available: http://www.umass.edu/landeco/research/fragstats/fragstats.html. Accessed 22 July 2013.

43. Araújo ED, Costa M, Chaud-Netto J, Fowler HG (2004) Body size and flight distance in stingless bees (Hymenoptera: Meliponini): inference of flight range and possible ecological implications. Braz J Biol 64: 563–568. doi:10.1590/S1519-69842004000400003

44. Greenleaf SS, Williams NM, Winfree R, Kremen C (2007) Bee foraging ranges and their relationship to body size. Oecologia 153: 589–596. doi:10.1007/s00442-007-0752-9

45. Saville NM, Dramstad WE, Fry GLA, Corbet SA (1997) Bumblebee movement in a fragmented agricultural landscape. Agric Ecosyst Environ 61: 145–154. doi:10.1016/S0167-8809(96)01100-0

46. Walther-Hellwig K, Frankl R (2000) Foraging distances of *Bombus muscorum, Bombus lapidarius, and Bombus terrestris* (Hymenoptera, Apidae). J Insect Behav 13: 239–246. doi:10.1023/A:1007740315207

47. Osborne Jl, Clark SJ, Morris RJ, Williams IH, Riley JR, et al. (1999) A landscape-scale study of bumble bee foraging range and constancy, using harmonic radar. J Appl Ecol 36: 519–533. doi:10.1046/j.1365-2664.1999.00428.x

48. Osborne JL, Williams IH (2001) Site constancy of bumble bees in an experimentally patchy habitat. Agric Ecosyst Environ 83: 129–141. doi:10.1016/S0167-8809(00)00262-0

49. Pasquet RS, Peltier A, Hufford MB, Oudin E, Saulnier J, et al. (2008) Long-distance pollen flow assessment through evaluation of pollinator foraging range suggests transgene escape distances. Proc Natl Acad Sci 105: 13456–13461. doi:10.1073/pnas.0806040105

50. Taylor PD, Fahrig L, Henein K, Merriam G (1993) Connectivity is a vital element of landscape structure. Oikos 68: 571–573. doi:10.2307/3544927

51. Colwell RK (2013) Estimates: statistical estimation of species richness and shared species from samples. Version 9. Available: Persistent URL < purl.oclc.org/estimates>.

52. R Development Core Team (2010) R: A language and environment for statistical computing. Version 2.12.0. R Foundation for Statistical. Vienna, Austria. Available: http://www.r-project.org/. Accessed 25 July 2013.

53. Kindt R, Coe R (2005) Tree diversity analysis: A manual and software for common statistical methods for ecological and biodiversity studies. World Agroforestry Centre Eastern and Central Africa Program. Available: http://books.google.com/books?hl=en&lr=&id=zn-xYQoG7ZgC&oi=fnd&pg=PP4&dq=%22manual+and+software+for+common+statistical+methods%22+%22and+biodiversity%22&ots=giZ4Wu5m7A&sig=Gx_p2gUDBS1RUKHdn-2aDMfYVWI. Accessed 14 January 2013.

54. Walther BA, Morand S (1998) Comparative performance of species richness estimation methods. Parasitology 116: 395–405.

55. Neame LA, Griswold T, Elle E (2013) Pollinator nesting guilds respond differently to urban habitat fragmentation in an oak-savannah ecosystem. Insect Conserv Divers 6: 57–66. doi:10.1111/j.1752-4598.2012.00187.x

56. Westrich P (1989) Die Wildbienen Baden-Württembergs Spezieller Teil. Germany: Eugen Ulmer. 536 p.

57. Amiet F, Müller A, Neumeyer R (1999) Apidae 2: Colletes, Dufourea, Hylaeus, Nomia, Nomioides, Rhophitoides, Rophites, Sphecodes, Systropha. 219 p.

58. Amiet F, Herrmann M, Müller A, Neumeyer R (2001) Apidae 3: *Halictus, Lasioglossum.* 208 p.

59. Amiet F, Herrmann M, Müller A, Neumeyer R (2004) Apidae 4: *Anthidium, Chelostoma, Coelioxys, Dioxys, Heriades, Lithurgus, Megachile, Osmia, Stelis.* Centre Suisse de Cartographie de la Faune. 273 p.

60. Amiet F, Herrmann M, Müller A, Neumeyer R (2007) Apidae 5: Ammobates, Ammobatoides, Anthophora, Biastes, Ceratina, Dasypoda, Epeoloides, Epeolus, Eucera, Macropis, Melecta, Melitta, Nomada, Pasites, Tetralonia, Thyreus, Xylocopa. 356 p.

61. Amiet F, Herrmann M, Müller A, Neumeyer R (2010) Apidae 6: *Andrena, Melitturga, Panurginus, Panurgus.* 316 p.

62. Oertli S, Mueller A, Dorn S (2005) Ecological and seasonal patterns in the diversity of a species-rich bee assemblage (Hymenoptera: Apoidea: Apiformes). Eur J Entomol 102: 53–63.

63. Cane JH (1987) Estimation of bee size using intertegular span (Apoidea). J Kans Entomol Soc 60: 145–147.

64. Rice WR (1989) Analyzing tables of statistical tests. Evolution 43: 223–225.

65. Williams NM, Minckley RL, Silveira FA (2001) Variation in native bee faunas and Its implications for detecting community changes. Conserv Ecol 5: [online] URL: http://www.consecol.org/vol5/iss1/art7/.

66. Leonhardt SD, Gallai N, Garibaldi LA, Kuhlmann M, Klein A-M (2013) Economic gain, stability of pollination and bee diversity decrease from southern to northern Europe. Basic Appl Ecol: http://dx.doi.org/10.1016/j.baae.2013.06.003. doi:10.1016/j.baae.2013.06.003

67. Lelièvre F, Sala S, Volaire F (2010) Climate change at the temperate-Mediterranean interface in southern France and impacts on grasslands production. Option Méditerranéennes: 187–192.

68. Michener CD (1979) Biogeography of the bees. Ann Mo Bot Gard 66: 277–347. doi:10.2307/2398833

69. Sheffield CS, Pindar A, Packer L, Kevan PG (2013) The potential of cleptoparasitic bees as indicator taxa for assessing bee communities. Apidologie: 10. doi:10.1007/s13592-013-0200-2

70. Combes C (1996) Parasites, biodiversity and ecosystem stability. Biodivers Conserv 5: 953–962. doi:10.1007/BF00054413

71. Morand S, Gonzalez EA (1997) Is parasitism a missing ingredient in model ecosystems? Ecol Model 95: 61–74. doi:10.1016/S0304-3800(96)00028-2

72. Horwitz P, Wilcox BA (2005) Parasites, ecosystems and sustainability: an ecological and complex systems perspective. Int J Parasitol 35: 725–732. doi:10.1016/j.ijpara.2005.03.002

73. Marcogliese DJ (2004) Parasites: small players with crucial roles in the ecological theater. EcoHealth 1: 151–164. doi:10.1007/s10393-004-0028-3

74. Hudson PJ, Dobson AP, Lafferty KD (2006) Is a healthy ecosystem one that is rich in parasites? Trends Ecol Evol 21: 381–385. doi:10.1016/j.tree.2006.04.007

75. Wood CL, Byers JE, Cottingham KL, Altman I, Donahue MJ, et al. (2007) Parasites alter community structure. Proc Natl Acad Sci 104: 9335–9339. doi:10.1073/pnas.0700062104

76. Moretti M, De Bello F, Roberts SPM, Potts SG (2009) Taxonomical vs. functional responses of bee communities to fire in two contrasting climatic regions. J Anim Ecol 78: 98–108.

77. Sheffield CS, Kevan PG, Pindar A, Packer L (2012) Bee (Hymenoptera: Apoidea) diversity within apple orchards and old fields in the Annapolis Valley, Nova Scotia, Canada. Can Entomol 145: 94–114. doi:10.4039/tce.2012.89

78. Baum KA, Wallen KE (2011) Potential bias in pan trapping as a function of floral abundance. J Kans Entomol Soc 84: 155–159. doi:10.2317/JKES100629.1

79. Kolasa J, Rollo CD (1991) Introduction: The Heterogeneity of Heterogeneity: A Glossary. In: Kolasa J, Pickett STA, editors. Ecological Heterogeneity. Ecological Studies. Springer New York. pp. 1–23. Available: http://link.springer.com/chapter/10.1007/978-1-4612-3062-5_1. Accessed 14 October 2013.

80. Porter EE, Forschner BR, Blair RB (2001) Woody vegetation and canopy fragmentation along a forest-to-urban gradient. Urban Ecosyst 5: 131–151. doi:10.1023/A:1022391721622

81. Gaston KJ, Smith RM, Thompson K, Warren PH (2005) Urban domestic gardens (II): experimental tests of methods for increasing biodiversity. Biodivers Conserv 14: 395–413.

82. Hernandez JL, Frankie GW, Thorp RW (2009) Ecology of urban bees: a review of current knowledge and directions for future study. Cities Environ CATE 2: 3.

83. Kirk WD, Howes F (2012) Plants for bees: a guide to the plants that benefit the bees of the British Isles. Cardiff: International Bee Research Association. 312 p.

84. Gathmann A, Tscharntke T (2002) Foraging ranges of solitary bees. J Anim Ecol 71: 757–764. doi:10.1046/j.1365-2656.2002.00641.x

85. Klein A-M, Cunningham SA, Bos M, Steffan-Dewenter I (2008) Advances in pollination ecology from tropical plantation crops. Ecology 89: 935–943. doi:10.1890/07-0088.1

86. Henle K, Davies KF, Kleyer M, Margules C, Settele J (2004) Predictors of species sensitivity to fragmentation. Biodivers Conserv 13: 207–251. doi:10.1023/B:BIOC.0000004319.91643.9e

87. Winfree R, Griswold T, Kremen C (2007) Effect of human disturbance on bee communities in a forested ecosystem. Conserv Biol 21: 213–223. doi:10.1111/j.1523-1739.2006.00574.x

88. Steffan-Dewenter I, Klein A-M, Gaebele V, Alfert T, Tscharntke T (2006) Bee diversity and plant–pollinator inter- actions in fragmented landscapes. In: Waser NM, Ollerton J, editors. Plant–pollinator interactions: from specialization to generalization. Illinois, USA: University of Chicago Press. pp. 387–407.

89. Niemelä J (1999) Ecology and urban planning. Biodivers Conserv 8: 119–131. doi:10.1023/A:1008817325994

90. Rebele F (1994) Urban ecology and special features of urban ecosystems. Glob Ecol Biogeogr Lett 4: 173–187. doi:10.2307/2997649

91. McIntyre NE (2000) Ecology of urban arthropods: a review and a call to action. Ann Entomol Soc Am 93: 825–835. doi:10.1603/0013-8746(2000)093[0825:EOUAAR]2.0.CO;2

92. Witte R, Diesing D, Godde M (1985) Urbanophobe, urbanoneutral, urbanophile–behavior of species concerning the urban habitat. Flora 177: 265–282.

93. Blair RB (2001) Birds and Butterflies Along Urban Gradients in Two Ecoregions of the United States: Is Urbanization Creating a Homogeneous Fauna? In: Lockwood JL, McKinney ML, editors. Biotic Homogenization. Springer US. pp. 33–56. Available: http://link.springer.com/chapter/10.1007/978-1-4615-1261-5_3. Accessed 14 October 2013.

94. Kuhn I, Brandl R, Klotz S (2004) The flora of German cities is naturally species rich. Evol Ecol Res 6: 749–764.

95. Goulson D, Lye GC, Darvill B (2008) Decline and conservation of bumble bees. Annu Rev Entomol 53: 191–208. doi:10.1146/annurev.ento.53.103106.093454

96. Tscharntke T, Klein AM, Kruess A, Steffan-Dewenter I, Thies C (2005) Landscape perspectives on agricultural intensification and biodiversity – ecosystem service management. Ecol Lett 8: 857–874. doi:10.1111/j.1461-0248.2005.00782.x

97. Samways MJ, Stork NE, Cracraft J, Eeley HA, Foster M, et al. (1995) Scales, planning and approaches to inventorying and monitoring. In: Heywood V, Watson RT, editors. Global biodiversity assessment. Cambridge, UK. pp. 475–517.

98. Kremen C, Ullman KS, Thorp RW (2011) Evaluating the quality of citizen-scientist data on pollinator communities: citizen-scientist pollinator monitoring. Conserv Biol 25: 607–617. doi:10.1111/j.1523-1739.2011.01657.x

99. Entwistle AC, Dunstone N, Mickleburgh S (2000) Mammal conservation: current contexts and opportunities. In: Entwistle AC, Dunstone N, editors. Priorities for the Conservation of Mammalian Diversity: Has the Panda Had Its Day? Cambridge UK: Cambridge University Press. pp. 1–7.

100. Danforth B (2007) Bees. Curr Biol 17. Available: http://cat.inist.fr/?aModele = afficheN&cpsidt = 18586301. Accessed 26 August 2013.

101. Steffan-Dewenter I, Potts SG, Packer L (2005) Pollinator diversity and crop pollination services are at risk. Trends Ecol Evol 20: 651–652.

102. Clucas B, McHugh K, Caro T (2008) Flagship species on covers of US conservation and nature magazines. Biodivers Conserv 17: 1517–1528. doi:10.1007/s10531-008-9361-0

103. Home R, Keller C, Nagel P, Bauer N, Hunziker M (2009) Selection criteria for flagship species by conservation organizations. Environ Conserv 36: 139–148. doi:10.1017/S0376892909990051

104. United Nations (2007) World population prospects. the 2006 revision. New York: United Nations.

105. Ramalho CE, Hobbs RJ (2012) Time for a change: dynamic urban ecology. Trends Ecol Evol 27: 179–188. doi:10.1016/j.tree.2011.10.008

19

The Value of Countryside Elements in the Conservation of a Threatened Arboreal Marsupial *Petaurus norfolcensis* in Agricultural Landscapes of South-Eastern Australia—The Disproportional Value of Scattered Trees

Mason J. Crane*, David B. Lindenmayer, Ross B. Cunningham

Fenner School of Environment and Society, The Australian National University, Canberra, ACT, Australia

Abstract

Human activities, particularly agriculture, have transformed much of the world's terrestrial environment. Within these anthropogenic landscapes, a variety of relictual and semi-natural habitats exist, which we term countryside elements. The habitat value of countryside elements (hereafter termed 'elements') is increasingly recognised. We quantify the relative value of four kinds of such 'elements' (*linear roadside remnants*, *native vegetation patches*, *scattered trees* and *tree plantings*) used by a threatened Australian arboreal marsupial, the squirrel glider (*Petaurus norfolcensis*). We examined relationships between home range size and the availability of each 'element' and whether the usage was relative to predicted levels of use. The use of 'elements' by gliders was largely explained by their availability, but there was a preference for *native vegetation patches* and *scattered trees*. We found home range size was significantly smaller with increasing area of *scattered trees* and a contrasting effect with increasing area of *linear roadside remnants* or *native vegetation patches*. Our work showed that each 'element' was used and as such had a role in the conservation of the squirrel glider, but their relative value varied. We illustrate the need to assess the conservation value of countryside elements so they can be incorporated into the holistic management of agricultural landscapes. This work demonstrates the disproportional value of *scattered trees*, underscoring the need to specifically incorporate and/or enhance the protection and recruitment of *scattered trees* in biodiversity conservation policy and management.

Editor: Paul Adam, University of New South Wales, Australia

Funding: The authors have no support or funding to report.

Competing Interests: The authors have declared that no competing interests exist.

* Email: mason.crane@anu.edu.au

Introduction

Humans have transformed most of the earth's terrestrial biosphere into highly modified biomes [1], resulting in the loss and decline of many species [2]. To counter this, much focus has been on the establishment of formal conservation reserves. While these are a critical component of biodiversity conservation worldwide, reserve networks rarely provide comprehensive, adequate and representative coverage of ecosystems [3–6]. It is often the case that those natural ecosystems associated with agricultural land are the most poorly represented in conservation reserve networks. Australian box gum woodlands are a classic example of this [7]. Despite comparatively low levels of reservation in agricultural landscapes, many species associated with such ecosystems still persist there. Such species have been shown to use the relictual and semi-natural habitats of these landscapes (e.g. [8,9–15]). These habitats or countryside elements ('elements') [16] often have very different and distinct characteristics, each offering a different range of resources for wildlife.

The role that countryside elements and matrix habitats play in biodiversity conservation is increasingly appreciated. This has been demonstrated by work conducted in Australia [16], Central and South America [10,17,18] and West Africa [19]. However, to date there have been few works comparing different kinds of 'elements' and quantifying their relative usage for a single species. In this study, we examined the relative value of four kinds of wooded countryside elements in the conservation of a threatened arboreal marsupial, the squirrel glider *Petaurus norfolcensis*, in agricultural landscapes of south-eastern Australia. These 'elements' were: *scattered trees*, *tree plantings*, *linear roadside remnants* and small *native vegetation patches*. These four broad kinds of 'elements' can be found in many agricultural landscapes around the world (e.g. [20,21–24]). While their use as wildlife habitat has been reported (e.g. [12,20,21,25–28]), there are few unequivocal examples of their relative value in the conservation of different species or taxa [16,29]. To achieve this we address the following questions:

1. Is the use of an 'element' for denning and feeding proportional to their availability? If not, is there evidence of preferential usage of some kinds of 'elements'?

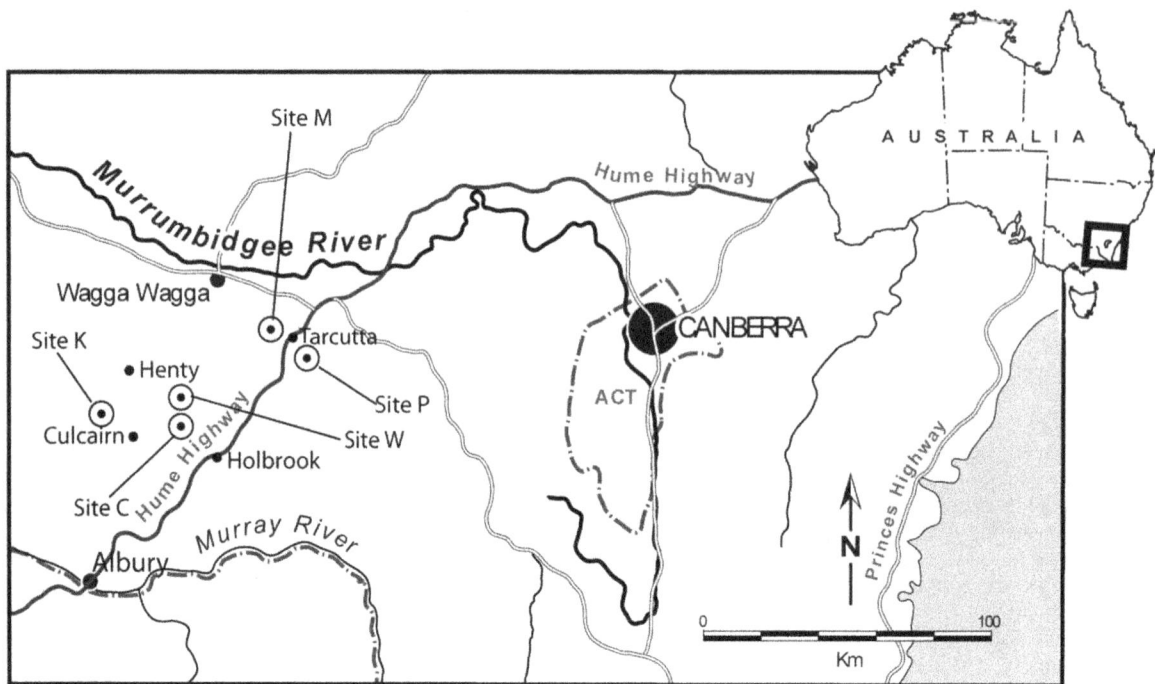

Figure 1. Location of the five study areas (C, K, M, P and W) used in radio-tracking the Squirrel Glider.

2. Does home range size depend on the proportional availability of an 'element'? Differences in home range size within species have often been attributed to habitat quality [30–32], with smaller home range indicating better quality habitat. If an 'element' offers a significantly higher or lower quality habitat relative to others, it would be expected that their availability would have a significant influence on home range size.

Our work is the first to attempt to quantify the relative contribution of different countryside elements for the squirrel glider and is one of the few that attempt to compare the ecological value of countryside elements, using empirical data. An understanding of the value of such 'elements' for biodiversity is essential for better integration of biodiversity conservation and broader management of agricultural landscapes. Without such data, managers are in danger of undervaluing certain 'elements'. This may, in turn, lead to the loss of critical habitats and further threaten associated species in these landscapes.

Methods

The squirrel glider

The squirrel glider is a nocturnal, arboreal, gliding marsupial in the Family Petauridae. It is a medium-sized possum weighing between 190–300 g and which feeds on invertebrates, insect exudates, sap of trees and shrubs, and pollen and nectar [33–35]. The species is listed as threatened in three of the four Australian states in which it is found (Victoria – *Flora and Fauna Guarantee Act 1988*; South Australia – *National Parks and Wildlife Act 1972*; New South Wales – *Threatened Species Conservation Act 1995*; and Queensland – no formal listing).

Study area

Our investigation encompassed five study areas within the south-west slopes of New South Wales, Australia [36] (Figure 1).

The region is the most extensively and intensively disturbed of the 13 botanical regions of NSW, with an estimated 85% of the original cover of native vegetation removed in the past 200 years [37]. The five study areas were located in heavily modified agricultural landscapes, used predominantly for livestock grazing and dryland cropping. Study areas were approximately 3 km×3 km. Woody vegetation occurred primarily as relictual scattered paddock trees, native vegetation plantings and remnant temperate *Eucalyptus* woodlands on private lands, road reserves and travelling stock reserves.

Ethics statement

We conducted trapping and radio tracking under The Australian National University Animal Ethics Committee protocol number C.RE.39.05. Squirrel Gliders are a native species and therefore protected. Relevant permits to handle the animals were obtained from New South Wales Government agencies. Animals were captured in wire mesh cage traps covered with non-transparent, heavy-duty plastic sleeves to minimise stress on animals and protect from cold and wet weather. Qualified wildlife veterinarians anaesthetised captured animals using isofluorane gas delivered via a portable gas anaesthesia machine. Isofluorane anaesthesia ensures recovery of animals within minutes, which is considerably faster than injectable agents, thus minimising holding time and stress. Anaesthetising animals ensured that accurate measurements of body size and reproductive status could be made without undue stress to the animals. It also enabled the veterinarians to properly fit radio-collars and microchip each animal. Gliders not fitted with collars were ear tagged (hamster ear tags, Sieper & Co., Sydney, Australia) in each ear. When animals had recovered from the anaesthetic, they were released at the exact point of capture.

Land accessed was a mix of privately owned farmland, local government managed road reserves, and travelling stock reserves managed by the Livestock Health and Pest Authority and relevant access permissions were obtained from the respective land managers.

Radio-tracking

We captured gliders using drop-door, wire mesh cage traps (170 mm×200 mm×500 mm) over a three night period at each site in March 2005 [38]. We fitted 32 gliders with a single stage brass loop radio transmitter, weighing 4.5 grams (Sirtrack, New Zealand). When selecting which gliders were to be collared, we preferred adult gliders and attempted to achieve an equal sex ratio and an equal spatial coverage of animals within and between sites. We tracked gliders to their diurnal denning site at least twice a week and to a nocturnal location at least 1–3 times every 14 days, over a 4–5 month period [38,39]. For each fix, we recorded the countryside element in which the glider was located.

Home range

We derived home range estimates by using both diurnal and nocturnal fixes. The woody vegetation was scattered or represented in irregularly shaped patches, surrounded by cleared agricultural matrix. Common parametric approaches such as the minimum convex polygon method were therefore unsuitable. This is because they would have (inappropriately) included large areas of unused cleared agricultural land – as has been found by Martin et al. [40] and van der Ree and Bennett [15]. We estimated home range by using the non-parametric, grid cell method [41]. The size of the grid cell is arbitrary and can have a major influence on either underestimating or over-estimating the den range size [41]. Three sized cells were tested 40×40 m, 50×50 m and 60×60 m and compared to the minimum convex polygon (MCP) method (where MCP could be appropriately used, i.e. for some animals in areas dominated by remnant vegetation patches or large clusters of scattered trees). We calculated the minimum convex polygon using Home Range extension within Arc View GIS (ESRI, California, USA). We selected a 50×50 m grid cell as it aligned more closely with the commonly used MCP method. We connected disjointed cells by including cells that were crossed by the most direct line joining to consecutive locations, taking into consideration gap-crossing ability ([i.e. gaps in canopy <70 m; see [42]). We estimated the home range size of each squirrel glider on 95% of fixes. This was done to give an objective, repeatable method of comparison of normal home range [41]. We deleted 5% of fixes at the extremities of each animal's range to reduce the influence of exploratory movements or outlying fixes outside the 'normal' home range [15].

Countryside elements

Within our study area, we recognised four categories of countryside elements that contained woody vegetation.

(1) **Linear roadside remnants.** These were linear strips of remnant vegetation along roads. Remnant roadside vegetation is a major feature across agricultural landscapes providing a network of remnant vegetation corridors across what are otherwise generally heavily cleared landscapes. The width of the roadside

reserves in this study ranged from 40–60 m. Vegetation along road reserves is subject to high levels of disturbance such as road construction and maintenance. However, grazing pressure by domestic livestock is often low and irregular. These areas regularly contain regeneration of overstorey species. Native understorey species are generally present but their dominance and diversity varies.

(2) **Native vegetation patches.** These were patches of remnant vegetation where the understorey was dominated by a diverse array of native plants. In this study, these areas were mostly on travelling stock reserves, and some patches of remnant native vegetation on freehold land. The travelling stock route network was established more than 150 years ago to facilitate the movement of domestic livestock between properties and to markets [43]. The network is made up of travelling stock routes (which today are often incorporated into the road reserves) and holding paddocks which are generally referred to as travelling stock reserves (TSRs). Because of their reservation for these purposes, these areas generally escaped clearing and continuous high-intensity livestock grazing. The *native vegetation patches* varied in size with the largest patches occurring on TSRs (≈100 ha) and the smallest patches on freehold land (>5 ha).

(3) **Scattered trees.** These were scattered, (mostly large, old) relictual trees remaining on land used for grazing or cropping. They include dead and living trees, are often widely spaced (more widely spaced than expected in their natural state) and contain a simple understorey, generally dominated by introduced grasses and forbs, with a low diversity of native plant species.

(4) **Tree plantings.** These were Australian native vegetation plantings, generally containing dense stands of trees (predominately *Eucalyptus*) and shrubs (such as *Acacia* and *Melaleuca*). The species composition was predominantly locally indigenous species, but often included species naturally found outside the region. *Tree plantings* vary in their shape and size as they were planted for various purposes such as shelterbelts or to reduce rising water tables [44]. The level of grazing by livestock within them also varied.

The vegetation was assigned to the countryside element of best fit, using the above descriptions as a guide. Where elements were adjacent to each other, boundaries were defined by differences in vegetation structure and composition, and/or management practices. We calculated the area of each countryside element available to an individual squirrel glider by measuring the total area of woody vegetation attributed to that 'element', within a 1000 m radius of the centre point of all fixes for each individual glider. We used a 1000 m radius, as 2000 m is approximately the maximum home range length that has been reported for our study species [15]. We measured the area of woody vegetation using geographical information systems software (ArcGIS 9.2-esri) to draw polygons over the canopy of woody vegetation interpreted from satellite imagery (spot 5-Astrium). We deemed that woody vegetation isolated by a gap distance of greater than 70 metres was unavailable to gliders [42].

Table 1. Number of squirrel gliders captured/collared at each site.

	Site C	Site K	Site M	Site P	Site W	Total
No. captures	5	8	25	3	11	52
No. collars fitted	5	4	12	3	8	32

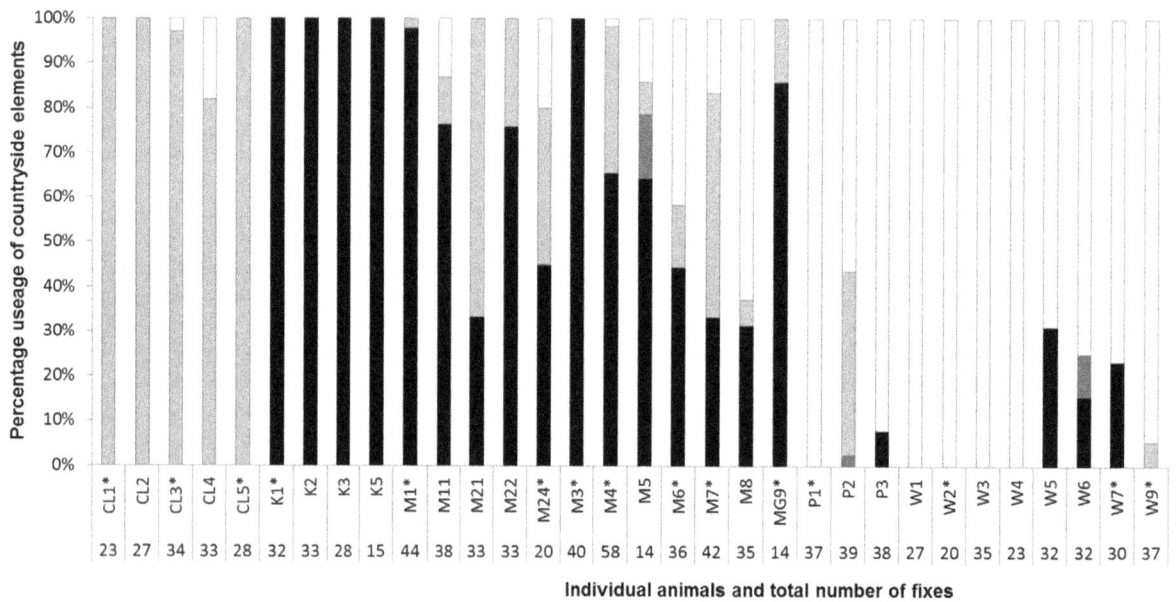

Figure 2. The percentage usage of countryside elements by individual squirrel gliders (* = female). *Remnant vegetation patches* (black shading), *tree planting* (dark grey shading), *linear roadside remnant* (intermediate grey shading) and *scattered trees* (light grey shading).

Data analyses

For each squirrel glider, our data consisted of a count of the number of fixes in each 'element', classified by circadian time - day or night. Associated data were the total area of woody vegetation in each 'element', available within a 1000 m radius of the centre point of all fixes for each individual glider. From these data, we can obtain the expected number of fixes by apportioning total fixes according to the relative area of each 'element'. Thus, for day and night data separately, we have a 2-way contingency table cross-classified by animal ID and 'element' type, where the cells are the observed count of the number of fixes and a concomitant variable is the expected number of fixes based on the relative availability of each of the 'elements'.

The Model

Considered as a two-way contingency table, our data can be modelled by

$$E\left[\log\left(y_{ij}\right)\right] = u + b\,\log\left(f_{ij}\right) + animal_i + CSE_j$$

where y_{ij} are the expected frequencies, u is the grand mean, f_{ij} are the 'expected' frequencies derived from the percentage are occupied by '*element*'$_j$ *animal*$_i$ and '*element*'$_j$ are constants to

account for the marginal distribution of the two-way animal by 'element' table.

The above model is a particular case of a class of models commonly known as log-linear models, and which are often used in the analyses of multiway contingency tables. The log-linear model belongs to the class of generalised linear models [45]. These models can be fitted by the use of maximum likelihood methods available in many statistical software packages, such as GENSTAT Version 15. The goodness-of-fit of the model and the significance of individual terms can be assessed by examining an analysis of deviance.

We completed further linear regression analysis to quantify relationships between home range size and total woody vegetation and each of the countryside elements. We avoided statistical issues associated with intrinsic collinearity of countryside elements by considering each of the 'elements' separately.

Results

Radio-tracking

We captured 52 individual gliders and fitted 32 with radio-transmitting collars (see Table 1). The numbers of fixes for individuals varied as signals were lost for some animals throughout

Table 2. Accumulated analysis of deviance Table after fitting a sequence of log-linear Poisson models (see section 'The Model').

Term	d.f.	Night use	Day use
+ Animal	31	27	76
+ CSE	3	186	391
Residual	91	576	949
+ log(f)	1	367	675
Residual (final)	92	209	286
Total	127	789	1416

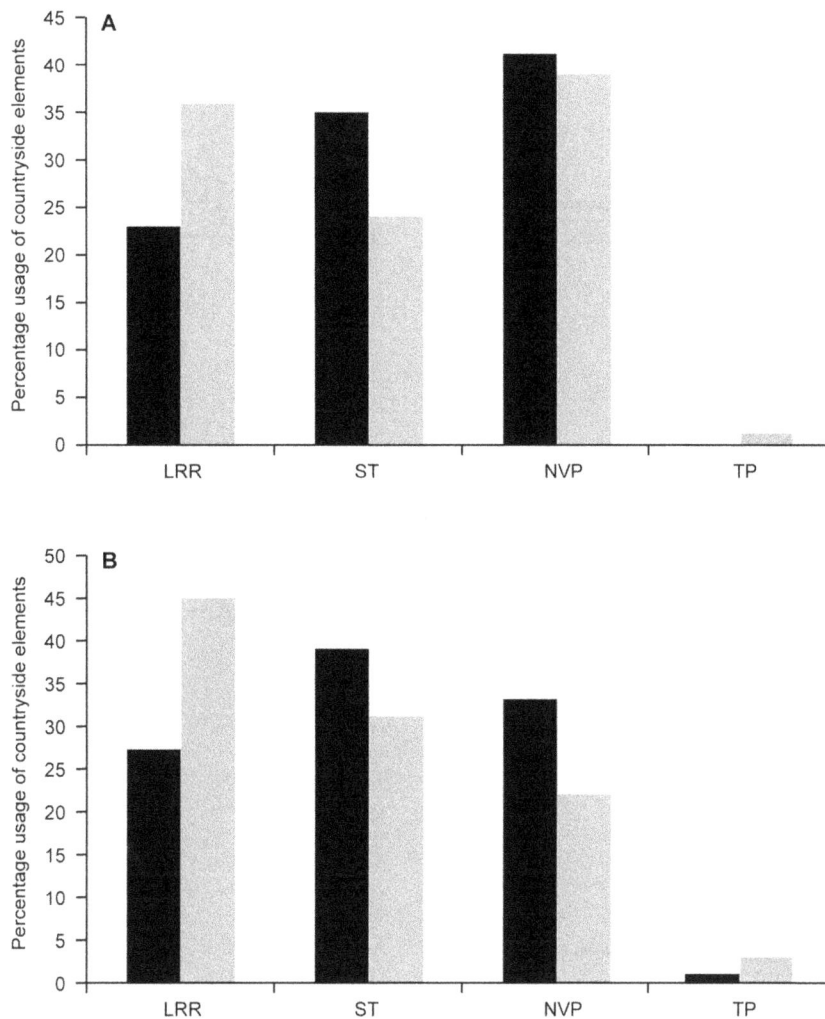

Figure 3. Actual usage verse the predicted usage of countryside elements. LRR = Linear roadside remnants, ST = Scattered trees, NVP = Native vegetation patches, TP = Tree plantings. Actual percentage usage is shown as black shading and the predicted percentage usage is shown as grey shading. Panel (a) is based on diurnal fixes, and Panel (b) is based on nocturnal fixes.

the study. Over a five month period, we tracked gliders to 1027 independent locations (655 diurnal and 372 nocturnal locations). We tracked individual gliders to an average of 21 ± 1.16 (mean $\pm$ s.e.) diurnal locations and to 12 ± 0.6 (mean $\pm$ s.e.) nocturnal locations.

Usage of countryside elements

We found that gliders used all four wooded countryside elements nocturnally and all but one element (*tree plantings*) for diurnal denning. Some individual gliders exclusively used one category of 'element', *scattered trees*, *linear roadside remnants* or *native vegetation* (Figure 2).

The fit of the model as defined above is summarised by the analysis of deviance (Table 2). As can be seen from Table 2, 64% (367/576) and 71% of the residual deviance was accounted for by the 'expected' count based on the percentage of available 'elements' for each animal, for night and day use respectively. The final residual deviance, based on 92 degrees of freedom after fitting the terms for animal, 'element' and expected frequency was 209 and 286, for night and day, respectively. This suggested that there remains non-random unexplained variation.

A further breakdown of the components of this residual variation revealed evidence of preferential selection of *native vegetation patches* and *scattered trees* nocturnally, and a strong preference to *scattered trees* for diurnal use, as is shown in Figure 3.

Relationship between home range size and availability of CSEs

Home ranges varied from 2.5 to 12 ha and averaged 4.9 ± 0.45 ha (mean ±1 s.e.). Home range size increased with addition fixes, but plateaued towards the end of our study (Figure 4).

We found no significant relationship between total woody vegetation cover and home range size. However, there were significant relationships between home range size and area of woody vegetation within three 'elements'. We found a significant ($p<0.001$) negative relationship between home range size and the available area of *scattered trees* and significant ($p=0.009$ and $p=0.022$) positive relationships with the available area of *native vegetation patches* (Figure 5) and *linear roadside remnants*.

Site M

Site K

Site C

Site W

Site P

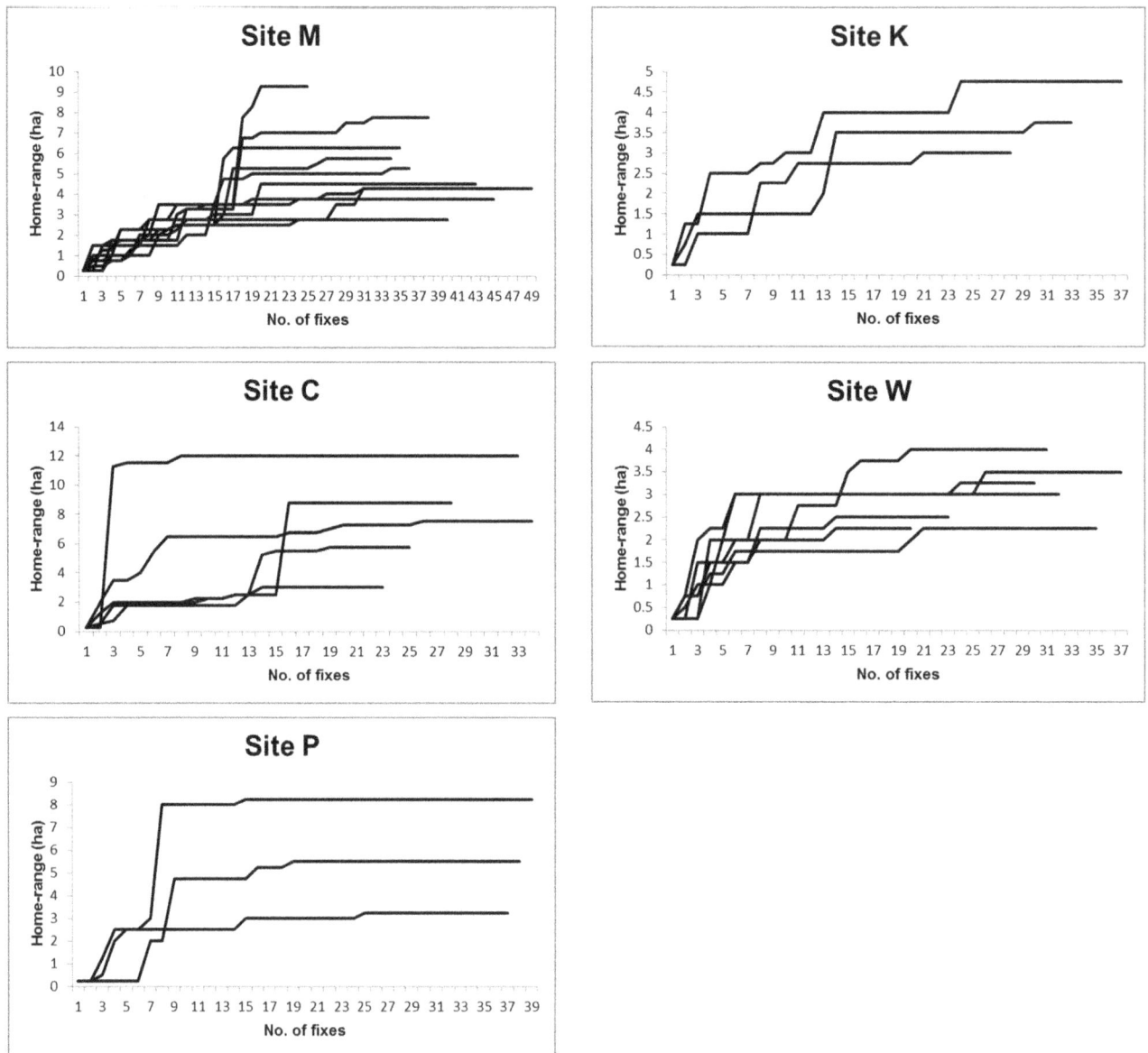

Figure 4. Home range estimates plotted against the number of fixes in sequential tracking order. The fixes are for all animals (with >20 fixes) at each of the five study sites.

Discussion

The contribution that countryside elements make towards biodiversity conservation is increasingly recognised in many ecosystems around the world [16–19]. The challenge for conservationists is to recognise different 'elements' and to understand the role each play in the conservation of different species. Our work is one of the few studies that quantify the relative value of different countryside elements for a single species. We provide empirical evidence of the disproportionate value of *scattered trees* in agricultural landscapes. Our work also highlights the need to examine the roles and values of countryside elements in biodiversity conservation, particular for species such as the squirrel glider which has a distribution largely confined to highly modified agricultural landscapes.

Across our five study sites, gliders relied entirely on countryside elements (*scattered trees, linear remnant vegetation, native*

vegetation patches and *tree plantings*) located within road reserves, traveling stock reserves and freehold land. Our study demonstrates that the squirrel glider will use all of these four wooded 'elements', with availability being a key factor in determining usage.

In agricultural landscapes, roadside vegetation provides important habitat for many species of wildlife, particularly in facilitating migration and dispersal [20,46–48]. We found that the squirrel glider would commonly use *linear roadside remnants*, with some individuals relying entirely on this 'element'. The ability of *linear roadside remnants* to support squirrel glider populations has been previously reported [28]. While it is clear this kind of countryside element is of conservation importance to this species, we found evidence that suggests *linear roadside remnants* may provide inferior quality habitat compared *to native vegetation patches* and *scattered trees*. Larger home range sizes associated with increased area *of linear roadside remnants* and the under-utilisation of this 'element' is evidence of this. This may be explained by structural

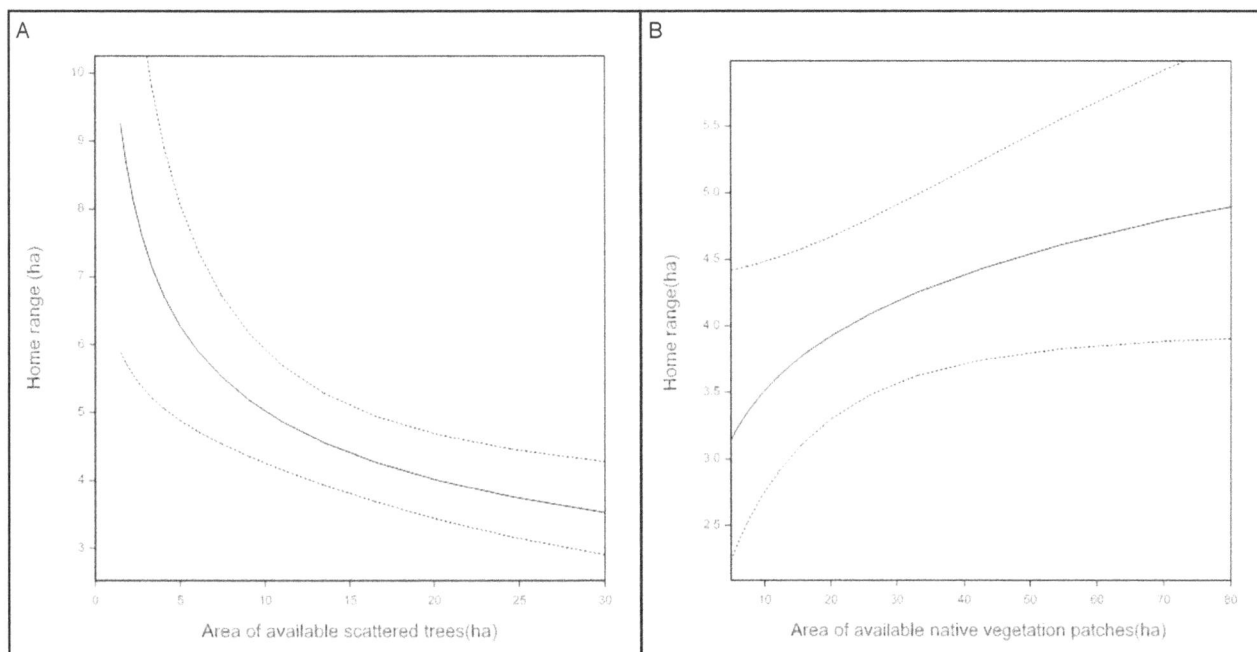

Figure 5. Influence of available area of three countryside elements on squirrel glider home range size. Panel (a) scattered trees, Panel (b) native vegetation patches. Based on 95% of fixes and associated 95% confidence intervals.

differences in the vegetation between the 'elements', such as a *linear roadside remnants* containing a higher proportion of small regrowth trees (which have been shown to offer poorer quality habitat than large trees [34,38,39]). Issues associated with the geometry of linear habitats also may explain our findings. Lindenmayer et al. [49] and Recher et al. [50] highlight the problems for species inhabiting linear habitats, such as disrupted social behaviour and additional expenditure of energy in obtaining food.

While *native vegetation patches* in agricultural landscapes are often small and highly fragmented, they have been shown to be important habitat for many species [24,51,52]. In our study, the majority (over 90%) of *native vegetation patches* occurred on traveling stock reserves. These reserves themselves have been shown to have high conservation value (higher than remnant vegetation on private land, particularly for arboreal marsupials) [53]. We found that *native vegetation patches* were heavily used by gliders, with a number of individuals using this 'element' exclusively. The use of *native vegetation patches* was higher than predicted for both diurnal and nocturnal activity. While *native vegetation patches* were preferentially used, gliders in areas dominated by *native vegetation patches* had larger home ranges. This result may suggest that while quality habitat exists within *native vegetation patches*, it is often dispersed among poorer habitats, resulting in gliders having to cover larger areas to gather resources.

Tree plantings have been shown to play an important role in the conservation of some species in agricultural landscapes, particularly birds [9,27]. For arboreal marsupial conservation, *tree plantings* may have a more limited role, at least in the short to medium term [54]. Our data indicated that *tree plantings* were used less than predicted based on availability. When compared to other treed 'elements', plantings would seem to be of minor importance to the species. However, *tree plantings* were nevertheless still used and should not be discounted as their value may

increase over time as trees mature, given squirrel gliders' preference for large trees [34,38,39]. *Tree plantings* are the only countryside element examined here that can be readily introduced into these highly modified landscape. Our data suggests that for *tree plantings* to be used, they must be located in association with other countryside elements containing remnant trees.

Scattered trees are widely recognised as a critically important countryside element in many agricultural landscapes globally and are considered a keystone structure in such landscapes [23]. Numerous studies have highlighted the ecological value of *scattered trees* for various taxa of wildlife [12,16,25,26,52]. Fischer et al. [29] have shown that *scattered trees* have a disproportionate value for birds and bats. Our work provides further evidence of this. The home ranges of gliders were significantly smaller with increasing coverage of *scattered trees*, an opposite pattern to all other 'elements'. We also found that *scattered trees* were used at higher rates than predicted, particularly as diurnal den sites. Both results indicate that scattered trees have a comparatively higher habitat value for the squirrel glider than the other three kinds of countryside elements we examined.

Scattered trees are often the oldest living structures in disturbed landscapes [23]. This was the case in our study, with *scattered trees* represented by predominately large mature relictual *Eucalyptus* trees. These trees generally had well developed hollows and a large canopy, both kinds of resources have been shown to be preferentially sought by the squirrel glider [34,38,39]. Despite this, *scattered trees* are arguably under the greatest threat of all countryside elements and populations of such trees are rapidly diminishing resource in many agricultural landscapes [55,56]. It is estimated that tens of millions of *scattered trees* will be lost in grazed landscapes in Australia over the next 50 years, due to factors such as natural attrition, clearing, and a lack of regeneration [55] – a problem common to many agricultural landscapes globally [56]. Current conservation strategies offer little protection for *scattered trees* [55], and they are cleared for cropping and irrigation [57]. Without a concerted shift in conservation policy and awareness in general of the value of *scattered trees*, there is a real risk of losing this critical countryside element.

Conclusions

Agricultural landscapes are dynamic. Changing technologies, land uses and attitudes continually transform these systems [58]. There are significant threats to the integrity and extent of various kinds of countryside elements within agricultural landscapes. Therefore, understanding the contribution they make to biodiversity conservation is essential. In the case of the squirrel glider, we found that four kinds of countryside elements are important, but their relative values vary. We suggest that these countryside elements need to be better integrated into landscape management.

References

1. Ellis CE (2011) Anthropogenic transformation of the terrestrial biosphere. Philos T R Soc B 369: 1010–1035.
2. Chapin FS III, Zavaleta ES, Eviner VT, Naylor RL, Vitousek PM, et al. (2000) Consequences of changing biodiversity. Nature 405: 234–242.
3. Aung UM (2007) Policy and practice in Myanmar's protected area system. J Environ Manage 84: 188–203.
4. Ioja CI, Patroescu M, Rozylowicz L, Popescu VD, Verghelet M, et al. (2010) The efficacy of Romania's protected areas network in conserving biodiversity. Biol Conserv 143: 2468–2476.
5. Pressey RL, Tully SL (1994) The cost of adhoc reservation: A case study in western New South Wales. Aust J Ecol 19: 375–384.
6. Rouget M, Richardson DM, Cowling RM (2003) The current configuration of protected areas in the Cape Floristic Region, South Africa - reservation bias and representation of biodiversity patterns and processes. Biol Conserv 112: 129–145.
7. Prober SM (1996) Conservation of the grassy white box woodlands: Rangewide floristic variation and implications for reserve design. Aust J Bot 44: 57–77.
8. Burel F (1996) Hedgerows and their role in agricultural landscapes. Critical Rev Plant Sci 15: 169–190.
9. Cunningham RB, Lindenmayer DB, Crane M, Michael DR, MacGregor C, et al. (2008) The combined effects of remnant vegetation and tree planting on farmland birds. Conserv Biol 22: 742–752.
10. Daily GC, Ceballos G, Pacheco J, Suzan G, Sanchez-Azofeifa A (2003) Countryside biogeography of neotropical mammals: Conservation opportunities in agricultural landscapes of Costa Rica. Conserv Biol 17: 1814–1826.
11. Hazell D, Cunnningham R, Lindenmayer D, Mackey B, Osborne W (2001) Use of farm dams as frog habitat in an Australian agricultural landscape: factors affecting species richness and distribution. Biol Conserv 102: 155–169.
12. Lumsden LF, Bennett AF (2005) Scattered trees in rural landscapes: foraging habitat for insectivorous bats in south-eastern Australia. Biol Conserv 122: 205–222.
13. Michael DR, Cunningham RB, Lindenmayer DB (2008) A forgotten habitat? Granite inselbergs conserve reptile diversity in fragmented agricultural landscapes. J Appl Ecol 45: 1742–1752.
14. Öckinger E, Smith HG (2007) Semi-natural grasslands as population sources for pollinating insects in agricultural landscapes. J Appl Ecol 44: 50–59.
15. van der Ree R, Bennett AF (2003) Home range of the squirrel glider (*Petaurus norfolcensis*) in a network of remnant linear habitats. J Zool 259: 327–336.
16. Haslem A, Bennett AF (2008) Countryside elements and the conservation of birds in agricultural environments. Agr Ecosyst Environ 125: 191–203.
17. Daily GC, Ehrlich PR, Sanchez-Azofeifa GA (2001) Countryside biogeography: use of human dominated habitats by the avifauna of southern Costa Rica. Ecol Appl 11: 1–13.
18. Renjifo LM (2001) Effect of natural and anthropogenic landscape matrices on the abundance of sub-Andean bird species. Ecol Appl 11: 14–31.
19. Söderström B, Kiema S, Reid RS (2003) Intensified agricultural landuse and bird conservation in Burkina Faso. Agr Ecosyst Environ 99: 113–124.
20. Bennett AF (1999) Linkages in the landscape: the role of corridors and connectivity in wildlife conservation. Gland, Switzerland: IUCN.
21. Fischer J, Lindenmayer DB (2002) Small patches can be valuable for biodiversity conservation: two case studies on birds in southeastern Australia. Biol Conserv 106: 129–136.
22. Forman RTT (1995) Land Mosaics: The Ecology of Landscapes and Regions. New York: Cambridge University Press.
23. Manning AD, Fischer J, Lindenmayer DB (2006) Scattered trees are keystone structures - implications for conservation. Biol Conserv 132: 311–321.
24. Pacheco R, Vasconcelos HL, Groc S, Camacho GP, Frizzo TLM (2013) Importance of remnants of natural vegetation for maintain ant diversity in Brazilian agricultural landscapes. Biodivers Conserv 22: 983–997.
25. Fischer J, Lindenmayer DB (2002) The conservation value of paddock trees for birds in a variegated landscape in southern New South Wales. 1. Species composition and site occupancy patterns. Biodivers Conserv 11: 807–832.
26. Manning AD, Lindenmayer DB, Barry SC (2004) The conservation implications of bird reproduction in the agricultural "matrix": a case study of the vulnerable superb parrot of south-eastern Australia. Biol Conserv 120: 363–374.
27. Munro N, Fischer J, Barrett G, Wood J, Leavesley A, et al. (2011) Bird's response to revegetation of different structure and floristics – are "restoration plantings" restoring bird communities? Restor Ecol 19: 223–235.
28. van der Ree R (2002) The population ecology of the squirrel glider (Petaurus norfolcensis) within a network of remnant linear habitats. Wildlife Res 29: 329–340.
29. Fischer J, Stott J, Law BS (2010) The disproportionate value of scattered trees. Biol Conserv 143: 1564–1567.
30. McLoughlin PD, Cluff HD, Gau RJ, Mulders R, Case RL, et al. (2003) Effect of special differences in habitat on home ranges of grizzly bears. Ecoscience 10: 11–16.
31. Mitchell MS, Powell RA (2004) A mechanistic home range model for optimal use of spatially distributed resources. Ecol Model 15: 209–232.
32. van Beest FM, Rivrud IM, Loe LE, Milner JM, Mysterud A (2011) What determines variation in home range size across spatiotemporal scales in a large browsing herbivore. J Anim Ecol 80: 771–785.
33. Ball T, Adams A, Goldingay RL (2009) Diet of the squirrel glider in a fragmented landscape near Mackay, central Queensland. Aust J Zool 57: 295–304.
34. Holland GJ, Bennett AF, van der Ree R (2007) Time-budget and feeding behaviour of the squirrel glider (Petaurus norfolcensis) in remnant linear habitat. Wildlife Res 34: 288–295.
35. Sharpe DJ, Goldingay RL (1998) Feeding behaviour of the squirrel glider at Bungawalbin Nature Reserve north-eastern New South Wales. Wildlife Res 25: 243–254.
36. Anderson RH (1961) Introduction, flora series Nos. 1–8: 1–6. Contributions from the New South Wales National Herbarium. Sydney: Royal Botanic Gardens.
37. Benson JS (2008) New South Wales vegetation classification and assessment: Part 2 Plant communities of the NSW South-western Slopes bioregion and update of the NSW Western Plains plant communities, version 2 of the NSWVCA database. Cunninghamia 10: 599–673.
38. Crane M, Montague-Drake RM, Cunningham RB, Lindenmayer DB (2008) The characteristics of den trees used by the Squirrel Glider (Petaurus norfolcensis) in temperate Australian woodlands. Wildlife Res 35: 663–675.
39. Crane MJ, Lindenmayer DB, Cunningham RB (2012) Use and characteristics of nocturnal habitats of the squirrel glider (Petaurus norfocensis) in Australian temperate woodlands. Aust J Zool 60: 320–329.
40. Martin JK, Handasyde KA, Taylor AC (2007) Linear roadside remnants: Their influence on the den-use, home range and mating system in bobucks (Trichosurus cunninghami). Austral Ecol 32: 686–696.
41. White GC, Garrott RA (1990) Analysis of Wildlife Radio-tracking Data. San Diego, USA: Academic Press.
42. van der Ree R, Bennett AF, Gilmore DC (2004) Gap-crossing by gliding marsupials: thresholds for use of isolated woodland patches in an agricultural landscape. Biol Conserv 115: 241–249.
43. Spooner PG (2005) On squatters, settlers and early surveyors: Historical development of country road reserves in southern New South Wales. Australian Geographer 36: 55–73.
44. Stirzaker R, Vertessey R, Sarre A, editors (2002) Trees, Water and Salt. An Australian Guide to Using Trees for Healthy Catchments and Productive Farms. Canberra: Joint Venture Agroforestry Program.
45. McCullagh P, Nelder JA (1989) Generalised Linear Models. New York: Chapman and Hall.

Acknowledgments

We thank the Hume Livestock Health and Pest Authority (formally Hume and Wagga Wagga Rural Lands Protection Boards), Wagga Wagga City and Greater Hume Shire (formerly Culcairn Shire Council) Councils, and private landholders for allowing us access to their properties, Cody Keys for assistance with trapping and radio tracking, and Damian Michael, Chris MacGregor, Ben MacDonald, Nicki Munro, and Jake Gillen for assistance with trapping. We thank our two wildlife veterinarians Karen Viggers and Arianne Lowe. We also thank Paul Adams (Academic Editor) and the two anonymous reviewers for their thoughtful contributions.

Author Contributions

Conceived and designed the experiments: MJC DBL RBC. Performed the experiments: MJC. Analyzed the data: MJC DBL RBC. Contributed to the writing of the manuscript: MJC DBL RBC.

46. Arnaud J (2003) Metapopulation genetic structure and migration pathways in the land snail *Helix aspersa*: influence of landscape heterogeneity. Landscape Ecol 18: 333–346.

47. Cummings JR, Vessey SH (1994) Agricultural influences on the movement patterns of white-footed mice (*Peromyscus leucopus*). Am Midl Nat 132: 209–218.

48. Getz LL, Cole FR, Gates DL (1978) Interstate roadside as dispersal routes for *Microtus pennnsylvanicus*. J Mammal 59: 208–212.

49. Lindenmayer DB, Cunningham RB, Donnelly CF (1993) The conservation of arboreal marsupials in the montane ash forests of the Central Highlands of Victoria, south-east Australia: IV. The presence and abundance of arboreal marsupials in retained linear habitats (wildlife corridors) within logged forest. Biol Conserv 66: 207–221.

50. Recher HF, Shields J, Kavanagh RP, Webb G (1987) Retaining remnant mature forest for nature conservation at Eden, New South Wales: A review of theory and practice. In: Saunders DA, Arnold GW, Burbidge AA, Hopkins AJM, editors. Nature Conservation: the Role of Remnants of Vegetation. Chipping Norton, N.S.W.: Surrey Beatty and Sons. pp.177–194.

51. Collard S, Le Broque A, Zammit C (2009) Bird assemblages in fragmented agricultural landscapes: the role of a small brigalow remnants and adjoining land uses. Biodivers Conserv 18: 1649–1670.

52. Fischer J, Lindenmayer DB (2002) The conservation value of paddock trees for birds in a variegated landscape in southern New South Wales. 2. Paddock trees as stepping stones. Biodivers Conserv 11: 833–849.

53. Lindenmayer DB, Cunningham RB, Crane M, Montague-Drake R, Michael D (2010) The importance of temperate woodland in travelling stock reserves for vertebrate biodiversity conservation. Ecol Manage Restor 11: 27–30.

54. Cunningham RB, Lindenmayer DB, Crane M, Michael D, MacGregor C (2007) Reptile and arboreal marsupial response to replanted vegetation in agricultural landscapes. Ecol Appl 17: 609–619.

55. Gibbons P, Lindenmayer DB, Fischer J, Manning AD, Weinberg A, et al. (2008) The future of scattered trees in agricultural landscapes. Conserv Biol 22: 1309–1319.

56. Lindenmayer DB, Laurance W, Franklin WF, Likens GE, Banks SC, et al. (2014) New policies for old trees: averting a global crisis in a keystone ecological structure. Conserv Lett 7: 61–69.

57. Maron M, Fitzsimons JA (2007) Agricultural intensification and loss of matrix habitat over 23 years in the West Wimmera, south-eastern Australia. Biol Conserv 135: 587–593.

58. Lindenmayer DB, Cunningham SA, Young A, editors (2012) Land Use Intensification: Effects on Agriculture, Biodiversity and Ecological Processes. Melbourne: CSIRO Publishing.

20

Biogeographic Distribution Patterns and Their Correlates in the Diverse Frog Fauna of the Atlantic Forest Hotspot

Tiago S. Vasconcelos[1]*, Vitor H. M. Prado[2], Fernando R. da Silva[3], Célio F. B. Haddad[2]

1 Departamento de Ciências Biológicas, Universidade Estatual Paulista, Bauru, São Paulo, Brazil, **2** Departamento de Zoologia, Universidade Estadual Paulista, Rio Claro, São Paulo, Brazil, **3** Departamento de Ciências Ambientais, Universidade Federal de São Carlos, Sorocaba, São Paulo, Brazil

Abstract

Anurans are a highly diverse group in the Atlantic Forest hotspot (AF), yet distribution patterns and species richness gradients are not randomly distributed throughout the biome. Thus, we explore how anuran species are distributed in this complex and biodiverse hotspot, and hypothesize that this group can be distinguished by different cohesive regions. We used range maps of 497 species to obtain a presence/absence data grid, resolved to 50×50 km grain size, which was submitted to k-means clustering with v-fold cross-validation to determine the biogeographic regions. We also explored the extent to which current environmental variables, topography, and floristic structure of the AF are expected to identify the cluster patterns recognized by the k-means clustering. The biogeographic patterns found for amphibians are broadly congruent with ecoregions identified in the AF, but their edges, and sometimes the whole extent of some clusters, present much less resolved pattern compared to previous classification. We also identified that climate, topography, and vegetation structure of the AF explained a high percentage of variance of the cluster patterns identified, but the magnitude of the regression coefficients shifted regarding their importance in explaining the variance for each cluster. Specifically, we propose that the anuran fauna of the AF can be split into four biogeographic regions: a) less diverse and widely-ranged species that predominantly occur in the inland semideciduous forests; b) northern small-ranged species that presumably evolved within the Pleistocene forest refugia; c) highly diverse and small-ranged species from the southeastern Brazilian mountain chain and its adjacent semideciduous forest; and d) southern species from the Araucaria forest. Finally, the high congruence among the cluster patterns and previous eco-regions identified for the AF suggests that preserving the underlying habitat structure helps to preserve the historical and ecological signals that underlie the geographic distribution of AF anurans.

Editor: Carlos A. Navas, University of Sao Paulo, Brazil

Funding: This study was supported by Fundação de Amparo a Pesquisa no Estado de São Paulo (FAPESP, grants: 2008/50928-1; 2010/50125-6; 2011/18510-0; 2012/07765-0). CFBH research is also supported by Conselho Nacional de Desenvolvimento Científico e Tecnológico (CNPq). The funders had no role in study design, data collection and analysis, decision to publish, or preparation of the manuscript.

Competing Interests: The authors have declared that no competing interests exist.

* Email: zoologia@ig.com.br

Introduction

Dividing the world or large geographical regions into meaningful biological units has long been of general interest for macroecologists and biogeographers. For instance, the evaluation of the world's zoogeographical regions proposed by A. R. Wallace more than 100 years ago is still a subject of recent studies (e.g., [1]). While early biogeographical regions were generated based on researchers' knowledge of species distribution (e.g., the original zoogeographical regions proposed by Wallace), recent regionalization proposals have been performed by considering a large amount of species information available on digital databases coupled with the use of one or several quantitative statistical methods (e.g., [1,2]). Irrespective of what method is used, a species assemblage within a determined biogeographic region can be expected to share a large amount of history with other assemblages within the region, but relatively little with those in other biogeographic regions [3]. For this reason, biogeographic regions may be viewed as operational species pools [3], which provide fundamental abstractions of the geographical organization of life in response to past or current physical and biological forces.

Regionalization schemes thus provide spatially explicit frameworks for answering many basic and applied questions in historical and ecological biogeography, evolutionary biology, systematics, and conservation [2,4].

Biogeographic regionalizations in South America have mainly been performed at a global scale perspective, and have relied on a variety of methods and biological models (see examples and references in [1,2]). These schemes either consider the whole continent as a distinct biogeographic unit (e.g., Neotropical region *sensu* Wallace's zoogeographical classification) or split the continent into two or three regions depending on either the methodological approaches or the biological traits among taxa (e.g., dispersal capability) [1,2,5]. All else being equal, the scale of analysis is an important factor in determining the final number of regions. For instance, global analyses using similar clustering methods always identify Europe as part of the Palaearctic region [1,2], but scaling down the analysis to the continent level generates a more refined identification of sub-regions [6]. In South America, a cluster analysis was performed in order to devise a regionalization system based on amphibian distribution. In the analysis, the authors recognized four biogeographic regions for the group [7].

Specifically, although some areas of the seasonally dry Atlantic Forest were grouped within the savanna-like vegetation cluster, the authors found that most of the area encompassed by the Atlantic Forest hostspot (*sensu* [8]) is considered to be a biogeographic unit for the South American amphibians [7]. Here, we devise a regionalization scheme for the current original extent of the Atlantic Forest hotspot (i.e., without considering habitat loss by recent deforestation that occurred during the last century) in order to explore how amphibians are distributed throughout this complex and biodiverse domain, and then to generate a map of amphibian diversity focused on the composition of regional faunas within the hotspot.

Amphibian species of the Atlantic Forest (hereafter AF) are a highly diverse group, and their morphological structures, behavioral repertoires, and breeding strategies are greatly diversified as well. For instance, there are approximately 550 anuran species from the AF that exhibit 39 different reproductive modes, most of which are endemic at the species, genus, or even family level [9,10]. This high diversity of reproductive modes is attributed to the successful utilization of the diversified and humid microhabitats present in this biome [9]. Yet, the gradient of species richness and number of reproductive modes is not randomly distributed, so there is a parallel of increased species richness and number of reproductive modes between dry/seasonal and evergreen humid forests [11,12]. There is also a great number of micro-endemic species associated with the Atlantic coast, some of which have been reported at only one location [9]. Because of this high anuran diversity associated with different patterns of species richness and concentrations of micro-endemic species, we hypothesize that the anuran distribution within the AF can be distinguished by different cohesive regions, thus consisted of different species pools.

Specifically, our first goal is to determine the number and the spatial position of these regions using a cluster analysis. Then, based on preview studies that showed that richness gradients and range size of species are differently distributed throughout the hotspot [9,11–13], we hypothesize that gradients of climatic conditions, topographic variations, and habitat structures are non-mutually exclusive conditions that determine cohesive regions within the AF. Thus, because the patterns of species distributions are ultimately determined by the rates of speciation, extinction, and dispersal [14,15], and because physiological constraints and limited dispersal are two key characteristics of most amphibian species, we hypothesize three potential explanations for the cohesive anuran regions in the AF. The first hypothesis considers the well-known fact that larger ranges in elevation promote speciation through habitat specialization and altitudinal isolation, which increases endemism and, consequently, the discrepancy in species richness between sites within a region [9,16–18]. Therefore, we first hypothesize that topography could be one of the determinants of the anuran biogeographic regions, because regions with extensive variation in topography would harbor small-ranged species due to historically limited dispersal capabilities and would thus increase the chance of higher speciation rates at these areas. The second hypothesis considers the fact that energy- and humidity-related variables have been shown to be key environmental determinants of the richness and composition of amphibian communities [11,12,19]. Due to the wide latitudinal variation in the AF, our second hypothesis is that climate may be a strong predictor of the anuran biogeographic regions identified by our cluster analysis. Finally, Rueda and collaborators [6] showed that the habitat structure in Europe has a strong influence on the identification of biogeographic regions for different taxa (including anurans). Thus, considering the fact that the habitat provides the templet on which evolution forges animal life-history strategies (the

concept of habitat templet [20,21]), our third hypothesis is that the anuran's cohesive regions can be recognized as a consequence of the vegetation distribution within the AF. We also used deviance partitioning techniques to disentangle the relative influence of each predictor and to identify the independent and shared influences of topography, climate, and vegetation structure on the identified anuran biogeographic sub-regions within the AF hotspot.

Materials and Methods

Study area

Characterized by a complex topography (elevation varies from sea level to 2,000 m a.s.l.) and a wide latitudinal distribution along the Brazilian Atlantic coast (latitudinal distribution of *c.* 25°), the AF hotspot is considered one of the world's most species-rich, yet notoriously endangered and understudied ecosystems [8,22]. There are many classifications attributed to the AF (e.g., [23]), and one of the most commonly used [24] divides the domain in terms of its floristic composition, landscape, and climatic attributes into the categories of open, dense, and mixed ombrophilous/evergreen forest, which are widely distributed throughout the Brazilian coast, but the mixed forest (also known as the Araucaria forest) is mainly found along the southern rim of the hotspot [23,24]; the seasonally dry forest is also known as semideciduous and deciduous forests, and it is characterized by the partial and total loss of leaves, respectively, as a result of the pronounced precipitation seasonality over the year (Figure 1A). Although they also have wide latitudinal distributions, deciduous and semideciduous forests are located in inland areas that are mostly located in northeastern and southeastern Brazil [23,24].

To devise a new amphibian regionalization scheme, we considered the current original extent of the AF (i.e., deforestation has not been considered herein, see [24]) provided by the Conservation International portal (http://www.conservation.org/where/priority_areas/hotspots/Documents/CI_Biodiversity-Hotspots_2011_ArcView-Shapefile-and-Metadata.zip). We then divided the AF into 469 grids at *c.* 50×50 km grain size, considering that each grid was covered by at least 50% of the AF. Finally, we were able to construct a presence/absence matrix based on the anuran distribution, which was then submitted to cluster analysis.

Species distribution data

There are currently 543 amphibian species in the AF hotspot, 529 of which are anurans [10]. Here, we excluded the grids in which small natural patches of the AF do not cover at least 50% of the biome. Consequently, we were unable to consider the species restricted to these small patches, such as the narrowly-ranged species *Adelophryne baturitensis*, *A. maranguapensis*, *Bokermannohyla diamantina*, and *B. itapoty*. Island-endemic species, such as *Scinax alcatraz* and *S. faivovichi*, have also been excluded from the analysis. In the end, a total of 496 species (~94% of AF anurans) were considered for the regionalization process (Table S1).

Almost all species range maps were obtained from the International Union for Conservation of Nature (IUCN) portal (http://www.iucnredlist.org/technical-documents/spatial-data), and the amphibian nomenclature was updated according to the Amphibian Species of the World 5.6 portal [25]. The species that were not available in the database (e.g., recently described species, such as *Brachycephalus pulex* and *B. toby*) had their maps created in ArcGIS 10.1 considering their original descriptions. The rasterized range maps were overlaid onto each grid cell to generate a presence/absence matrix. Although they possess some

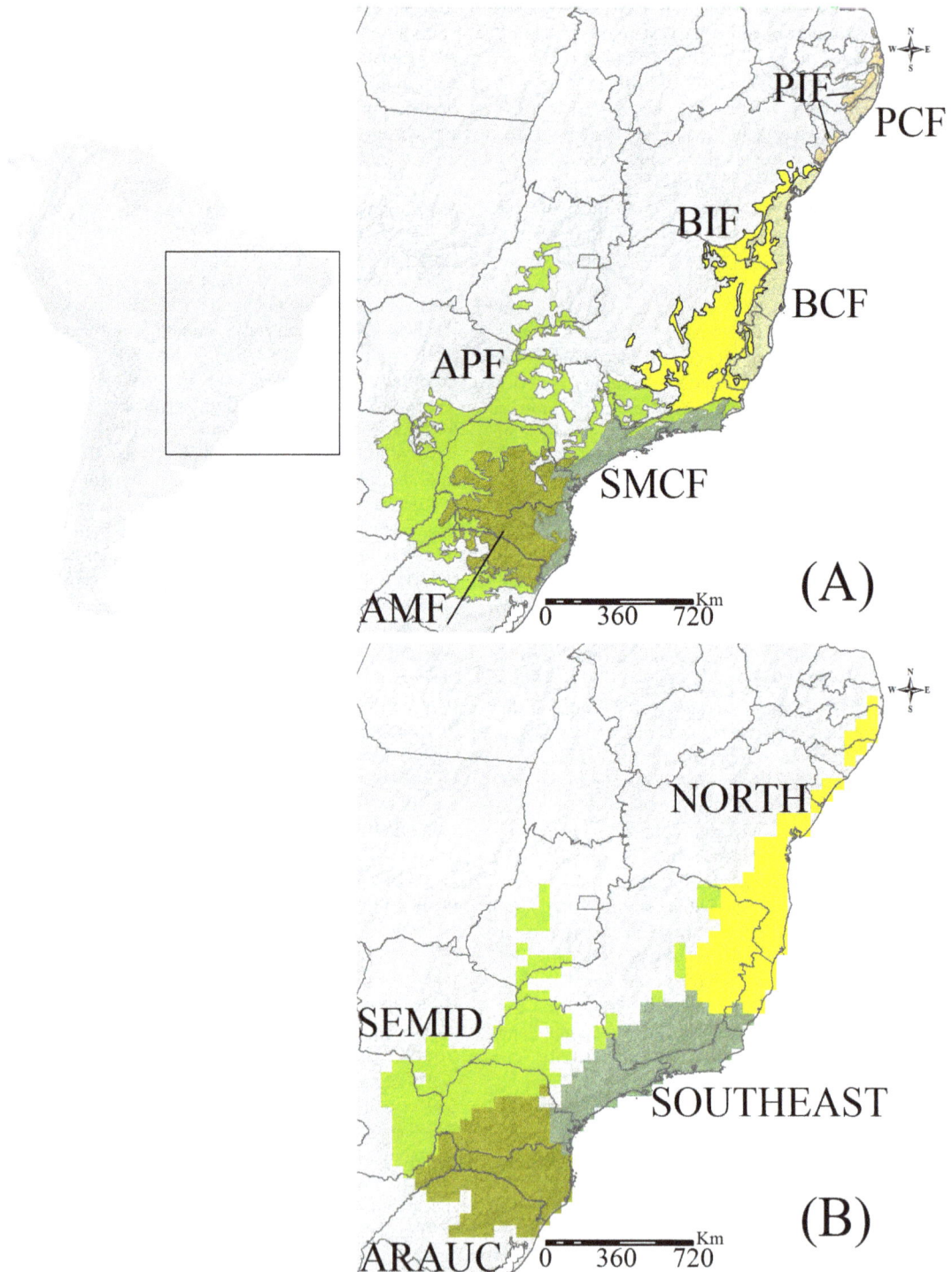

Figure 1. Major Atlantic Forest eco-regions, modified from the World Wildlife Fund designations (A): AMF = Araucaria Moist Forests; APF = Alto Paraná (semideciduous/deciduous) Forests; BCF = Bahia Coastal (moist) Forests; BIF = Bahia Interior (semidecidual/decidual) Forests; PCF = Pernambuco Coastal (moist) Forests; PIF = Pernambuco Interior (semideciduous) Forests; and SMCF = Serra do Mar Coastal (moist) Forests. Biogeographic regions based on the anuran fauna generated through k-means clustering with v-fold cross-validation (B).

level of error [26], range maps represent the areas where a particular species can be expected to occur, and it will be expected to be found only in suitable habitats within these areas [6]. Thus, overprediction is an inherent methodological limitation of these kinds of range maps [27]. Within a macroecological perspective, however, they may function very well at grains greater than 50×50 km [28]. That is to say, although the IUCN amphibian maps might include either over- or underpredictions [26], using range maps is presumed to be as reliable as more refined information regarding a given species distribution (e.g., point occurrence records from survey data and/or herpetological collections) if the goal is to document broad-scale patterns of species distribution [28].

Environmental and topographic variables

Five abiotic variables (one topographic and four climatic variables) were gathered and averaged for each grid cell. Annual precipitation (P), precipitation seasonality (PS), and mean annual temperature (T) were obtained from the WorldClim database at a 10×10 km resolution [29]. Annual actual evapotranspiration (AET), a measure of water-energy balance, was also obtained at a 10×10 km resolution at http://www.fao.org/geonetwork/srv/en/metadata.show?id = 37233. The standard deviation of elevation, a measure of topographic heterogeneity (TOP), was calculated for each grid cell based on elevation data (~1×1 km resolution) available at https://lta.cr.usgs.gov/GTOPO30. All of these variables are known to represent either potential physiological limits for amphibians or as barriers to dispersal, and they are closely associated with species richness patterns of both plants and animals [11–12,19,30].

Regionalization procedure

First, k-means clustering combined with v-fold cross-validation was applied to the presence/absence matrix [31,32]. The classical k-means clustering algorithm requires the number of clusters (k) to be established in advance, and utilizes a subset of k random initialization cells that are treated as the initial cluster centers, and then proceeds as a two-step iterative procedure in which cluster centres and clusters are successively recalculated. The first step starts with the assignment of each cell to its nearest cluster center in terms of species compositional distance, herein considered as the Hellinger distances [33]. In the second step, each cluster center is updated by making it equal to the mean of the cells assigned to it. The process is repeated (we used 50 iterations) so that the cluster and cluster centers change in each replicate, and they converge to a locally optimal position in the data space [6,7]. The k-means clustering technique was combined with v-fold cross-validation in order to obtain the optimal number of clusters based on species composition without regard to the spatial proximity of the grids (see [6]). In summary, the algorithm determines the "best" number of clusters within a range of pre-determined cluster numbers (we set these from two to 25 clusters). The k-means clustering technique with v-fold cross-validation was performed using Statistica (StatSoft).

Correlates of cluster patterns

Considering the fact that the regionalization procedure we used is designed to generate biotic regions based on differences in species assemblages affected by complex interacting factors [5,6], we explore the extent to which topography, climate, and the vegetation structure of the AF are expected to identify the cluster patterns recognized by k-means clustering. Because the climatic and topographic variables were gathered at higher resolutions than those of the ~50×50 km AF grid, we averaged all values of these variables within each AF grid cell, thus balancing out the different data scales inherent in each independent variable. The vegetation structure was considered based on the major AF eco-regions from the World Wildlife Fund designations [34] (Figure 1A) and was used as a multinomial variable to evaluate the extent to which animal species composition is associated with the AF habitat structure (e.g., [6]).

Because the dataset is linearly distributed (visually checked by means of partial residual plots graphic, [35]; Figure S1), we followed Rueda and collaborators' study [6] and performed

Generalized Linear Models (GLMs) with multinomial logit-link for modelling a multinomial response variable (i.e., the present k-means cluster solution) as a function of one or more continuous predictors. Due to different magnitudes of measurement of each predictor and in order to facilitate the interpretation of the regression coefficients, all predictor variables were standardized to have a mean of zero and a standard deviation of 1.0 prior to analysis [36]. Collinearity among the predictors was verified by the Variance Inflation Factor (VIF; [37]) and we considered them to be not strongly collinear (VIF<5.1).

We generated several single- and multiple-variable explanatory models that could potentially explain the cluster patterns: a) full multiple-variable model that considered all predictors; b) a climatic and eco-regional multiple-variable model that considered the AF eco-regions and climatic variables; c) a climatic and topographic multiple-variable model that considered climate and the standard deviation of elevation, which is presumed to generate high levels of endemism in the AF (see [9]); d) a topographic and eco-regional multiple-variable model that considered the AF eco-regions and the standard deviation of elevation; e) a climatic multiple-variable model that considered only climatic variables; f) an eco-regional multiple-variable model that considered only the AF eco-regions; and g) a topographic single-variable that considered only the standard deviation of elevation. The model selection approach was based on the lowest Akaike Information Criterion (AIC; [38]). We considered the best model to be the one based on the lowest AIC required to partition the deviance of each response variable into independent effects of a particular predictor and co-varying effects of two or more predictors that cannot be disentangled [6,39].

We also performed a Principal Component Analysis (PCA) of all 469 grid cells in order to visualize the patterns of distribution of the abiotic characteristics and the clusters that represented each grid cell [40].

Results

The cluster analysis identified four biogeographic regions in the AF based on anuran species composition (Figure 1B). Cluster 1 (hereafter SEMID) is located in AF inland areas, and it encompasses most of the semideciduous forest and transitional areas to the Cerrado (i.e., the Brazilian savanna-like vegetation). Cluster 2 (hereafter SOUTHEAST) is comprised of the coastal AF in southeastern Brazil, where most of the area falls within the ombrophilous forest and adjacent areas of semideciduous forest. Cluster 3 (hereafter NORTH) encompasses the northeastern Brazilian semideciduous/deciduous and ombrophilous forests, and cluster 4 (hereafter ARAUC) is mostly congruent with the Araucaria forest in southern Brazil (Figure 1B).

Among all models of predictor variables, the full model (with all variables included) was the best one for explaining the cluster patterns (Table 1). This model explained higher levels of variance (80.10%) than the models comprised solely of climatic, topographic, or eco-regional variables (Table 1). However, the magnitude of the regression coefficients of the full model shifted in terms of their importance in explaining the variance for each cluster patterns (Table 2). The variable temperature and precipitation seasonality were the strongest predictors of the SEMID and ARAUC clusters, but this relationship was positively and negatively associated with these clusters, respectively (Table 2). That is, while warmer temperatures and precipitation seasonality predict the former cluster, cooler temperatures and more homogeneous rainfall predict the ARAUC cluster. Positive temperature is the only predictor of the NORTH cluster, whereas

Table 1. Generalized Linear Models of amphibians' *k*-means group in the Atlantic Forest.

Rank	Model	k	AIC	w$_i$	Pseudo R$_2$
1	Full Model	6	328.2	0.972	80.10
2	Climatic & Eco-regional Model	5	335.3	0.028	79.08
3	Climatic & Topographic Model	5	470.5	0	66.26
4	Climatic Model	1	576.4	0	57.56
5	Topographic & Eco-regional Model	2	615.5	0	55.93
6	Eco-regional Model	1	689.1	0	49.74
7	Topographic Model	1	1090.1	0	16.27

The models are sorted according to the lowest Akaike Information Criterion (AIC). k = number of the predictor variables included in the model; Pseudo R$_2$ = coefficient of determination; w$_i$ = evidence of 0.972 for the Model 1. See Methods for predictors' abbreviations.

cooler temperatures, rough topography, AET, and precipitation seasonality are significant predictors of the SOUTHEAST cluster (Table 2).

The PCA results are represented in Figure 2 and shows that the first axis segregates both the SEMID and NORTH clusters from the other ones, since the SEMID and NORTH clusters are characterized by having higher precipitation seasonality than SOUTHEAST and ARAUC, which, in turn, have lower values of annual precipitation. The clusters overlap greatly at the second axis, but SOUTHEAST is slightly more commonly associated with rough topography than the other clusters are (Figure 2).

The deviance partitioning indicates that a combined effect of the climate and vegetation structure of the AF (eco-regions) accounted for the largest fraction (25.8%) of the variability of the anuran cluster patterns identified herein (Figure 3). However, the largest independent effect is accounted for climate (24.2%), followed by the vegetation structure of the AF (13.8%) and topography (1.02%) (Figure 3).

Discussion

The biogeographic patterns found for amphibians are broadly congruent with ecoregions identified in the AF, but their edges, and sometimes the whole extent of some clusters, present much less resolved patterns compared to the previous classification (e.g., [23,24]). The SEMID and ARAUC clusters are broadly congruent with the southeastern Brazilian semideciduous and southern Araucaria forests, respectively (Figure 1A and 1B). On the other hand, the SOUTHEAST and NORTH clusters are consistent with a combination of subregions, mostly composed by ombrophilous and their adjacent semideciduous/deciduous forests.

The present study identified that climate, topography (i.e., the endemism-related variable), and the vegetation structure of the AF explained a high percentage of variance of the cluster patterns identified, a finding which agrees with previous studies that defined biogeographic regions for diverse taxa, including amphibians (e.g., [6,7]). For instance, climate is well known to be strongly associated with broad-scale geographic patterns of species distributions [19,30]. Therefore, it is reasonable that climatic gradients determined by the latitudinal variation in the AF are important forces in determining the present clusters (Table 2). Furthermore, as reported previously in Europe [6], the underlying vegetation structure of the AF is also considerably important for predicting the present cluster patterns, in which some of the clusters represent specific AF eco-regions (e.g., the Araucaria forest and the ARAUC cluster), while others represent a combination of eco-regions (e.g., SOUTHEAST). Indeed, it is well known that the water-energy balance is a strong correlate of plant distribution [6,19], so it is not surprising that the shared effect of climate and AF vegetation distribution on the anuran biogeographic patterns was relatively high in the present study. Finally, although the topography accounts for only a small fraction of the variance of the identified cluster patterns, it is particularly important in predicting the SOUTHEAST cluster, which harbors the complex mountain chain in this region (see discussion ahead).

Considering a previous regionalization performed for South American amphibians [7], the identification of the SEMID cluster was already expected. The frog fauna from the inland semideciduous forest is made up of a mix of typical Cerrado and AF species, most of which are widely-ranged species [7,41]. Hence, the most common feature shared by the SEMID species is the fact that they are both less diverse [11] and more widely distributed compared to

Table 2. Regression coefficients of determination of the full multiple-variable Generalized Linear Model of amphibians' *k*-means group in the Atlantic Forest (eco-regional variables are omitted due to the lack of statistical significance with any cluster).

Clusters	TOP	AET	T	P	PS	ECOR
pFM	**0.004**	**0.023**	**0.000**	0.511	**0.000**	**0.000**
SEMID	−0.599	0.004	**2.120**	0.453	**1.360**	-
SOUTHEAST	**0.637**	**1.235**	−0.950	−0.197	**3.511**	-
NORTH	0.397	−0.349	**1.973**	−0.203	−0.423	-
ARAUC	−0.435	−0.890	**−3.143**	−0.053	**−4.449**	-

TOP = topography; AET = annual actual evapotranspiration; T = temperature; P = precipitation; PS = precipitation seasonality; ECOR = AF eco-regions; pFM = ANOVA p values of the full model. See Results for clusters' abbreviations. Significant coefficients (p≤0.01) are highlighted in bold.

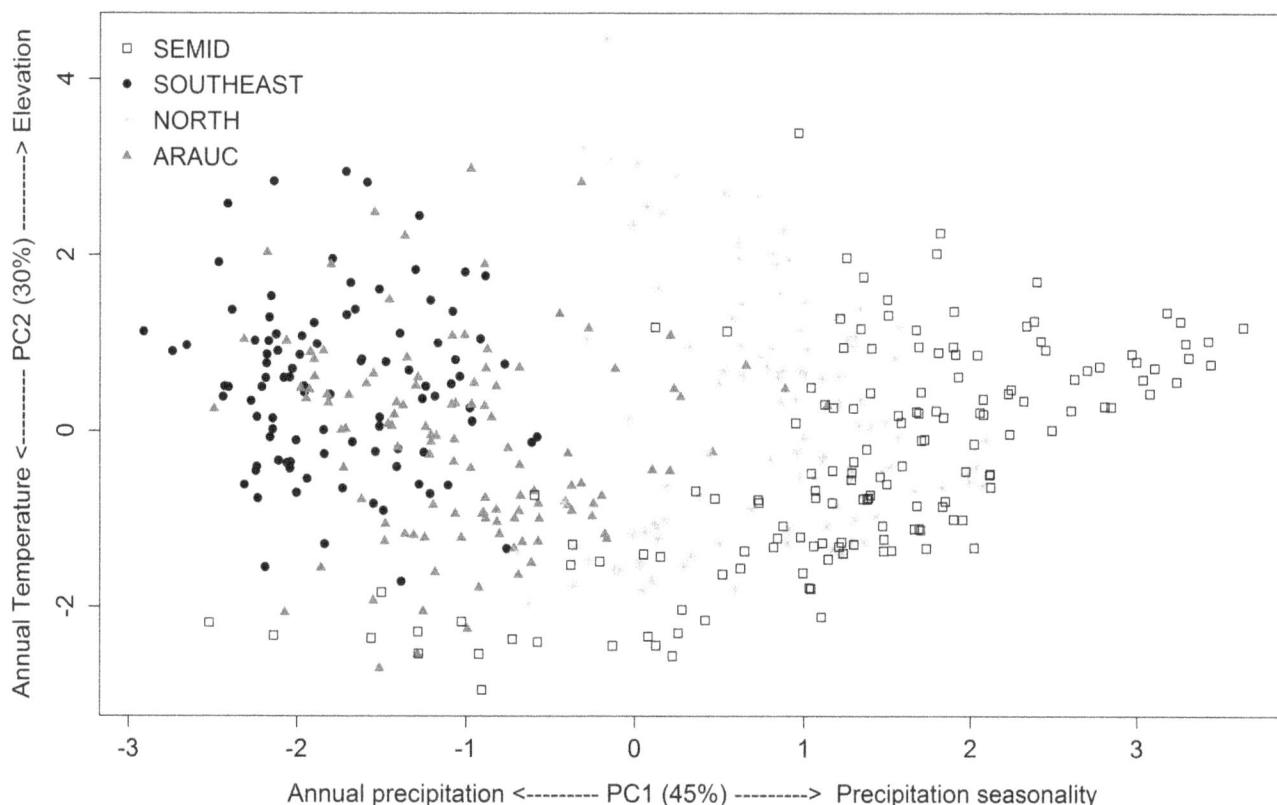

Figure 2. Principal component analysis (PCA) based on abiotic (annual precipitation, precipitation seasonality, mean annual temperature, annual actual evapotranspiration, and standard deviation of elevaton) and biotic (species richness and range size) variables of all 469 grid cells in the Atlantic Forest. Different symbols represent biogeographic regions based on the anuran fauna generated through *k*-means clustering with *v*-fold cross-validation.

species from other clusters (see also Figure S2 and Figure S3). This factor might be related to the fact that areas with minor topographic variation, such as the area encompassed by the SEMID cluster, favor population dispersal, and consequently, low speciation rates are expected in this area. This finding is reinforced by the fact that anurans from the semidecidual AF are more similar to the adjacent Cerrado anuran assemblages, which have more similar homogeneous topography and a harsher environment (higher temperatures and precipitation seasonality) than other AF ecoregions, such as the ombrophilous forest, which is more humid and which presents a rough topography [41]. Conversely, more homogeneous rainfall over the year and cooler temperatures are the strongest correlates of the ARAUC cluster. This is expected because, while variation in precipitation decreases, temperature markedly increases its seasonality at higher latitudes [42].

In NORTH, the only correlate identified was positive temperature. This is expected in a way, because the influence of positive temperatures becomes evident in the more northerly regions, closer to the Equator, where the climate is hotter. The SOUTHEAST cluster was correlated with almost all climatic variables analyzed, but precipitation seasonality was the strongest correlate. This finding was not expected, because this cluster mostly includes the ombrophilous forest, which is characterized by moist weather over the year, with no well-defined dry season [10,23]. Thus, this unexpected correlation is more influenced by the presence of transitional areas of semidecidual forest in SOUTHEAST (see Figures 1A and 1B). The semidecidual forest

considered in SOUTHEAST was likely not clustered with the SEMID because the anuran fauna of the semideciduous forest closer to coastal mountains includes some species that typically reside in the ombrophilous forest, and which is usually absent from those more distant and inland semideciduous forests (see examples in [41]). All other correlates in SOUTHEAST are expected: AET is known to be highly correlated with animal distribution (see [19]); and the negative correlation found between temperature and SOUTHEAST is probably due to cooler climate in this cluster than in NORTH and SEMID ones, particularly in areas where the mean temperature tends to decrease as the altitude increases (e.g., the southeastern Brazilian mountain chain).

Although the NORTH and SOUTHEAST clusters have different environmental predictors, these regions share interesting features in terms of anuran biogeographic patterns. In fact, they are recognized as "rich and rare" regions in South America for their amphibian diversity (i.e., they possess high species richness with restricted ranges; [13], see also Figure S2 and Figure S3). Due to different aims and methodological approaches, Villalobos and collaborators [13] considered the entire extent of the SOUTH-EAST and NORTH clusters to be a continuous "rich and rare" region, but the identification of two distinct biogeographic species pools in the present study raises interesting questions regarding the evolution of amphibians in the AF. Although we found that climate, topography, and the vegetation structure of the AF are important in determining the present cluster patterns, we hypothesize that the recognition of two distinct micro-endemic species pools should result, at least partially, from the past climate

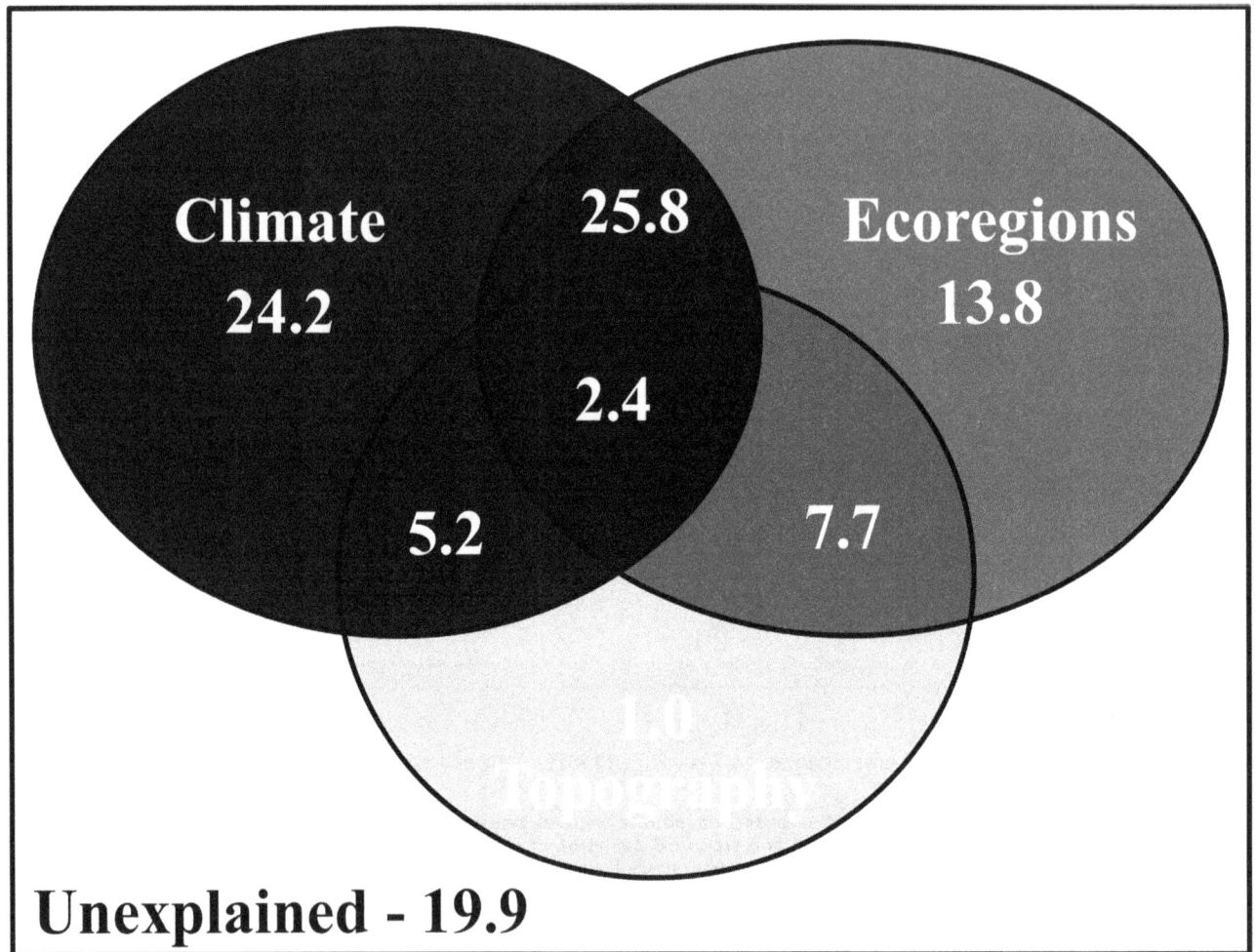

Figure 3. Deviance partitioning analysis representing the deviance in cluster configurations explained by climate (annual precipitation, precipitation seasonality, mean annual temperature, and annual actual evapotranspiration), vegetation structure of the AF (eco-regions considered in Figure 1A), and topography (range in elevation). The light-dark gradient of the figure represents the low-high deviance explained by the predictors.

history (e.g., the persistence of historically stable areas during the Pleistocene glaciations) and also from differences in topography along the extent of the AF (such as the presence of the complex mountain chains of Serra do Mar and Serra da Mantiqueira in the SOUTHEAST cluster). The extent of the NORTH and SOUTHEAST clusters agree with the predicted historical Pleistocene forest refugia (21,000 years before present; 21 ka BP), so these historically stable areas are expected to retain high levels of endemism for diverse taxa ([43] and references therein), including amphibians [13,22,44]. Moreover, the mountain chain in the southeastern Brazil is expected to favor the genetic diversification of amphibians, since it breaks the AF biome up into many small humid microhabitats and ultimately promotes speciation through geographic isolation [9]. This phenomenon has been found to be the case for amphibians [22], and high levels of endemism have also been reported for diverse taxa in SOUTH-EAST ([43] and references therein). Therefore, although both NORTH and SOUTHEAST are recognized by their large numbers of small-ranged anuran species, we hypothesize that the different topography and the persistence of these areas over the course of the climate history (since the late Pleistocene) experi-

enced by these clusters ultimately resulted in the evolution of two distinct species pools.

Conclusions

In summary, we propose that the anuran fauna of the AF can be split into four biogeographic regions characterized by: a) less diverse and widely-ranged species that predominantly occur in the inland semideciduous forests, where the weather is hot and seasonally dry (SEMID); b) northern small-ranged species that presumably evolved/survived to extinction within the Pleistocene forest refugia, where the climate nowadays is hot (NORTH); c) highly diverse and small-ranged species from the southeastern ombrophilous and its adjacent semidecidous forest, where the climate is cooler (except when compared to ARAUC) and the topography is rough (SOUTHEAST); and d) southern species from the Araucaria forest, where the weather is cooler and the rains are well distributed throughout the year (ARAUC). The high congruence among the cluster patterns and previous eco-regions identified for the AF (Figure 1A and 1B) corroborates the habitat templet concept [20,21], and suggests that preserving the underlying habitat structure (i.e., natural forest formations) helps to preserve the historical and ecological signals that underlie the geographic distribution of species [6], including the AF anurans. Nonetheless, it is important to emphasize that our regionalization scheme did not consider the human-induced deforestation that reduced the AF extension to ~7% of its original distribution [10]. In addition, the herpetological literature is dynamic regarding the updating of the geographic ranges of the species and/or the description of new ones (e.g., [45,46]). Thus, future regionalization schemes that consider the current remnants of the AF and the updated geographic ranges of the species after deforestation would be of interest for conservation biogeographers, who would be able to assess how much habitat loss can erase and/or maintain the broad-scale biogeographic patterns of AF anurans, mainly in areas with higher deforestation rates, such as the area encompassed by the NORTH cluster [22]. Finally, although other biological data (e.g., species traits and phylogenetic relationships) and the congruence of biogeographic patterns across multiple taxonomic groups are undoubtedly necessary for properly establishing conservation actions [7,13], the regionalization process is an important step for identifying biogeographic regions that contain centers of origin, have been colonized by dispersing organisms, or have been subjected to large-scale forces such as the Pleistocene glaciations [47].

Author Contributions

Conceived and designed the experiments: TSV VHMP FRdS CFBH. Performed the experiments: TSV VHMP FRdS. Analyzed the data: TSV VHMP FRdS. Contributed reagents/materials/analysis tools: TSV VHMP FRdS CFBH. Wrote the paper: TSV VHMP FRdS CFBH.

References

1. Rueda M, Rodríguez MÁ, Hawkins BA (2013) Identifying global zoogeographical regions: lessons from Wallace. J Biogeogr 40: 2215–2225.
2. Kreft H, Jetz W (2010) A framework for delineating biogeographical regions based on species distributions. J Biogeogr 37: 2029–2053.
3. Carstensen DW, Lessard J-P, Holt BG, Krabbe M, Rahbek C (2013) Introducing the biogeographic species pool. Ecography 36: 1–9. DOI: 10.1111/j.1600-0587.2013.00329.x
4. Morrone JJ (2009) Evolutionary biogeography: an integrative approach with case studies. New York: Columbia University Press. 301 p.
5. Proches Ş (2005) The world's biogeographical regions: cluster analysis based on bat distributions. J Biogeogr 32: 607–614.
6. Rueda M, Rodríguez MÁ, Hawkins BA (2010) Towards a biogeographic regionalization of the European biota. J Biogeogr 37: 2067–2076.
7. Vasconcelos TS, Rodríguez MÁ, Hawkins BA (2011) Biogrographic distribution patterns of South American amphibians: a regionalization based on cluster analysis. Natureza & Conservação 9: 67–72.
8. Myers N, Mittermeier RA, Mittermeier CG, Fonseca GAB, Kent J (2000) Biodiversity hotspots for conservation priorities. Nature 403: 853–858.
9. Haddad CFB, Prado CPA (2005) Reproductive modes in frogs and their unexpected diversity in the Atlantic Forest of Brazil. BioScience 55: 207–217.
10. Haddad CFB, Toledo LF, Prado CPA, Loebmann D, Gasparini JL, et al. (2013) Guide to the amphibians of the Atlantic Forest: diversity and biology. São Paulo: Anolis Books. 544 p.
11. Vasconcelos TS, Santos TG, Haddad CF, Rossa-Feres DC (2010) Climatic variables and altitude as predictors of anuran species richness and number of reproductive modes in Brazil. J Trop Ecol 26: 423–432.
12. da Silva FR, Almeida-Neto M, Prado VHM, Haddad CFB, Rossa-Feres DC (2012) Humidity levels drive reproductive modes and phylogenetic diversity of amphibians in the Brazilian Atlantic Forest. J Biogeogr 39: 1720–1732.
13. Villalobos F, Dobrovolski R, Provete DB, Gouveia SF (2013) Is rich and rare the common share? Describing biodiversity patterns to inform conservation practices for South American anurans. PLoS ONE: doi:10.1371/journal.pone.0056073.
14. Ricklefs RE (1987) Community diversity: relative roles of local and regional processes. Science 235: 167–171.
15. Wiens JJ, Donoghue MJ (2004) Historical biogeography, ecology and species richness. Trends Ecol Evol 19: 639–644.
16. Lomolino MC (2001) Elevation gradients of species-density: historical and prospective views. Global Ecol Biogeogr 10:3–13.
17. Rahbek C, Graves GR (2001) Multiscale assessment of patterns of avian species richness. Proc Natl Acad Sci USA 98: 4534–4539.
18. Ruggiero A, Hawkins BA (2008) Why do mountains support so many species of birds? Ecography 31: 306–315.
19. Hawkins BA, Field R, Cornell HV, Currie DJ, Guégan J-F, et al. (2003) Energy, water, and broad-scale geographic patterns of species richness. Ecology 84: 3105–3117.
20. Southwood TRE (1977) Habitat, the templet for ecological strategies? J Anim Ecol 46: 337–365.
21. Southwood TRE (1988) Tactics, strategies and templets. Oikos 52: 3–18.
22. Carnaval AC, Hickerson MJ, Haddad CFB, Rodrigues MT, Moritz C (2009) Stability predicts genetic diversity in the Brazilian Atlantic forest hotspot. Science 323: 785–789.
23. IBGE (Instituto Brasileiro de Geografia e Estatística) (2012) Manual técnico da vegetação brasileira. Rio de Janeiro: IBGE, Brazil. 275 p.
24. MMA (Ministério do Meio Ambiente), IBAMA (Instituto Brasileiro do Meio Ambiente e dos Recursos Naturais Renováveis) (2010) Monitoramento do desmatamento nos biomas brasileiros por satélite: monitoramento do bioma Mata Atlântica. Brasília: MMA, Brasil. 42 p.
25. Frost DR (2013) Amphibian Species of the World: an Online Reference. Version 5.6 (9 January 2013). Available: http://research.amnh.org/herpetology/amphibia/index.html. American Museum of Natural History, New York, USA. Accessed 01 August 2013.
26. Ficetola GF, Rondinini C, Bonardi A, Katariya V, Padoa-Schioppa E, et al. (2013) An evaluation of the robustness of global amphibian range maps. J Biogeogr Early view: doi:10.1111/jbi.12206
27. Graham CH, Hijmans RJ (2006) A comparison of methods for mapping species range and species richness. Global Ecol Biogeogr 15: 578–587.
28. Hawkins BA, Rueda M, Rodríguez MÁ (2008) What do range maps and surveys tell us about diversity patterns? Diversity 43: 345–355.
29. Hijmans RJ, Cameron SE, Parra JL, Jones PG, Jarvis A (2005) Very high resolution interpolated climate surfaces for global land areas. Int J Climatol 25: 1965–1978.
30. Field R, Hawkins BA, Cornell HV, Currie DJ, Diniz-Filho JAF, et al. (2009) Spatial species-richness gradients across scales: a meta-analysis. J Biogeogr 36: 132–147.
31. Bishop CM (1995) Neural networks for pattern recognition. Oxford: Clarendon Press. 482 p.
32. Molinaro A, Simon R, Pfeiffer RM (2005) Prediction error estimation: a comparison of resampling methods. Bioinformatics 21: 3301–3307.
33. Legendre P, Gallagher ED (2001) Ecologically meaningful transformations for ordination of species data. Oecologia 129: 271–280.
34. Olson DM, Dinerstein E, Wikramanayake ED, Burgess ND, Powell GVN, et al. (2001) Terrestrial ecoregions of the world: a new map of life on Earth. Bioscience 51: 933–938.
35. Zuur AF, Ieno EN, Walker NJ, Saveliev AA, Smith GM (2009) Mixed Effects Models and Extensions in Ecology with R. New York: Springer, New York. 574p.
36. Gotelli NJ, Ellison AM (2004) A primer of ecological statistics. Sunderland: Sinauer Associates. 492 p.
37. Zuur AF, Ieno EN, Elphick CS (2010) A protocol for data exploration to avoid common statistical problems. Method Ecol Evol 1:3–14.
38. Burnham KP, Anderson DR (2002) Model selection and multimodel inference: a practical information-theoretical approach. 2 ed, New York: Springer. 488 p.
39. Lobo JM, Lumaret JP, Jay-Robert P (2002) Modelling the species richness distribution of French dung beetles (Coleoptera, Scarabaeidae) and delimiting the predictive capacity of different groups of explanatory variables. Global Ecol Biogeogr 11: 265–277.

40. Legendre P, Legendre L (2012) Numerical Ecology, 3rd English edn. Amsterdam: Elsevier Science BV. 990 p.

41. Santos TG, Vasconcelos TS, Rossa-Feres DC, Haddad CFB (2009) Anurans of a seasonally dry tropical forest: the Morro do Diabo State Park, São Paulo State, Brazil. J Nat Hist 43: 973–993.

42. Canavero A, Arim M, Brazeiro A (2009) Geographic variations of seasonality and coexistence in communities: the role of diversity and climate. Austral Ecol 34: 741–750.

43. Carnaval AC, Moritz C (2008) Historical climate modelling predicts patterns of current biodiversity in the Brazilian Atlantic forest. J Biogeogr 35: 1187–1201.

44. Thomé MTC, Zamudio KR, Giovanelli JGR, Haddad CFB, Baldissera FA Jr, et al. (2010) Phylogeography of endemic toads and post-Pliocene persistence of the Brazilian Atlantic Forest. Mol Phylogenet Evol 55: 1018–1031.

45. Brusquetti F, Thomé MTC, Canedo C, Condez TH, Haddad CFB (2013) A new species of *Ischnocnema parva* species series (Anura, Brachycephalidae) from northern State of Rio de Janeiro, Brazil. Herpetologica 69:175–185.

46. Condez TH, Clemente-Carvalho RBG, Haddad CFB, dos Reis SF (2014) A new species of *Brachycephalus* (Anura: Brachycephalidae) from the highlands of the Atlantic Forest, southeastern Brazil. Herpetologica 70: 89–99.

47. Mackey BG, Berry SL, Brown T (2008) Reconciling approaches to biogeographic regionalization: a systematic and generic framework examined with a case study of the Australian continent. J Biogeogr 35: 213–229.

21

Species Richness and Assemblages in Landscapes of Different Farming Intensity – Time to Revise Conservation Strategies?

Erik Andersson[1]*, **Regina Lindborg**[2]

1 Stockholm Resilience Centre, Stockholm University, Stockholm, Sweden, **2** Dept. of Physical Geography and Quaternary Geology, Stockholm University, Stockholm, Sweden

Abstract

Worldwide conservation goals to protect biodiversity emphasize the need to rethink which objectives are most suitable for different landscapes. Comparing two different Swedish farming landscapes, we used survey data on birds and vascular plants to test whether landscapes with large, intensively managed farms had lower richness and diversity of the two taxa than landscapes with less intensively managed small farms, and if they differed in species composition. Landscapes with large intensively managed farms did not have lower richness than smaller low intensively managed farms. The landscape types were also similar in that they had few red listed species, normally targeted in conservation. Differences in species composition demonstrate that by having both types of agricultural landscapes regional diversity is increased, which is seldom captured in the objectives for agro-environmental policies. Thus we argue that focus on species richness or red listed species would miss the actual diversity found in the two landscape types. Biodiversity conservation, especially in production landscapes, would therefore benefit from a hierarchy of local to regional objectives with explicit targets in terms of which aspects of biodiversity to focus on.

Editor: Zhigang Jiang, Institute of Zoology, China

Funding: The study was financially supported by the Swedish Research Council for Environment, Agricultural Sciences and Spatial Planning (FORMAS) to Ekoklim. The funders had no role in study design, data collection and analysis, decision to publish, or preparation of the manuscript.

Competing Interests: The authors have declared that no competing interests exist.

* Email: erik.andersson@su.se

Introduction

Efforts to conserve biodiversity have traditionally focused mainly on natural habitats and pristine environments. During the last decades conservation focus has somewhat changed towards human dominated landscapes, especially farmlands [1,2]. In regions like Europe and Asia, long traditions of low-intensive management have favoured biodiversity, and many species occurring in those landscapes are now dependent on human management for their survival (e.g. [1,3,4]). However, agricultural practices are changing and more modern landscapes of intensive food production often have little or no biodiversity left and in these landscapes biodiversity is mostly protected within national parks and nature reserves [5,6]. This has resulted in polarized views on conservation management in agricultural landscapes, where land sparing separates intensively used agricultural land with high production from larger permanent preservation areas [6,7] and land sharing incorporating small natural habitats within the agricultural system, often with lower yield per unit area. However, biodiversity conservation in regions with mixed land use intensities may need new strategies and targets that go species richness or red listed species.

The most common approach to survey biodiversity is still to collect binary data like presence/absence to get a measure of species richness [8]. This measure has been criticized for being crude and not capturing the differences between environments or the actual change when monitoring habitats over time as the link between species richness and e.g. phylogenetic diversity, functional diversity and community structure and identity of species is still controversial (e.g. [9]). Recent studies suggest that knowledge about species identity and community composition is especially important for maintaining ecosystem functions within a landscape [10,11]. As a measure of the variation in species identity between sites, β diversity provides a link that connects diversity measures across scales, that is, between α (small scale) and γ (broad scale) diversities [12,13]. Understanding the factors driving each of these components and their interrelationships provide useful insights for understanding the mechanisms that structure diversity and community composition in landscapes. The relationships between farming systems and biodiversity have so far mainly been studied at a very broad scale (e.g. [14]), focusing on the contrasting effects of the main farming systems, or at local scale investigating the relation between different crops and biodiversity (e.g. [15,16]). As the most appropriate scale for investigating species movement and population dynamics most often is the intermediate landscape scale, studies on farming effects on biodiversity should be conducted at landscape scale where both the agricultural land and the surrounding matrix shape local biodiversity patterns [17]. This is also a relevant spatial scale for practical aspects of conservation biology since farms are the units for implementing agri-environmental policies.

This article examine the effect of two different Swedish landscapes types, one dominated by large-scale intensive farming

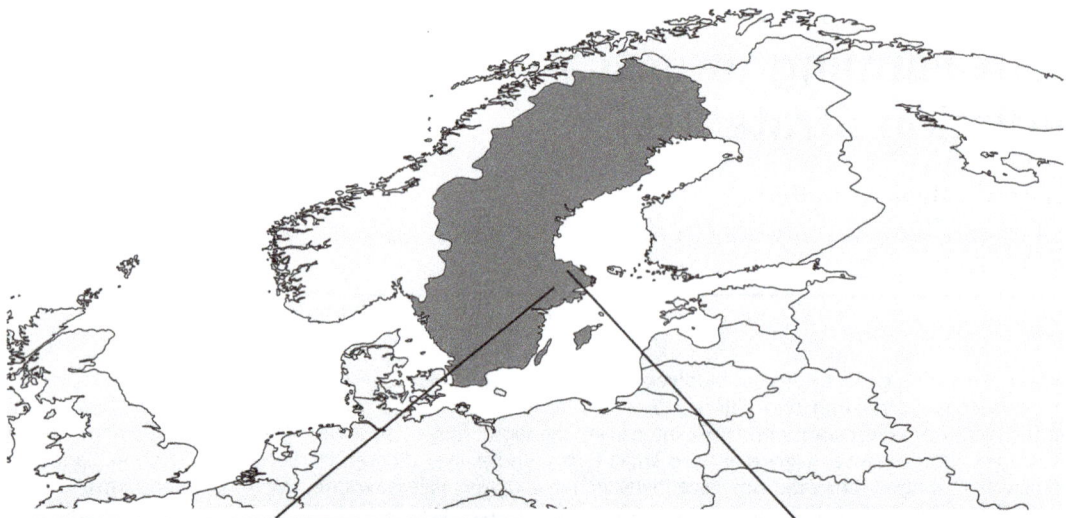

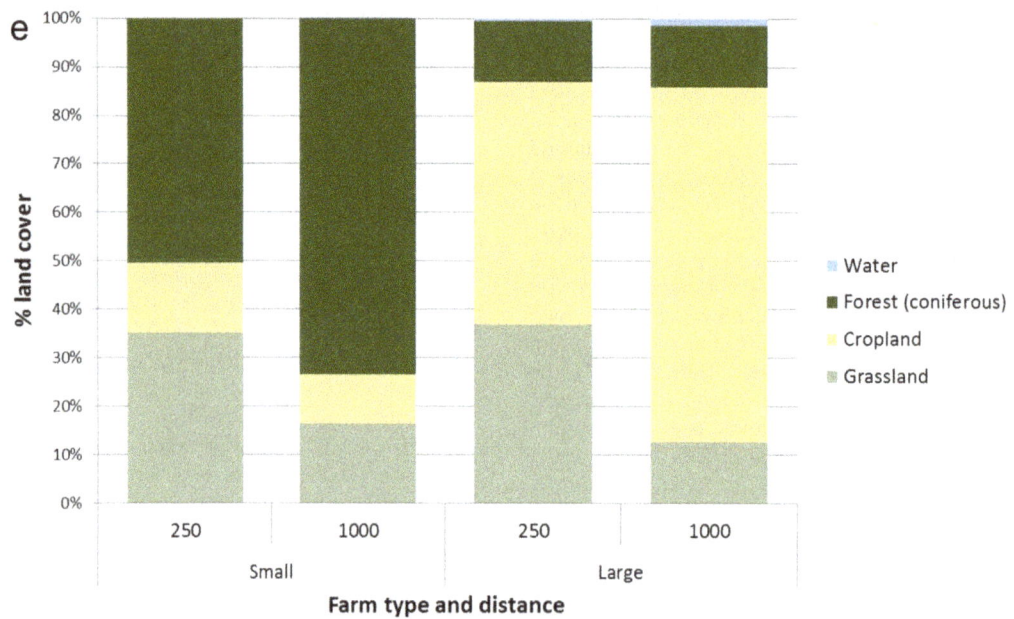

Figure 1. The two different landscape types. Pictures a) and b) illustrate the large farm landscapes and c) and d) the small farm landscapes. Picture e) shows the average land cover composition within 250 meters and 1000 meters, respectively, from each farmhouse in the different landscape types.

and one by small-scale, low intensive farming, on richness, diversity, abundance and identity of bird and plant species. These two taxa can be expected to highlight different aspects of the same landscapes due to differences in the scales and environmental drivers they respond to. Specifically, we test the hypotheses that landscapes with intensively managed farms should have 1) lower richness, diversity and abundance of species, and 2) be different in species composition, i.e. have fewer rare and specialist species, than the low intensity, small scale farm landscapes. Further, based on our empirical findings, we discuss the broader issue of conservation objectives in relation to different dimensions of biodiversity and how different landscapes and scales may be more or less suited to different measures of and targets for biodiversity.

Study Area

The study area is situated in south-central Sweden in the county of Uppsala (Fig. 1). In this area 16 farm-based sites were selected to represent the two landscape types; eight of the farms were large (mean size 336 ha; hereafter "large farms"), and found in an intensively managed agricultural area around Uppsala-Enköping-Västerås (59.680°N, 16.866°E); and eight were small scale and less intensively managed farms (on average 13 ha; hereafter "small farms") located on the Hållnäs peninsula (60.553°N, 17.868°E; all farm locations can be found in Table S1 in Appendix S1). The farms were selected from County board data, where we chose the largest and smallest farms in the county, respectively. The county has homogeneous climate with July being the warmest month (average maximum temperature of 21°C), and January the coldest (with an average minimum of −8°C), with freezing spells that can last a number of consecutive days. Rainfall is higher during the summer months of the year (up to 60 mm/day), while less abundant in winter (up to 25 mm/day), accumulating around 530 mm at the end of the year. The two landscape types mainly differ in the relative proportion of agricultural land and forest, with near farmhouse environments being more similar in their composition than the larger surroundings (Fig. 1), and in the intensity of land use. As an effect of land use changes since the 1950s the landscapes types have developed along different trajectories. The landscapes surrounding the small farms have seen a trend of reforestation and a shift away from full-time family farming, although they still hold fragments of managed, semi-natural grasslands acknowledged for their high biodiversity. The large scale farming landscape has in turn seen a development towards larger farm units and intensified use.

However, due to different physical environment in the region, larger farms are always located in areas with richer, clay dominated soils and flatter topography, while small farms are mostly found in remote areas with poorer, sandy soils [18,19]. This trend is evident all over Europe, where geophysical constrains like low soil fertility, steep slopes or stony grounds is a major limitation for mechanization and increased productivity (e.g. [20,21]).

Methods

Field observational data were collected at two levels of detail, strictly standardized (in space and time) abundance or frequency counts, and as presence - absence (standardized only in terms of time spent at a site) per site. The latter was used only for species lists. No legal permits were required for the data collection

(Sweden has a general right to access also to private land). However, land owners were contacted before the surveys to obtain their informal approval.

Bird survey

We used the point count method of surveying bird species and their abundances at each site [22]. Five survey points were located at and around each farmhouse. A central point was located next to the farmhouse and the other four in the NW, NE, SE and SW corners, respectively, 150 to 200 meters from the central point. With a mobile taxon like birds (e.g. [23]), this approach was deemed sufficient to capture the landscape variation and habitat diversity found at each site. Surveys were conducted two times: in early May and late May/early June 2012. We chose survey periods to coincide with the annual peak in singing activity. Surveys were begun at first light, at approximately 4:30am in early May and 3:30am late May/early June, and finished at latest 3 hours later. This period overlapped with the daily peak in bird vocal activity. Surveys were only conducted in mornings with favourable weather conditions, i.e. low winds and no heavy rain. At each point all birds seen or heard during five minutes and within two distance bands, 0–50 meters and 50–100 meters, were recorded, except birds that were detected flying over the counting station without landing. Birds that were farther away than 100 meters or seen or heard when walking between the points were recorded only as present, as were over-flying birds. In total, approximately 45 minutes were spent at each site per visit.

Plant survey

Plant inventories for each farm were done in June-July 2011 and 2012 in four habitat types representative for Swedish rural landscapes and present in both landscapes, i.e. forest, semi-natural pasture, grazed ex-arable field and field margin, to compensate for plans being sessile. We selected habitat patches located adjacent (within a radius of 1 km) to the farmhouses. For each habitat and farm we inventoried presence/absence of vascular plants in 10 randomly distributed 1 m^2 plots, totalling 40 plots per farm. Based on this binary data we calculated the frequency of each plant species. In addition, a walking inventory of all species was also conducted [24] in one hectare of each habitat at each farm, noting all plants present but not found in the plots, spending approximately 20 minutes per habitat.

Statistical analyses

Differences and/or similarities in community structure between the two types of landscapes were tested statistically using Bray-Curtis similarity and one-way analysis of similarities (ANOSIM) randomization test [25], a non-parametric analogue to the standard univariate 1- and 2-way ANOVA tests. Bray-Curtis dissimilarities were also used to estimate β-diversity and relative contribution of different species to these differences, the latter done through the SIMPER procedure [25]. Differences in species richness were tested with Welch's t-test. As indicated above, each site has two data sets, one with species frequencies from the point/ plot counts and a more inclusive species list including all species seen or heard at a site. The primary analyses were done on frequency data from the sample points/plots, complemented by analyses of the total species lists from each site. Frequency data were untransformed, using the relative abundances of different

Table 1. Descriptive statistics for birds and plants based on a survey in large and small scale farming systems (n = 16).

		Mean species richness	STDEV	Shannon diversity (log e)	STDEV
Birds	Small scale, frequencies	24.5	±3.207	2.79	±0.17
	Large scale, frequencies	20.75	±4.334	2.57	±0.32
	Small scale, presence	29.25	±3.412	n/a	n/a
	Large scale, presence	28.375	±5.236	n/a	n/a
Vascular plants	Small scale, frequencies	107.875	±5.489	4.28	±0.054
	Large scale, frequencies	114.5	±6.459	4.36	±0.068
	Small scale, presence	170.125	±14.74	n/a	n/a
	Large scale, presence	177.125	±13.163	n/a	n/a

The table includes both survey plots (frequency) and overall (presence).

species, species lists were only presence-absence. Tests were done using PRIMER 6 software [26] and R version 3.01 (http://www.r-project.org/). To further analyse potential patterns we partitioned data into different classes based on species data such as red listed status [27] and primary habitat association. The habitats were: forest, open land, woodland, wetland and mixed/generalist for birds [28] and ruderal or grassland for plants [29]. Plant groups were chosen to reflect the hypothesis, i.e. capture the species normally in focus for conservation and species associated with poor environments. To test for spatial auto-correlation we used the RELATE test in the PRIMER 6 [26], a non-parametric form of Mantel test [30], usually employed to assess trends in time or space, but with good capacity to detect auto-correlation in a multivariate context. In this case, the Spearman rank matrix correlation (r) was computed between two resemblance matrices: one constructed as Bray-Curtis dissimilarities between the samples of species abundance and the other as (non-normalized) Euclidean distances calculated from the spatial coordinates (X and Y) determining the location of the 16 sites.

Results

In total we found 70 bird species, 55 in the small farm landscapes and 58 in the large, and 434 species of vascular plants,

of which 316 species occurred in large farm landscapes and 300 species in small (Tables S2 & S3 in Appendix S1). Species richness was similar across farm types (Tables 1 & 2). However, bird and plant communities, respectively, showed statistically significant differences between large and small farm landscapes (Table 2 & Fig. 2), both in terms of species composition and abundance. Also the within group variation for birds and plants were different, with a higher similarity among small farm landscapes; ranging from 50% to 65% for birds and 70% to 75% for plants, compared to 50% to 70% for birds and 60% to 70% for plants on large farms (Fig. 2). Differences in bird communities were primarily caused by a smaller subset of species (seven species accounted for close to 50% of the difference between the landscape types, see Table 3). Plant communities did not have the same species driven differences (48 species to account for 50% of the difference). When analysed separately, two habitat guilds of birds demonstrated clear differences both in number of species and in species composition. Small scale farm landscapes had significantly more forest associated species while large farm landscapes had more species preferring open land. Plants did not signal such clear habitat related differences between the two farming landscapes; there were significantly more ruderal species in the large farm landscapes but we found no difference in the number of grassland

Table 2. Statistical significance of the differences in community compositions and species richness (within sample plots) between the two landscape types based on frequency data.

	ANOSIM, differences in community composition		t-test, differences in species richness		
	Sample statistics (Global R)	p- level	t	df	p-level
Bird species	0.775	0.001**	−1.967	12.897	0.071
Plant species	0.972	0.001**	1.002	13.825	0.334
Ruderal plant species	0.156	0.029*	2.966	11. 616	0.012*
Grassland plant species	0.694	0.001**	−0.222	13.792	0.828
Birds associated with forest	0.501	0.001**	−7.92	11.58	5.2e-6***
Birds associated with open land	0.79	0.001**	3.591	13.912	0.003**
Birds associated with woodland	0.04	0.237	0.5627	2.366	0.6224
Generalist bird species	0.126	0.066	−0.6333	9.939	0.5408
Birds, other	n/a	n/a	n/a	n/a	n/a

The ANOSIM analyses all used 999 permutations and the t-tests all had 1 degree of freedom. * indicates a p-level between 0.05 and >0.01, **between 0.01 and 0.001, and *** below 0.001.

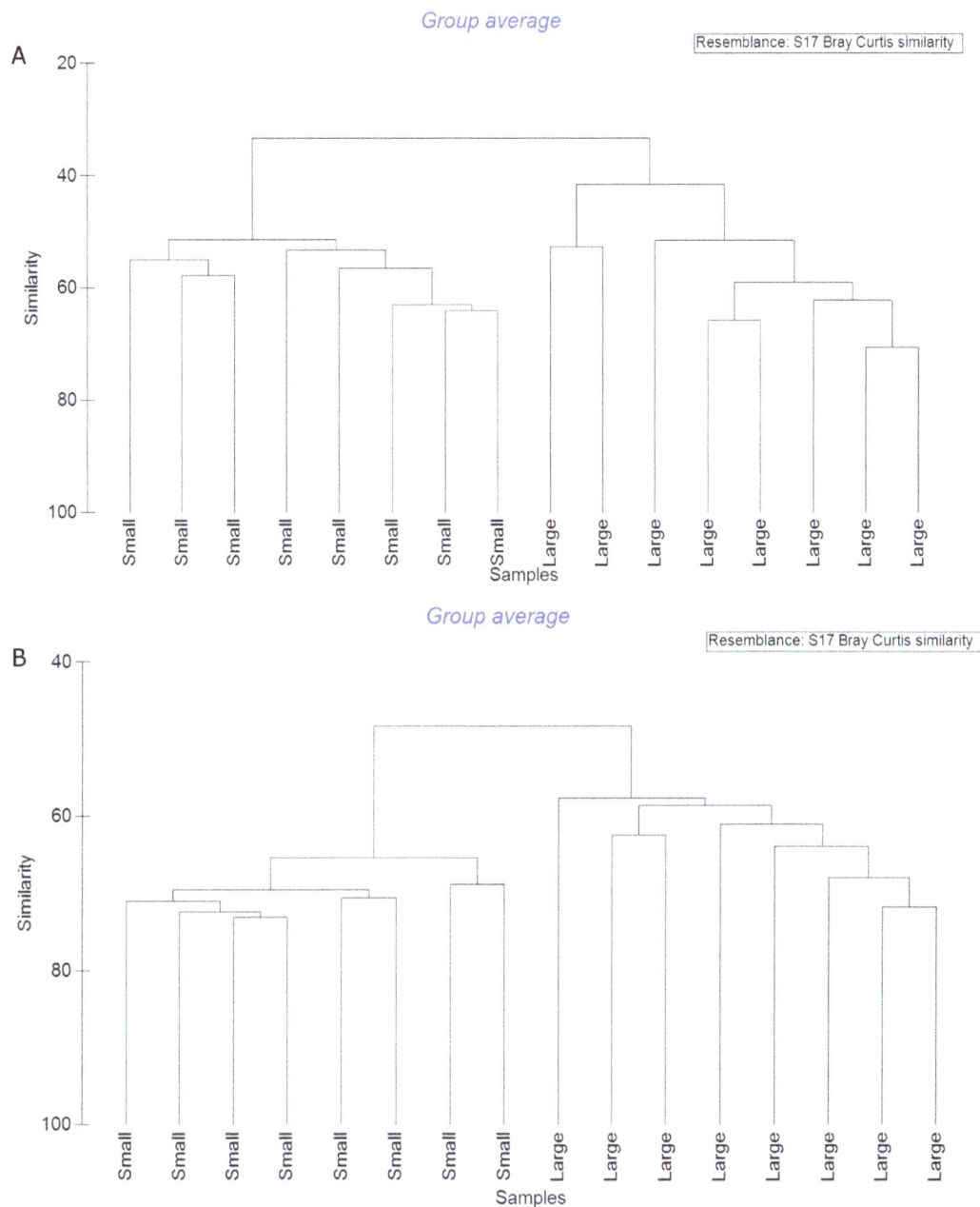

Figure 2. Within and between group similarities for a) birds and for b) plants in small and large farm landscapes. Similarities are based on Bray-Curtis dissimilarities.

plants (Table 4). However, both groups had significantly different species composition. Overall, the results showed that for both birds and plants the level of α-diversity was similar in all farms, whereas β-diversity was higher among the large farms (Fig. 2).

Red listed species were few and infrequent (Table 1 & 3), and while data were insufficient for statistical analysis, the actual species were not the same in the two landscape types (Tables S2 and S3 in Appendix S1). Among the red listed birds we found primarily species associated with open land, and two out of three of these were absent from the small scale farm landscapes. The small farm landscapes had one species associated with woodlands or mixed habitats but neither landscape type had any forest specialists. A majority of the red listed plants were associated to grassland habitats, and small farms held a few more species than

large farms, but most often they occurred in both types of farms (Table S3 in Appendix S1).

The RELATE test show a potentially significant spatial effect in the comparison between the landscape types (p = 0.001 for birds and p = 0.001 for plants). However, it did not detect any spatial auto-correlation within the two landscape types (p = 0.522 and 0.216 for birds and p = 0.17 and 0.233 for plants, large:small).

Discussion

Small farms with long traditional and extensive management have been shown to harbour higher species diversity compared to more intensively managed larger farms, not only as an effect of lower nutrient and pesticide input [31], but also due to higher

Table 3. Bird species contributing the most to the difference between small and large farm landscapes, with 50% as the cut-off point for cumulative contribution.

Species	Av.Abund small farm	Av.Abund Large farm	Av.Diss	Diss/SD	Contrib%	Cum.%	Habitat association
Sky Lark	0.13	13.38	7.51	2.01	11.27	11.27	Open
Willow Warbler	11.13	1.63	5.41	1.89	8.12	19.39	Forest
Chaffinch	13.25	9.13	4.81	1.51	7.22	26.61	Forest/Woodland
House Sparrow & Tree Sparrow	0	7.38	4.1	1.39	6.15	32.76	Open
Great Tit	8	10.75	3.8	1.65	5.7	38.45	Forest/Woodland
Jackdaw	0.75	7.13	3.54	1.25	5.31	43.76	Open
Starling	1.5	6.63	3.17	1	4.76	48.52	Open

Dissimilarity measures are based on Bray-Curtis dissimilarities between communities.

landscape heterogeneity surrounding smaller farms (e.g. [7]). Therefore, we hypothesized that large farm landscapes should have lower richness, diversity and abundance of species and have fewer rare and specialist species than the small farm landscapes. Our results did not fully support these hypotheses. Although there were no statistical differences in the total number of species and number of red listed species the landscape types supported, the combined species lists for small and large farm landscapes, respectively, showed higher total number of birds and plants in the surroundings of the large farms. For plants, the landscape types were also similar in terms of the number of species associated with grasslands, which is surprising since grassland specialists are especially favoured by low intensive management [4,19,32]. Part of the explanation may be that the habitats where grassland specialists are found are managed in similar ways in the two landscape types. We did find more ruderal plant species in the large farm landscapes, which indirectly supports our hypothesis that these landscapes should have fewer specialists and rare species, i.e. the relative proportion rather than the absolute number of rare and specialized species was somewhat lower.

Given the similarities in species richness, it was somewhat surprising to find such big differences in community composition for both plants and birds, and in the identity of the few red listed species present. Although similar in α diversity, community differences indicate high β diversity between the two landscape types [33] (understood as the inverse of the community similarities presented in Fig. 2). Within the landscape types, β-diversity has been suggested to be higher in intensively managed agricultural landscapes, with larger distances between habitats [34,35], a pattern in line with our findings. The more heterogeneous conditions could also explain the larger variation among bird communities. In addition, small scale farming has a longer, and perhaps more uniform, management tradition that might have acted as a stable environmental filter for community composition. The landscape composition at a regional scale, with large and small scale farming being separated in space, made it impossible to completely avoid spatial auto-correlation as a potential explanation for differences between the landscape types. However, we take the lack of spatial structuring within each landscape type as an indication that it is environmental differences rather than farm location that create the patterns we found.

For birds our richness estimates demonstrated clear differences between the landscape types in terms of the number of habitat specific species, differences that are congruent with regional differences in terms of the forest to open land ratio. The composition and land use of the larger surrounding landscape can be expected to influence organisms differently depending on their mobility, turnover time and resilience/persistence to change [36]. Thus, it was not unexpected that we found bird communities to reflect the larger landscape more clearly than plants. Mobile animals can benefit from complementary or supplementary resources available in the surrounding matrix outside the study site and hence be less confined by local resource limits (e.g. [37,38]) compared to sessile organisms like plants [17]. Habitat specialists can be expected to have tolerance thresholds below which the amount of habitat is insufficient to attract them to a site [39,40]. Although our study was not set up to identify neither the relevant spatial scale for different organisms nor the actual thresholds, our results indicate that despite being qualitatively similar (judging from the plant communities), open land in the small farm landscapes is clearly not extensive enough to attract many of the bird species associated with open land found in the large farm landscapes. Similarly, and more surprising since it does contain approximately 41% forest, the large farm landscapes are

Table 4. Descriptive statistics for different groups of vascular plants (ruderal, grassland, red listed species) and birds (forest, open and red listed birds) in large and small scale farming systems (n = 16).

	Classification	Landscape type	Mean number of species	Standard deviation
Birds	Forest	Small scale	12.375	±1.302
		Large scale	5.375	±2.134
	Open	Small scale	4.375	±2.387
		Large scale	8.5	±2.204
	Red listed	Small scale	0.75	±0.707
		Large scale	1.25	±0.707
Plants	Ruderal	Small scale	12.857	±1.345
		Large scale	15.625	±2.134
	Grassland	Small scale	34.375	±4.779
		Large scale	33.875	±4.224
	Red listed	Small scale	1.375	±1.506
		Large scale	1	±1.195

missing many of the forest associated species. Despite the relatively high proportion of forest the landscape may be less suited to support large populations due to non-optimal location in the landscape [41].

Placement of the semi-natural habitats within a landscape is crucial in the land sharing vs. land sparing debate. In our case, both small and large farm landscapes seem best described as land sharing since semi-natural habitats are patchy and dispersed. Locally based conservation strategies focusing on rare or threatened species seem less suited to the two landscape types and near farm environments in this study. Red listed species were few, and while a focus on them can be expected to support many other species and certain characteristics of the landscapes it would miss the much of the diversity actually there, not least on the regional scale. Land sharing, with habitats and resources more evenly distributed, may instead be well suited to ensure access to ecosystem services [42]. However, it may be that some land sharing landscapes are sinks for biodiversity and depend on a larger scale regional pool of species for their long term ability to provide ecosystem services [41,43].

Ecosystem services are connected to biodiversity through functional traits, and thus related to community composition [11]. If biodiversity conservation would focus on habitats and community profiles our two taxa tell different stories, as is often the case [23,44]. Bird communities, opposite to plants, clearly indicate that regional β-diversity is linked to large scale differences in the relative ratio between forest and open land, suggesting that large scale land use distributions would be the most suitable target for interventions and policies. Together, birds and plants show that conservation strategies must target several scales without compromising each other, and that the creation/maintenance of habitat alone will not necessarily predict or preserve differences or similarities between communities.

Concluding Remark

Strategies targeting biodiversity conservation in agricultural landscapes should be couched differently, depending on the desired outcomes. If, as in our study, landscapes have relatively high species richness but few rare and specialist species, the most appropriate conservation measure is to focus on the more common species and how to secure ecological functions and the landscape-wide delivery of biodiversity related ecosystem services. This strategy corresponds best with a land sharing approach, i.e. maintaining small semi-natural biotopes interspersed in the landscape, which is also a good strategy for local biodiversity conservation. Furthermore, as we found the different landscapes to be similar in terms of α-diversity but different in their species communities, we argue that β-diversity should be a regional concern. At least for birds this means maintaining the different forest to open land ratios at a spatial scale larger than the farms. We argue that production landscapes should benefit from a nested hierarchy of scale specific conservation objectives, each with its explicit targets in terms of which aspects of biodiversity to focus on.

Supporting Information

Appendix S1   Location of all farms and all bird and plant species found in the inventory of small and large scale farms, related to habitat type and red list. Table S1, Location of all 16 surveyed farms. Coordinates are given in GCS_SWEREF99, prime meridian is Greenwich and angular units are in in degrees. Table S2, All bird species found in the surveys, their red list status and habitat associations. Habitat associations are open, forest, woodland (scattered trees, otherwise not specific), water/forest (need both; not present in the point count), wetlands (including reeds) and generalist. Table S3, All vascular plant species found in the surveys, their red list status and whether they are associated with either ruderal land or traditional grasslands.

Acknowledgments

Plant surveys were done by S. Jakobsson and I. Bränning.

Author Contributions

Conceived and designed the experiments: EA RL. Analyzed the data: EA RL. Wrote the paper: EA RL.

References

1. Sutherland WJ (2002) Restoring a sustainable countryside. Trends Ecol Evol 17: 148–150. Available: http://dx.doi.org/10.1016/S0169-5347(01)02421-1. Accessed 5 April 2013.
2. Butler SJ, Vickery JA, Norris K (2007) Farmland Biodiversity and the Footprint of Agriculture. Science (80–) 315: 381–384. Available: http://www.sciencemag.org/cgi/content/abstract/315/5810/381.
3. Báldi A, Batáry P (2011) The past and future of farmland birds in Hungary. Bird Study 58: 365–377. Available: http://dx.doi.org/10.1080/00063657.2011.588685. Accessed 5 April 2013.
4. Eriksson O, Cousins SAO, Bruun H-H (2002) Land-use history and fragmentation of traditionally managed grasslands in Scandinavia. J Veg Sci 13: 743–748.
5. Phalan B, Onial M, Balmford A, Green RE (2011) Reconciling food production and biodiversity conservation: land sharing and land sparing compared. Science 333: 1289–1291. Available: http://www.sciencemag.org/content/333/6047/1289.abstract. Accessed 30 October 2012.
6. Green RE, Cornell SJ, Scharlemann JPW, Balmford A (2005) Farming and the fate of wild nature. Science 307: 550–555. Available: http://www.sciencemag.org/content/307/5709/550.short. Accessed 10 November 2012.
7. Benton TG, Vickery JA, Wilson JD (2003) Farmland biodiversity: is habitat heterogeneity the key? Trends Ecol Evol 18: 182–188. Available: http://www.sciencedirect.com/science/article/B6VJ1-47S6M9J-1/2/6373457cabe152314e0ab54f31ed6908.
8. Balvanera P, Pfisterer AB, Buchmann N, He J-S, Nakashizuka T, et al. (2006) Quantifying the evidence for biodiversity effects on ecosystem functioning and services. Ecol Lett 9: 1146–1156. Available: http://www.blackwell-synergy.com/doi/abs/10.1111/j.1461-0248.2006.00963.x.
9. Naeem S, Duffy JE, Zavaleta E (2012) The functions of biological diversity in an age of extinction. Science (80–) 336: 1401–1406. Available: http://www.sciencemag.org/content/336/6087/1401.abstract. Accessed 4 March 2013.
10. Mace GM, Norris K, Fitter AH (2012) Biodiversity and ecosystem services: a multilayered relationship. Trends Ecol Evol 27: 19–26. Available: http://dx.doi.org/10.1016/j.tree.2011.08.006. Accessed 21 May 2013.
11. Cardinale BJ, Duffy JE, Gonzalez A, Hooper DU, Perrings C, et al. (2012) Biodiversity loss and its impact on humanity. Nature 486: 59–67. Available: http://dx.doi.org/10.1038/nature11148. Accessed 19 September 2013.
12. Jost L (2007) Partitioning diversity into independent alpha and beta components. Ecology 88: 2427–2439. Available: http://www.esajournals.org/doi/abs/10.1890/06-1736.1. Accessed 19 December 2013.
13. Anderson MJ, Crist TO, Chase JM, Vellend M, Inouye BD, et al. (2011) Navigating the multiple meanings of β diversity: a roadmap for the practicing ecologist. Ecol Lett 14: 19–28. Available: http://www.ncbi.nlm.nih.gov/pubmed/21070562. Accessed 11 December 2013.
14. Robinson RA, Sutherland WJ (2002) Post-war changes in arable farming and biodiversity in Great Britain. J Appl Ecol 39: 157–176. Available: http://doi.wiley.com/10.1046/j.1365-2664.2002.00695.x. Accessed 17 December 2013.
15. Bommarco R, Kleijn D, Potts SG (2013) Ecological intensification: harnessing ecosystem services for food security. Trends Ecol Evol 28: 230–238. Available: http://www.sciencedirect.com/science/article/pii/S016953471200273X. Accessed 13 January 2014.
16. Garibaldi LA, Steffan-Dewenter I, Winfree R, Aizen MA, Bommarco R, et al. (2013) Wild pollinators enhance fruit set of crops regardless of honey bee abundance. Science 339: 1608–1611. Available: http://www.sciencemag.org/content/339/6127/1608.short. Accessed 13 January 2014.
17. Tscharntke T, Clough Y, Wanger TC, Jackson L, Motzke I, et al. (2012) Global food security, biodiversity conservation and the future of agricultural intensification. Biol Conserv 151: 53–59. Available: http://www.sciencedirect.com/science/article/pii/S0006320712000821. Accessed 26 October 2012.
18. Strijker D (2005) Marginal lands in Europe—causes of decline. Basic Appl Ecol 6: 99–106. Available: http://www.sciencedirect.com/science/article/pii/S1439179105000022. Accessed 8 November 2013.
19. Lindborg R, Bengtsson J, Berg Å, Cousins SAOO, Eriksson O, et al. (2008) A landscape perspective on conservation of semi-natural grasslands. Agric Ecosyst Environ 125: 213–222. Available: http://www.sciencedirect.com/science/article/pii/S0167880908000066. Accessed 22 January 2014.
20. Firmino A (1999) Agriculture and landscape in Portugal. Landsc Urban Plan 46: 83–91. Available: http://www.sciencedirect.com/science/article/pii/S0169204699000493. Accessed 20 December 2013.
21. Cousins SAO (2009) Landscape history and soil properties affect grassland decline and plant species richness in rural landscapes. Biol Conserv 142: 2752–

2758. Available: http://www.sciencedirect.com/science/article/pii/S0006320709002961. Accessed 20 December 2013.
22. Bibby CJ, Burgess ND, Hill DA (2000) Bird Census Techniques. London, UK: Academic Press.
23. Gonthier DJ, Ennis KK, Farinas S, Hsieh H-Y, Iverson AL, et al. (2014) Biodiversity conservation in agriculture requires a multi-scale approach. Proc R Soc B Biol Sci 281: 20141358–20141358. Available: http://rspb.royalsocietypublishing.org/content/281/1791/20141358.short. Accessed 6 August 2014.
24. Sutherland WJ (2000) The conservation handbook. Research, Management and Policy. Oxford, UK: Blackwell Publishing.
25. Clarke KR (1988) Detecting change in benthic community structure. Proceedings XIVth International Biometric Conference, Namur: Invited Papers. . Société Adophe Quélét, Gemblous, Belgium. pp. 131–142.
26. Clarke KR, Gorley RN (2006) PRIMER v6.
27. Gärdenfors U (2010) Rödlistade arter i Sverige 2010 - The 2010 Red List of Swedish Species. Uppsala, Sweden: ArtDatabanken.
28. Cramp S (1977–1994) Handbook of the birds of Europe, the Middle East, and North Africa: the birds of the western Palearctic. Oxford University Press.
29. Plue J, Cousins SAO (2013) Temporal dispersal in fragmented landscapes. Biol Conserv 160: 250–262. Available: http://www.sciencedirect.com/science/article/pii/S000632071300058X. Accessed 31 July 2014.
30. Mantel N (1967) The Detection of Disease Clustering and a Generalized Regression Approach. Cancer Res 27: 209–220. Available: http://cancerres.aacrjournals.org/content/27/2_Part_1/209.short. Accessed 7 August 2014.
31. Stoate C, Báldi A, Beja P, Boatman ND, Herzon I, et al. (2009) Ecological impacts of early 21st century agricultural change in Europe – A review. J Environ Manage 91: 22–46. Available: http://www.sciencedirect.com/science/article/pii/S0301479709002448. Accessed 20 December 2013.
32. Pykälä J (2005) Plant species responses to cattle grazing in mesic semi-natural grassland. Agric Ecosyst Environ 108: 109–117. Available: http://www.sciencedirect.com/science/article/pii/S0167880905000447. Accessed 20 December 2013.
33. Anderson MJ, Ellingsen KE, McArdle BH (2006) Multivariate dispersion as a measure of beta diversity. Ecol Lett 9: 683–693. Available: http://www.ncbi.nlm.nih.gov/pubmed/16706913. Accessed 17 December 2013.
34. Tscharntke T, Tylianakis JM, Rand TA, Didham RK, Fahrig L, et al. (2012) Landscape moderation of biodiversity patterns and processes - eight hypotheses. Biol Rev Camb Philos Soc 87: 661–685. Available: http://www.ncbi.nlm.nih.gov/pubmed/22272640. Accessed 18 October 2013.
35. Lindborg R, Plue J, Andersson K, Cousins SAO (2014) Function of small habitat elements for enhancing plant diversity in different agricultural landscapes. Biol Conserv 169: 206–213. Available: http://www.sciencedirect.com/science/article/pii/S0006320713003947. Accessed 20 December 2013.
36. Prevedello JA, Vieira MV (2010) Does the type of matrix matter? A quantitative review of the evidence. Biodivers Conserv 19: 1205–1223. Available: http://link.springer.com/10.1007/s10531-009-9750-z. Accessed 27 February 2013.
37. Dunning JB, Danielson BJ, Pulliman HR (1992) Ecological processes that affect populations in complex landscapes. Oikos 65: 169–175.
38. Debinski DM (2006) Forest fragmentation and matrix effects: the matrix does matter. J Biogeogr 33: 1791–1792. Available: http://doi.wiley.com/10.1111/j.1365-2699.2006.01596.x. Accessed 7 March 2013.
39. Fahrig L (2001) How much habitat is enough? Biol Conserv 100: 65–74. Available: <Go to ISI>://000169481300008.
40. Andrén H (1994) Effects of Habitat Fragmentation on Birds and Mammals in Landscapes with Different Proportions of Suitable Habitat - a Review. Oikos 71: 355–366. Available: <Go to ISI>://A1994QA75600002.
41. Bengtsson J, Angelstam P, Elmqvist T, Emanuelsson U, Folke C, et al. (2003) Reserves, Resilience and Dynamic Landscapes. Ambio 32: 389–396.
42. Blitzer EJ, Dormann CF, Holzschuh A, Kleind A-M, Rand TA, et al. (2012) Spillover of functionally important organisms between managed and natural habitats. Agric Ecosyst & Environ 146: 34–43. Available: http://dx.doi.org/10.1016/j.agee.2011.09.005. Accessed 21 November 2011.
43. Fahrig L, Baudry J, Brotons L, Burel FG, Crist TO, et al. (2011) Functional landscape heterogeneity and animal biodiversity in agricultural landscapes. Ecol Lett 14: 101–122.
44. Grenyer R, Orme CDL, Jackson SF, Thomas GH, Davies RG, et al. (2006) Global distribution and conservation of rare and threatened vertebrates. Nature 444: 93–96. Available: http://www.nature.com.ezp.sub.su.se/nature/journal/v444/n7115/full/nature05237.html. Accessed 14 December 2013.

Permissions

All chapters in this book were first published in PLOS ONE, by The Public Library of Science; hereby published with permission under the Creative Commons Attribution License or equivalent. Every chapter published in this book has been scrutinized by our experts. Their significance has been extensively debated. The topics covered herein carry significant findings which will fuel the growth of the discipline. They may even be implemented as practical applications or may be referred to as a beginning point for another development.

The contributors of this book come from diverse backgrounds, making this book a truly international effort. This book will bring forth new frontiers with its revolutionizing research information and detailed analysis of the nascent developments around the world.

We would like to thank all the contributing authors for lending their expertise to make the book truly unique. They have played a crucial role in the development of this book. Without their invaluable contributions this book wouldn't have been possible. They have made vital efforts to compile up to date information on the varied aspects of this subject to make this book a valuable addition to the collection of many professionals and students.

This book was conceptualized with the vision of imparting up-to-date information and advanced data in this field. To ensure the same, a matchless editorial board was set up. Every individual on the board went through rigorous rounds of assessment to prove their worth. After which they invested a large part of their time researching and compiling the most relevant data for our readers.

The editorial board has been involved in producing this book since its inception. They have spent rigorous hours researching and exploring the diverse topics which have resulted in the successful publishing of this book. They have passed on their knowledge of decades through this book. To expedite this challenging task, the publisher supported the team at every step. A small team of assistant editors was also appointed to further simplify the editing procedure and attain best results for the readers.

Apart from the editorial board, the designing team has also invested a significant amount of their time in understanding the subject and creating the most relevant covers. They scrutinized every image to scout for the most suitable representation of the subject and create an appropriate cover for the book.

The publishing team has been an ardent support to the editorial, designing and production team. Their endless efforts to recruit the best for this project, has resulted in the accomplishment of this book. They are a veteran in the field of academics and their pool of knowledge is as vast as their experience in printing. Their expertise and guidance has proved useful at every step. Their uncompromising quality standards have made this book an exceptional effort. Their encouragement from time to time has been an inspiration for everyone.

The publisher and the editorial board hope that this book will prove to be a valuable piece of knowledge for researchers, students, practitioners and scholars across the globe.

LIST OF CONTRIBUTORS

Maike Abbas, Robert Ptacnik and Helmut Hillebrand
Institute for Chemistry and Biology of the Marine Environment, Carl von Ossietzky University of Oldenburg, Wilhelmshaven, Germany

Anne Ebeling
Institute of Ecology, Friedrich- Schiller-University Jena, Jena, Germany

Yvonne Oelmann
Geoecology/Geography, Eberhard Karls-University Tübingen, Tübingen, Germany

Christiane Roscher
UFZ, Helmholtz Centre for Environmental Research, Department of Community Ecology, Halle, Germany

Alexandra Weigelt
Institute of Biology, University of Leipzig, Leipzig, Germany

Wolfgang W. Weisser
Institute of Ecology, Friedrich-Schiller-University Jena, Jena, Germany
Terrestrial Ecology, Department of Ecology and Ecosystem Management, Center of Life and Food Sciences Weihenstephan; Technische Universität München, Freising, Germany

Wolfgang Wilcke
Geographic Institute, University of Berne, Berne, Switzerland

Nubia França da Silva Giehl
Programa de Pós-graduação em Ecologia e Conservação, Universidade do Estado de Mato Grosso (UNEMAT), Nova Xavantina, Mato Grosso, Brasil
Laboratório de Entomologia, Universidade do Estado de Mato Grosso (UNEMAT), Nova Xavantina, Mato Grosso, Brasil

Karina Dias-Silva
Programa de Pósgraduação em Ciências Ambientais, ICB1, Universidade Federal de Goiás, Goiânia, Goiás, Brasil

Leandro Juen
Laboratório de Ecologia e Conservação, Instituto de Ciências Biológicas, Universidade Federal do Pará (UFPA), Belém, Pará, Brasil

Joana Darc Batista
Laboratório de Entomologia, Universidade do Estado de Mato Grosso (UNEMAT), Nova Xavantina, Mato Grosso, Brasil

Helena Soares Ramos Cabette
Laboratório de Entomologia, Universidade do Estado de Mato Grosso (UNEMAT), Nova Xavantina, Mato Grosso, Brasil
Departamento de Biologia, Universidade do Estado de Mato Grosso (UNEMAT), Nova Xavantina, Mato Grosso, Brasil

Anne Ebeling, Nico Eisenhauer and Anja Vogel
Institute of Ecology, University of Jena, Jena, Germany

Sebastian T. Meyer
Department of Ecology and Ecosystem Management, Center for Food and Life Sciences Weihenstephan, Technische Universität München, Freising, Germany

Maike Abbas and Helmut Hillebrand
Institute for Chemistry and Biology of the Marine Environment, Carl-von-Ossietzky-University Oldenburg, Wilhelmshaven, Germany

Markus Lange
Max Planck Institute for Biogeochemistry, Jena, Germany

Christoph Scherber
DNPW, Agroecology, Georg-August University Göttingen, Göttingen, Germany

Alexandra Weigelt
Department for Systematic Botany and Functional Biodiversity, University of Leipzig, Leipzig, Germany

Wolfgang W. Weisser
Institute of Ecology, University of Jena, Jena, Germany
Department of Ecology and Ecosystem Management, Center for Food and Life Sciences

Maria M. Romeiras
Tropical Botanical Garden, Tropical Research Institute (IICT), Lisbon, Portugal
Centre for Biodiversity, Functional and Integrative Genomics (BIOFIG), Faculty of Sciences, University of Lisbon, Lisbon, Portugal

Rui Figueira and Maria Cristina Duarte
Tropical Botanical Garden, Tropical Research Institute (IICT), Lisbon, Portugal

Pedro Beja
CIBIO - Research Center in Biodiversity and Genetic Resources/InBIO, University of Porto, Vairão, Portugal

Iain Darbyshire
Royal Botanic Gardens, Kew. Richmond, United Kingdom

Jitka Horáčková
Department of Ecology, Charles University in Prague, Faculty of Science, Prague 2, Czech Republic
Department of Zoology, Charles University in Prague, Faculty of Science, Prague 2, Czech Republic

Lucie Juřičková
Department of Zoology, Charles University in Prague, Faculty of Science, Prague 2, Czech Republic

Arnošt L. Šizling
Center for Theoretical Study, Charles University and the Academy of Sciences of the Czech Republic, Prague 1, Czech Republic

Vojtěch Jarošík and Petr Pyšek
Department of Ecology, Charles University in Prague, Faculty of Science, Prague 2, Czech Republic
Department of Invasion Ecology, Institute of Botany, Academy of Sciences of the Czech Republic, Průhonice, Czech Republic

Ana Novoa
Departamento de Bioloxía Vexetal e Ciencias do Solo, Facultade de Bioloxía, Universidade de Vigo, Vigo, Spain
Centre for Invasion Biology, Department of Botany and Zoology, Stellenbosch University, Stellenbosch, South Africa

Luís González
Departamento de Bioloxía Vexetal e Ciencias do Solo, Facultade de Bioloxía, Universidade de Vigo, Vigo, Spain

Ana Villarroya
The Nature Conservancy, Boulder, Colorado, United States of America

Ana Cristina Barros
The Nature Conservancy, Brasilia, Brazil

Joseph Kiesecker
The Nature Conservancy, Fort Collins, Colorado, United States of America

Nadja K. Simons, Martin M. Gossner, Esther Pašalić, Wolfgang W. Weisser and Manfred Türke
Terrestrial Ecology Research Group, Department of Ecology and Ecosystem Management, School of Life Sciences Weihenstephan, Technische Universität München, Freising, Germany

Thomas M. Lewinsohn
Department of Animal Biology, Institute of Biology, University of Campinas, Campinas, Sao Paulo, Brazil

Steffen Boch and Stephanie A. Socher
Institute of Plant Sciences, University of Bern, Bern, Switzerland

Markus Lange
Max-Planck-Institute for Biogeochemistry, Jena, Germany

Jörg Müller
Institute of Biochemistry and Biology, University of Potsdam, Potsdam, Germany

Markus Fischer
Institute of Plant Sciences, University of Bern, Bern, Switzerland
Biodiversity and Climate Research Centre, Senckenberg Gesellschaft für Naturforschung, Frankfurt/Main, Germany

Anna Scott, Jannah M. Hardefeldt and Karina C. Hall
National Marine Science Centre and Marine Ecology Research Centre, School of Environment, Science and Engineering, Southern Cross University, Coffs Harbour, New South Wales, Australia

Yi Yu, Qiang Fan, Rujiang Shen, Jianhua Jin and Wenbo Lia
Guangdong Key Laboratory of Plant Resources and Key
Laboratory of Biodiversity Dynamics and Conservation of Guangdong Higher Education Institutes, School of Life Sciences, Sun Yat-Sen University, Guangzhou, China

Wei Guo
Department of Horticulture and Landscape Architecture, Zhongkai University of Agriculture and Engineering, Guangzhou, China

Dafang Cui
College of Forestry, South China Agriculture
University, Guangzhou, China

Bernard W. T. Coetzee
Centre for Invasion Biology, Department of Botany
and Zoology, Stellenbosch University, Stellenbosch,
Western Cape, South Africa
School of Biological Sciences, Monash University,
Melbourne, Victoria, Australia

Kevin J. Gaston
Environment and Sustainability Institute, University
of Exeter, Penryn, Cornwall, United Kingdom

Steven L. Chown
School of Biological Sciences, Monash University,
Melbourne, Victoria, Australia

Benjamin Blonder
Sky School, University of Arizona, Tucson, Arizona,
United States of America

Lindsey Sloat
Department of Ecology and Evolutionary Biology,
University of Arizona, Tucson, Arizona, United States
of America

Brian J. Enquist
Department of Ecology and Evolutionary Biology,
University of Arizona, Tucson, Arizona, United States
of America
Santa Fe Institute, Santa Fe, New Mexico, United
States of America

Brian McGill
School of Biology and Ecology, University of Maine,
Orono, Maine, United States of America

Xuan Fang and Guoan Tang
Key Laboratory of Virtual Geographic Environment,
Ministry of Education, School of Geography Science,
Nanjing Normal University, Nanjing, China

Bicheng Li
Research Center of Soil and Water Conservation and
Ecological Environment, Chinese Academy of Sciences,
Yangling, Shaanxi, China

Ruiming Han
School of Geography Science, Nanjing Normal
University, Nanjing, China

Rachel M. Mitchell
School of Environmental and Forest Sciences, University
of Washington, Seattle, Washington, United States of
America

Jonathan D. Bakker
School of Environmental and Forest Sciences, University
of Washington, Seattle, Washington, United States of
America
Smithsonian Environmental Research Center,
Edgewater, Maryland, United States of America

Qiuyuan Huang
Department of Geology and Environmental Earth
Science, Miami University, Oxford, Ohio, United States
of America
Department of Microbiology, University of Georgia,
Athens, Georgia, United States of America

Brandon R. Briggs
Department of Geology and Environmental Earth
Science, Miami University, Oxford, Ohio, United
States of America

Hailiang Dong
Department of Geology and Environmental Earth
Science, Miami University, Oxford, Ohio, United States
of America
State Key Laboratory of Biogeology and Environmental
Geology, China University of Geosciences, Beijing,
China
State Key Laboratory of Biogeology and Environmental
Geology, China University of Geosciences, Wuhan,
China

Hongchen Jiang and Geng Wu
State Key Laboratory of Biogeology and Environmental
Geology, China University of Geosciences, Wuhan,
China

Christian Edwardson
Department of Microbiology, University of Georgia,
Athens, Georgia, United States of America

Iwijn De Vlaminck and Stephen Quake
Departments of Bioengineering and Applied Physics,
Stanford University and the Howard Hughes Medical
Institute, Stanford, California, United States of America

Rebecca Ostertag
Department of Biology, University of Hawai'i at Hilo,
Hilo, Hawai'i, United States of America

Faith Inman-Narahari
Department of Natural Resources and Environmental
Management, University of Hawai'i at Mānoa,
Honolulu, Hawai'i, United States of America

Susan Cordell and Christian P. Giardina
Institute of Pacific Islands Forestry, Pacific Southwest
Research Station, USDA Forest Service, Hilo, Hawai'i,
United States of America

Lawren Sack
Department of Ecology and Evolutionary Biology, University of California Los Angeles, Los Angeles, California, United States of America

Michal Horsák and Milan Chytry
Department of Botany and Zoology, Masaryk University, Brno, Czech Republic

Laura Fortel, Laurent Guilbaud and Anne Laure Guirao
INRA, UR 406 Abeilles et Environnement, Avignon, France

Mickaë l Henry and Bernard E. Vaissière
INRA, UR 406 Abeilles et Environnement, Avignon, France
UMT Protection des Abeilles dans l'Environnement, Avignon, France

Michael Kuhlmann
Department of Life Sciences, Natural History Museum, London, United Kingdom

Hugues Mouret
Arthropologia, Ecocentre du Lyonnais, La Tour de Salvagny, France

Orianne Rollin
UMT Protection des Abeilles dans l'Environnement, Avignon, France
ACTA, Site Agroparc, Avignon, France

Mason J. Crane, David B. Lindenmayer and Ross B. Cunningham
Fenner School of Environment and Society, The Australian National University, Canberra, ACT, Australia

Tiago S. Vasconcelos
Departamento de Ciências Biológicas, Universidade Estatual Paulista, Bauru, São Paulo, Brazil

Vitor H. M. Prado and Célio F. B. Haddad
Departamento de Zoologia, Universidade Estadual Paulista, Rio Claro, São Paulo, Brazil

Fernando R. da Silva
Departamento de Ciências Ambientais, Universidade Federal de São Carlos, Sorocaba, São Paulo, Brazil

Erik Andersson
Stockholm Resilience Centre, Stockholm University, Stockholm, Sweden

Regina Lindborg
Dept. of Physical Geography and Quaternary Geology, Stockholm University, Stockholm, Sweden

Index

www.ingramcontent.com/pod-product-compliance
Lightning Source LLC
Chambersburg PA
CBHW080530200326
41458CB00012B/4391